Assessing Progress towards Sustainability

Frameworks, Tools and Case Studies

Assessing Progress towards Sustainability

Frameworks, Tools and Case Studies

Edited by

Carmen Teodosiu

Professor and Director of the Department of Environmental Engineering and Management from „Gheorghe Asachi" Technical University Iasi (TUIASI), Romania

Silvia Fiore

Associate Professor at the Department of Environment, Land and Infrastructure Engineering (DIATI) at Politecnico di Torino, Italy

Almudena Hospido

Associate Professor at the Department of Chemical Engineering at Universidade de Santiago de Compostela (USC), Spain

ELSEVIER

Elsevier
Radarweg 29, PO Box 211, 1000 AE Amsterdam, Netherlands
The Boulevard, Langford Lane, Kidlington, Oxford OX5 1GB, United Kingdom
50 Hampshire Street, 5th Floor, Cambridge, MA 02139, United States

ISBN: 978-0-323-85851-9

For information on all Elsevier publications
visit our website at https://www.elsevier.com/books-and-journals

Publisher: Candice Janco
Editorial Project Manager: Aleksandra Packowska
Production Project Manager: Paul Prasad Chandramohan
Cover designer: Christian Bilbow

Typeset by STRAIVE, India

Contents

SECTION II Assessment tools for sustainability-methodological issues

SECTION III Case studies for sustainability assessments

Contributors

Feni Agostinho
Programa de Pós-graduação em Engenharia de Produção, Laboratório de Produção e Meio Ambiente, Universidade Paulista (UNIP), São Paulo, Brazil

Rubén Aldaco
Department of Chemical and Biomolecular Engineering, University of Cantabria, Santander, Spain

Cecilia M.V.B. Almeida
Programa de Pós-graduação em Engenharia de Produção, Laboratório de Produção e Meio Ambiente, Universidade Paulista (UNIP), São Paulo, Brazil

Jacopo Bacenetti
Department of Environmental Science and Policy, Università degli Studi di Milano, Milan, Italy

George Barjoveanu
Department of Environmental Engineering and Management, "Gheorghe Asachi" Technical University of Iasi, Iasi, Romania

Florin Bucatariu
Department of Environmental Engineering and Management, "Gheorghe Asachi" Technical University of Iasi; "Petru Poni" Institute of Macromolecular Chemistry, Iasi, Romania

Diana M. Byrne
Department of Civil Engineering, University of Kentucky, Lexington, KY, United States

Roberto Chirone
Department of Chemical, Materials and Production Engineering, University of Naples Federico II; eLoop S.r.l., Naples, Italy

Roland Clift
Centre for Environment and Sustainability (CES), University of Surrey, Guildford, United Kingdom; Department of Chemical and Biological Engineering, University of British Columbia,Vancouver, BC, Canada

Brett Cohen
The Green House; Department of Chemical Engineering, University of Cape Town, Cape Town, South Africa

Mauro Cordella
TECNALIA, Basque Research and Technology Alliance (BRTA), Derio, Spain

Lluís Corominas
Catalan Institute for Water Research (ICRA); Universitat de Girona, Girona, Spain

Jorge Cristóbal
Department of Chemical and Biomolecular Engineering, University of Cantabria, Santander, Spain

Sabino De Gisi
Department of Civil, Environmental, Land, Building Engineering and Chemistry (DICATECh), Polytechnic University of Bari, Bari, Italy

Ana Belén de Isla
LKS Krean, KREAN Group, Mondragón, Spain

Tiziano Distefano
DEM (Department of Economics and Management), University of Pisa, Pisa, Italy

Andrew Ferdinando
LKS Krean, KREAN Group, Mondragón, Spain

Daniela Fighir
Department of Environmental Engineering and Management, "Gheorghe Asachi" Technical University of Iasi, Iasi, Romania

Silvia Fiore
Department of Environment, Land and Infrastructure Engineering (DIATI), Politecnico di Torino, Torino, Italy

Daniela Gavrilescu
Department of Environmental Engineering and Management, "Gheorghe Asachi" Technical University of Iasi, Iasi, Romania

Shabbir H. Gheewala
The Joint Graduate School of Energy and Environment (JGSEE), King Mongkut's University of Technology Thonburi (KMUTT); Centre of Excellence on Energy Technology and Environment (CEE), PERDO, Ministry of Higher Education, Science, Research and Innovation, Bangkok, Thailand

Biagio F. Giannetti
Programa de Pós-graduação em Engenharia de Produção, Laboratório de Produção e Meio Ambiente, Universidade Paulista (UNIP), São Paulo, Brazil

Sara González-García
CRETUS, Department of Chemical Engineering, Universidade de Santiago de Compostela, Santiago de Compostela, Spain

Almudena Hospido
CRETUS, Department of Chemical Engineering, Universidade de Santiago de Compostela, Santiago de Compostela, Spain

Brandon Kuczenski
Institute for Social, Behavioral, and Economic Research University of California, Santa Barbara, CA, United States

Claudia Labianca
Department of Civil and Environmental Engineering, Hong Kong Polytechnic University, Hung Hom, Kowloon, Hong Kong, China; Department of Civil, Environmental, Land, Building Engineering and Chemistry (DICATECh), Polytechnic University of Bari, Bari, Italy

Francesco Laio
Department of Environment, Land and Infrastructure Engineering (DIATI), Politecnico di Torino, Torino, Italy

Jara Laso
Department of Chemical and Biomolecular Engineering, University of Cantabria, Santander, Spain

Paola Lettieri
Department of Chemical Engineering, University College London, London, United Kingdom

Yvonne Lewis
The Green House, Cape Town, South Africa

Eléonore Loiseau
ITAP, Univ Montpellier, INRAE, Institut Agro; Elsa, Research Group for Environmental Lifecycle and Sustainability Assessment, Montpellier, France

Simon Mair
CES, University of Surrey, Guildford; Circular Economy and Data Analytics, School of Management, University of Bradford, Bradford, United Kingdom

María Margallo
Department of Chemical and Biomolecular Engineering, University of Cantabria, Santander, Spain

Carolin Märker
Institute of Energy and Climate Research - Systems Analysis and Technology Evaluation (IEK-STE), Forschungszentrum Jülich, Jülich; University of Bonn, Bonn, Germany

George Martin
Department of Sociology, Montclair State University, Montclair, NJ, United States; CES, University of Surrey, Guildford, United Kingdom

Jose Miguel Martínez
LKS Krean, KREAN Group, Mondragón, Spain

Antonio Marzocchella
Department of Chemical, Materials and Production Engineering, University of Naples Federico II, Napoli, Italy

Miguel Mauricio-Iglesias
CRETUS, Department of Chemical Engineering, Universidade de Santiago de Compostela, Santiago de Compostela, Spain

Marcela Mihai
Department of Environmental Engineering and Management, "Gheorghe Asachi" Technical University of Iasi; "Petru Poni" Institute of Macromolecular Chemistry, Iasi, Romania

Irina Morosanu
Department of Environmental Engineering and Management, "Gheorghe Asachi" Technical University of Iasi, Iasi, Romania

Anuska Mosquera-Corral
CRETUS, Department of Chemical Engineering, Universidade de Santiago de Compostela, Santiago de Compostela, Spain

Rattanawan Tam Mungkung
Centre of Excellence on enVironmental strategy for GREEN business (VGREEN); Department of Environmental Technology and Management, Faculty of Environment, Kasetsart University, Bangkok, Thailand

Michele Notarnicola
Department of Civil, Environmental, Land, Building Engineering and Chemistry (DICATECh), Polytechnic University of Bari, Bari, Italy

Andrea Paulillo
eLoop S.r.l., Napoli, Italy; Department of Chemical Engineering, University College London, London, United Kingdom

Oana Plavan
Department of Environmental Engineering and Management, "Gheorghe Asachi" Technical University of Iasi, Iasi, Romania

Olatz Pombo
LKS Krean, KREAN Group, Mondragón, Spain

Beatriz Rivela
Inviable Life Cycle Thinking, Madrid, Spain

Alba Roibás-Rozas
CRETUS, Department of Chemical Engineering, Universidade de Santiago de Compostela, Santiago de Compostela, Spain

Philippe Roux
ITAP, Univ Montpellier, INRAE, Institut Agro; Elsa, Research Group for Environmental Lifecycle and Sustainability Assessment, Montpellier, France

Mateo Saavedra del Oso
CRETUS, Department of Chemical Engineering, Universidade de Santiago de Compostela, Santiago de Compostela, Spain

Serenella Sala
European Commission, Joint Research Centre, Ispra, VA, Italy

Piero Salatino
Department of Chemical, Materials and Production Engineering, University of Naples Federico II, Napoli, Italy

Thibault Salou
ITAP, Univ Montpellier, INRAE, Institut Agro; Elsa, Research Group for Environmental Lifecycle and Sustainability Assessment, Montpellier, France

Brindusa Sluser
Department of Environmental Engineering and Management, "Gheorghe Asachi" Technical University of Iasi, Iasi, Romania

Dolores Sucozhañay
Department of Space and Population, University of Cuenca, Cuenca, Ecuador

Carmen Teodosiu
Department of Environmental Engineering and Management, "Gheorghe Asachi" Technical University of Iasi, Iasi, Romania

Marta Tuninetti
Department of Environment, Land and Infrastructure Engineering (DIATI), Politecnico di Torino, Torino, Italy

Ana-Maria Vasiliu
"Petru Poni" Institute of Macromolecular Chemistry, Iasi, Romania

Ian Vázquez-Rowe
Peruvian LCA and Industrial Ecology Network (PELCAN), Department of Engineering, Pontificia Universidad Católica del Perú, San Miguel, Lima, Peru

Sandra Venghaus
Institute of Energy and Climate Research - Systems Analysis and Technology Evaluation (IEK-STE), Forschungszentrum Jülich, Jülich; RWTH Aachen University, School of Business and Economics, Aachen, Germany

Giulia Zarroli
CRETUS, Department of Chemical Engineering, Universidade de Santiago de Compostela, Santiago de Compostela, Spain; Department of Environment, Land and Infrastructure Engineering (DIATI), Politecnico di Torino, Torino, Italy

About the editors

Carmen Teodosiu is Professor and Director of the Department of Environmental Engineering and Management from 'Gheorghe Asachi' Technical University of Iasi (TUIASI), Romania. She obtained her chemical engineering degree at TUIASI (1981); post-graduate diploma (1995) and a Master of Science degree with distinction in Environmental Science and Technology (1996), both at IHE Institute of Hydraulics and Environmental Engineering, UNESCO-IHE Delft, The Netherlands; and her PhD in chemistry at TUIASI (1998). From 2008 to 2016, she was Vice-Rector for research and Director of the Doctoral Schools at TUIASI. Her research interests are in the fields of advanced wastewater treatment, integrated water resources management, and environmental and sustainability assessments.

Prof. Carmen Teodosiu published more than 198 scientific papers in international peer-reviewed journals, 29 book chapters, 8 patents, Hirsch index = 26 (Scopus and Web of Science or WoS), and H = 30 (Google Scholar). She was involved, as a coordinator/principal investigator, in 224 research projects funded by European and National agencies or industries. She is a PhD supervisor in the domains of chemical and environmental engineering, and 15 candidates received their PhD titles under her supervision. Prof. Teodosiu received the Doctor Honoris Causa award from Pannonia University Veszprem, Hungary. She is the initiator and chairperson of the International Conference in Environmental Engineering and Management (ICEEM), with its 11th edition in 2021 (four previous editions were organised in Hungary, Austria, Italy, and Switzerland, www.iceem.ro).

Prof. Carmen Teodosiu is a member of editorial boards and subject editor for international peer-reviewed journals such as *Sustainable Production and Consumption, Journal of Cleaner Production, Water,* and *Environmental Engineering and Management Journal.* She is also a guest editor of special issues of journals such as *Process Safety and Environmental Protection, Sustainable Production and Consumption, Water, Sustainability,* and *Environmental Engineering and Management Journal.*

Silvia Fiore is Associate Professor at Department of Environment, Land and Infrastructure Engineering (DIATI), Politecnico di Torino, Italy, since 2014. She obtained her chemistry degree at the University of Turin (1997) and a PhD in Geoenvironmental Engineering at Politecnico di Torino (2003). She instructs the courses 'Environmental Chemistry' and 'Circular Economy and Environmental Sustainability' at the Politecnico di Torino. She is responsible of the Circular Economy laboratory. Her current research interests are water and wastewater treatments, waste management and circular economy, and environmental and economic assessments of full-scale processes. She has published about 80 papers in peer-reviewed journals (H-index 22 on Scopus, 23 on Google Scholar), has supervised 10 PhD students (3 ongoing) and 12 research fellows, and was the principal coordinator of 11 research projects funded by European/National agencies and industries. She is currently involved in three Horizon 2020 projects (BEST4Hy, HYDRA, and NICE) and in one ERA-MIN2 project (BASH-TREAT). She has been Visiting Scholar at McGill University, Canada, in 2016. She was awarded Doctor Honoris Causa by '*Gheorghe Asachi*' Technical University of Iasi in 2019 and has been appointed Adjunct Research Professor (2018–24) in the Department of Chemical and Biochemical Engineering of Western University of Ontario, Canada. She is an editorial board member for *Environmental Engineering and Management Journal* and the *Sustainability journal*. She is also a guest editor for *Sustainable Production and Consumption, Water,* and *Sustainability* journals.

Almudena Hospido is Associate Professor at the Department of Chemical Engineering and a member of the Center for Interdisciplinary Research in Environmental Technologies (CRETUS), both at the Universidade de Santiago de Compostela (USC), Spain. Her current research interests are related to the application and development of life cycle assessment and environmental footprints for monitoring progress and support decisions, with a special focus on the wastewater sector (both treatment and valorisation) and on the food production and processing industry.

International collaboration has always been a key pillar of her research activities, with research stays in Sweden, Canada, and the United Kingdom, and an active role in European projects: at present participating in two H2020 actions (USABLE Packaging and

PROTECT) and in the past as coordinator of the H2020 CAS ENERWATER that developed, tested, and validated a method to measure, benchmark, and communicate the energy efficiency of wastewater treatment plants. The method has been adopted by the CEN/TC165 and approved in January 2021 as the *CEN/TR 17614: Standard method for assessing and improving the energy efficiency of wastewater treatment plans*.

She has published more than 70 papers in peer-reviewed journals (H-index = 35, Scopus) and 10 book chapters. She has supervised eight PhD theses (three of them ongoing). She is an editorial board member for the *International Journal of Life Cycle Assessment*, responsible for the wastewater section.

Almudena Hospido was also Vice-Rector of Internationalisation (September 2015–June 2018), CEO of Cursos Internacionais da USC (October 2015–June 2018), member of the Executive Committee of the Compostela Group of Universities (September 2015–June 2018), and USC International Relations Coordinator (September 2013–April 2014).

Acknowledgements

We acknowledge the important contributions of all authors to this book elaboration. Their scientific expertise, involvement, and dedication are highly appreciated, especially considering the time constraints related to the submissions and revisions of the chapters. It was a privilege for us to work on the planning, review, and development of the book. It was a long journey, and we hope that the readers will enjoy reading it and find useful insights for their research and professional development.

We express our gratitude to the Elsevier representatives, the Editorial Project Manager, Miss Aleksandra Packowska, and the Production Manager, Mr. Paul Prasad Chandramohan; their support for the book finalisation was constant and very important.

CHAPTER

An integrated approach to assess the sustainability progress

1

Carmen Teodosiu[a], Almudena Hospido[b], and Silvia Fiore[c]

[a]*Department of Environmental Engineering and Management, "Gheorghe Asachi" Technical University of Iasi, Iasi, Romania* [b]*CRETUS, Department of Chemical Engineering, Universidade de Santiago de Compostela, Santiago de Compostela, Spain* [c]*Department of Environment, Land and Infrastructure Engineering (DIATI), Politecnico di Torino, Torino, Italy*

1 The challenges of sustainable development and sustainability progress

During the last 40 years, the concepts of *Sustainable development* and *Sustainability* contributed to a paradigm shift of the economic and social development due to the finite character of the Earth resources that cannot support the actual population growth and the massive impacts of pollution events and climate change, with major effects on human health, ecosystems, and biodiversity preservation. Sustainable development was defined as the "*development that meets the needs of the present without compromising the ability of future generations to meet their own needs*" (WCED, 1987). This vision has been incorporated in international agreements and national legislation of many countries, and influenced industry, agriculture, services, and the urban and regional development (Ruggerio, 2021). Sustainable development is a cross-cutting subject with three dimensions: social, economic, and environment, which should be kept under dynamic equilibrium. Social sustainability refers to poverty reduction and social equity; economic sustainability refers to long-term sustainability of renewable and non-renewable resources providing long-term economic benefits; and environmental sustainability refers to preservation and maintenance of life forms that exist on earth, whilst reducing the pollution impacts (Kwatra, Kumar, & Sharma, 2020).

The concepts of sustainable development and sustainability are sometimes used as synonyms, although theoretical and methodological aspects have recently pointed out difference in their definitions (Ruggerio, 2021). According to this study, sustainability must comply with the following criteria: (a) account for the complexity of socio-ecological systems (SESs) by encompassing economic, ecological, social,

Assessing Progress towards Sustainability. http://doi.org/10.1016/B978-0-323-85851-9.00020-1

and political factors; (b) account for intergenerational and intragenerational equity; (c) address the hierarchical organisation of nature, by acknowledging the feedback between the SESs and their surroundings (Ruggerio, 2021).

The problems of sustainable development and sustainability are very challenging since they approach all compartments of development (economic, social, and environmental), they are based on scientific research and innovation in various fields and disciplines (formal, natural, social, and applied sciences) but also on various management, legislation, and governance concepts and practices. The indicator Earth Overshoot Day (Global Footprint Network, n.d) is a suggestive way to describe the rapid consumption of resources at national scales, and together with other pollution problems, major accidents/disasters and climate events (heatwaves, heavy precipitation, droughts, tropical cyclones) provide an overview of how unsustainable is the current format of our development.

The last Intergovernmental Panel on Climate Change (IPCC) scientific report (2021) formulated "a red code for humanity" and showed that "*it is unequivocal that human influence has warmed the atmosphere, ocean, land and widespread and rapid changes in the atmosphere, ocean, cryosphere and biosphere have occurred*". Moreover, the report has proven, by means of various scenarios, that to limit the future climate changes and to improve air quality it is absolutely necessary to decrease the human-induced global warming to a specific level which requires reducing cumulative CO_2 emissions, reaching at least net zero CO_2 emissions, along with strong reductions in other greenhouse gas emissions and CH_4 emissions (IPCC, 2021).

The 2030 Agenda for Sustainable Development, adopted by all United Nations (UN) Member States in 2015, provides a blueprint for peace and prosperity for people and the planet, for the present and future, having at its central place the 17 Sustainable Development Goals (SDGs), which are an urgent call for action by all countries—developed and developing—in a global partnership (UN-SDGs, 2019). The SDGs build on decades of work by countries and the UN, and has started in June 1992, at the Earth Summit in Rio de Janeiro, Brazil, and recognise that ending poverty and other deprivations must go hand in hand with strategies that improve health and education, reduce inequality, and spur economic growth—all the whilst tackling climate change and working to preserve our oceans and forests. Whilst also recognising the need to quantify sustainability progresses, in July 2017, the UN General Assembly adopted a global indicator list, including 232 different indicators. These indicators cover all the 169 targets of the 2030 Agenda (as some indicators are used to monitor more than 1 target, the list overall includes 244 indicators).

However, to monitor progresses towards sustainability, apart from the SDG indicators, other indicators, indices, and models have been developed and adapted for various systems to provide objective information that enable decision making and track the progress towards the proposed goals. Such indicators have been used globally to assess the progress made by regions towards a sustainable economy, society, and environment and indicate whether progress has been made, in the long term and short term, or if there is no significant progress at all (Kwatra et al., 2020). Various sustainability indicators have been used to assess, quantify, and evaluate performances of countries (Pitkänen et al., 2016), industries, e.g. mining (Fuentes, Negrete, Herrera-León, &

Kraslawski, 2021), bio-based chemicals (Van Schoubroeck, Van Dael, Van Passel, & Malina, 2018), real estate (Rogmans & Ghunaim, 2016), building industry (Gholami Rostam & Abbasi, 2021), contaminated site remediation (Li, Cundy, Chen, & Lyu, 2021), eco-industrial parks (Valenzuela-Venegas, Salgado, & Díaz-Alvarado, 2016), energy, conventional and renewables (Gunnarsdottir, Davidsdottir, Worrell, & Sigurgeirsdottir, 2020; Liu, 2014), agriculture (Nadaraja, Lu, & Islam, 2021), urban water systems (Spiller, 2016), climate change (Barry & Hoyne, 2021), etc.

Another approach to save the planet's environment was proposed by Rockström et al. (2009a, 2009b) through the concept of planetary boundaries (PBs) for critical interlinked biophysical processes, including climate change, biodiversity loss, biogeochemical flows, stratospheric ozone depletion, ocean acidification, global freshwater use, change in land use, atmospheric aerosol loading, and chemical pollution. PBs use various approaches to determine the environmental boundaries at sub-global scales and are increasingly integrated with the Environmental footprints and Life cycle assessment (LCA), with the potential to monitor the progress and gaps of the SDGs (Chen, Li, Li, & Fang, 2021).

The preoccupation to implement these concepts at various scales (for economic and social activities, at national/regional/local level, for all the development projects that influence the society–nature relationship), as well as for the assessment of the progress that is encountered in terms of sustainability, has been a consistent subject of researchers, as it may be observed by the number of research and review articles, book chapters, encyclopaedia that have been published in the last 10 years in the Science Direct database (Fig. 1). It is also important to note that almost half of this research is directed towards the assessment part that is associated also with

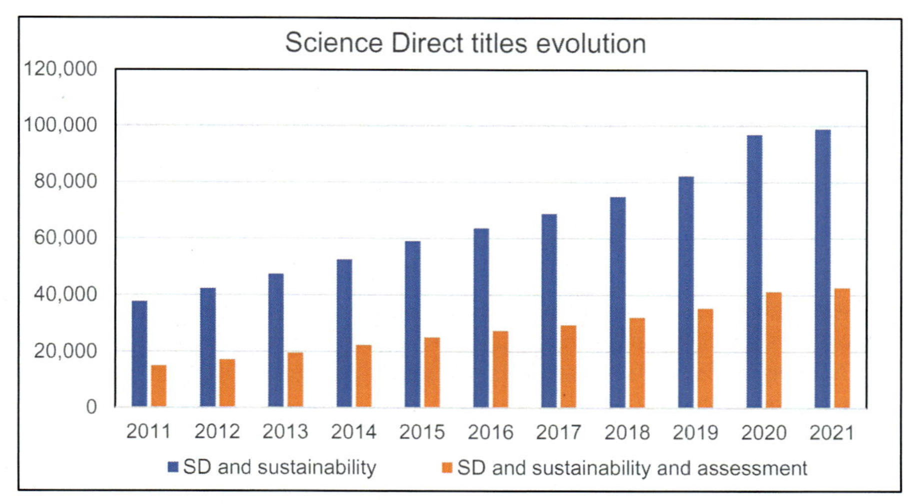

FIG. 1

Science Direct search by using the keywords: sustainable development, sustainability, and assessment (the number of titles represents research and review articles, book chapters, and encyclopaedia).

the practical applications of these concepts in various fields and geographical scales.

Although many books that have been published deal with sustainable development, sustainability, and case studies, the focus on a quantitative viewpoint of sustainability concepts and assessment tools as well as its materialisation in diverse case studies to illustrate the decrease of the environmental impacts and risks and the increase in resources and energy efficiency is, to the best of our knowledge, lacking.

2 The book vision and approach

This book provides an integrated approach of sustainability progress measurement, by considering both the frameworks and methodological developments of various tools, and their implementation in assessing sustainability of processes, products, and services, through a selection of case studies that brings together diverse sectors and a wide international perspective. By doing so, this book provides examples and incentives for introducing sustainability concepts and assessment tools, for diminishing environmental impacts and risks, and for increasing resources and energy efficiency, and is addressing a large audience such as environmental engineers, environmental scientists, managers, consultants, academics, researchers, chemical engineers, NGO representatives, etc.

2.1 Frameworks used to evaluate the sustainability progress

A framework is defined as "a system of rules, ideas, or beliefs that is used to plan or decide something" (Cambridge Dictionary, n.d). For the sustainability assessments, the frameworks are very important because they are associated to specific tools, principles, indicators, policies, and therefore all these issues might influence the implementation of sustainable development principles. The integrated approach proposed by this book and its sections is presented in Fig. 2.

In this figure, the geographic boundaries refer to the implementation at various levels (country, regional, local), whilst the production and consumption refer to the assessment subject, i.e. industrial/agricultural processes, services, products. The policies and specific legislation, and the institutional development capacity are also part of the decisional context.

Sustainable development and sustainability have been the conceptual background of frameworks and approaches such as Green economy (Pitkänen et al., 2016; Wanner, 2015), Circular economy (Schroeder, Anggraeni, & Weber, 2019; Suárez-Eiroa, Fernández, Méndez-Martínez, & Soto-Oñate, 2019; Geissdoerfer, Savaget, Bocken, & Hultink, 2017), nexus (Parsa, Van De Wiel, & Schmutz, 2021), Sustainable consumption and production (UNEP, 2015), European green deal

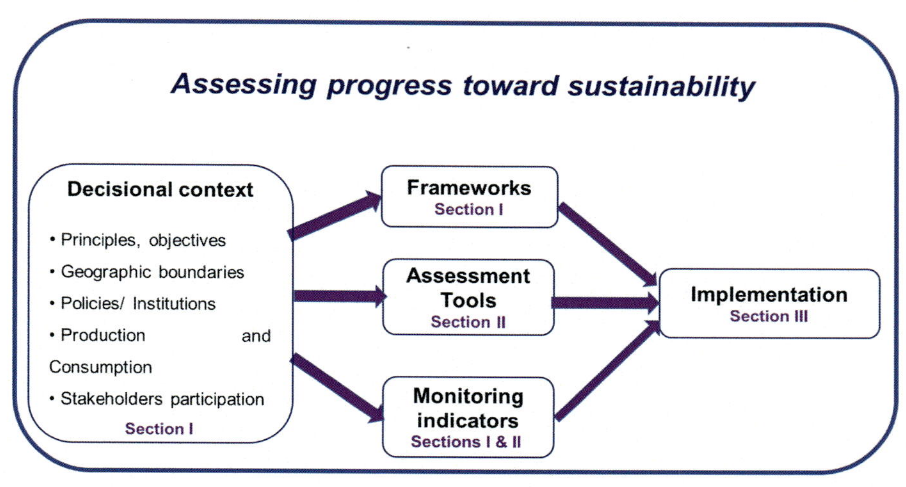

FIG. 2

Integrated approach to assess the sustainability progress.

(https://ec.europa.eu/info/strategy/priorities-2019-2024/european-green-deal_en), Planetary boundaries (Chen et al., 2021). So, this book presents in Section I (i.e. Chapters 2–5) a selection of the most important frameworks to assess the sustainability progress described as follows:

- **Chapter 2. Sustainable development and its goals** (SDGs) presents the concept and describes the relationships between its objectives and sustainability. The relationships are discussed along with the potential contributions from the academia for their application, in the absence of a scientific model linking the SDGs and sustainability.
- **Chapter 3. Sustainability and the circular economy** considers an approach to close the material flows in order to increase economic activity or market share, without overlooking the equity dimension, by introducing the Performance Economy model. This model concentrates on making best use of stocks and extending product life through reuse, remanufacturing, and reprocessing and shifting from non-renewable inputs (including energy) to renewable inputs (including labour).
- **Chapter 4. The Food–Energy–Water Nexus Approach** is a research framework containing five different spaces: the natural, technical, economic, political, and social. In line with this research focus, the important system elements are defined and three consecutive rules allowing a structured nexus analysis are presented.
- **Chapter 5. The European Green Deal (EGD) in the Global Sustainability Context.** This framework builds on the sustainability concept and the related scientific support, pointing out the need to decouple economic growth from

resources use and environmental impacts to stay within the planetary boundaries. The EGD, together with the lessons of the 2008–2009 financial crisis, offers a global reference for shaping sustainable and resilient economies in the recovery from the Covid-19 crisis.

2.2 Assessment tools for sustainability—Methodological issues

Section II (i.e. Chapters 6–11) introduces some of the methodologies that are more frequently used to assess, quantify, and communicate sustainability. A common structure has been defined for all the chapters to facilitate the reading of the whole section, considering: (i) goal and target audience, (ii) key concepts, (iii) tool description and software used, (iv) data requirements, (v) scale of analyses, and (vi) future developments.

- **Chapter 6. Life Cycle Sustainability Assessment-based tools** establishes a baseline of ideas of what Life Cycle Thinking means: going beyond the traditional focus, including the whole environmental, social, and economic implications of decision-making processes to identify potential conflicts, synergies, and trade-offs. The life cycle tools (life cycle assessment, life cycle costing, and social life cycle assessment) are described, offering an overview of the available software, databases, and ongoing challenges.
- **Chapter 7. Footprint tools** presents a variety of methods that have been developed over the last 30 years to better understand the environmental impacts of human activities. The "family of footprint tools" covers a broad range of environmental sustainability issues and various scales of application, whether they focus on resources or emissions, impacts or inventories, single or multiple issues.
- **Chapter 8. The combined use of Life Cycle Assessment** (LCA) **and Data Envelopment Analysis** (DEA) may be used to analyse the environmental efficiency of a wide spectrum of production systems. This chapter presents a critical review on current practices of the joint application of LCA + DEA, as well as point towards the methodological challenges and opportunities for future applications.
- **Chapter 9. Territorial Life Cycle Assessment** presents the principles and methodology of LCA adapted to be applied to territories, such as cities, metropolitan areas, agricultural areas, or regions. Territorial LCA approaches have two main goals: (i) providing an environmental diagnosis and (ii) comparing the performances of land planning scenarios. Combinations with other tools such as metabolism studies, Geographic Information System (GIS), or economic modelling are also discussed.
- **Chapter 10. Environmental Impact and Risk Assessment** presents the main goals, key concepts, assumptions, and target audience for the impact quantification but also in association with the risk assessment methodology and

integrated methods. New indices developed for impact and risk quantification are also presented.

- **Chapter 11. Multi-Criteria Decision Making** (MCDM) method offers a scientifically sound decision support, which can provide a comprehensive and transparent basis for sustainability assessments. The chapter aims at highlighting the main steps, advantages, and the risks of developing a MCDM framework for the resolution of complex environmental problems. Further information about software applications, data requirements, scale of the analyses, target audience, and future trends are provided as well.

2.3 Implementation concepts and case studies

A selection of case studies is compiled in the last section (Section III: Chapters 12–20) of the book with the aim of accomplishing an in-depth knowledge of some of the tools presented before (i.e. learning by doing). The sustainability of several products, processes, and services, covering from material development to food and energy production, to urban water systems or bioplastic production and processing, is evaluated by the application of different tools, methods, or frameworks. These cases are presented in a unitary format, strengthening the actual challenges related to applications and the practical usage of sustainability assessments (environmental, social, economic), providing also policy recommendations.

- **Chapter 12. LCA for eco-design in product development** explores the environmental performance of novel materials (organic/inorganic composites and sorbents obtained from waste rapeseed biomass) used in water/wastewater treatment. LCA is used in support to eco-design to identify and quantify potential environmental impacts, which may appear during the early stages (laboratory synthesis and testing) of the development of these products.
- **Chapter 13. LCA for the design of a pilot recovery plant in the European context** presents the eco-design of a low-impact environmental recovery plant to value brewery by-products for aquaculture feed ingredients. LCA was used from the preliminary design steps for decision making, considering different passive energy saving measures and material alternatives to reduce the impacts of the building phase.
- **Chapter 14. LCA and food and personal care products sustainability: Case studies of Thai Riceberry rice products** investigates the use of LCA and Eco-efficiency for evaluating the environmental and economic performances of various kinds of riceberry rice products (food and personal care products).
- **Chapter 15. Environmental and economic sustainability of cocoa production in West sub-Saharan Africa** assesses the water use related to the cultivation of cocoa from 2010 to 2017. The analysis of the water footprint was made at unitary (crop's water use efficiency) and aggregated (annual production's water) scale. Also, the economic aspects related to the production and distribution of cocoa beans and derivatives were evaluated.

- **Chapter 16. Environmental assessment of urban water systems: LCA case studies** provides an overview of LCA studies applied to these systems and describes the related challenges, firstly as a state of the art and secondly through three case studies, describing construction and operation inventories and including different elements of the UWS.
- **Chapter 17. Environmental sustainability in energy production systems** presents the LCA of bioenergy production from pruning and forest residues. The cradle-to-energy factory gate approach was considered for medium/small installations, discussing the methodological issues, identifying the environmental hotspots, and proposing future research activities.
- **Chapter 18. Sustainability assessment of biotechnological processes: LCA and LCC of second-generation biobutanol production** is focused on the production of biobutanol—a second-generation biofuel—from renewable resources (agro-food waste) via the biotechnological route. The comparison of the environmental performances with the fossil-based alternative was presented and the key hotspots identified.
- **Chapter 19. Footprint Assessment of Solid Waste Management Systems** analyses the carbon footprint of different solid waste management systems (SWMS). The waste categories under investigation were municipal solid waste (MSW), waste of electrical and electronic equipment (WEEE), packaging waste (PW), and biowaste. Four aggregated models were used to calculate the carbon footprint and GHG efficiency indicator.
- **Chapter 20. How can we validate the environmental profile of bioplastics? Towards the introduction of Polyhydroxyalkanoates (PHA) in the value chains.** This chapter presents the state of the art of the LCA of biodegradable plastics produced from renewable sources like PHA. The literature published in the last 20 years was analysed and the methodological features (functional unit and allocation choices, data sources) discussed.

The book reaches its end with Chapter 21 that presents the main conclusions and future research directions.

References

Barry, D., & Hoyne, S. (2021). Sustainable measurement indicators to assess impacts of climate change: Implications for the New Green Deal Era. *Current Opinion in Environmental Science & Health*, *22*, 100259. https://doi.org/10.1016/j.coesh.2021.100259.

Cambridge Dictionary. https://dictionary.cambridge.org/.

Chen, X., Li, C., Li, M., & Fang, K. (2021). Revisiting the application and methodological extensions of the planetary boundaries for sustainability assessment. *Science of the Total Environment*, *788*. https://doi.org/10.1016/j.scitotenv.2021.147886.

Fuentes, M., Negrete, M., Herrera-León, S., & Kraslawski, A. (2021). Classification of indicators measuring environmental sustainability of mining and processing of copper. *Minerals Engineering*, *170*. https://doi.org/10.1016/j.mineng.2021.107033.

Geissdoerfer, M., Savaget, P., Bocken, N., & Hultink, E. J. (2017). The circular economy—A new sustainability paradigm? *Journal of Cleaner Production*, *143*, 757–768. https://doi.org/10.1016/j.jclepro.2016.12.048.

Gholami Rostam, M., & Abbasi, A. (2021). A framework for identifying the appropriate quantitative indicators to objectively optimize the building energy consumption considering sustainability and resilience aspects. *Journal of Building Engineering*, *44*, 102974. https://doi.org/10.1016/j.jobe.2021.102974.

Global Footprint Network. https://www.footprintnetwork.org/.

Gunnarsdottir, I., Davidsdottir, B., Worrell, E., & Sigurgeirsdottir, S. (2020). Review of indicators for sustainable energy development. *Renewable and Sustainable Energy Reviews*, *133*, 110294. https://doi.org/10.1016/j.rser.2020.110294.

IPCC. (2021). Climate change 2021: The physical science basis. In *Contribution of working group I to the sixth assessment report of the intergovernmental panel on climate change [Masson-Delmotte, V., P. Zhai, A. Pirani, S. L. Connors, C. Péan, S. Berger, N. Caud, Y. Chen]* (p. 3949). Cambridge Univ. Press.

Kwatra, S., Kumar, A., & Sharma, P. (2020). A critical review of studies related to construction and computation of sustainable development indices. *Ecological Indicators*, *112*, 106061. https://doi.org/10.1016/j.ecolind.2019.106061.

Li, X., Cundy, A. B., Chen, W., & Lyu, S. (2021). Systematic and bibliographic review of sustainability indicators for contaminated site remediation: Comparison between China and western nations. *Environmental Research*, *200*, 111490. https://doi.org/10.1016/j.envres.2021.111490.

Liu, G. (2014). Development of a general sustainability indicator for renewable energy systems: A review. *Renewable and Sustainable Energy Reviews*, *31*, 611–621. https://doi.org/10.1016/j.rser.2013.12.038.

Nadaraja, D., Lu, C., & Islam, M. M. (2021). The sustainability assessment of plantation agriculture—A systematic review of sustainability indicators. *Sustainable Production and Consumption*, *26*, 892–910. https://doi.org/10.1016/j.spc.2020.12.042.

Parsa, A., Van De Wiel, M. J., & Schmutz, U. (2021). Intersection, interrelation or interdependence? The relationship between circular economy and nexus approach. *Journal of Cleaner Production*, *313*, 127794. https://doi.org/10.1016/j.jclepro.2021.127794.

Pitkänen, K., Antikainen, R., Droste, N., Loiseau, E., Saikku, L., Aissani, L., et al. (2016). What can be learned from practical cases of green economy? –Studies from five European countries. *Journal of Cleaner Production*, *139*, 666–676. https://doi.org/10.1016/j.jclepro.2016.08.071.

Rockström, J., Steffen, W., Noone, K., Persson, Å., Chapin, F. S., III, Lambin, E. F., et al. (2009a). A safe operating space for humanity. *Nature*, *461*, 472–475.

Rockström, J., Steffen, W., Noone, K., Persson, Å., Chapin, F. S., III, Lambin, E., et al. (2009b). Planetary boundaries: Exploring the safe operating space for humanity. *Ecology and Society*, *14*(2), 32.

Rogmans, T., & Ghunaim, M. (2016). A framework for evaluating sustainability indicators in the real estate industry. *Ecological Indicators*, *66*, 603–611. https://doi.org/10.1016/j.ecolind.2016.01.058.

Ruggerio, C. A. (2021). Sustainability and sustainable development: A review of principles and definitions. *Science of the Total Environment*, *786*, 147481. https://doi.org/10.1016/j.scitotenv.2021.147481.

Schroeder, P., Anggraeni, K., & Weber, U. (2019). The relevance of circular economy practices to the sustainable development goals. *Journal of Industrial Ecology*, *23*, 77–95. https://doi.org/10.1111/jiec.12732.

Spiller, M. (2016). Adaptive capacity indicators to assess sustainability of urban water systems—Current application. *Science of the Total Environment*, *569–570*, 751–761. https://doi.org/10.1016/j.scitotenv.2016.06.088.

Suárez-Eiroa, B., Fernández, E., Méndez-Martínez, G., & Soto-Oñate, D. (2019). Operational principles of circular economy for sustainable development: Linking theory and practice. *Journal of Cleaner Production*, *214*, 952–961.

UN-SDGs. (2019). *United Nations sustainable development goals platform*. https://sdgs.un.org/goals. Accessed 17 August 2021.

United Nations Environment Programme (UNEP). (2015). *Sustainable consumption and production, a handbook for policy makers*. ISBN: 978-92-807-3364-8.

Valenzuela-Venegas, G., Salgado, J. C., & Díaz-Alvarado, F. A. (2016). Sustainability indicators for the assessment of eco-industrial parks: Classification and criteria for selection. *Journal of Cleaner Production*, *133*, 99–116. https://doi.org/10.1016/j.jclepro.2016.05.113.

Van Schoubroeck, S., Van Dael, M., Van Passel, S., & Malina, R. (2018). A review of sustainability indicators for biobased chemicals. *Renewable and Sustainable Energy Reviews*, *94*, 115–126. https://doi.org/10.1016/j.rser.2018.06.007.

Wanner, T. (2015). The new 'passive revolution' of the green economy and growth discourse: Maintaining the 'sustainable development' of neoliberal capitalism. *New Political Economy*, *20*, 21–41. https://doi.org/10.1080/13563467.2013.866081.

WCED. (1987). *Our common future, Brundtland report*. https://sustainabledevelopment.un.org/content/documents/5987our-common-future.pdf.

SECTION

I

Frameworks for assessing sustainability

CHAPTER

Sustainable development and its goals

2

Biagio F. Giannetti, Feni Agostinho, and Cecilia M.V.B. Almeida
Programa de Pós-graduação em Engenharia de Produção, Laboratório de Produção e Meio Ambiente, Universidade Paulista (UNIP), São Paulo, Brazil

1 Introduction

For almost 50 years, a sequence of World Summits has been creating awareness that succeeding sustainable development in the 21st century is not a choice but an imperative (Table 1).

In September 2000, prior to the Johannesburg Summit, political leaders from around the world set the targets for Millennium Development Goals related to the challenges of sustainable development. Poverty, hunger, education, gender, health, environmental sustainability, and a global partnership for development were set as priorities.

Finally, in 2015, the member countries of the United Nations (UN) Organisation signed the document "Transforming Our World: the 2030 Agenda for Sustainable Development." This agreement provides that all signatories strengthen actions for the planet to move towards sustainable development, on the basis of poverty eradication, economic growth, and environmental protection in an integrated and transversal way. Within the scope of the 2030 Agenda, the 17 Sustainable Development Goals (SDGs), which provide elements to guide global development, governments, the private sector, the civil society, and the academia to develop aligned initiatives with the Agenda.

The SDGs are universal, which means that they should apply to all countries in the world. However, its realisation will depend on the ability to make them reality at different scales, depending on individual capabilities and development priorities. There is a process for taking into account scale contexts in carrying out the 2030 Agenda, from the setting of objectives and goals until the determination of the means of implementation, as well as the use of indicators to measure and track progress. It thus refers both to the way in which local and regional governments can support the achievement of the SDGs through 'bottom-up' shares, as to how the SDGs can provide a framework for a local development policy.

Assessing Progress towards Sustainability. https://doi.org/10.1016/B978-0-323-85851-9.00009-2

Table 1 Sequence of World Summits, main calls and features, with the Brundtland Report highlighted in *green*.

Year		Highlights	Main features
1972	UN conference Stockholm	Conserving biodiversity to ensure human rights to a healthful world	• Developing countries argued that their priority was development • Developed countries defended environmental protection and conservation.
1982	Nairobi Summit	Called upon intensifying efforts to protect the environment and the necessary international cooperation	• Tensions between Western Governments and the former Soviet Union flawed progress towards action
1987	Brundtland Report	Concurrent equity, growth, and environmental maintenance should be possible Each nation should be capable of achieving full economic potential while enhancing its resource base	
1992	Earth Summit	Conveyed governments to negotiate an agenda for environment and development Nongovernmental organisations discussed strategies and actions towards sustainable development	• The Rio Declaration established broad principles to guide national conduct on environmental protection and development
2002	Johannesburg World Summit	The low progress in turning Agenda 21 into actions led to the thrust on public–private partnerships for sustainable development	• Most of partnerships agreements failed to be implemented

Within academia, where each researcher has such different characteristics and capabilities, the process is of fundamental importance. This chapter, in addressing the introduction of the SDGs in several and heterogeneous research areas, brings the discussion about SDGs to another level. Furthermore, it promotes reflection on how researchers can make a critical contribution for the general scope of the SDGs and provide a framework for future research, in the sense of seeking strategies for sustainable development that can be adapted to specific contexts.

2 The relationships between the SDGs and sustainability

Before exploring the relationships between the SDGs and sustainability, it is important to note that the Sustainable Development Goals have a strong normative component and, therefore, will always have a significant ideological component. With regard to the UN, there are two predominant approaches. The first, a neorealist approach that considers the international system anarchic with unregulated competition predominating between states, always leading to the maximisation of the national interest of each country in detriment of an international cooperation. Within this approach, regulation and standardisation produced by the UN would be of very low relevance and effectiveness. The second, a liberal-institutionalist approach that considers that, gradually multilateral international organisations would have created a system of global governance, placing limits on national sovereignty. In this case, the complex and laborious elaboration of the SDGs, with a wide participation of multiple governmental, private, philanthropic, and social movements, could result in the evolution of humanity.

By 2030 it is expected that the SDGs will have produced a new minimum civilising consensus on almost all the relevant issues of human life, from the eradication of poverty to the development of smart cities and the protection of the oceans. Of course, some relevant issues have been left out, either because they would not reach consensus or because they are not yet in the general public radar, such as the nature of the political regimes, reduced military spending, or the regulation of disruptive technologies such as artificial intelligence and synthetic biology.

The 2030 Agenda declared the 17 Sustainable Development Goals highlighting their cohesive and indissoluble nature (UN, 2015), and trying to overcome the deficiencies of integration amongst policies, strategies, and applications that have been noticed as key drawbacks of the prior efforts towards sustainable development, such as the preceding Millennium Development Goals (Le Blanc, 2015; Liu et al., 2018; Obersteiner et al., 2016). The sustainable development targets are supposed to cover the scopes of sustainable development—economic, environmental, and social—and their institutional and/or governance structures (Costanza et al., 2016). Therefore integrated approaches are required to clarify and evidence the interdependencies amongst the SDGs and to expedite their effective implementation.

The understanding of the SDGs embodies a foremost accomplishment for evaluating environmental, social, and economic improvements (UN, 2017a) and directing forthcoming developments (Ripple et al., 2017—with 15,364 scientist signatories from 184 countries). Worldwide, authorities have committed to safeguard the environment, foster equity, and encourage sustainable development, whilst concurrently acknowledging the connections amongst these objectives for achieving social well-being. In this regard, several authors provided models to understand the SDGs to help policy makers and society to identify the SDG connections and priorities. The wedding cake model (Fig. 1A) was developed by the Stockholm Resilience Centre (Stockholm Resilience Centre, 2016) and set the priority on the biosphere—which holds economy and society—and biosphere health as a precondition to economic development and human well-being. However, the goals were also organised in a way in which people, environment, and spiritual issues are linked to determine the pathway to sustainability (SDG pyramid, 2019) prioritising human welfare and human actions. Whatever the mental model considered, the role of the scientific community in understanding and reaching the SDGs is vital. Encouragement to research on sustainability—and on the tools to achieve it—allows the protection

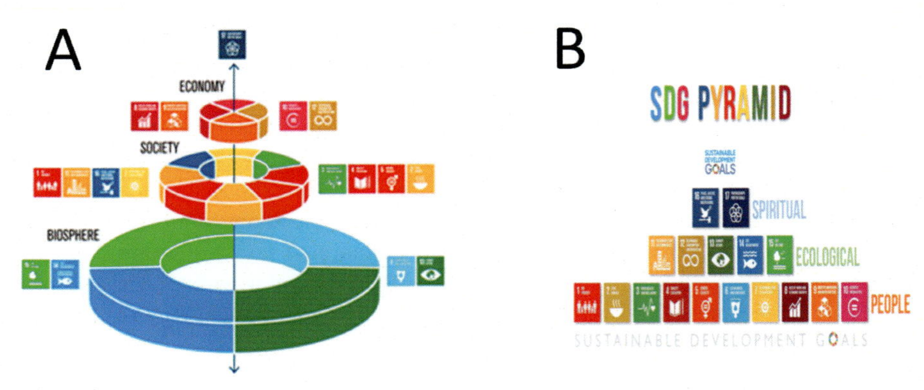

FIG. 1

Proposed models for the Sustainable Development Goals. (A) The wedding cake model and (B) the pyramid model.

of the environment, the economy, and equity, and act for a first step of the coalition between sustainable development and current policies—which still do not prevent society from surpassing the limits of natural resources exploitation—that is, to succeed in reaching the SDGs (Leal Filho et al., 2018).

The relationships amongst the 17 SDGs have also been explored in recent publications, since prioritising one or other SDG may trigger conflicting results (Weitz, Carlsen, Nilsson, & Skånberg, 2017). Pradhan, Costa, Rybski, Lucht, and Kropp (2017) analysed the SDG interactions identifying synergies and trade-offs using data from 227 countries. Taking SDG couples, synergies and trade-offs were ranked on national and global scales, and amongst them SDG 1 was found synergic with most goals, whilst SDG 12 was mostly connected to trade-offs. For these authors, the accomplishment of the SDGs will deeply depend on leveraging the identified negotiating trade-offs. van Soest et al. (2019) proposed the use of integrated assessment models to analyse the interactions/connections amongst the SDGs showing how modelling can contribute to policy logic and coherence.

This section briefly describes each of the 17 objectives, starting from different angles and theoretical perspectives, but which complement each other and enrich the proposed debate.

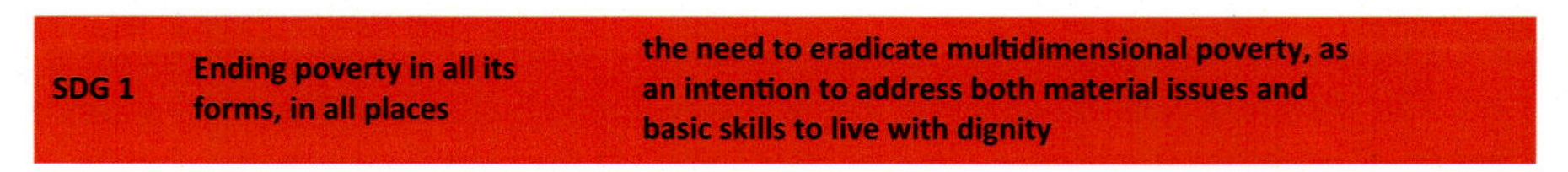

The first SDG recognises that the eradication of poverty, in all its forms, is the greatest global challenge to achieve sustainable development. There is a need for creating solid political frameworks, at national, regional, and international levels, based on pro-poor and gender-sensitive development strategies to support accelerated investments in poverty eradication actions. The academia has an important role in developing new ideas and new tools to build the resilience of the poor and those in a vulnerable situation and reduce their exposure and vulnerability to extreme events related to climate and other economic, social, and environmental blows and calamities.

SDG 2	Ending hunger, achieving food security and improved nutrition and promoting sustainable agriculture	To deal with hunger as a political problem and the questioning of alternative development models, as well as the intrinsic relationship between food habits, organization of the territory and the critique of the hegemonic economic model

During the past two decades, rapid economic growth and the development of agriculture have been responsible for reducing the proportion of malnourished people in the world. SDG 2 aims to end all forms of hunger and malnutrition by 2030, to ensure that everyone—especially children—has sufficient access to nutritious food every year. To achieve this goal, it is necessary to promote sustainable agricultural practices, through research and extension of agricultural services, technology development, in order to increase agricultural production capacity, in particular in the least developed countries. Scientists may help to maintain the genetic diversity of seeds, cultivated plants, domesticated animals and their respective wild species and develop ways to ensure equitable sharing of the resulting benefits. In this area, the academia can effectively contribute to the development of sustainable food production systems and the implementation of robust agricultural practices, which strengthen the capacity to adapt to climate change and progressively improve the quality of land and soil and help to maintain the ecosystems health.

SDG 3	Ensuring a healthy life and promoting well-being for all	the promotion of integral health of individuals and social groups depends on cross-cutting and innovative strategies

Historic progress has been made in reducing child mortality and fighting the most diverse diseases, such as HIV and malaria, but despite progress, chronic diseases and those resulting from disasters—and now pandemics—continue to be the main factors contributing to poverty and the deprivation of the most vulnerable. The SDG 3 addresses the promotion of health and well-being as essential to the promotion of human capacities.

The obvious contribution of science is the research and development of vaccines and medicines for communicable and non-communicable diseases, which mainly affect developing countries, but it is also important to strengthen the capacity of all countries, particularly developing countries, to maintain educational institutions to develop ways to substantially reduce the number of deaths and illnesses from dangerous chemicals and from contamination and pollution of air, water, and soil to avoid and pandemics avoiding preventable deaths.

SDG 4	Quality education	Ensuring inclusive and equitable and quality education, and promoting lifelong learning opportunities for all

Progress has been made in promoting universal access to primary education for children around the world, but all levels of education are included in the SDG 4. The promotion of training and empowerment of individuals is at the heart of this objective, which aims to expand the opportunities of the most vulnerable people on the path of development, also contributing to SDG 3. There is a need to substantially increase not only the number of qualified teachers, but also the quantity of higher education institutions to promote the acquisition of knowledge and skills necessary to foster sustainable development. The achievement of SDG 4 would substantially increase the number of people who have relevant technical and professional skills, for employment, decent work, and entrepreneurship, contributing, then, to SDG 8.

SDG 5	**Achieving gender equality and empowering women**	**considering that the many obstacles are also diversified according to the different cultural, and social contexts**

Gender equality is not only a fundamental human right but also the necessary basis for building a peaceful, prosperous, and sustainable world. The effort to reach SDG 5 runs across all the other SDGs and reflects the growing evidence that gender equality has multiplier effects on sustainable development and that promoting the empowerment of women can help emphatically in promoting development sustainable, through their participation in politics, the economy, and in various areas of decision-making. Social sciences may help to strengthen policies and legislation to promote gender equality and empower all women ensuring equal opportunities for leadership at all levels of decision-making in political, economic, and public life.

SDG 6	**Clean water and sanitation**	**Ensuring availability and sustainable management water and sanitation for all**

Water is at the basis of sustainable development. Water resources, as well as the services associated with them, underpin efforts to eradicate poverty, economic growth, and environmental sustainability. Access to water and sanitation matters for all aspects of human dignity: from food and energy security to human and environmental health. For SDG 6, several areas of science can make a direct contribution, such as international cooperation for sharing of water treatment technologies, protection restoration of water-related ecosystems, implementation of integrated water resources management, including via cross-border cooperation, increase the efficiency of water use in all sectors, and ensure sustainable withdrawals and fresh water supply to address water scarcity. Science and technology can also contribute to improving water quality by reducing pollution, eliminating waste and minimising the release of hazardous materials, and substantially increasing safe recycling and reuse globally at a crossing with SDGs 9, 11, and 12. In achieving access to adequate and equitable sanitation and hygiene for all, and universal and equitable access to safe, accessible, and safe drinking water for all crossing SDG 2.

SDG 7	**Affordable and Clean Energy**	**reliable, sustainable, modern and affordable access to energy for all, focusing on sustainability-related challenges and energy justice.**

Fossil fuels, still affordable, cause drastic changes in the climate, and for the next few years, the trend is to increase the demand for cheap and renewable energy. Meeting the needs of the economy and protecting the environment is one of the greatest challenges for sustainable development. In this sense, SDG 7 recognises the importance and outlines goals focused on energy transition, from non-renewable and polluting sources, to clean renewable sources, with special attention to the needs of people and countries in the most vulnerable situations. In this area, academia has contributed very strongly both in the theoretical and practical fields by expanding infrastructure and modernising technology to provide sustainable/renewable energy, strengthen international cooperation to facilitate access to clean energy research and technologies, including renewable energy, energy efficiency, and advanced and cleaner fossil fuel technologies, and substantially increasing the share of renewable energy in the global energy matrix.

SDG 8	Decent Work and Economic Growth	Promote sustained, inclusive and sustainable economic development, full and productive employment and decent work for all

Economic revitalisation contributes to creating better conditions for stability and sustainability. It is well known that inequality undermines economic growth and the achievement of sustainable development, since the most vulnerable often have lower life expectancies, in opposition to SDG 3, and cause a vicious cycle of school failure and low qualifications (SDG 4) and few prospects for quality jobs. SDG 8 recognises the urgency to eradicate indecent labour (forced, slavery) to guarantee everyone the full reach of their potential and capabilities. In this case, mainly the economic and social sciences can assist in strengthening the capacity of financial institutions to increase the access to banking, financial, and insurance services for all. Moreover, it is important to develop and implement policies to protect labour rights and promote safe and protected work environments reducing the proportion of people without jobs, education, or training (SDGs 4 and 5). Economic growth should turn into economic development by reassessing the consumption and production efficiencies of global resources (SDG 12) and decoupling economic growth from environmental degradation. This SDG in particular depends on the realisation of several other indicators, which in turn depend on the integration of various areas of science such as social sciences to promote development-oriented policies that support decent job creation, entrepreneurship, creativity and innovation, and engineering to provide clean water (SDG 6), affordable energy (SDG 7), or industry innovation and technological modernisation (SDG 9).

SDG 9	Industry, Innovation and Infrastructure	Promoting inclusive and sustainable industrialization, fostering technological innovation and enabling transformative, disruptive and regenerative replacements.

Infrastructure and innovation are basic conditions for sustainable economic development through the promotion of energy efficiency (SDG 7), social inclusion (SDGs 4,

5, and 8), and technological progress (SDG 12). Ensuring equal access to technologies is crucial to promoting knowledge for all, and the SDG 9 includes the development of resilient and modern structures (SDG 11), efficient and productive systems (SDG 12). Infrastructure includes the access to information and communication technologies and scientific research (SDG 4).

SDG 10	Reducing inequality within countries and among them	confronting territorial inequalities

Income inequality and wealth distribution within countries have crippled efforts to achieve development and expand people's opportunities. The strategic idea of this SDG is to build on SDG 1 in all its dimensions, reducing socioeconomic inequalities and combating discrimination of all kinds (SDG 5). The scope depends on all sectors for the promotion of opportunities for the most excluded people on the path of development. An important focus of SDG 10 is the contemporary challenge ensuring a stronger representation of developing countries in global economic decision-making and improving regulation and monitoring of global markets adopting wage and social protection policies to progressively achieve greater equality. This includes the elimination of discriminatory laws, policies, and practices to empower and promote social, economic, and political inclusion for all.

SDG 11	Sustainable Cities and Communities	Making cities and human settlements inclusive, safe, resilient and sustainable

Significantly transforming the construction and management of urban spaces is essential if sustainable development is to be achieved. Themes intrinsically related to urbanisation, such as mobility, solid waste management, and sanitation, are included in the goals of SDG 11, as well as the planning and resilience of human settlements. SDG 11 includes technical and financial assistance for sustainable and robust buildings, using local materials, and adopting and implementing integrated policies for inclusion (SDGs 5, 8, and 10), resource efficiency (SDGs 6, 7, and 12), mitigation and adaptation to climate change (SDG 13), and resilience to disasters (SDGs 14 and 15). Support the balance amongst economic, social, and environmental relations, strengthening national and regional development planning, and providing access to safe, inclusive, accessible, and green public spaces paying special attention to air quality (SDG 130), municipal waste management (SDGs 12 and 14), guaranteeing access to safe, adequate, and affordable housing, and to basic services (SDGs 6 and 7).

SDG 12	Responsible Consumption and Production	Ensuring sustainable production and consumption patterns

SDG 12 calls for changing consumption and production patterns as essential measures to reduce the ecological footprint on the environment. These measures are the basis for sustainable economic and social development and must promote the efficient use of energy (SDG 7) and natural resources (SDG 6), sustainable infrastructure (SDG 9 and 11), and access to basic services (SDG 3). In addition, SDG

12 prioritises information, coordinated management, transparency, and accountability on consuming natural resources as tools for achieving more sustainable standards of production and consumption. The key idea here is to deal with the inefficient structures and practices in dealing with resources and energy usage, whilst promoting awareness about sustainable consumption and encouraging lifestyles in harmony with nature. Scientists from all areas can contribute to this goal by promoting sustainable public procurement practices and integrating sustainability information prevention, reduction, recycling and reuse practices at the industrial, retail, and consumer levels, to reduce losses along production and supply chains.

SDG 13	Action Against Global Climate Change	Take urgent action to combat climate change and its impacts

As a transnational event, climate change requires immediate action and the establishment of SDG 13 just to deal with climate issues is strategic for promoting the necessary changes to prevent projections from becoming a reality. In order to achieve SDG 13, integration and collaboration are needed between various areas of science that will allow the design of mechanisms for effective management of climate change causes and effects with a focus on women (SDG 5) and marginalised communities (SDGs 10 and 11), improved education (SDG 4), increased human and institutional awareness and capacity on global climate mitigation (SDG 12), adaptation (SDG 11) reinforcing resilience and the ability to adapt to risks related to climate and natural disasters in all countries.

SDG 14	Life on the Water	Conserving and promoting the sustainable use of oceans, seas and marine resources for sustainable development

The role of the oceans both in supporting life and in regulating various global biosphere processes is well known. Oceans give support to food security (SDG 2), transportation (SDG 12), energy supply (SDG 7), tourism (SDG 8), amongst others. In addition, by regulating their temperature, chemistry, currents, and life forms, the oceans regulate many of the most critical ecosystem services on the planet, such as the carbon and nitrogen cycle, climate regulation, and oxygen production. Oceans also contribute to the global economy. Social sciences may help by ensuring implementation of international laws. Economic sciences may help in providing small-scale fishing with access to markets and increased scientific knowledge in several areas will contribute to develop research capacities/technologies to improve the health of the oceans and preserve marine biodiversity, minimise oceans' acidification, whilst ensuring healthy and productive oceans.

SDG 15	Earth Life	Protect, recover and promote the sustainable use of terrestrial ecosystems

SDG 15 aims to promote the sustainable use of terrestrial ecosystems by managing forests, combating desertification, stopping and reversing land loss and degradation. Earth, like oceans, provides food and helps to maintain clean air, clean water, and to combat climate change. Moreover, Earth is home to millions of species. Thus promoting sustainable use of natural resources in production chains (SDG 12), in community subsistence activities (SDGs 1, 2, and 3) and integrating them into public policies is a central task for achieving SDG 15. Global policies should reinforce efforts to combat trafficking of protected species and increase the capacity of local communities to pursue sustainable livelihood opportunities. To achieve this objective, the mobilisation and integration of all areas of science will be necessary to promote sustainable forests management (SDGs 2, 6, 9, and 12) and conservation, preserve biodiversity and ecosystems, whilst ensuring a fair and equitable sharing of benefits derived from the use of genetic resources. The most significant and radical change is expected from the economic sciences to integrate the values of ecosystems and biodiversity into global, national, and local development strategies, ensuring the conservation, recovery, and sustainable use of terrestrial and freshwater ecosystems and their services.

SDG 16	Peace, Justice and Effective Institutions	Promote peaceful and inclusive societies for sustainable development

SDG 16 aims to promote strong, inclusive, and transparent institutions in order to keep peace and respect for human rights as the basis for sustainable human development. Amongst the principles that support SDG 16 there are sensitive topics, such as combating sexual exploitation, human trafficking, and torture, and also by ending corruption, terrorism, and criminal practices that violate human rights. The social and economic sciences have a decisive role in achieving this objective by promoting non-discriminatory laws and policies for sustainable development, strengthening institutions, including through international cooperation, for the prevention of violence and the fight against terrorism and crime, and expanding the participation of developing countries in institutions of global governance. But there is also the need to ensure public access to information to protect fundamental freedoms and reduce the flow of financial and illegal weapons and stolen resources to reduce all forms of economic-induced violence and related mortality rates.

SDG 17	Partnerships and Means of Implementation	Strengthen the means of implementation and revitalize the global partnership for sustainable development

Only a broad global partnership that includes all interested sectors and people affected by development processes will allow to achieve the 17th sustainable development goal. SDG 17 calls for the coordination of efforts in the international arena as essential for cooperation, the transfer of technology, and the exchange of data and human capital. To reach this goal, it will be necessary to build on existing initiatives to progress towards sustainable development by organising high-quality and reliable data, disaggregated by income (SDGs 8 and 10), gender, age, race, and ethnicity (SDGs 5 and 10) and other relevant characteristics in (inter)national contexts. It is crucial to encourage and promote effective public, public–private, private, and civil society awareness/partnerships based on resource mobilisation, data, monitoring and sharing of knowledge, experiences, technologies, and financial resources to support the achievement of sustainable development goals globally.

3 Indicators and the perception of scientists on their ability to reach each SDG

The description of the SDGs highlights the plurality of approaches and perspectives that may be adopted for their achievement. This also explains the essentially critical positions found in the current literature that deserves to be explored, whether in the field of territorial planning, public policies, environmental sciences, international relations, economics, amongst so many other areas of knowledge.

There are actually several studies trying to establish indicators to measure how the SDGs are implemented and introducing quantification to proposed procedures (Alleyne, Beaglehole, & Bonita, 2015), how to discern progress (Bebbington & Unerman, 2018), and how to evaluate accomplishments in different countries and regions (BCCIC, 2016; European Union, 2016, 2017). At global level, two reports (UN, 2017a) were published underscoring both benefits and threats to sustainability as the societies move towards the 2030 Agenda for Sustainable Development Goals (UN, 2017a, 2017b), but there are also reports that are dedicated to the performance of countries (Sachs, Schmidt-Traub, Kroll, Durand-Delacre, & Teksoz, 2017), continents, and regions (World Bank, 2017). Europe, for example, has been advancing in enhancing energy efficiency and the portion of renewable sources used. There were also progresses in resource use, sustainable and responsible consumption, decrease in waste generation and CO_2 emissions, but indicators still showed that the conditions of ecosystems have not satisfactorily improved (European Union, 2017). In Latin America and the Caribbean, the major challenges were identified as required improvements in the health and education systems (Sachs et al., 2017; World Bank, 2017). Progresses related to environmental aspects were related to industrial innovation and increasing energy access, whilst inequality and employment were still critical issues to be resolved (Nicolai, Bhatkal, Hoy, & Aedy, 2016).

There are still few publications explicitly directed to the degree to which the SDGs are being pursued/reached through the numerous cleaner production actions and practices, although cleaner production is considered worldwide as a tool to reach sustainable development. Cleaner production combines the use of preventive environmental methods to processes, products, and services targeting to increase efficiency and to diminish the threats to society and environment. These preventive practices contribute to achieve economic savings as well as a better quality environment for society, which are fundamentally emphasised in the SDG description. The advances in cleaner production are accounted for by processes (saving energy and raw materials, reducing the use of toxic substances, diminishing the quantity of waste and emissions) and products (decreasing impacts along the whole life cycle), as well as for services (Shereni, 2019). Cleaner production can be particularly advantageous to developing countries and those experiencing economic transition by facilitating innovative advances in the reuse, remanufacturing, and recycling. Amongst the studies that relate cleaner production practices with the SDGs, research was focused on resource use (Fader, Cranmer, Lawford, & Engel-Cox, 2018; Mugagga & Nabaasa, 2016), energy efficiency (Chirambo, 2018), and the role of environmental assessment methods to meet the goals (Hoekstra, Chapagain, & van Oel, 2017; Laurent et al., 2019; Sala & Castellani, 2019). The SDG guidance for the operation of supply chains (Russell, Lee, & Clift, 2018), industrial sectors (Oliveira-Neto, Rodrigues-Pinto, Castro-Amorim, Giannetti, & Almeida, 2018), individual companies (Fonseca & Carvalho, 2019), and services (Avrampou, Skouloudis, Iliopoulos, & Khan, 2019; Di Vaio & Varriale, 2019) was also explored.

In the pursuit of the SDGs, society, and in particular science, should use a multidisciplinary and transformative approach bringing together different disciplines that would allow policy development built upon a consistent scientific approach (Schmalzbauer & Visbeck, 2016). In this context, scientific guidance was pointed out as the required basis to obtain effective solutions to complex problems, projecting and modelling future scenarios, and offering a bottom line for policy making establishing priorities for action (ICSU, 2017).

One of the responsibilities of the scientific community is to help in the interpretation of the large-scale goals into the real-world local level agenda. Another responsibility of the scientific community is to help to reaffirm the required support of education (Annan-Diab & Molinari, 2017; Dlouhá, Henderson, Kapitulcinová, & Mader, 2017; Storey, Killian, & O'Regan, 2017) and research (Bebbington & Unerman, 2018; Leal Filho et al., 2018). Research is a constructive route to foster sustainability, and a sustainable future can be only achieved by relying on scientific guidance (Schmalzbauer & Visbeck, 2016). Therefore the academia is striving to fill out a number of knowledge gaps on how to achieve the SDGs and, consequently, contribute to sustainability (Décamps, Barbat, Carteron, Hands, & Parkes, 2017; ICSU, 2017; Soini, Jurgilevich, Pietikäinen, & Korhonen-Kurki, 2018).

4 The lack of a universal scientific model linking the SDGs and sustainability

It is not hard to find scientific works on sustainability in which a conceptual model supporting indicators choice, interpretation of results, and discussion is missing (Agostinho, Silva, Almeida, Liu, & Giannetti, 2019). The definition of sustainability by the United Nations Brundtland Report (1987) allows different interpretations and the proposition of different conceptual models, including 'weak', 'medium', and 'strong' relationships amongst the three main forms of capital (social, economic, and environmental). The weak sustainability model (traditional) is characterised by the interdependent relationship amongst capitals, allowing their total substitutability. For example, the environmental capital can be reduced as long as this reduction causes an equal increase in economic or social capital, and vice versa. On the other hand, the strong sustainability model implies that the different capitals cannot be substituted amongst each other, since they provide different and unique contributions. Although several different models exist, it is recognised that strong sustainability must be pursued for medium- to long-term strategic political planning, since it considers the Earth's biophysical limits in providing resources and receiving waste (Rockström, Steffen, & Noone, 2009) (Fig. 2).

As a contribution to the scientific discussion on sustainability assessment, Giannetti et al. (2019) proposed a different interpretation of sustainability, which is represented by the five sectors sustainability model (5SenSu; Fig. 3). This model stems upon (i) a multi-dimensional approach embracing social, economic, and environmental dimensions or capitals; (ii) multiple perspectives, by assuming both the donor and receiver side of different forms of capital; (iii) a multi-metric approach allowing the use of different indicators; (iv) a multi-criteria approach that requires the use of weighting techniques and composite indicators. All these characteristics together with the consideration of all the three forms of capital make the 5SenSU a more holistic model than others available in literature.

Accommodating the 17 SDGs within the 5SenSu model is a relevant exercise to verify: (1) if and how the SDGs cover aspects of each one of the different forms of capital, (2) if the SDGs are distributed in a balanced way amongst the forms of capital, and (3) if sustainability experts have similar understandings of each SDGs.

To test the perception of scientists on their ability to reach each SDG and accommodate them in a structured scientific model, there is no better room than a scientific meeting with a high quality audience—scientists from different disciplines and countries—dealing with sustainability issues in their research activities (Giannetti et al., 2020). Authors from the seventh International Workshop on Advances in Cleaner Production, IWACP (www.advancesincleanerproduction.net) were requested to choose amongst the 17 SDGs the closest related to the main theme of their work. Having this information, and knowing the scientific profile of the invited key speakers and the number of participants (including listeners and presenters), the workshop's organising committee selected four SDGs (Responsible consumption

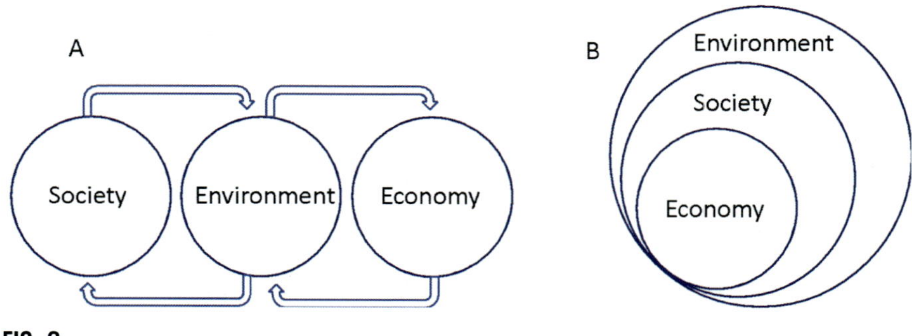

FIG. 2

The weak sustainability model (A, traditional) and the strong sustainability model (B).

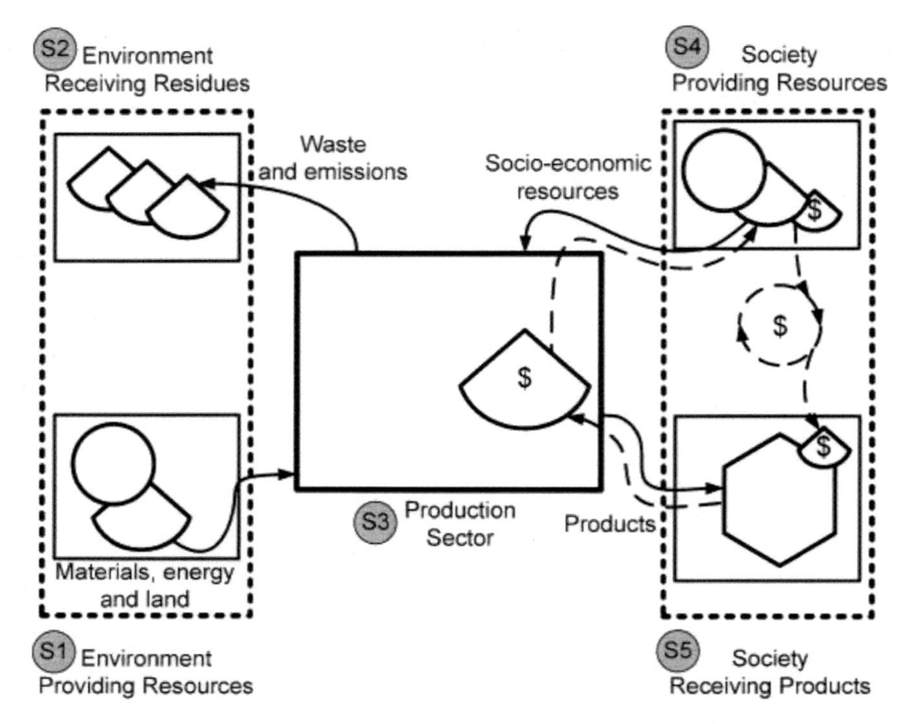

FIG. 3

Five sectors sustainability model (5SenSu). *S*, sectors. *Dashed lines* indicate monetary flows in exchange to energy flows.

(Model adapted from Giannetti, B.F., Sevegnani, F., Almeida, C.M.V.B., Agostinho, F., Garcı́a, R.R.M., Liu, G. (2019). Five sector sustainability model: A proposal for assessing sustainability of production systems. Ecological Modelling, 406, *98–108.)*

and production, SDG 12; Climate action and affordable clean energy, SDG 7; Sustainable cities and communities, SDG 11; and Industry innovation and infrastructure, SDG 9) with the aim of reuniting experts that, potentially, could provide different perspectives on one particular SDG during the workshop discussions. All the thematic sections were carried out under a similar structure by including a half hour exposition of concepts and ideas behind the SDG, followed by an exercise in which each participant provided his/her thoughts on the SDGs as related to two different conceptual models of sustainability: the 5SenSU and the traditional models.

The four following key questions were considered by the thematic sections' chairs of the seventh IWACP to steer the discussions:

Question #1: Is the use of the 5SEnSU model helpful in allocating the SDGs amongst the different forms of capital (or sectors)?
Question #2: Do you identify any disadvantage or misunderstanding about the 5SenSU as a conceptual model for sustainability?
Question #3: Would you suggest any improvement?

Question #4: Should the SDGs come first as a conceptual model representing sustainability or should a scientific model supporting the SDGs come first?

In regard to the traditional model, the highest share of participants understood most of the SDGs as directly related to social capital (Fig. 4). Specifically, considering the higher percentage of participants' perception for each SDG as a criterion, four of them were related to the environmental capital, three to the economic capital, and 10 to the social capital. This perception of the SDGs as mostly related to the social dimension is in line with further groupings of the SDGs as environmental, social, and economic. For example, Costanza et al. (2016) identified eight SDGs as social (1–5, 10, 16, and 17), four as economic (7–9, 11, and 12), and four as environmental (6, 13–15). Similarly to Claret, Metzger, Kettunen, and ten Brink (2018) referred to a classification by the Stockholm Resilience Centre recalled by Fioramonti, Coscieme, and Mortensen (2019), which related eight SDGs specifically to social capital, whilst Mortensen (2018) only distinguished three SDGs (13, 14, and 15) as core environmental.

In this context, a lack of balance emerges amongst the SDGs in terms of how they relate with the three main forms of capital, with social capital over-represented, followed by the other ones. A country or a region that achieves all the established SDG targets could be labelled as sustainable even if the SDGs mostly focus on social aspects. Assuming that a 'weak' sustainability conceptual model allows the substitution of any form of capital with another one without compromising sustainability, this unbalance is acceptable, but it calls for the implementation of a 'strong' sustainability thinking for monitoring progress towards sustainability in the long term.

Few SDGs were related to only one form of capital (SDG 5 as a social indicator, and SDGs 3, 10, and 16 as social) and the classification of most of the SDGs was heterogeneous amongst the participants. For example, SDG 7 (affordable and clean energy) was related to environmental capital by about 22% of participants, to

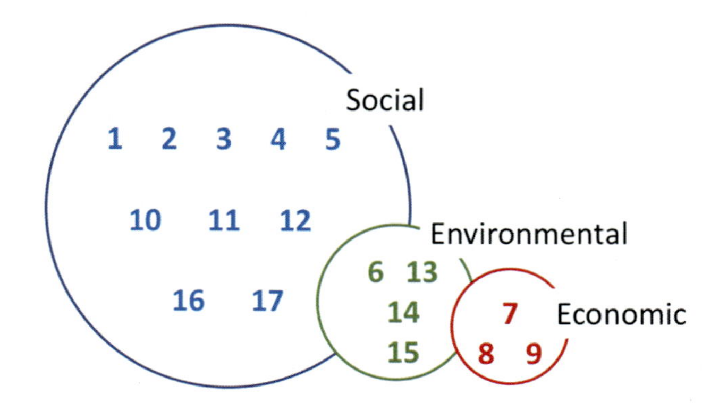

FIG. 4

The perception of the SDGs in approaching social, economic, and environmental dimensions or capitals.

economic capital by about 42% of participants, and to social capital by about 36%. Similar results were found for SDGs 6, 8, 9, 11–15. This heterogeneous interpretation sends a negative signal for the SDGs as a set of goals that could lead to sustainability and call for a scientific model to align and balance targets with practical policy purposes towards sustainability. That is, a scientific model should come first in supporting the SDGs.

Considering the same criterion of higher percentage of participants' perception for each SDG, a similar analysis was performed considering SDGs' allocation under the 5SenSUS model (Fig. 3). In this case, SDGs 13, 14, and 15 were related to the environment receiving waste; SDG 6 to the environment providing resources; SDGs 7, 8, 9, and 12 to the production sector; SDGs 5, 10, 11, 16, and 17 to society as provider; and SDGs 1, 2, 3, and 4 to society as receiver of products. As well as for the traditional sustainability model, there is an unbalanced distribution of the SDGs across the five sectors of the 5SenSU model. In particular, there is a concentration of SDGs related to society as providers (five SDGs), whilst only one SDG is related to the environment providing resources. As also observed by Wackernagel, Hanscom, and Lin (2017), our results suggest that according to the participants perception, the biophysical limits of the natural environment in providing resources to support societal development at the basis of the 'strong' sustainability approach (Meadows, Meadows, Randers, & Behrens III, 1972; Neumayer, 2003; Odum, 1996; Odum & Odum, 2006; Wackernagel & Rees, 1996) are under-considered in the UN 2030 Agenda and the SDGs. All other sectors seem to be balanced and/or well represented with a similar number of SDGs.

With the exception of SDGs 8 and 9 (decent work and economic growth; industry, innovation, and infrastructure), with about 70% of participants relating them to the production sector, a heterogeneous allocation of the SDGs across the five sectors is evident. For example, focusing on SDG 2 (zero hunger), 12% understand it as belonging to the environment as supplier sector, 2% to the environment as receiver, 28% to the production sector, 24% to the society as supplier, and 34% to the society as receiver. A similar heterogeneous allocation emerges for all other SDGs. Should not the SDGs be related to one form of capital or sector within a sustainability model in order to allow a homogeneous interpretation and quantification by researchers for informing policy?

According to the United Nations, the SDGs provide "a shared blueprint for peace and prosperity for people and the planet, now and into the future," by recognising "that ending poverty and other deprivations must go hand-in-hand with strategies that improve health and education, reduce inequality, and spur economic growth – all while tackling climate change and working to preserve our oceans and forests" (UN-SDGs, 2019). These statements have important key aspects related to a 'strong' sustainability vision: (i) they recognise the importance of people but also of the planet; (ii) they recognise that 'time' is an important variable, and short-, medium-, and long-term actions and policies should be strategically considered; (iii) they recognise the society, economy, and environment as important forms of capital relevant to sustainability. On the other hand, our results show that the environment as a provider of resources, and its biophysical limits are only partially addressed by the SDGs.

Considering now the initial questions that guided the discussions in the thematic sections, the main outcomes can be summarised as follows: (1) the 5SenSu provides a good representation/conceptual model for sustainability, although it lacks the consideration of biophysical boundaries as limits to growth; (2) most of the SDGs are not clearly allocated to one specific sector or form of capital, which, although reflecting the complexity of environmental, social, and economic contributions to sustainability, might imply that SDGs are difficult to implement in sectoral policies; (3) the SDGs are perceived as the result of a more practical rather than scientific-based approach to sustainability, a strong conceptual model is thus fundamental to ensure that sustainable development policies reflect scientific knowledge.

The unbalanced distribution of the SDGs as relevant to the three main forms of capital (environmental, social, and economic) indicates that both environmental and economic capital should receive further attention in terms of the number of goals in post-2030 agendas for sustainable development. By considering the 5SenSU model, we were able to understand further which specific capital or sector of sustainability are less represented by the SDGs. In particular, the environment acting as a provider of resources was related to only one of the SDGs (SDG 6, clean water and sanitation). Since 'strong' sustainability must be understood within the Earth's biophysical limits, it is urgent to rethink about the SDGs and their targets, making some of them more relevant for the environment as a source. This aspect is strongly aligned to cleaner production concepts and practices towards resource savings.

In spite of the inspirational nature of this kind of global initiative, the heterogeneous interpretation of the SDGs calls for further efforts by policy making for improving the understanding and scientific resonance of future SDG-like initiatives.

5 Concluding remarks

With less than 10 years to reach the ambitious goals of Agenda 2030 and in the UN's own assessment, the SDG actions were advancing at a slow and insufficient pace, in the current scenario, especially in face of the coronavirus pandemic with its devastating consequences, accentuating further social inequalities. The prospects for a recovery within the expectations expressed in the 2030 Agenda goals, despite several academic and political efforts that have recently emerged in addressing the need to conduct this process recovery with a strong sustainability bias, demanding a realignment of SDGs in view of pandemic risks (Gulseven, Al, Al Falasi, & ALshomali, 2020), are nothing encouraging.

Pandemics and other disasters caused by human action must become more and more frequent if we do not start to recognise that it is not possible to imagine that simply through new technological inventions, for example, through a Screen New

Deal (Klein, 2020), we will be able to face such threats, without rethinking and reinventing our ways of life and without imagining a new future.

The seriousness of the ecological crisis and the extreme difficulty that public and private agents, operating at the local, regional, and national levels are increasingly amplifying and transforming in economic, social, and political crises, make it difficult to believe in the possibility of achieving the objectives of sustainable development. But, in any case, there seems to be no doubt that in order to achieve a fair and equitable sustainable development, there is a need to strengthen the contributions of science, which will be decisive to renew the development process towards the goals of the SDGs.

The authors of this chapter share the conviction of the need for deeper changes and, consequently, also share the hope that the empirical studies and critical discussions can contribute to an increasingly broad debate in society about strategies required to make real progress in implementing the 2030 Agenda.

Acknowledgments

The Erasmus KA107 programme enabling the collaboration between "Gheorghe Asachi" Technical University of Iaşi and Universidade Paulista, is gratefully acknowledged.

References

Agostinho, F., Silva, T. R., Almeida, C. M. V. B., Liu, G., & Giannetti, B. F. (2019). Sustainability assessment procedure for operations and production processes (SUAPRO). *Science of the Total Environment*, *685*, 1006–1018.

Alleyne, G., Beaglehole, R., & Bonita, R. (2015). Quantifying targets for the SDG health goal. *Lancet*, *385*, 208–209.

Annan-Diab, F., & Molinari, C. (2017). Interdisciplinarity: Practical approach to advancing education for sustainability and for the sustainable development goals. *The International Journal of Management Education*, *15*, 73–83.

Avrampou, A., Skouloudis, A., Iliopoulos, G., & Khan, N. (2019). Advancing the sustainable development goals: Evidence from leading European banks. *Sustainable Development*, *27*(4), 743–757.

BCCIC. (2016). *Where Canada stands: A sustainable development goals Progress report the British Columbia council for international cooperation* (p. 94). Vancouver.

Bebbington, J., & Unerman, J. (2018). Achieving the United Nations sustainable development goals: An enabling role for accounting research account. *Accounting Auditing & Accountability Journal*, *31*, 2–24.

Chirambo, D. (2018). Towards the achievement of SDG 7 in sub-Saharan Africa: Creating synergies between Power Africa, Sustainable Energy for All and climate finance in-order to achieve universal energy access before 2030. *Renewable and Sustainable Energy Reviews*, *94*, 600–608.

Claret, C., Metzger, M. J., Kettunen, M., & ten Brink, P. (2018). Understanding the integration of ecosystem services and natural capital in Scottish policy. *Environmental Science and Policy*, *88*, 32–38.

Costanza, R., Daly, L., Fioramonti, L., Giovannini, E., Kubiszewski, I., Mortensen, L. F., et al. (2016). Modelling and measuring sustainable wellbeing in connection with the UN Sustainable Development Goals. *Ecological Economics*, *130*, 350–355.

Décamps, A., Barbat, G., Carteron, J. C., Hands, V., & Parkes, C. (2017). Sulitest: A collaborative initiative to support and assess sustainability literacy in higher education. *The International Journal of Management Education*, *15*, 138–152.

Di Vaio, A., & Varriale, L. (2019). SDGs and airport sustainable performance: Evidence from Italy on organizational, accounting and reporting practices through financial and non-financial disclosure. *Journal of Cleaner Production*, *249*, 119431.

Dlouhá, J., Henderson, L., Kapitulcinová, D., & Mader, C. (2017). Sustainability-oriented higher education networks: Characteristics and achievements in the context of the UN DESD. *Journal of Cleaner Production*, *112*, 3464–3478.

European Union. (2016). *Sustainable development in the European Union: A statistical glance from the viewpoint of the UN sustainable development goals*. http:/ec.europa.eu/eurostat/documents/3217494/7745644/KS-02-16-996-EN-N.pdf. Accessed 5th Jan 2020.

European Union. (2017). *Sustainable development in the European Union: monitoring report on progress towards the SDGs in an EU context*. http:/ec.europa.eu/eurostat/documents/3217494/8461633/KS-04-17-780-EN-N.pdf/f7694981-6190-46fb-99d6-d092ce04083f. Accessed 5th Jan 2020.

Fader, M., Cranmer, C., Lawford, R., & Engel-Cox, J. (2018). Toward an understanding of synergies and trade-offs between water, energy, and food SDG targets. *Frontiers in Environmental Science*, *6*(Nov), 112.

Fioramonti, L., Coscieme, L., & Mortensen, L. F. (2019). From gross domestic product to wellbeing: How alternative indicators can help connect the new economy with the sustainable development goals. *The Anthropocene Review*, 1–16.

Fonseca, L., & Carvalho, F. (2019). The reporting of SDGs by quality, environmental, and occupational health and safety-certified organizations. *Sustainability*, *11*(20), 5797.

Giannetti, B. F., Agostinho, F., Almeida, C. M. V. B., Liu, G., Contreras, L. E. V., Vandercaeele, C., et al. (2020). Insights on the United Nations sustainable development goals scope: Are they aligned with a 'strong' sustainable development? *Journal of Cleaner Production*, *252*, 119574.

Giannetti, B. F., Sevegnani, F., Almeida, C. M. V. B., Agostinho, F., García, R. R. M., & Liu, G. (2019). Five sector sustainability model: A proposal for assessing sustainability of production systems. *Ecological Modelling*, *406*, 98–108.

Gulseven, O., Al, H. F., Al Falasi, M., & ALshomali, I. (2020). *How the COVID-19 pandemic will affect the UN sustainable development goals? SSRN*. https:/ssrn.com/abstract=3592933. Accessed 9th Dec 2020.

Hoekstra, A. Y., Chapagain, A. K., & van Oel, P. R. (2017). Advancing water footprint assessment research: Challenges in monitoring progress towards sustainable development goal 6. *Water*, *9*, 438.

ICSU. (2017). *A guide to SDG interactions: From science to implementation International Council for Science (ICSU), Paris*. https:/www.icsu.org/cms/2017/05/SDGs-GuidetoInteractions.pdf. Accessed 8th Jan 2020.

Klein, N. (2020). *Screen new deal: Under cover of mass death, Andrew Cuomo calls in the billionaires to build a high-tech dystopia. The intercept. 2020*. https:/theintercept.com/2020/05/08/andrew-cuomo-eric-schmidt-coronavirus-tech-shock-doctrine/. Accessed 9th Dec 2020.

Laurent, A., Molin, C., Owsianiak, M., Frantke, P., Dewulf, W., Herrmann, C., et al. (2019). The role of life cycle engineering (LCE) in meeting the sustainable development goals—Report from a consultation of LCE experts. *Journal of Cleaner Production*, *230*, 378–382.

Le Blanc, D. (2015). Towards integration at last? The sustainable development goals as a network of targets. *Sustainable Development*, *23*(3), 176–187.

Leal Filho, W., Azeiteiro, U., Alves, F., Pace, P., Mifsud, M., Brandli, L., et al. (2018). Reinvigorating the sustainable development research agenda: The role of the sustainable development goals (SDG). *International Journal of Sustainable Development & World Ecology*, *25*(2), 131–142.

Liu, J., Hull, V., Godfray, H. C. J., Tilman, D., Gleick, P., Hoff, H., et al. (2018). Nexus approaches to global sustainable development. *Nature Sustainability*, *1*, 466–476.

Meadows, D. H., Meadows, D. L., Randers, J., & Behrens, W. W., III. (1972). *The limits to growth: A report for the Club of Rome's project on the predicament of mankind*. New York: Universe Books.

Mortensen, L. F. (2018). *EESC hearing on civil society indicators for the sustainable development goals*. Copenhagen, Denmark: European Environment Agency. www.eesc.europa.eu.

Mugagga, F., & Nabaasa, B. B. (2016). The centrality of water resources to the realization of sustainable development goals (SDG). A review of potentials and constraints on the African continent. *International Soil and Water Conservation Research*, *4*, 215–223.

Neumayer, E. (2003). *Weak versus strong sustainability: Exploring the limits of two opposing paradigms Edward Elgar, Northampton*.

Nicolai, S., Bhatkal, T., Hoy, C., & Aedy, T. (2016). *Projecting Progress: The SDGs in Latin America and the Caribbean*. London: Overseas Development Institute. https:/www.odi.org/sites/odi.org.uk/files/resource-documents/11376.pdf. Accessed 27th Dec 2019.

Obersteiner, M., Walsh, B., Frank, S., Havlik, P., Cantele, M., Liu, J., et al. (2016). Assessing the land resource–Food price nexus of the sustainable development goals. *Science Advances*, *2*(9), e1501499.

Odum, H. T. (1996). *Environmental accounting—Emergy and environmental decision making*. John Wiley & Sons, Inc.

Odum, H. T., & Odum, E. C. (2006). The prosperous way down. *Energy*, *31*, 21–32.

Oliveira-Neto, G. C., Rodrigues-Pinto, L. F., Castro-Amorim, M. P., Giannetti, B. F., & Almeida, C. M. V. B. (2018). A framework of actions for strong sustainability. *Journal of Cleaner Production*, *196*, 1629–1643.

Pradhan, P., Costa, L., Rybski, D., Lucht, W., & Kropp, J. P. (2017). A systematic study of sustainable development goal (SDG) interactions. *Earth's Future*, *5*, 1169–1179.

Ripple, W. J., Wolf, C., Newsome, T. M., Galetti, M., Alamgir, M., Crist, E., et al. (2017). World scientists' warning to humanity: A second notice. *Bioscience*, *67*, 1026–1028.

Rockström, J., Steffen, W., & Noone, K. (2009). A safe operating space for humanity. *Nature*, *461*(2009), 472–475.

Russell, E., Lee, J., & Clift, R. (2018). Can the SDGs provide a basis for supply chain decisions in the construction sector? *Sustainability*, *10*(3), 629.

Sachs, J., Schmidt-Traub, G., Kroll, C., Durand-Delacre, D., & Teksoz, K. (2017). *SDG index and dashboards report 2017 Bertelsmann Stiftung and sustainable development solutions network (SDSN), New York*. http://www.sdgindex.org/assets/files/2017/2017-SDG-Index-and-Dashboards-Report–full.pdf. Accessed 6th Jan 2020.

Sala, S., & Castellani, V. (2019). The consumer footprint: Monitoring sustainable development goal 12 with process-based life cycle assessment. *Journal of Cleaner Production*, *240*, 118050.

Schmalzbauer, B., & Visbeck, M. (2016). *The contribution of science in implementing the sustainable development goals German committee future earth, Stuttgart/Kiel*. http:/www.

dkn-future-earth.org/data/mediapool/2016_report_contribution_science_v8_light_final_fin.pdf. Accessed 7th Jan 2020.

SDG pyramid. (2019). https://www.sdgpyramid.org/about-sdg-pyramid/. Accessed 8th Jan 2020.

Shereni, N. C. (2019). The tourism sharing economy and sustainability in developing countries: Contribution to SDGs in the hospitality sector. *African Journal of Hospitality, Tourism and Leisure*, *8*(5).

Soini, K., Jurgilevich, A., Pietikäinen, J., & Korhonen-Kurki, K. (2018). Universities responding to the call for sustainability: A typology of sustainability centres. *Journal of Cleaner Production*, *170*, 1423–1432.

Stockholm Resilience Centre. (2016). https:/www.stockholmresilience.org/research/research-news/2017-02-28-contributions-to-agenda-2030.html. Accessed 5th Feb 2020.

Storey, M., Killian, S., & O'Regan, P. (2017). Responsible management education: Mapping the field in the context of the SDGs. *The International Journal of Management Education*, *15*, 93–103.

UN. (1987). *United Nations Brundtland Report. Report of the world commission on environment and development: our common future*. http:/www.un-documents.net/our-common-future.pdf. Accessed 9th Feb 2019.

UN. (2015). *Transforming our world: The 2030 Agenda for Sustainable Development. Resolution A/RES/70/1. New York*. https:/www.un.org/en/development/desa/population/migration/generalassembly/docs/globalcompact/A_RES_70_1_E.pdf. Accessed 4th Jan 2020.

UN. (2017a). *United Nations Statistical Commission. Report on the Forty-eighth Session*. https:/unstats.un.org/unsd/statcom/48th-session/documents/Report-on-the-48th-Session-of-the-Statistical-Commission-E.pdf. Accessed 6th Apr 2018.

UN. (2017b). *United Nations the sustainable development goals report 2017*. https:/unstats.un.org/sdgs/files/report/2017/TheSustainableDevelopmentGoalsReport2017.pdf (Accessed 5th Apr 2018).

UN-SDGs. (2019). *United Nations sustainable development goals platform*. https:/sustainabledevelopment.un.org/?menu=1300 (Accessed 9th Apr 2019).

van Soest, H. L., van Vuuren, D. P., Hilaire, J., Minx, J. C., Harmsen, M. J. H. M., Krey, V., et al. (2019). Analysing interactions among sustainable development goals with integrated assessment models. *Global Transitions*, *1*, 210–225.

Wackernagel, M., Hanscom, L., & Lin, D. (2017). Making the sustainable development goals consistent with sustainability. *Frontiers in Energy Research*. https:/doi.org/10.3389/fenrg.2017.00018.

Wackernagel, M., & Rees, W. E. (1996). *Our ecological footprint—Reducing human impact on the earth new solutions publishers*.

Weitz, N., Carlsen, H., Nilsson, M., & Skånberg, K. (2017). Towards systemic and contextual priority setting for implementing the 2030 Agenda. *Sustainability Science*, *13*, 531–548.

World Bank. (2017). *Atlas of sustainable development goals 2017—From world development indicators*. https:/openknowledge.worldbank.org/handle/10986/26306. Accessed 5th Jan 2020.

CHAPTER

Sustainability and the circular economy

3

Roland Clift[a,b], George Martin[c,d], and Simon Mair[d,e]

[a]*Centre for Environment and Sustainability (CES), University of Surrey, Guildford, United Kingdom* [b]*Department of Chemical and Biological Engineering, University of British Columbia, Vancouver, BC, Canada* [c]*Department of Sociology, Montclair State University, Montclair, NJ, United States* [d]*CES, University of Surrey, Guildford, United Kingdom* [e]*Circular Economy and Data Analytics, School of Management, University of Bradford, Bradford, United Kingdom*

1 Introduction

Of the Sustainable Development Goals discussed in the previous chapter (UN, 2015), this chapter is concerned mainly with no. 12—Responsible Consumption and Production. We explore how changes in infrastructure, business models and practices, supply systems, and consumer expectations can all lead to a more sustainable economy.

1.1 A brief history of sustainability

Sustainability is an emergent property of a complex system, whereas *sustainable development* is a process of change towards a more sustainable state. These concepts were introduced by the Brundtland report (UN, 1987), which conceived sustainable development as meeting "the needs of the present without compromising the ability of future generations to meet their needs" and built on the earlier Brandt Commission (Brandt, 1982, p. 23) which had pointed out firmly that:

> *One must avoid the persistent confusion of growth with development, and we strongly emphasise that the prime objective of development is to lead to self-fulfilment and creative partnership in the use of a nation's productive forces and its full human potential.*

Amongst the literature exploring these concepts and the related question of how happiness and quality of life can best be promoted, Jackson (2010, p. 20) has provided one of the most succinct articulations of sustainability:

> *Sustainability is the art of living well, within the ecological limits of a finite planet.*

Assessing Progress towards Sustainability. https://doi.org/10.1016/B978-0-323-85851-9.00001-8

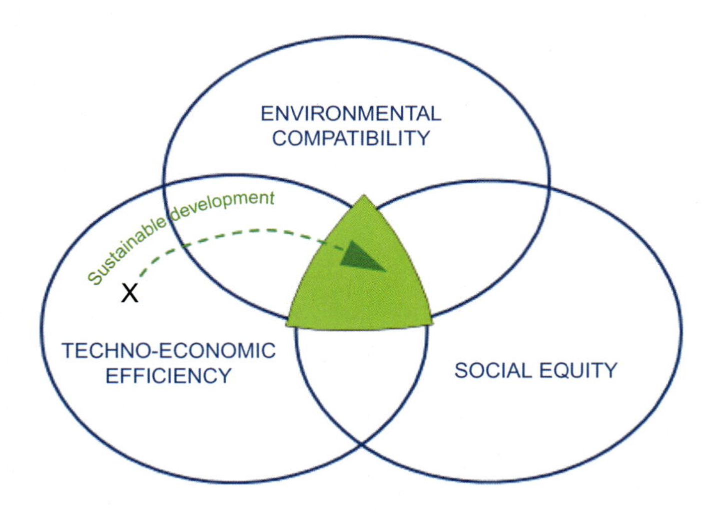

FIG. 1

Sustainability and sustainable development.

Adapted from Clift, R. (1995). The challenge for manufacturing. In J. McQuaid (Ed.) Engineering for sustainable development. *London: The Royal Academy of Engineering and Clift, R., et al. (2013). Sustainable consumption and production: Quality, luxury and supply chain equity. In I.S. Jawahir et al. (Eds.)* Treatise on sustainability science and engineering. *[Online]. Dordrecht: Springer Netherlands. Available at: https://doi.org/10.1007/978-94-007-6229-9_17 [Accessed 5 May 2021].*

'Living well' means flourishing, with a decent level of material comfort, security, and dignity, but also has a moral sense: not living at the expense of the well-being of others and thus feeling your life is good in ethical terms.

Recognition of limits is fundamental: if economic activity could be expanded without limits, sustainability of development would not be a concern. Sustainability is commonly represented as having three sets of limits: techno-economic, environmental or ecological, and social or societal (e.g. Clift, 1995; Mitchell et al., 2004; Purvis et al., 2019). The relationship between them is frequently represented as a Venn diagram (Fig. 1) in which the three lobes represent decision spaces, bounded by 'hard' or 'soft' constraints.[a] Showing them overlapping represents our hope that a sustainable economy satisfying all three limits can be found. Sustainable development implies finding a path from current activities (point X in Fig. 1) to sustainability, i.e. the region where all three sets of constraints are respected.

'Techno-economic efficiency' represents the ranges of possible activities limited by technical skills and ingenuity, the laws of thermodynamics, and the need for efficiency as defined by the prevailing economic system. It is important that the laws of thermodynamics are hard-wired into the universe whereas the 'laws' of economics are human constructs and therefore mutable, for example by changes to the fiscal system. 'Environmental compatibility' represents activities consistent with the

[a]Authors who argue that the lobes should be concentric (e.g. Mitchell, 2000; Velenturf & Purnell, 2021) are illustrating a different metaphor in which they represent fields of interest or governance.

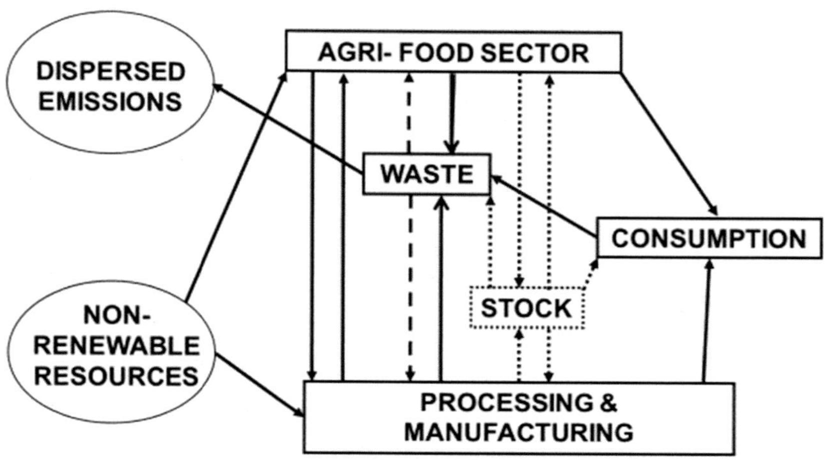

FIG. 2

Linear, circular, and performance economies.

Adapted from Clift, R. (2013). Clean technology and industrial ecology. In R.M. Harrison (Ed.) Pollution- causes, effects and control *(5th ed.) [Online]. Available at: https://doi.org/10.1039/9781847551719-00411 [Accessed 5 May 2021].*

resource and carrying capacity of the planet. Recent attempts to define these constraints and apply them in environmental management tools have focussed on the Planetary Boundaries approach (Clift et al., 2017; Rockström et al., 2009; Steffen et al., 2015). 'Social equity', emphasised in the Brandt report (1982), is an ethical principle at the heart of sustainable development. Dearing et al. (2014) and Raworth (2017), amongst others, have considered how to reconcile the social foundations of development with the Planetary Boundaries.

1.2 Circularity and sustainability

Publications concerned with the idea of the circular economy have expanded ten-fold since 2015 (Velenturf & Purnell, 2021). The circular economy, with 'closed loop' flows within the economy, is usually presented as an alternative to the 'linear' use of resources that has prevailed since the industrial revolution. Fig. 2 shows the distinction. Flows in the 'linear' economy are shown as full lines. Production and post-consumer waste are inevitable—a 'waste-free' economy is thermodynamically impossible (Georgescu-Roegen, 2017)—but waste can remain part of the economy even in a linear economy[b] because material is only truly lost when it is dispersed into the environment.

A commonly cited definition (McKinsey & Co., 2012, p. 14) states:

> *The circular economy refers to an industrial economy that is restorative by intention…and eradicates waste through careful design.*

[b]For example, old landfills are increasingly mined for their content of relatively scarce metals.

This interpretation emphasises immediate re-use and recycling of 'waste' materials and used products: 'circularity' refers to flows within the economy, shown by the broken lines in Fig. 2. This improves resource efficiency by reducing demand for fresh products, waste from processing and manufacturing, and resource extraction.

However, 'circular economy' has a broad range of meanings. Kirchherr et al. (2017) reviewed 114 proposed definitions and concluded "the circular economy is most frequently depicted as a combination of reduce, reuse and recycle activities...(with)...few explicit linkages...to sustainable development." The usual focus is economic prosperity, sometimes with environmental quality, rarely including social or intergenerational equity, and even more rarely recognising systemic changes or the agents who might drive them.

Rather than introducing yet another definition, we use that given by the (UK) Waste and Resources Action Programme (WRAP, 2021):

> *A circular economy is an alternative to a traditional linear economy (make, use, dispose) in which we keep resources in use for as long as possible, extract the maximum value from them whilst in use, then recover and regenerate products and materials at the end of each service life.*

This definition focuses on the stock of materials in use (infrastructure, buildings, plant, vehicles, appliances, etc.) and considers the resource flows and interactions indicated by the dotted arrows in Fig. 2. Although the link to sustainability is not explicit, the need for systemic change is recognised, emphasising the longevity, utility, and efficiency of the stock of products and materials in use. Thus economic restructuring for better resource efficiency goes beyond 'circular' flows: 'reduce, reuse and recycle activities' are secondary, one of several approaches to improving resource efficiency in building up, maintaining, and using stock and infrastructure and in reducing the impacts of consumption. It includes the renewable stock of labour and so also enables the neglected social lobe to be included.

The focus on efficient use of stock rather than just 'closing loops' in material flows is becoming more widely accepted. It is embodied in the Action Plan of the European Commission (2020). The Ellen MacArthur Foundation, long an advocate of the 'circular flows' approach (Webster, 2015), has now (Ellen MacArthur Foundation, 2021) articulated its policy goals as follows:

- Stimulate design for the circular economy;
- Manage resources to preserve value;
- Make the economics work;
- Invest in innovation, infrastructure, and skills;
- Collaborate for system change.

Following Stahel (2010, 2019), we refer to the stock-centred model as the *Performance Economy*. The argument that living and interacting with high-quality stock, including the built environment, is at least as important as 'consumption' flows in maintaining quality of life (Clift et al., 2013; Jackson, 2017) further supports focussing on the most resource-efficient ways to use, maintain, and improve capital stock. However, even though thinking has moved away from circular material flows back to the emphasis on resource efficiency that is central to industrial ecology (Clift, 2013), the term 'Circular Economy' has become so familiar that we retain it here as a general heading.

2 Material resource efficiency

2.1 Re-use, remanufacturing, and recycling

'Stock' includes 'manufactured capital': buildings, goods, and infrastructure. Fig. 3 shows the three 'loops' that summarise the approach to reducing material intensity in the performance economy (King et al., 2006; Stahel, 2010) with their geographical scales:

Re-use includes direct re-use such as refilling beverage containers or cleaning and refurbishing garments. It also includes cases where ownership changes, for example via a second-hand market or a mechanism such as eBay. These activities are usually carried out locally.

Remanufacturing (Loop 1) includes repair and remanufacture of used goods and 'upgrading' to meet new performance standards or fashion. It is typically more labour intensive and less energy intensive than recycling, and may be a small-scale local activity such as repairing appliances or vehicles (Stahel, 2010; Stahel & Clift, 2016). Complex items may be taken to regional service centres, but global movement of goods for remanufacturing is rare.

Recycling and reprocessing (Loop 2) refers to reprocessing or dismantling to recover materials or components. The recovered materials may be recycled into the same product system ('closed-loop recycling'), as shown in Fig. 3, or may pass to Loop 2 of a different product system ('open-loop recycling'), often with lower performance requirements ('down-cycling'). Recycling is usually carried out at a larger scale than remanufacturing, in a regional or global product system. Production and use of both virgin and secondary materials are typically more energy intensive than remanufacturing (Allwood, 2014; Stahel, 2010; Stahel & Clift, 2016).

To both reduce non-renewable resources use and generate employment, re-use and remanufacturing (Loop 1) are usually preferable to reprocessing (Loop 2), with primary production the least desirable (Stahel, 2010, 2019). More detailed analysis (Stahel & Clift, 2016) shows that the priority order for changing design and practice is as follows:

(1) Extend service life, to reduce material throughput;
(2) Intensify use of stock, to reduce stock needed;
(3) Increase the proportion of post-use products remanufactured;
(4) Increase the proportion of post-use products and materials reprocessed.

These priorities underpin the structural changes needed. Product longevity is specifically emphasised in the policies of the European Union (EC, 2020), although making it a legal requirement is problematic (Tonner & Malcolm, 2017).

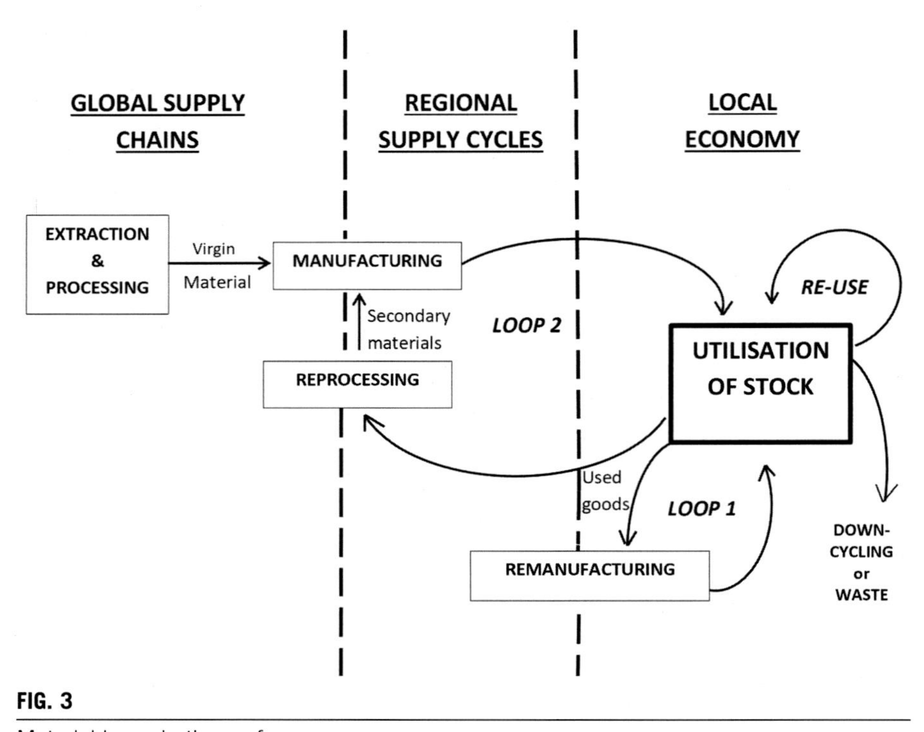

FIG. 3

Material loops in the performance economy.

Adapted from Stahel, W., & Clift, R. (2016). Stocks and flows in the performance economy. In Taking stock of industrial ecology *(pp. 137–158). Online: Springer International Publishing.*

2.2 Material reprocessing and energy recovery

Fig. 4 shows an idealised system to obtain maximum value from a material as it passes through the economy, through a ***cascade of uses*** with successively lower requirements on purity and material properties. It is complementary to Fig. 3, which refers to a manufactured product.

Following primary production, the material is processed for USE 1. It may be re-used several times in this form and also reprocessed for the same use. Eventually, because of contamination or degradation, the material must leave the first use loop. It may then be ***downcycled*** to a second use with lower performance specifications; re-used and reprocessed several times in this use, before being downcycled again; and so on. As a specific example, aluminium may be used in beverage containers, then passed to use in mechanical components (e.g. in vehicles), and then structural items (e.g. building components). A system for plastics is shown in Fig. 5 and explained in the Text Box, to illustrate that multiple re-use and recycling is more complex than merely closing material loops.

Ultimately, the material must become so contaminated or degraded that there is no demand for it or the resource input or cost to reprocess it becomes larger than for virgin material. End-of-life waste is inevitable even in a fully developed circular

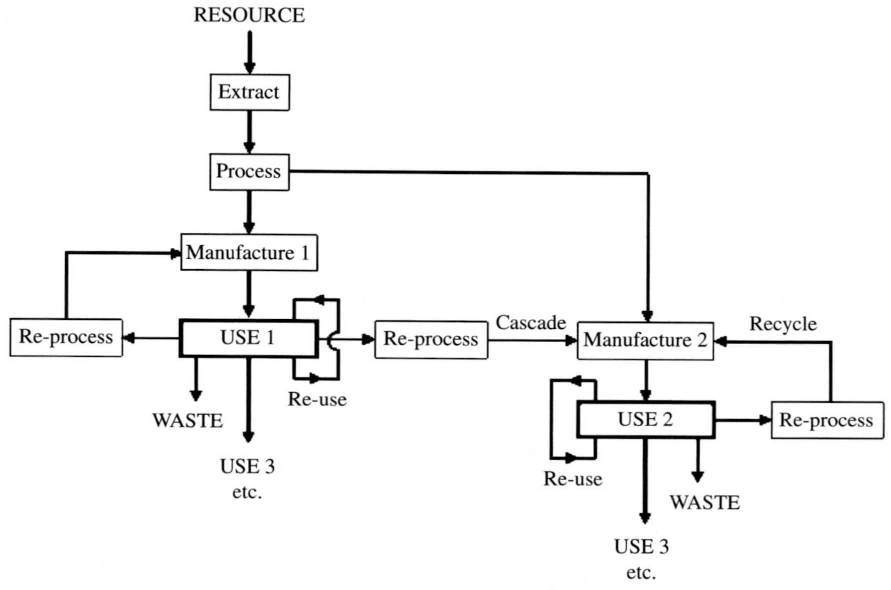

FIG. 4

Re-use, reprocessing, and cascaded use of a material (Clift, 2013).

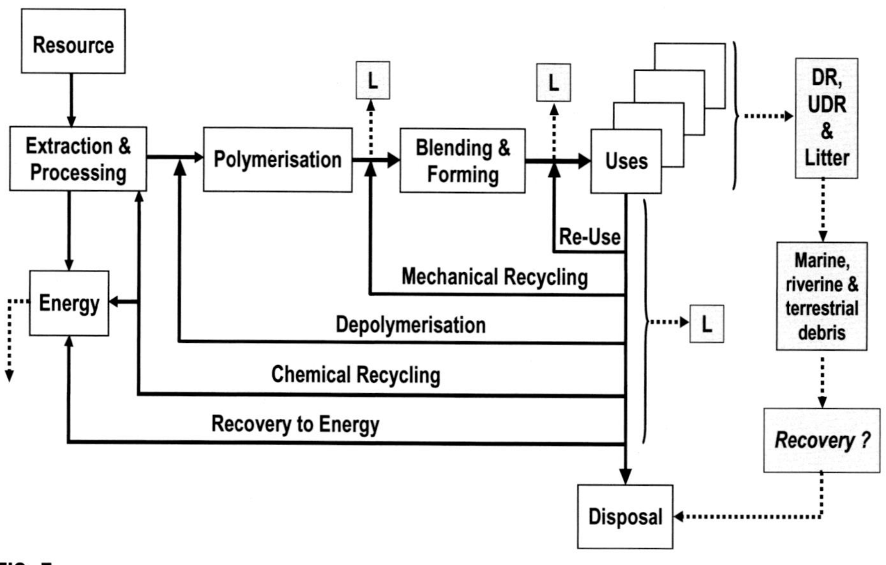

FIG. 5

The industrial ecology of plastics.

Adapted from Clift, R. (1997). Clean technology—The idea and the practice. Journal of Chemical Technology & Biotechnology, 68(4), *347–350 and Clift, R. et al. (2019). Managing plastics: Uses, losses and disposal special issue on designing law and policy towards managing plastics in a circular economy.* Law, Environment and Development Journal (LEAD Journal), 15(2), *93–107.*

economy so that waste management will always be needed. Plastic packaging illustrates that energy recovery from waste is necessary even in a circular economy (see Text Box).

Text Box: Circular economy for plastics

Fig. 5 maps the industrial ecology of plastic produced from fossil hydrocarbons, i.e. it shows the principal flows and uses of plastics in the economy.

Starting at the top left, the raw material is extracted and processed to produce the monomer (e.g. ethylene) which is polymerised to make the polymer (e.g. polyethylene). Before forming into a plastic product, additives including plasticisers, fillers, and pigments must be added to give the required material properties. Thus plastics put to different uses are already differentiated at this point.

Possible routes for re-use, reprocessing, and recycling start at the top right of Fig. 5. Plastic formed into a durable product, such as a toner cartridge for a printer, may be re-used. However, these items eventually become damaged or contaminated so that they must be remanufactured or reprocessed. Provided the plastic is not contaminated or mingled with different materials, it may be possible to re-form it into a replacement or a similar product (closed-loop 'mechanical recycling' in Fig. 5). The demand for items made from mixed waste plastic is small, so open-loop mechanical recycling is not shown. Some plastics may be 'depolymerised' back to the monomer, which can be recycled to polymerisation. More commonly, mixed waste plastic must be recycled chemically, i.e. converted thermally into a chemical product which is reprocessed or used as a fuel. Failing this, the waste can be used directly in energy from waste plants, or else landfilled.

The further the loops are from the corner in Fig. 5, the less environmentally beneficial they are. Keeping material in the inner loops requires it to be kept clean and not mixed with plastics containing different polymers or additives. This option is not available for single-use plastic, including most packaging. Thus use as an energy source is a necessary part of any truly closed-loop system for plastics. The alternative is landfilling to sequester carbon, but energy recovery is usually preferable on environmental grounds (Evangelisti et al., 2015; Gear et al., 2018).

Fig. 5 also shows the principal points in the industrial ecology where plastics can 'leak' from the economy into the environment to contribute to the problem of plastic pollution (L). Once released, a plastic item is carried by environmental flows to end up in the ocean. Ocean plastic is so mixed and contaminated that it cannot realistically be recycled and may not even be usable as a fuel (Clift et al., 2019): it is a legacy that must be collected and landfilled, and also a source of social inequity because the impacts affect people other than those who used the plastic. To limit future plastic pollution, it is essential to avoid 'leakage' by ensuring that all the loops in Fig. 5 are in place, including energy recovery.

3 Structural changes

Pursuing sustainability entails restructuring to achieve a more resource-efficient and equitable economy. Developing the approaches in Section 2 is only part of this restructuring: circularity is not an end in itself. We now consider other factors determining the performance of an economy.

3.1 Redeployment: Labour in the performance economy

To understand the structural changes needed to move to a more sustainable performance economy, 'stock' must be extended to include the 'five capitals': manufactured, social, human, financial, and natural (Forum for the Future, 2014; Porritt, 2005).

In addition to 'natural capital' (i.e. renewable and non-renewable natural resources), we focus on 'social capital' (i.e. the formal and informal structures and relationships on which a society depends) and 'human capital' (i.e. the knowledge, skills, and experience of individuals and groups in the society) which we conflate as 'labour'. Natural capital relates to the 'Environmental Compatibility' lobe of Fig. 1 whilst labour relates mainly to Social Equity. Restructuring involves more than changes in processes to maintain physical stock: the performance economy implies an integrative framework that links the life cycles of materials to their social webs (Laurenti et al., 2018), realigning social relationships and improving human skills.

Labour has both stock and flow dimensions. Labour as non-material stock represents the available resource of social and human capital. Labour also denotes the use of this stock in economic and social activities, i.e. a flow into these activities. The stock is not depleted when labour is used; therefore it is a renewable resource, in contrast with non-renewable stocks of materials and fossil energy.

Non-renewable resources, primarily deposits of minerals and fossil energy, are key parts of natural capital. Unemployment and low wages are features of socio-economic inequality. Therefore substituting non-renewable resource use by renewable labour addresses both the 'Environmental Compatibility' and 'Social Equity' dimensions of sustainability (Fig. 1). In aiming for higher employment and therefore lower labour efficiency, the performance economy represents a reversal of the structural economic changes during and following the industrial revolution. Seeing employment as a social good represents a reversal of conventional economic thinking, which sees labour as a cost and labour efficiency as an indicator of desirable social outcomes such as wages and leisure time.[c]

Stahel's (2010, 2019) concept of the Performance Economy also addresses policy measures to promote the shift from energy use to labour: taxing use of non-renewable resources, rather than renewable resources like labour, provides a practical application of the idea that the economic part of 'Techno-economic Efficiency' (Fig. 1) can be changed by fiscal measures. However, achieving meaningful shifts of taxation from labour to materials is complex and politically difficult.

3.2 Socialising business practices: Beyond the profit motive

The economics laureate Milton Friedman (1970) famously declared "the social responsibility of business is to increase its profits" to maximise short-term shareholder value. However, obsession with short-term profit is incompatible with the environmental and equity goals inherent in sustainability. COVID-19's social problems have intensified critiques of the Friedman paradigm, making it "sound emptier than ever" (Ward, 2020, p.1) and amplifying expectations that business should provide societal benefits as well as profits.

[c]See Isham et al. (2021) for a discussion of this tension.

The idea of ***Corporate Social Responsibility*** **(CSR)** developed from the 1980s as a move to engage external *stake*holders who are outside a business but affected by its actions (customers, suppliers, creditors, the local community, society, government) in addition to the internal *share*holders (employees, owners, investors) recognised by Friedman (Freeman & Reed, 1983). Attempts to measure CSR performance have achieved limited success (Korhonen, 2003). More transparent ***Environmental, Social, and Governance*** **(ESG)** ratings have evolved to provide investors with data for assessing the performance of corporations in these domains (e.g. Sustainalytics, 2021). However, they are still intended to measure economic risks rather than sustainability. Although current CSR and ESG initiatives embody a move away from Friedman's exclusive focus on profit, they do not fully address the social and environmental components of sustainability.

Socialising business practices involves a more radical approach. It was noted in Section 1 that the circular economy approach focuses on material flows, whilst the performance economy approach recognises both social and physical stocks. Labour as social stock is an underappreciated factor for promoting sustainability in the circular economy: "reconsidering labor is essential to tackling the large share of dissipated material and energy flows that cannot be recovered economically" (Moreau, 2017, p. 497). Circular economy transitions are generally predicted to increase employment (Wiebe et al., 2019; Wijkman & Skanders, 2015) and hence reduce the structural socio-economic disparities that have a negative effect on quality of life for all (Wilkinson & Pickett, 2010).

The more radical approach requires a different conception of ***value chains***. The term conventionally refers to the activities and functions within businesses to deliver products and services and provide competitive economic advantages (Chopra, 2020). The concept is a development of the Friedman business model: it maximises shareholder profit (Porter, 1985) but has been updated to incorporate CSR (Porter & Kramer, 2011). By contrast, the Performance Economy model challenges businesses to develop the labour component of their value chains as a goal in itself, thereby helping to fulfil their neglected social remit; a value chain is seen as a set of relationships that delivers benefits in both directions along the chain (Clift et al., 2013; Makower, 2021). The social benefit of providing employment can lead to the environmental benefit of reducing non-renewable inputs. Under an appropriate fiscal system (see Section 4), this can translate into economic advantages.

Education serves as an illustration. Social benefits accrue from individuals acquiring new craft and digital skills (Tilak, 2008). Businesses generally rely on outside institutions to provide workers with the required knowledge and skills to provide the necessary input, i.e. to build up and maintain the stock of labour. Furthermore, educational programmes provide occupations for their staffs and teachers. This illustrates the way in which the performance economy recognises and benefits from its social embeddedness: education and training to maintain the labour stock are amongst the two-way benefits of value chains. As Laurenti et al. (2018) put it: "this important feedback loop needs to be explicitly taken into account in circular economy initiatives."

In the current circular economy discourse, little attention is paid to the wider structural framework of social systems within which all economies operate, but social sustainability requires the broader perspective (Ede, 2017). ***Social enterprises*** provide opportunities to integrate the economic, environmental, and social domains of sustainability and sustainable development (Sahakian, 2016). A social enterprise is a business with monetary return but without profit maximisation. The intent is to address social or environmental problems, using business practices as a goal to support this purpose (Gomez, 2016; Vickers et al., 2017). Social enterprises have been emerging around the world, involving major bridging institutions: markets, civil organisations, states, and international aid programmes (Kerlin, 2010). The model is rooted in a value proposition that 'embeds' business activity in social and environmental performance (Stratan, 2017). Social enterprises can link social causes with environmental performance and include both in accounts along with financial performance. Profits are retained for expansion and improvement. Investors receive back what they put in but there are no dividends.

3.3 Dematerialisation and employment

The focus on stock in the Performance Economy model leads to the priorities identified in Section 2.1 to better align economic profitability with sustainability by decreasing material throughput and non-renewable resource use and increasing the use of renewable labour. Remanufacturing to extend the service lives of products is generally more labour intensive than primary extraction and processing, as well as less intensive in use of physical inputs and the associated environmental impacts (Jensen et al., 2019; Schau et al., 2012). It is less routine and repetitive than primary manufacture, and therefore less open to automation. The European Commission (2020) estimates that job-intensive reforms in circular economies could generate about 700,000 jobs in the EU by 2030. Servicing and remanufacturing require humans with flexible skills, not programmed robots, countering the possibility that a new wave of automation could displace workers, particularly in 'blue-collar' craft occupations (WEF, 2020). Automobiles are a prime example: whilst virgin models are made on large centralised assembly lines, their lives are extended by local enterprises employing human labour (Stahel, 2019). Businesses based on quality goods, produced and serviced by highly skilled labourers, can similarly improve quality of life for both consumers and workers (Clift et al., 2013). The associated shifts in labour and social relations meet the MacArthur Foundation policy goal to invest in technical innovation, infrastructure, and human skill (see Section 1.2).

Dematerialisation is associated with a broader change: ***servicisation***—i.e. moving away from material products to service provision (Stahel, 2013, 2019). Stahel (2016) sees this as "a new relationship with our goods and materials". Service provision as a business model has been growing since the 1980s (Vandermerwe & Rada, 1988), illustrated by leasing of equipment like photocopiers. The value chain for a servicised business can lead to greater resource efficiency and increased employment (Heiskanen & Jalas, 2000) by promoting the changes identified in Section 2.1.

However, servicisation has not yet produced a widespread sustainability-driven paradigm shift in business practices; rather, as in the interpretation of 'circularity' discussed in Section 1.2, it has been seen as a way to make continuing profits beyond product sales. Servicisation is most likely to succeed commercially if a product has high use value relative to production and disposal costs (Örsdemir et al., 2018); it has resulted in profit growth particularly for producers of complex high-value products, such as aircraft, automobiles, and ICT equipment, whose use is accompanied by service contracts (Lightfoot, 2021; Stahel & Clift, 2016). However, reluctance to shift from product sales to service is widespread: "most companies either don't know how or don't care to provide after-sales services" (Cohen et al., 2006).

Extended Producer Responsibility (EPR) is a complementary approach in which the producer retains liability for material products at the end of their service life but does not usually retain ownership or liability in use. EPR is sometimes seen as a driver for stock preservation similar to servicisation (Malcolm, 2019; Stahel, 2010), requiring manufacturers to incorporate end-of-life remanufacturing and reprocessing and longevity in product design. A few companies have adopted EPR for commercial reasons, particularly where their business is part service so that the return system already exists; the Xerox Asset Recovery Operation is a notable early example (Clift, 2013). However, EPR is rarely economically attractive (Clift & Wright, 2000): only in a few instances are the prices of scarce materials (notably some rare earths) high enough to promote recovery, and technological change often removes the possibility of re-using components in 'new' products. Environmental taxes have not generally been effective in promoting re-use or recycling because, with the exception of disposal charges for some hazardous materials, they are not high enough to offset the expense of recovery and disassembly. In the absence of economic incentives aligned with environmental benefits, EPR has developed mainly as a regulatory approach rather than a voluntary business strategy: companies are mandated to 'take back' their products at end of life.

EPR legislation has two complementary objectives: ensuring that substances presenting risks to human and environmental health are recycled or at least managed, and inducing manufacturers to design their products to facilitate re-use of components and recovery of materials. The latter objective requires legislation to be framed so that items are returned to the original manufacturer, to build-in the link between take-back and product design (Lindhqvist & Lifset, 2003). However, due to a combination of confused legislative processes and industry push-back, implementation of EPR has generally been cumbersome (Castell et al., 2008; Mayers et al., 2011). Mayers (2016) concluded:

> *While EPR was developed with good motives, in practice its implementation is both administratively and logistically complex, and to date the main purpose to incentivise design is largely unfulfilled.*

EPR is not a driver for some of the other benefits of servicisation, and specifically does not promote longevity. Tonner and Malcolm (2017) suggested statutory lifespan guarantees for specific groups of products. However, this would be resisted as a move towards a command-and-control approach. Furthermore, it would be difficult to frame legislation that recognised that optimal service life differs between different products.

3.4 Localisation of activities: 'Small is beautiful'

Formal and informal ***co-operative associations*** at the local level provide frameworks for socialising the circular economy. Remanufacturing—'Loop 1' in Fig. 3—is typically carried out in small-scale local sites or even in the home, and is labour intensive. This makes remanufacturing less subject than capital-intensive activities to 'economies of scale'. Efficiencies result from loops that are more proximate (Stahel & Clift, 2016). Local enterprises can benefit from co-operation in developing and deploying the human capital required for remanufacturing. This, in turn, encourages local networks to grow into stand-alone regional enterprises (Stahel & Clift, 2016). A social economy incorporating these elements can become one of the agents transforming the biophysical-based economy (North, 1999; Sahakian, 2016).

Mending and restoring goods as they wear and breakdown is essential to extend product life but can also be an antidote to globalised mass production of low cost throwaway goods. Local businesses, as well as individual consumers, can contribute to the necessary development of repair shops. For example, there are more than a thousand Repair Cafe volunteer organisations in Western Europe, helping people to fix broken goods (Charter, 2018). The emergence of a 'right-to-repair' movement, embodying the principle of "keeping resources in use for as long as possible" (WRAP, 2021), indicates the growing public demand for product life extension (Duvall, 2016). It challenges control of the repair function by manufacturers that can increase the lifetime cost (Stahel, 2019) and limit independent repair shops by denying access to spare parts, specialised tools, and technical manuals.

Industrial symbiosis (IS) is a different type of co-operative association, linking productive activities through flows of physical by-products, especially flows that would otherwise be wastes. IS provides an opportunity to focus on labour stocks as well as material flows because labour is involved in connecting material flows and the repair and re-use of necessary equipment. IS depends on close relationships between the participants (Chertow & Park, 2016). Proximity favours both the development of these relationships and the physical exchanges.

3.5 Implications for global trade

Global trade is one of the characteristics of the linear economy. In the decades before the 2008 financial crisis, global supply chains became progressively more complex (Los et al., 2014). The pace of globalisation has subsequently slowed, with trade as a percentage of GDP remaining roughly constant at 60% of world GDP (Wang & Sun, 2020). Primary resources are an important part of this trade: in 2019 accounting for around 25% of international trade by value (UNCTAD, n.d.).

Transition to a performance economy, with local re-use and remanufacturing, would reduce production and trade of primary resources (OECD, 2018). The consequences would differ between the 'Global North' and the 'Global South'. Wiebe et al. (2019) modelled a range of circular economy policies and concluded that, where circularity can be implemented by shrinking primary production and growing secondary sectors within the same nation, distributional impacts on employment and wages would be minimal: reducing production and trade of primary materials by around 10% would increase in global employment by less than 5%, primarily driven by increases in remanufacturing. The impact is small because sectors producing secondary resources (e.g. metals from scrap) do not currently have systematically higher employment intensities. This highlights the importance of understanding the performance economy in terms of an economy-wide shift away from resource-intensive processes towards labour-intensive processes, rather than a scaling up of some sectors as others recede.

However, circular economy policies, such as those in the EU, focus on changing the global distribution of primary and secondary production (Gregson et al., 2015). If economies in the North reduce material intensity, fewer imports will be needed from low-income countries, removing the livelihoods of workers in the Global South dependent on primary production and exports, for example from their agricultural and textile sectors (e.g. cotton and cloth) (Schröder et al., 2020). Other rebound effects are also likely, such as consumers using monetary savings to purchase more goods and so expand output. A meta-analysis of 300 CE scenarios to 2050 (Aguilar-Hernandez et al., 2021) found that they pay insufficient attention to potential rebounds.

Delivering a global economy that reduces unsustainability for both the 'North' and 'South' is the subject of a nascent literature highlighting the power relationships inherent in the linear economy (dominated by the Global North) and how they influence circular economy practice and thinking (Schröder, 2019). However, there is a possibility to shift these power relations. Measures to curtail export of waste plastic to the 'South' illustrate the shift (Clift et al., 2019). The concept of a sustainable value chain as a set of relationships that delivers benefits in both directions along the chain (see Section 3.2) is essential here, exemplified by movements like ***Fair Trade*** that promote ethical consumer choices (Reed, 2009) and link sustainable consumption to sustainable production (Clift et al., 2013; Sahakian, 2016).

The dependence of many workers in the Global South on primary resource exports is a feature of the vertical specialisation increasingly evident in global trade from around 1970 (Pahl & Timmer, 2019; Timmer et al., 2019): countries become specialised in particular stages of the supply chain. A performance economy, maintaining stocks as close to the point of use as possible, favours a more resilient system in which countries have a broad basis in multiple stages of production. Relocalisation can promote more equitable distribution of social benefits by avoiding trapping countries in the parts of the supply chain with low economic value and disproportionately high environmental impact (Clift & Wright, 2000): it does not merely mean returning blue-collar industrial jobs to the North; it can also generate white-collar jobs in the Global South.

4 Drivers for a sustainable economy

A sustainable economy must respect all three constraints in Fig. 1:

(1) Techno-economic efficiency: the scientific laws and economic constraints that shape the ways we provide for ourselves and seek to improve our societies. Whereas scientific laws are immutable, the 'laws' defining economic efficiency are not: 'economic efficiency' is a construct that can be modified, for example through changes to the fiscal system.

(2) Environmental compatibility: the environmental and resource constraints that bound a 'safe operating space' for humanity.

(3) Social equity: the ethical constraints that ensure we do not place our well-being ahead of the well-being of others (including future generations).

The performance economy aims to provide enhanced quality of life whilst recognising these three sets of constraints.

Linear economies define efficiency in terms of the monetary value associated with the flows of goods and services produced and consumed. The performance economy approach shifts the focus to maximising the value obtained from stock in use. This perspective shifts the view of the economic system to overlap with environmental and social concerns: a focus on stock preservation helps harness technological efficiency to satisfy environmental and equity constraints. For example, the stock of natural capital provides value by preserving environmental conditions in which human societies flourish. Likewise, the stock of skilled labour can provide value by providing meaningful and well rewarded employment. Earlier sections of this chapter introduced the idea that a focus on stock preservation leads to a greater focus on the role of skilled labour in producing value, recognising that labour is a renewable resource. However, shifting to a labour-based economy requires major structural shifts. We therefore consider measures that would promote such shifts.

It was noted in Section 1.2 that much of the literature on the circular economy emphasises its potential as a profit generating activity rather than a route to improving sustainability. However, studies of barriers to the implementation of circular economy strategies highlight that they are not always immediately profitable (Govindan & Hasanagic, 2018; Kirchherr et al., 2018). This is to be expected: the circular economy constrains economic goals to meet environmental and equity goals. Therefore promoting the performance economy may require changes to the techno-economic domain. Government actions are generally more effective than 'softer' private or behavioural pressure (Lehner et al., 2016; Mols et al., 2015), i.e. the state must play an active role. In addition to regulation, the state can intervene by ***fiscal measures***, which are the principal means by which governments can reshape the techno-economic domain of Fig. 1.

Ecological taxes on use of non-renewable stocks represent fiscal measures to promote resource efficiency by replacing non-renewable inputs by renewables (Lawn, 2000; Stahel, 2013). Ecological tax reforms are an important part of the circular economy policy landscape, particularly in Europe where tax reform is a key pillar of the resource efficiency roadmap (Domenech & Bahn-Walkowiak, 2019). The latest European circular economy action plan (European Commission, 2020) includes provisions for member states to use variable tax rates to incentivise consumers to use repair services. However, such strategies appear to have had limited impacts, in part because consumers may believe that some goods are not worth repairing even if it is cheap to do so (Milios, 2021). Domenech and Bahn-Walkowiak (2019) point also to the political context, arguing that taxes are used as elements of economic competition between states who resist substantive taxes on resources as they do not wish to harm their competitiveness. We return to this point later. Social equity can be promoted by ***personal taxation*** with a progressive rate structure that reduces income taxes for the lower paid. However, taxation policy is politically determined so that, in practice, there are limits on the extent to which personal taxation can be redistributive. Combining personal taxation with ecological taxes offers a potentially more 'acceptable' approach, improving both resource efficiency and equity.

The central idea of using taxes to increase the prices of non-renewable resources and decrease costs of renewables includes the shift from non-renewables to labour in the performance economy (Stahel, 2013, 2019; Stahel & Clift, 2016). Opposition to ecological taxes can be muted if the changes are presented as ***revenue-neutral***, i.e. as a shift in the tax base, not an additional tax. Taxes or levies on emissions of greenhouse gases—so-called carbon taxes—are a specific form of ecological tax.[d] The Canadian province of British Columbia illustrates this approach: carbon taxes are already high by global standards (CDN$45 per tCO_2e) but are being increased even further without widespread opposition. Some of the revenue is returned to lower-paid individuals and families as a 'carbon tax rebate', emphasising that carbon taxes represent redistribution rather than additional taxation. There are signs that this fiscal approach is driving restructuring towards greater resource efficiency (Murray & Rivers, 2015; Smart Prosperity Institute, 2021).

One of the concerns around fiscal measures is that increasing material or emission costs in one region will shift the associated activities elsewhere. This has been discussed primarily in terms of 'carbon leakage' (Cosbey, 2019; Keen & Kotsogiannis, 2014; Ward et al., 2015)—i.e. migration of industries emitting greenhouse gases to regions other than those where their products are consumed—but the principles apply more widely. Taxes on emissions or resource use make it more expensive to make a product than in regions where there is no such tax. Production, investment, and employment may then migrate to a lower tax region, from where the product will be imported into the region with the ecological taxes. ***Boundary or border taxes*** on imported goods are intended to negate the perverse competitive advantage of goods from a low-tax region by imposing tariffs equal to the ecological taxes. However,

[d]The efficacy of direct taxes and levies relative to an indirect system for 'pricing' emissions is not a topic we have space to discuss. We refer here to taxes or levies because that relates more directly to our analysis and because we believe such direct measures have proven more effective in practice.

World Trade Organisation rules raise questions over such taxes, and carbon border tax proposals in the US and Europe have been rejected by the relevant legislative bodies (Mehling et al., 2019; Truby, 2010).

5 Conclusions

Following Brandt (1982), sustainable development implies enhancing quality of life and well-being for current and future generations, rather than increasing consumption. Sustainability has economic, social, and environmental dimensions; quality of life does not equate to spending power; ethical imperatives override financial gain; and the well-being of the planet overrides everything. However, most interpretations of the circular economy promote it as a way to expand economic activity and corporate profits. Shifting to a more sustainable 'performance economy', respecting all three dimensions, requires major changes in business focus, away from material flows towards managing stocks of goods and labour. The circular economy approach potentially supports most of the 17 SDGs, particularly SDG 12: "Ensure Sustainable Consumption and Production Patterns" by "changing the way we produce and consume resources." However, the SDGs ignore the distinction between economic growth and sustainable development. SDG 12 includes "Achieving economic growth." SDG 8 explicitly links "Decent Work and Economic Growth," ignoring the performance economy approach of restructuring to generate more jobs without increasing material consumption. As Jackson (2017) puts it:

> *People can flourish without endlessly accumulating more stuff. Another world is possible.*

Acknowledgments

Simon Mair's time was partially supported by the Economic and Social Research Council funded Centre for the Understanding of Sustainable Prosperity (ES/M010163/1).

References

Aguilar-Hernandez, G. A., et al. (2021). Macroeconomic, social and environmental impacts of a circular economy up to 2050: A meta-analysis of prospective studies. *Journal of Cleaner Production*, *278*, 123421.

Allwood, J. M. (2014). Squaring the circular economy: The role of recycling within a hierarchy of material management strategies. In *Handbook of recycling: State-of-the-art for practitioners, analysts, and scientists*. [Online]. Available at http://publications.eng.cam.ac.uk/706968/. Accessed: 13 May 2021.

Brandt, W. (1982). *North-South, a programme for survival: Report independent commission on international development issues*. 11 London: Pan Books.

Castell, A., et al. (2008). Extended producer responsibility policy in the European Union: A horse or a camel? *Journal of Industrial Ecology*, *8*(1–2), 4–7.

Charter, M. (2018). Repair cafes. *The Journal of Peer Production*, *3*(12), 37–45.

Chertow, M., & Park, J. (2016). Scholarship and practice in industrial symbiosis: 1989–2014. In R. Clift, & A. Druckman (Eds.), *Taking stock of industrial ecology*. Cham: Springer International Publishing. https://doi.org/10.1007/978-3-319-20571-7_5. [Online]. Accessed: 13 May 2021.

Chopra, S. (2020). *Supply chain management: Strategy, planning, and operation*. New York: Pearson Education Limited.

Clift, R. (1995). The challenge for manufacturing. In J. McQuaid (Ed.), *Engineering for sustainable development*. London: The Royal Academy of Engineering.

Clift, R. (2013). Clean technology and industrial ecology. In R. M. Harrison (Ed.), *Pollution-causes, effects and control* (5th ed.). https://doi.org/10.1039/9781847551719-00411. [Online]. Accessed: 5 May 2021.

Clift, R., & Wright, L. (2000). Relationships between environmental impacts and added value along the supply chain. *Technological Forecasting and Social Change*, *65*(3), 281–295.

Clift, R., et al. (2013). Sustainable consumption and production: Quality, luxury and supply chain equity. In I. S. Jawahir, et al. (Eds.), *Treatise on sustainability science and engineering*. Dordrecht: Springer Netherlands. https://doi.org/10.1007/978-94-007-6229-9_17. [Online]. Accessed: 5 May 2021.

Clift, R., et al. (2017). Planetary boundaries as a basis for strategic decision-making in companies with global supply chains. *Sustainability*, *9*(2), 279.

Clift, R., et al. (2019). Managing plastics: Uses, losses and disposal special issue on designing law and policy towards managing plastics in a circular economy. *Law, Environment and Development Journal (LEAD Journal)*, *15*(2), 93–107.

Cohen, M. A., et al. (2006). Winning in the aftermarket. *Harvard Business Review*. 1 May. Available at https://hbr.org/2006/05/winning-in-the-aftermarket. Accessed: 5 May 2021.

Cosbey, A., et al. (2019). Developing guidance for implementing border carbon adjustments: Lessons, cautions, and research needs from the literature. *Review of Environmental Economics and Policy*, *13*(1), 3–22. The University of Chicago Press.

Dearing, J. A., et al. (2014). Safe and just operating spaces for regional social-ecological systems. *Global Environmental Change*, *28*, 227–238.

Domenech, T., & Bahn-Walkowiak, B. (2019). Transition towards a resource efficient circular economy in Europe: Policy lessons from the EU and the member states. *Ecological Economics*, *155*, 7–19.

Duvall, L., et al. (2016). The fixer movement: A key piece of the circular economy revolution. In Ellen Macarthur Foundation (Ed.), *Empowering repair*. Isle of Wight, UK: Ellen Macarthur Foundation.

Ede, S. (2017). *The real circular economy: How relocalizing production with not-for-profit business models helps build resilient and prosperous societies*. Online Commons Transition. Available at: https://commonstransition.org/the-real-circular-economy/. Accessed: 13 May 2021.

Ellen MacArthur Foundation. (2021). *Universal circular economy policy goals*. Available at: https://emf.thirdlight.com/link/kt00azuibf96-ot2800/@/preview/1?o. Accessed: 5 May 2021.

European Commission. (2020). *Circular economy action plan*. Available at: https://eur-lex.europa.eu/legal-content/EN/TXT/?qid=1583933814386&uri=COM:2020:98:FIN.

Evangelisti, S., et al. (2015). Integrated gasification and plasma cleaning for waste treatment: A life cycle perspective. *Waste Management*, *43*, 485–496.

Forum for the Future. (2014). *The five capitals—A framework for sustainability*. Available at: https://www.forumforthefuture.org/the-five-capitals. Accessed: 5 May 2021.

Freeman, R. E., & Reed, D. L. (1983). Stockholders and stakeholders: A new perspective on corporate governance. *California Management Review*, *25*(3), 88–106.

Friedman, M. (1970). *A Friedman doctrine—The social responsibility of business is to increase its profits*. The New York Times. 13 Sep. Available at https://www.nytimes.com/1970/09/13/archives/a-friedman-doctrine-the-social-responsibility-of-business-is-to.html. Accessed: 5 May 2021.

Gear, M., et al. (2018). A life cycle assessment data analysis toolkit for the design of novel processes—A case study for a thermal cracking process for mixed plastic waste. *Journal of Cleaner Production*, *180*, 735–747.

Georgescu-Roegen, N. (2017). *The entropy law and the economic process*. Cambridge (Mass.): Harvard University Press.

Gomez, E. (2016). *10 Social enterprise examples and the principles that guide them*. Conscious Connection. Available at https://www.consciousconnectionmagazine.com/2016/02/social-enterprise-examples-and-principles/. Accessed: 13 May 2021.

Govindan, K., & Hasanagic, M. (2018). A systematic review on drivers, barriers, and practices towards circular economy: A supply chain perspective. *International Journal of Production Research*, *56*(1–2), 278–311. Taylor & Francis.

Gregson, N., et al. (2015). Interrogating the circular economy: The moral economy of resource recovery in the EU. *Economy and Society*, *44*(2), 218–243.

Heiskanen, E., & Jalas, M. (2000). *Dematerialization through services: A review and evaluation of the debate*. Helsinki: Ministry of Environment : Edita, jakaja.

Isham, A., et al. (2021). Worker wellbeing and productivity in advanced economies: Re-examining the link. *Ecological Economics*, *184*, 106989.

Jackson, T. (2010). Keeping out the giraffes. In *Long horizons* (pp. 17–20). London: British Council.

Jackson, T. (2017). *Prosperity without growth: Foundations for the economy of tomorrow* (2nd ed.). London and New York: Routledge.

Jensen, J. P., et al. (2019). Creating sustainable value through remanufacturing: Three industry cases. *Journal of Cleaner Production*, *218*, 304–314.

Keen, M., & Kotsogiannis, C. (2014). Coordinating climate and trade policies: Pareto efficiency and the role of border tax adjustments. *Journal of International Economics*, *94* (1), 119–128. North-Holland.

Kerlin, J. A. (2010). A comparative analysis of the global emergence of social enterprise. *Voluntas: International Journal of Voluntary and Nonprofit Organizations*, *21*(2), 162–179.

King, A. M., et al. (2006). Reducing waste: Repair, recondition, remanufacture or recycle? *Sustainable Development*, *14*(4), 257–267.

Kirchherr, J., et al. (2017). Conceptualizing the circular economy: An analysis of 114 definitions. *Resources, Conservation and Recycling*, *127*, 221–232.

Kirchherr, J., et al. (2018). Barriers to the circular economy: Evidence from the European Union (EU). *Ecological Economics*, *150*, 264–272.

Korhonen, J. (2003). Should we measure corporate social responsibility? *Corporate Social Responsibility and Environmental Management*, *10*(1), 25–39.

Laurenti, R., et al. (2018). The socio-economic embeddedness of the circular economy: An integrative framework. *Sustainability*, *10*(7), 2129.

Lawn, P. A. (2000). Ecological tax reform: Many know why but few know how. *Environment, Development and Sustainability*, *2*(2), 143–164.

Lehner, M., et al. (2016). Nudging—A promising tool for sustainable consumption behaviour? *Journal of Cleaner Production*, *134*, 166–177.
Lightfoot, H. (2021). *Servitization: Manufacturing paradigm shift or simply a business model?* Si2 PARTNERS. Available at https://si2partners.com/resources/servitization-manufacturing-paradigm-shift-simply-business-model/. Accessed: 5 May 2021.
Lindhqvist, T., & Lifset, R. (2003). Can we take the concept of individual producer responsibility from theory to practice? *Journal of Industrial Ecology*, *7*(2), 3–6.
Los, B., et al. (2014). How global are global value chains? A new approach to measure international fragmentation. *Journal of Regional Science*, *55*(1), 66–92.
Makower, J. (2021). *The state of green business 2021*. Greenbiz. Available at: https://www.greenbiz.com/article/state-green-business-2021. Accessed: 14 May 2021.
Malcolm, R. (2019). Life cycle thinking as a legal tool: A codex Rerum. *Law, Environment and Development Journal*, *15*(2), 208. SOAS.
Mayers, K. (2016). Practical implications of product-based environmental legislation. In R. Clift, & A. Druckman (Eds.), *Taking stock of industrial ecology*. Cham: Springer International Publishing. https://doi.org/10.1007/978-3-319-20571-7_16. [Online].
Mayers, K., et al. (2011). Redesigning the camel. *Journal of Industrial Ecology*, *15*(1), 4–8.
McKinsey & Co. (2012). *Towards the circular economy: Economic and business rationale for an accelerated transition*. London: Ellen MacArthur Foundation.
Mehling, M. A., et al. (2019). Designing border carbon adjustments for enhanced climate action. *American Journal of International Law*, *113*(3), 433–481. Cambridge University Press.
Milios, L. (2021). Towards a circular economy taxation framework: Expectations and challenges of implementation. *Circular Economy and Sustainability*, *1*, 477–498.
Mitchell, C. (2000). Integrating sustainability in chemical engineering practice and education: Concentricity and its consequences. *Process Safety and Environmental Protection*, *78*(4), 237–242.
Mitchell, C. A., et al. (2004). The role of the professional engineer and scientist in sustainable development. In A. Azapagic, et al. (Eds.), *Sustainable development in practice*. Chichester, UK: John Wiley & Sons, Ltd. https://doi.org/10.1002/0470014202.ch2. [Online]. Accessed: 5 May 2021.
Mols, F., et al. (2015). Why a nudge is not enough: A social identity critique of governance by stealth. *European Journal of Political Research*, *54*(1), 81–98.
Moreau, V., et al. (2017). Coming full circle: Why social and institutional dimensions matter for the circular economy. *Journal of Industrial Ecology*, *21*, 497–506.
Murray, B., & Rivers, N. (2015). British Columbia's revenue-neutral carbon tax: A review of the latest "grand experiment" in environmental policy. *Energy Policy*, *86*, 674–683.
North, P. (1999). Explorations in heterotopia: Local exchange trading schemes (LETS) and the micropolitics of money and livelihood. *Environment and Planning D: Society and Space*, *17*(1), 69–86.
OECD. (2018). *The macroeconomics of the circular economy transition: A critical review of modelling approaches*. 130. Available at. https://doi.org/10.1787/af983f9a-en. Accessed: 10 December 2020.
Örsdemir, A., et al. (2018). Is servicization a win-win strategy? Profitability and environmental implications of servicization. *Manufacturing & Service Operations Management*, *21*(3), 674–691. INFORMS.
Pahl, S., & Timmer, M. P. (2019). Patterns of vertical specialisation in trade: Long-run evidence for 91 countries. *Review of World Economics*, *155*(3), 459–486.
Porritt, J. (2005). *Capitalism as if the world matters*. London ; Sterling, VA: Earthscan.

Porter, M. E. (1985). *Competitive advantage: Creating and sustaining superior performance*. New York : London: Free Press; Collier Macmillan.

Porter, M. E., & Kramer, M. R. (2011). Creating shared value. *Harvard Business Review*, *89*, 2–17.

Purvis, B., et al. (2019). Three pillars of sustainability: In search of conceptual origins. *Sustainability Science*, *14*(3), 681–695.

Raworth, K. (2017). *Doughnut economics: 7 ways to think like a 21st century economist*. London: Random House Business Books.

Reed, D. (2009). What do corporations have to do with Fair Trade? Positive and normative analysis from a value chain perspective. *Journal of Business Ethics*, *86*(1), 3–26.

Rockström, J., et al. (2009). A safe operating space for humanity. *Nature*, *461*(7263), 472–475. Nature Publishing Group.

Sahakian, M. (2016). The social and solidarity economy: Why is it relevant to industrial ecology? In R. Clift, & A. Druckman (Eds.), *Taking stock of industrial ecology*. Cham: Springer International Publishing. https://doi.org/10.1007/978-3-319-20571-7_10. [Online]. Accessed: 13 May 2021.

Schau, E., et al. (2012). Life cycle approach to sustainability assessment: A case study of remanufactured alternators. *Journal of Remanufacturing*, *2*(5), 1–14.

Schröder, P., et al. (2019). *The circular economy and the global south: Sustainable lifestyles and green industrial development* (Online). Routledge. https://doi.org/10.4324/9780429434006. Accessed: 10 December 2020.

Schröder, P., et al. (2020). Making the circular economy work for human development. *Resources, Conservation and Recycling*, *156*, 104686.

Smart Prosperity Institute. (2021). *Just the facts please: The true story of how BC's carbon tax is working*. Available at: https://institute.smartprosperity.ca/content/just-facts-please-true-story-how-bc-s-carbon-tax-working. Accessed: 12 May 2021.

Stahel, W. (2010). *The performance economy* (2nd ed.). Palgrave Macmillan.

Stahel, W. R. (2013). Policy for material efficiency—Sustainable taxation as a departure from the throwaway society. *Philosophical Transactions of the Royal Society A: Mathematical, Physical and Engineering Sciences*, *371*(1986), 20110567. Royal Society.

Stahel, W. R. (2016). Circular economy. *Nature*, *531*, 435–438.

Stahel, W. R. (2019). *The circular economy: A user's guide*. London; New York: Routledge, Taylor & Francis.

Stahel, W., & Clift, R. (2016). Stocks and flows in the performance economy. In *Taking stock of industrial ecology* (pp. 137–158). Springer International Publishing. Online.

Steffen, W., et al. (2015). Planetary boundaries: Guiding human development on a changing planet. *Science*, *347*(6223), 1259855.

Stratan, D. (2017). Success factors of sustainable social enterprises through circular economy perspective. *Visegrad Journal on Bioeconomy and Sustainable Development*, *6*(1), 17–23.

Sustainalytics. (2021). *ESG risk ratings*. Available at: https://www.sustainalytics.com/esg-data. Accessed: 5 May 2021.

Tilak, J. B. G. (2008). Higher education: A public good or a commodity for trade? *Prospects*, *38*(4), 449–466.

Timmer, M. P., et al. (2019). Functional specialisation in trade. *Journal of Economic Geography*, *19*(1), 1–30.

Tonner, K., & Malcolm, R. (2017). *How an EU lifespan guarantee model could be implemented across the European Union*. European Commission, Directorate General for Internal Policies.

Truby, J. M. (2010). Towards overcoming the conflict between environmental tax leakage and border tax adjustment concessions for developing countries. *Vermont Journal of Environmental Law*, *12*(1), 149–184.

UN. (1987). *Our common future. Report of the world commission on environment and development*. New York: United Nations.

UN. (2015). *Transforming our world: The 2030 agenda for sustainable development*. Available at: https://sdgs.un.org/2030agenda. Accessed: 14 May 2021.

UNCTAD (n.d.). Merchandise trade matrix—Exports of individual economies in thousands United States dollars, annual. Available at: https://unctadstat.unctad.org/wds/ReportFolders/reportFolders.aspx. Accessed: 10 December 2020.

Vandermerwe, S., & Rada, J. (1988). Servitization of business: Adding value by adding services. *European Management Journal*, *6*(4), 314–324.

Velenturf, A. P. M., & Purnell, P. (2021). Principles for a sustainable circular economy. *Sustainable Production and Consumption*, *27*, 1437–1457.

Vickers, I., et al. (2017). Public service innovation and multiple institutional logics: The case of hybrid social enterprise providers of health and wellbeing. *Research Policy*, *46*(10), 1755–1768.

Wang, Z., & Sun, Z. (2020). From globalization to regionalization: The United States, China, and the post-Covid-19 world economic order. *Journal of Chinese Political Science*, *26*, 69–87.

Ward, M. (2020). *As the pandemic, fires, and inequity all rage, free market icon Milton Friedman's declaration that the sole responsibility of business "is to increase its profits" sounds emptier than ever*. Business Insider. 13 September https://businessinsider.com/milton-friedman-theory-role-of-business-feels-emptier-than-ever-2020-9.

Ward, J., et al. (2015). *Carbon leakage*. Available at http://documents1.worldbank.org/curated/en/138781468001151104/pdf/100369-NWP-PUBLIC-ADD-SERIES-Partnership-for-Market-Readiness-technical-papers-Box393231B.pdf.

Webster, K. (2015). *The circular economy: A wealth of flows*. London: Ellen MacArthur Foundation.

WEF. (2020). *The future of jobs report 2020*. Available at: https://www.weforum.org/reports/the-future-of-jobs-report-2020/. Accessed: 13 May 2021.

Wiebe, K. S., et al. (2019). Global Circular Economy Scenario in a Multiregional Input–Output Framework. *Environmental Science & Technology*, *53*(11), 6362–6373.

Wijkman, A., & Skanders, K. (2015). *The circular economy and benefits for society: Jobs and climate clear winners in an economy based on renewable energy and resource efficiency*. Online The Club of Rome. Available at: https://clubofrome.org/wp-content/uploads/2020/03/The-Circular-Economy-and-Benefits-for-Society.pdf. Accessed: 13 May 2021.

Wilkinson, R. G., & Pickett, K. (2010). *The spirit level: Why equality is better for everyone; [with a new chapter responding to their critics]*. Published with revisions, published with a new postscript London: Penguin Books.

WRAP. (2021). *WRAP and the circular economy*. WRAP. Available at: https://wrap.org.uk/about-us/our-vision/wrap-and-circular-economy. Accessed: 5 May 2021.

CHAPTER

The food–energy–water nexus approach

4

Carolin Märker[a,b] and Sandra Venghaus[a,c]

[a]*Institute of Energy and Climate Research - Systems Analysis and Technology Evaluation (IEK-STE), Forschungszentrum Jülich, Jülich, Germany* [b]*University of Bonn, Bonn, Germany* [c]*RWTH Aachen University, School of Business and Economics, Aachen, Germany*

1 Introduction

Like the concepts of sustainable development and circular economy, also the food–energy–water (FEW) nexus has frequently been proposed and used as a framework for assessing sustainability progress. At its core, the nexus is strongly related to and intertwined with the earlier two frameworks. It emerged as part of the discussion on sustainable development on the one hand and proposes a more efficient use of natural resources on the other. The origins of the nexus concept trace back to the Bonn2011 Conference 'The Water, Energy and Food Security Nexus—Solutions for the Green Economy' (Hoff, 2011). At the time, the nexus was developed to provide a new approach to highlight and account for the interrelations between water, energy, and food resources. Specifically, it called for a more integrated management in order to reach security of water, energy, and food supply (Fig. 1).

This integrated approach aimed at a more holistic perspective that was intended to help avoid conflicts of interest and create synergies in managing these resources (Märker, Venghaus, & Hake, 2018). The FEW nexus was thus introduced with the objective to globally reduce vulnerabilities to access shortages of resources, while improving resilience. The idea of the nexus approach is to not prioritise a single resource, but to consider the different dimensions of water, energy, and land (food)—those resources most critical to sustaining life (Gain, Giupponi, & Benson, 2015)—equally and in their mutual, dynamic interrelationships (Food and Agriculture Organization of the United Nations (FAO), 2014).

The Bonn2011 conference was held in preparation for the 2012 UN Conference on Sustainable Development in Rio, which marked the beginning of the SDG development process. The nexus debate reflected the increasing pressures on natural resources as well as their complex interplay. The need for a more integrated management of these resources—the core idea of the nexus—was eventually included in the preamble to the SDGs. Adopting a nexus perspective accordingly implies pursuing

Assessing Progress towards Sustainability. https://doi.org/10.1016/B978-0-323-85851-9.00007-9

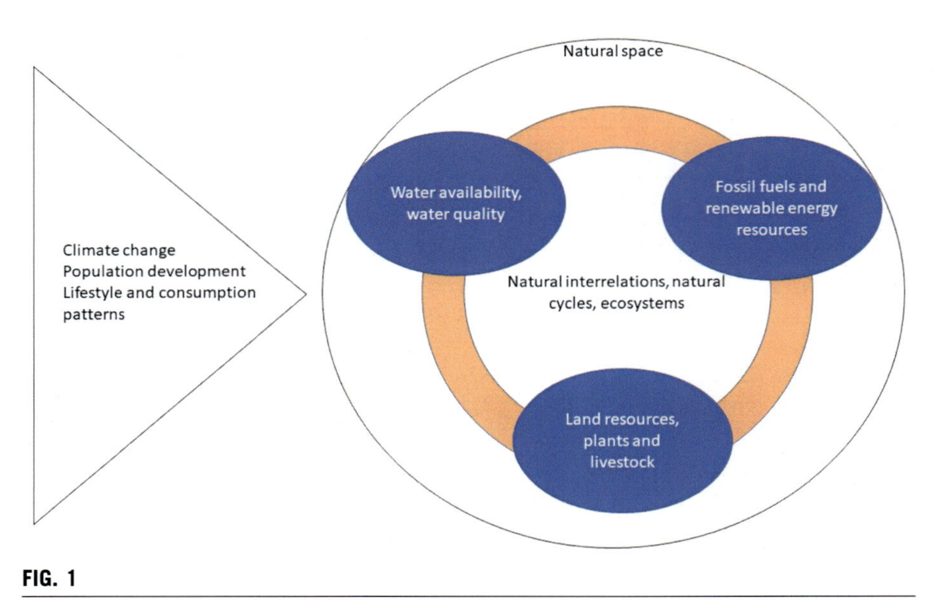

FIG. 1

The food–energy–water nexus of the natural space.

an approach that allows reaching these goals in complementarity with each other by avoiding negative side effects of one goal on another (Hoff, 2011). The FEW nexus, in particular, emphasises the interconnectedness of the goals that are related to the three resources, such as 'zero hunger' (goal 2), 'clean water and sanitation' (goal 6), and 'clean and affordable energy' (goal 7). From a slightly broader angle, the FEW nexus also relates to and addresses goals 13 to 15—'climate action', 'life below water', and 'life on land'.

The nexus concept was developed in response to the common criticism that the governance of water, energy, and food supply has widely occurred in separate 'silos', which in the past has led to inefficient use and conflicts of interest between the resources (Märker et al., 2018). Research further shows that ongoing global megatrends will likely and significantly aggravate these conflicts (Namany, Al-Ansari, & Govindan, 2019). A growing world population, on the one hand, will result in increasing demand, whereas the impacts of climate change will endanger the secure availability or production of these resources, on the other (Proctor, Tabatabaie, & Murthy, 2021). More frequent and/or more severe droughts, for example, are expected to threaten harvests and will lead to reduced water resources needed for drinking water, irrigation, or industrial uses. Capacity expansion of renewable energies, such as wind, solar, or bioenergy, requires land resources, thus competing with their need for food production or as natural habitats that need to be preserved. Many more of such interlinkages exist. With regard to these trends, an efficient management of water, energy, and food resources has become more and more important. Against this background, the nexus concept has not only emerged as a management concept but has also been transferred into academic research. Since then, research articles on the nexus can be found in countless journals and in just as many variations. According to a search in the Scopus database, the number of nexus-related articles increased by about a factor of over eight from 138 in 2011 to 1173 in 2020 (cf. Fig. 2, blue bar).

However, the use of the nexus concept has evolved along different pathways. The term 'nexus' appears not only in combination with water, energy, and food, but also in various other sector combinations, of which the water–energy nexus has been found to be the most prominent one (Fig. 3).

In light of this vast amount of nexus research, 'the nexus' as a precisely defined subject of investigation as often suggested does not exist. Instead, this chapter describes the nexus as a research framework for sustainability assessment that is open to many different applications. Defining the nexus as a framework allows lifting it to an overarching level capable of unifying the various different streams of research. This idea rests on the foundation of defining five different spaces of the FEW nexus—all of which are integrated in this framework and thus help to determine, which elements must be considered and accounted for in a comprehensive and systematic analysis. Before, however, the nexus framework is further introduced in Section 3, we will provide an overview of the most prominent streams of nexus research in order to briefly demonstrate the variety in which the nexus has been used and applied. As part of the discussion, the overarching rules for the framework application are described and discussed.

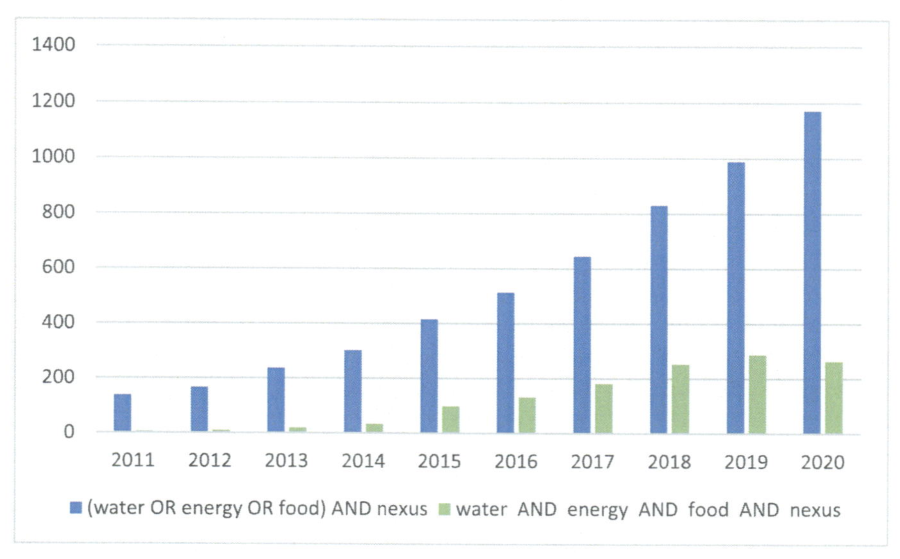

FIG. 2

Number of annually published nexus-related research articles.

Search within the Scopus database on July 21, 2021. Search in title, abstract, keywords.

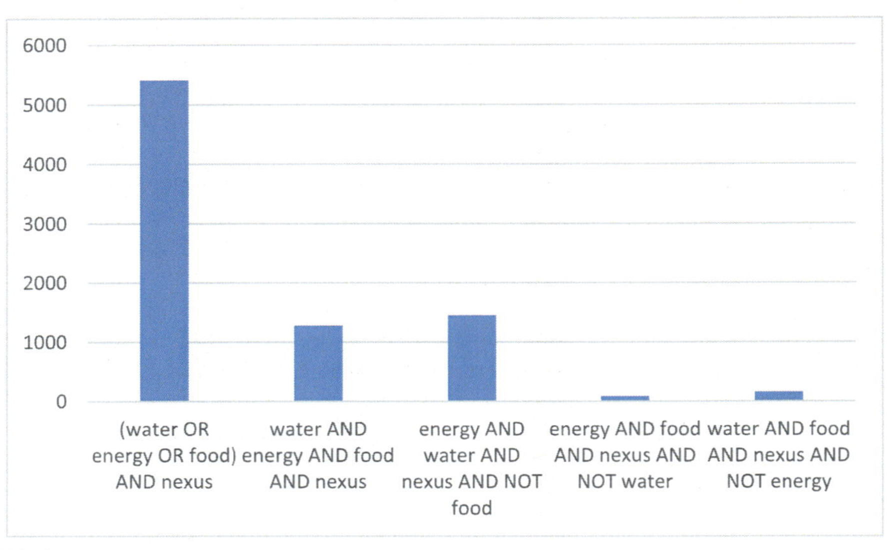

FIG. 3

Nexus-related research articles published in 2020.

Search within the Scopus database on July 21, 2021. Search in title, abstract, keywords.

2 Important streams of nexus research

The nexus-related scientific publications are categorised along the following categories: (1) analytical perspective, (2) sector combinations, and (3) application.

2.1 Analytical perspective

With respect to the analytical perspective of nexus research, two streams can be differentiated: techno-economic research and socio-economic research. Accordingly, depending on the perspective the used methods deviate between quantitative and qualitative research approaches. Techno-economic analyses often focus on resource flows, technology assessment or optimisation or simulation modelling. Examples of modelling approaches are provided, for example, by the following authors: Amjath-Babu et al. (2019) model the impacts of hydropower on the FEW nexus in the Himalaya. Brouwer et al. (2018) developed an energy systems modelling approach addressing the nexus, whereas Fouladi, AlNouss, and Al-Ansari (2021) and Gonzalez-Salazar et al. (2016) applied their modelling frameworks to the case of bioenergy. Khan et al. (2018) designed an integrated optimisation model for the energy–water nexus. Scenario development is also a commonly used method in this stream of nexus research (e.g. Fouladi et al., 2021; Moazeni, Khazaei, & Pera Mendes, 2020; Wang, Fath, & Chen, 2019). On the other hand, socio-economic approaches address, for example, topics such as policy development or the management and distribution of the resources related to the nexus. Some of the studies conducted stakeholder analyses or emphasised the need to engage them in nexus research (e.g. Ghodsvali, Krishnamurthy, & de Vries, 2019; Hoolohan et al., 2018; Hoolohan, McLachlan, & Larkin, 2019; White, Jones, Maciejewski, Aggarwal, & Mascaro, 2017; Zhang, Tan, Zhang, Zhang, & Zhang, 2021), whereas others provided policy analyses (e.g. Mercure et al., 2019; Portney et al., 2017; Venghaus & Hake, 2018; Venghaus, Märker, Dieken, & Siekmann, 2019). Another important stream of special relevance with respect to different analytical perspectives are conceptual thoughts on the nexus, such as those provided by, e.g. Al-Saidi and Elagib (2017), Weitz, Strambo, Kemp-Benedict, and Nilsson (2017), or Pahl-Wostl et al. (2020). Overall, however, a comprehensive review conducted by Simpson and Jewitt concludes "that much of the literature appertaining to the WEF nexus to date is apolitical and technical in nature" (Simpson & Jewitt, 2019).

2.2 Sector combinations

As mentioned before, the nexus does not only appear in the form of the food–energy–water nexus. With respect to the resource nexus and against the background of the Bonn2011 nexus conference, this resource combination of food, energy, and water, however, is generally considered to be the 'original' one and is widely used (e.g. Endo et al., 2020; Larkin, Hoolohan, & McLachlan, 2020; Simpson & Jewitt, 2019; Yu, Xiao, Zeng, Li, & Fan, 2020). One of the most prominent variations is the water–energy nexus (e.g. Abegaz, Datta, & Mahajan, 2018; Ahmad, Jia, Chen, Li, & Xu, 2020; Chen, Alvarado, & Hsu, 2018; Salehi, Ghannadi-Maragheh, Torab-Mostaedi, Torkaman, & Asadollahzadeh, 2020; Wang et al., 2019). Several other examples exist including, for example, the food–energy nexus (e.g. Cuberos Balda & Kawajiri, 2020; Schwoerer, Schmidt, & Holen, 2020) or the food–water nexus (e.g. Gephart et al., 2017; Li et al., 2021; Mrozik et al., 2019). Often and in many scientific publications further combinations are considered that include additional aspects and perspectives into the nexus, such as, e.g. economic growth, health, climate, environment, or carbon emissions (e.g. Aydin, 2018; Dale, Efroymson, & Kline, 2011; Ghani et al., 2019; Grobicki, 2016; Mohmmed et al., 2019; Ozturk & Bilgili, 2015; van den Heuvel, Blicharska, Masia, Sušnik, & Teutschbein, 2020). Security is another dimension that is repeatedly included (Ghodsvali et al., 2019; Hoolohan et al., 2018, 2019; White et al., 2017; Zhang et al., 2021). This broad variety of how the nexus is conceptualised to include different sectors and perspectives further strengthens the assumption that the core idea of emphasising interconnections between different sectors is more important than the sectors themselves.

2.3 Application

The modes of application are as diverse as the methods and combinations. Often, the nexus is applied to specific regions (e.g. Cuberos Balda & Kawajiri, 2020; Rodriguez et al., 2018; Saladini et al., 2018; Wang, Liu, & Chen, 2018) or river basins

Table 1 Categorisation of nexus research.

Categories	Specifications						
Sectors	Water	Energy	Agriculture/ Land/	Climate	Ecosystem/	Development policy	Security
Application	Regional application		Project application		Sectoral application	Theory application	
Analytical perspective and methods	Techno-economic dimension				Political and socio-economic dimension		
	Modelling approaches	Scenario development		Policy analysis	Resources management		Stakeholder analysis
	Quantitative				Mostly qualitative		

(e.g. Abegaz et al., 2018; Ahmad et al., 2020; Chen et al., 2018; Salehi et al., 2020; Wang et al., 2019). Others deal with sector-specific problems and address them from a nexus perspective, such as, for example, water supply or electricity production (e.g. Jin, Behrens, Tukker, & Scherer, 2019; Lechón, De La Rúa, & Cabal, 2018; Price et al., 2018; Vakilifard, Anda, Bahri, & P., & Ho, 2018). Table 1 summarises the different categories of nexus research.

These categorisations reflect that the use of the nexus concept in scientific research is not limited to the FEW nexus. It is used in many different combinations, with regard to many different applications, and by means of many different methods. At the same time, it is this wide applicability and relevance of the nexus that has led to a vast pool of related research papers, making it very difficult for researchers to keep track of developments in this research field. This further implicates that likewise also the number of analytical tools or instruments used and developed to analyse nexus-related research questions is equally broad and often a matter of individual choice. In order to systematically address this aspect, the following section serves to define the nexus as a research framework that can be used to analyse and address challenges related to natural resources management across natural, technical, economic, political, and social spaces. The nexus is thus conceptualised to provide a framework highlighting interrelations and connections among the natural resources themselves as well as between their consumption and management.

3 Defining the 'food–energy–water nexus' as a framework

The story of origin of the FEW nexus is somewhat different from the story of other sustainability concepts. The nexus first appeared as a concept representing the strong links that exist between the use of water, energy, and food resources. Against the background of globally growing demands, the World Economic Forum in 2011 brought the 'water–food–energy nexus' as one global risk into the focus of politics and science (World Economic Forum, 2011). Also, during the Bonn2011 nexus conference, the nexus was portrayed as a new perspective on water, energy, and food security as well as on the connections that constitute an essential precondition for reaching it (Hoff, 2011). In the background paper to the conference, Hoff already emphasised the need to understand the nexus as an approach that can be used to adopt a more holistic and integrated perspective in policy, governance, and management. Following this conference, the nexus concept was taken up by researchers from pretty much all disciplines, turning it into an easily adaptable research perspective used in countless applications and shapes. Even for recent papers the conference's background paper still serves as basic literature and as a point of entry (e.g. Albrecht, Crootof, & Scott, 2018; Benites-Lazaro, Giatti, Sousa Junior, & Giarolla, 2020; Mercure et al., 2019; van den Heuvel et al., 2020). However, despite the common basic assumption on which they rest, nexus research can barely be treated as one single research concern as the previous sections have shown.

Furthermore, resource management problems have existed and been addressed for many years. Reaching a globally secure supply with drinking water, energy, or food has long been a basic concern in development policy and environmental systems research, for example. The emergence of the nexus concept, however, has added a new drive to these questions (Pahl-Wostl et al., 2020). The hopes associated with the nexus are that this new, more integrated perspective will help to analyse as well as tackle these problems in a more efficient and effective way. The basic idea is that important conflicts of interest as well as specific challenges can only be revealed by a holistic perspective that takes interrelations between different resources, sectors, or policy fields into account (Howells et al., 2013). This new angle on challenges related to natural resources has resulted in an immense gain in knowledge on many different aspects of natural resources management across the world. Whereas this is certainly a great benefit, the dynamics and breadth in research on the FEW nexus—and thus also the pool of nexus-related publications—has become so large that scientists are hardly able to maintain an overview of ongoing nexus research relevant to their field of study. It thus becomes increasingly necessary to abandon the idea of treating the FEW nexus as a research topic in itself and, instead, understand it as a research framework. Defining the nexus as a framework for sustainability assessment allows keeping it open for any kind of application. According to Nobel laureate Elinor Ostrom "frameworks identify the elements and general relationships among these elements that one needs to consider [...]" (Ostrom, 2011). Frameworks thus support the researcher in achieving a structured approach of analysis (McGinnis, 2011). Looking at the different streams of nexus research helps setting the boundaries of the framework. The FEW nexus can be addressed from a sustainability perspective considering its focus on natural, technical, economic, social, as well as political aspects.

It becomes apparent that—depending on the research perspective—the food–energy–water nexus refers to entirely different aspects. At the same time, these aspects define the relevant elements of the framework that need to be considered for an analysis. However, the five spaces of the natural, technical, economic, political, and social environment are intertwined in many ways. Often, they cannot and should not be analysed separately from each other. When bringing these spaces together it becomes apparent that the natural space—or the resource availability—forms the outer boundary in which the other four spaces are embedded. Against the fact that the nexus concept emerged within the context of sustainable development, the outer ring of the framework—the natural space—can best be defined by the concept of planetary boundaries. Johann Rockström et al. used this concept to define a secure corridor for the use of natural resources (i.e. the planetary boundaries) with regard to nine crucial ecological risks—climate change, land use change, or freshwater use among them (Rockström et al., 2009). They concluded that the safe space (green zone) has already been transgressed within four areas. In the areas of land system change and climate change, humanity currently operates in a zone of uncertainty with increasing risks. With regard to biogeochemical flows (phosphorus and nitrogen) as well as biosphere integrity, genetic diversity in particular, humanity even operates in a zone of high risks. Continuing to exceed these boundaries may lead to severe negative impacts on nature and humanity. With regard to the nexus, the researchers emphasised the interdependencies that exist between the different areas calling for an integrated (nexus) approach in addressing these challenges (Steffen et al., 2015).

Fig. 4 shows the five spaces of the nexus framework. This system, as shall be further discussed, is built on the pillars of nature (i.e. the natural sphere, which

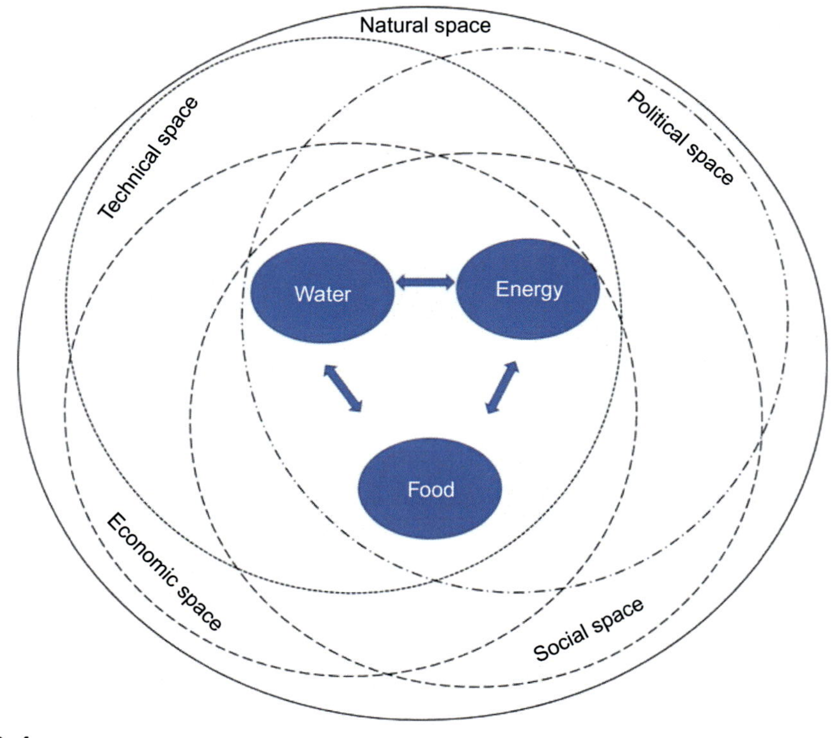

FIG. 4

Four spaces of the nexus framework.

determines the natural system boundaries), society and economy (i.e. the socio-technological sphere, which determines the technological, economic, and cultural (anthropocentric) system boundaries), and politics (i.e. the political sphere, which determines the political system boundaries). Furthermore, these pillars are externally influenced by different driving forces, subsumed under the three global trends of (a) climate change, (b) population development, and (c) lifestyles and consumption patterns (Chen et al., 2018; Gain et al., 2015; Hoff, 2011; Pahl-Wostl et al., 2020; Portney et al., 2017; Schwoerer et al., 2020; Venghaus et al., 2019).

4 Discussion

Opening up the framework by taking a deeper look into the different spaces implies focussing on different aspects of food, energy, and water including their interrelations. Depending on the application of the framework, different elements need to be considered. Table 2 summarises the key aspects of consideration for food, energy, and water with respect to both the different research perspectives as well as the related types of interconnections. Whereas any use of the framework with regard to the natural space deals with the natural resources themselves as well as their resource flows and natural cycles or ecosystems, political analyses need to focus on the laws and regulations related to the use and management of these resources. Thereby, the actual availability (the natural space) always forms the outer system boundary. No matter what perspective is taken, the complex feedbacks that exist between the sectors or policy fields make the development of an effective (and ideally optimised) management or policy approach a highly challenging task. With regard to the social perspective, it becomes important to understand that the nexus

Table 2 Key aspects within the five spaces of the food–energy–water nexus.

	Natural	Technical	Economical	Political	Social
Food	Land resources, plants, and livestock	Production and processing technologies	Costs	Laws, regulation, food policy	Access, affordability, distribution, quality, diet and nutrition
Energy	Fossil fuels, wind, sun, water, land	Infrastructure, energy technologies	Costs	Laws, regulation, energy policy	Access, affordability, distribution, sustainable
Water	Availability, quality	Infrastructure	Costs	Laws, regulations, water policy	Access, affordability, distribution
Type of interrelations	Natural interrelations, natural cycles, ecosystems	Resource flows	Input and output	Political side effects, impacts of management	Actor networks, cooperation, public debates

resources water, energy, and food are embedded in a larger surrounding system and thus their availability immediately depends on further resources and drivers such as land or climate. Although in previous nexus conceptualisations land has often been indiscriminately included as a nexus component, in our understanding land differs in a significant aspect from food or energy in that it exists as a natural resource with a finite physical quantity. The projected scarcities of food, water, and energy as well as the search for opportunities to meet these demands in the future have made the acquisition of 'land' a globally lucrative investment opportunity (Venghaus et al., 2019).

Using the nexus as a framework requires a specific set of rules to use: Firstly, the system boundaries have to be set. The natural availability always forms the outer boundary, in which the technical, economic, political, and social systems are embedded. Secondly, the relevant elements according to the key aspects in Table 2 need to be identified. Thirdly, the interrelations between these elements need to be defined. This process is especially important, since—as Verhoeven (2015) stated—never a single FEW nexus exists within a case study but a number of different nexuses instead. Defining the nexus as a framework instead of a research issue accounts for this fact and makes it possible to analyse a broad variety of different nexuses. Clearly identifying and defining the different elements presented in Table 2 helps setting a standard for analyses focusing on interconnections between water, energy, and food, which, in turn, would make nexus research more comparable and thus facilitate cooperation as well as further advancements in this field of research (Proctor et al., 2021). To do so, it will be necessary for every nexus researcher to clearly state the aim and purpose of the use of the nexus framework (Proctor et al., 2021; Schlör, Märker, & Venghaus, 2021). In practice, however, the different spaces cannot always be treated separately from each other. Often, the socio-technical-economic system forms an emergent system, where complex interdependencies determine the actual resource consumption Milchram et al. (2019). Adopting a socio-technical systems' perspective requires including elements from more than one space. In this perspective, the natural space determines the resource constraints whereas the political system sets the objectives for the sustainable use of resources. Instead of representing the framework by means of five different spaces (Fig. 4), the framework can also be illustrated as interrelated steps of a staircase that build on each other (Fig. 5).

5 Conclusions

The food–energy–water nexus has become a highly investigated and discussed research issue. Emerged as an idea born in policy and practice, the nexus concept has invited researchers from a wide range of different disciplines making it hard to frame a common understanding of this concept. The nexus has been applied in many different variations regarding sector combinations, methods, or its purpose. This chapter argues that the nexus provides a research framework rather than a subject of investigation. Defining the nexus as a framework provides a clear understanding of the nexus by including all different research perspectives at the

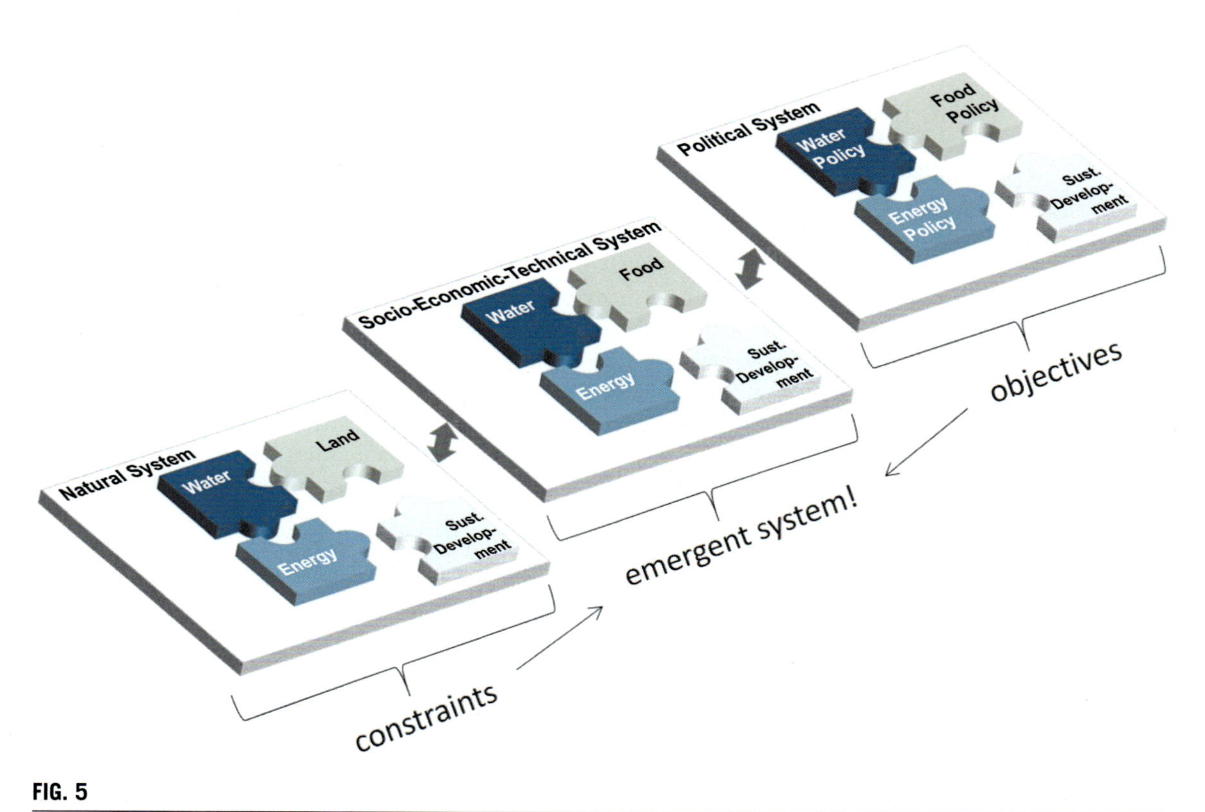

FIG. 5

Staircase FEW nexus framework.

same time. The food–energy–water nexus is not bound to any theory or method, it rather provides an overarching framework that helps to investigate the interrelations between different natural resource, resource sectors, or policy fields. Depending on the research focus, different aspects of food, energy, water as well as their interrelations determine the relevant elements of the framework. Thus, by using the nexus as a framework, the existing challenges in resource management can be analysed from various different points of view, which can reveal conflicts of interests or challenges that previously have not been recognised. The specific point of origin of the food–energy–water nexus as well as its understanding as a framework provides preconditions that make it a valuable contribution to solving existing resource management problems. Especially with regard to increasing impacts of climate change as well as other global megatrends, such as population growth, globalisation, or digitalisation, a more integrated perspective on the availability, use, and distribution of natural resources becomes more and more important. This offers some interesting and new linkages to practical approaches. Objective of a nexus-enabled policy framework, for example, is thus to "reconcile long-term and global objectives (e.g. climate protection, ecosystem stewardship and equity goals) not only with immediate economic benefits, but also with the need to secure local livelihoods and the non-negotiable human rights to water and food" (Hoff, 2011). However, market failures and largely disconnected regulatory frameworks for the different resources currently prevent an effective reconcilement of competing uses and a fair allocation of resources (Venghaus et al., 2019). In light of the ever-growing nexus research, however, the question remains: quo vadis nexus research? One key opportunity will certainly be an increasing focus on comparative case studies using the framework developed before. Its structured approach can increase the comparability of determining factors of resource management problems in specific contexts. Such analyses can bring forward transferable solutions that are essential for a fast progress with regard to current global problems. Aggravating impacts of climate change and current global trends call for quick and easily adaptable pilot projects or solutions that may work in several regions and different social, economic, and technical contexts.

References

Abegaz, B. W., Datta, T., & Mahajan, S. M. (2018). Sensor technologies for the energy-water nexus—A review. *Applied Energy*, *210*, 451–466. https://doi.org/10.1016/j.apenergy.2017.01.033.

Ahmad, S., Jia, H., Chen, Z., Li, Q., & Xu, C. (2020). Water-energy nexus and energy efficiency: A systematic analysis of urban water systems. *Renewable and Sustainable Energy Reviews*, *134*, 110381. https://doi.org/10.1016/j.rser.2020.110381.

Albrecht, T. R., Crootof, A., & Scott, C. A. (2018). The water-energy-food Nexus: A systematic review of methods for nexus assessment. *Environmental Research Letters*, *13*(4), 043002. https://doi.org/10.1088/1748-9326/aaa9c6.

Al-Saidi, M., & Elagib, N. A. (2017). Towards understanding the integrative approach of the water, energy and food nexus. *Science of the Total Environment*, *574*, 1131–1139. https://doi.org/10.1016/j.scitotenv.2016.09.046.

Amjath-Babu, T. S., Sharma, B., Brouwer, R., Rasul, G., Wahid, S. M., Neupane, N., et al. (2019). Integrated modelling of the impacts of hydropower projects on the water-food-energy nexus in a transboundary Himalayan river basin. *Applied Energy*, *239*, 494–503. https://doi.org/10.1016/j.apenergy.2019.01.147.

Aydin, M. (2018). Natural gas consumption and economic growth nexus for top 10 natural Gas–Consuming countries: A granger causality analysis in the frequency domain. *Energy*, *165*, 179–186. https://doi.org/10.1016/j.energy.2018.09.149.

Benites-Lazaro, L. L., Giatti, L. L., Sousa Junior, W. C., & Giarolla, A. (2020). Land-water-food nexus of biofuels: Discourse and policy debates in Brazil. *Environmental Development*, *33*, 100491. https://doi.org/10.1016/j.envdev.2019.100491.

Brouwer, F., Avgerinopoulos, G., Fazekas, D., Laspidou, C., Mercure, J. F., Pollitt, H., et al. (2018). Energy modelling and the Nexus concept. *Energy Strategy Reviews*, *19*, 1–6. https://doi.org/10.1016/j.esr.2017.10.005.

Chen, P. C., Alvarado, V., & Hsu, S. C. (2018). Water energy nexus in city and hinterlands: Multi-regional physical input-output analysis for Hong Kong and South China. *Applied Energy*, *225*, 986–997. https://doi.org/10.1016/j.apenergy.2018.05.083.

Cuberos Balda, M., & Kawajiri, K. (2020). The right crops in the right place for the food-energy nexus: Potential analysis on rice and wheat in Hokkaido using crop growth models. *Journal of Cleaner Production*, *263*, 121373.

Dale, V. H., Efroymson, R. A., & Kline, K. L. (2011). The land use-climate change-energy nexus. *Landscape Ecology*, *26*(6), 755–773. https://doi.org/10.1007/s10980-011-9606-2.

Endo, A., Yamada, M., Miyashita, Y., Sugimoto, R., Ishii, A., Nishijima, J., et al. (2020). Dynamics of water–energy–food nexus methodology, methods, and tools. *Current Opinion in Environmental Science and Health*, *13*, 46–60. https://doi.org/10.1016/j.coesh.2019.10.004.

Food and Agriculture Organization of the United Nations (FAO). (2014). *The water-energy-food nexus. A new approach in support of food security and sustainable agriculture*.

Fouladi, J., AlNouss, A., & Al-Ansari, T. (2021). Sustainable energy-water-food nexus integration and optimisation in eco-industrial parks. *Computers and Chemical Engineering*, *146*. https://doi.org/10.1016/j.compchemeng.2021.107229.

Gain, A. K., Giupponi, C., & Benson, D. (2015). The water–energy–food (WEF) security nexus: The policy perspective of Bangladesh. *Water International*, *40*(5–6), 895–910. https://doi.org/10.1080/02508060.2015.1087616.

Gephart, J. A., Troell, M., Henriksson, P. J. G., Beveridge, M. C. M., Verdegem, M., Metian, M., et al. (2017). The 'seafood gap' in the food-water nexus literature—Issues surrounding freshwater use in seafood production chains. *Advances in Water Resources*, *110*, 505–514. https://doi.org/10.1016/j.advwatres.2017.03.025.

Ghani, W. A. K., Salleh, M. A. M., Adam, S. N., Shafri, H. Z. M., Shaharum, S. N., Lim, K. L., et al. (2019). Sustainable bio-economy that delivers the environment–food–energy–water nexus objectives: The current status in Malaysia. *Food and Bioproducts Processing*, *118*, 167–186. https://doi.org/10.1016/j.fbp.2019.09.002.

Ghodsvali, M., Krishnamurthy, S., & de Vries, B. (2019). Review of transdisciplinary approaches to food-water-energy nexus: A guide towards sustainable development. *Environmental Science and Policy*, *101*, 266–278. https://doi.org/10.1016/j.envsci.2019.09.003.

Gonzalez-Salazar, M. A., Venturini, M., Poganietz, W. R., Finkenrath, M., Kirsten, T., Acevedo, H., et al. (2016). A general modeling framework to evaluate energy, economy, land-use and GHG emissions nexus for bioenergy exploitation. *Applied Energy*, *178*, 223–249. https://doi.org/10.1016/j.apenergy.2016.06.039.

Grobicki, A. (2016). Water-food-energy-climate: Strengthening the weak links in the nexus. In *The water, food, energy and climate nexus: Challenges and an agenda for action* (pp. 127–137). Taylor and Francis Inc. https://doi.org/10.4324/9781315640716.

Hoff, H. (2011). Understanding the nexus—Background paper for the Bonn2011 nexus conference. In *Bonn2011 conference the water, energy and food security nexus—Solutions for the green economy*.

Hoolohan, C., Larkin, A., McLachlan, C., Falconer, R., Soutar, I., Suckling, J., et al. (2018). Engaging stakeholders in research to address water–energy–food (WEF) nexus challenges. *Sustainability Science*, *13*(5), 1415–1426. https://doi.org/10.1007/s11625-018-0552-7.

Hoolohan, C., McLachlan, C., & Larkin, A. (2019). 'Aha' moments in the water-energy-food nexus: A new morphological scenario method to accelerate sustainable transformation.

Technological Forecasting and Social Change, *148*, 119712. https://doi.org/10.1016/j.techfore.2019.119712.

Howells, M., Hermann, S., Welsch, M., Bazilian, M., Segerström, R., Alfstad, T., et al. (2013). Integrated analysis of climate change, land-use, energy and water strategies. *Nature Climate Change*, *3*(7), 621–626. https://doi.org/10.1038/nclimate1789.

Jin, Y., Behrens, P., Tukker, A., & Scherer, L. (2019). Water use of electricity technologies: A global meta-analysis. *Renewable and Sustainable Energy Reviews*, *115*, 109391. https://doi.org/10.1016/j.rser.2019.109391.

Khan, Z., Linares, P., Rutten, M., Parkinson, S., Johnson, N., & García-González, J. (2018). Spatial and temporal synchronization of water and energy systems: Towards a single integrated optimization model for long-term resource planning. *Applied Energy*, *210*, 499–517. https://doi.org/10.1016/j.apenergy.2017.05.003.

Larkin, A., Hoolohan, C., & McLachlan, C. (2020). Embracing context and complexity to address environmental challenges in the water-energy-food nexus. *Futures*, *123*, 102612. https://doi.org/10.1016/j.futures.2020.102612.

Lechón, Y., De La Rúa, C., & Cabal, H. (2018). Impacts of decarbonisation on the water-energy-land (WEL) nexus: A case study of the Spanish electricity sector. *Energies*, *11*(5), 1203. https://doi.org/10.3390/en11051203.

Li, K., Feng, C., Liang, Y., Qi, J., Li, Y., Li, H., et al. (2021). Critical transmission sectors for provincial food-water nexus in China. *Journal of Cleaner Production*, *279*, 123886. https://doi.org/10.1016/j.jclepro.2020.123886.

Märker, C., Venghaus, S., & Hake, J. F. (2018). Integrated governance for the food–energy–water nexus—The scope of action for institutional change. *Renewable and Sustainable Energy Reviews*, *97*, 290–300. https://doi.org/10.1016/j.rser.2018.08.020.

McGinnis, M. D. (2011). An introduction to IAD and the language of the Ostrom workshop: A simple guide to a complex framework. *Policy Studies Journal*, *39*(1), 169–183. https://onlinelibrary.wiley.com/doi/pdf/10.1111/j.1541-0072.2010.00401.x.

Mercure, J. F., Paim, M. A., Bocquillon, P., Lindner, S., Salas, P., Martinelli, P., et al. (2019). System complexity and policy integration challenges: The Brazilian energy- water-food nexus. *Renewable and Sustainable Energy Reviews*, *105*, 230–243. https://doi.org/10.1016/j.rser.2019.01.045.

Milchram, C., Märker, C., Schlör, H., et al. (2019). Understanding the role of values in institutional change: the case of the energy transition. *Energy, Sustainability and Society*, *9*, 46. https://doi.org/10.1186/s13705-019-0235-y.

Moazeni, F., Khazaei, J., & Pera Mendes, J. P. (2020). Maximizing energy efficiency of islanded micro water-energy nexus using co-optimization of water demand and energy consumption. *Applied Energy*, *266*, 114863. https://doi.org/10.1016/j.apenergy.2020.114863.

Mohmmed, A., Li, Z., Olushola Arowolo, A., Su, H., Deng, X., Najmuddin, O., et al. (2019). Driving factors of CO_2 emissions and nexus with economic growth, development and human health in the top ten emitting countries. *Resources, Conservation and Recycling*, *148*, 157–169. https://doi.org/10.1016/j.resconrec.2019.03.048.

Mrozik, W., Vinitnantharat, S., Thongsamer, T., Pansuk, N., Pattanachan, P., Thayanukul, P., et al. (2019). The food-water quality nexus in periurban aquacultures downstream of Bangkok, Thailand. *Science of the Total Environment*, *695*, 133923. https://doi.org/10.1016/j.scitotenv.2019.133923.

Namany, S., Al-Ansari, T., & Govindan, R. (2019). Sustainable energy, water and food nexus systems: A focused review of decision-making tools for efficient resource management and governance. *Journal of Cleaner Production*, *225*, 610–626. https://doi.org/10.1016/j.jclepro.2019.03.304.

Ostrom, E. (2011). Background on the institutional analysis and development framework. *Policy Studies Journal*, *39*(1), 7–27. https://doi.org/10.1111/j.1541-0072.2010.00394.x.

Ozturk, I., & Bilgili, F. (2015). Economic growth and biomass consumption nexus: dynamic panel analysis for Sub-Sahara African countries. *Applied Energy*, *137*, 110–116. https://doi.org/10.1016/j.apenergy.2014.10.017.

Pahl-Wostl, C., Gorris, P., Jager, N., Koch, L., Lebel, L., Stein, C., et al. (2020). Scale-related governance challenges in the water–energy–food nexus: Toward a diagnostic approach. *Sustainability Science*, *16*, 615–629. https://doi.org/10.1007/s11625-020-00888-6.

Portney, K. E., Hannibal, B., Goldsmith, C., McGee, P., Liu, X., & Vedlitz, A. (2017). Awareness of the food–energy–water nexus and public policy support in the United States: Public attitudes among the American people. *Environment and Behavior*, *50*, 375–400. https://doi.org/10.1177/0013916517706531.

Price, J., Zeyringer, M., Konadu, D., Sobral Mourão, Z., Moore, A., & Sharp, E. (2018). Low carbon electricity systems for Great Britain in 2050: An energy-land-water perspective. *Applied Energy*, *228*, 928–941. https://doi.org/10.1016/j.apenergy.2018.06.127.

Proctor, K., Tabatabaie, S. M. H., & Murthy, G. S. (2021). Gateway to the perspectives of the food-energy-water nexus. *Science of the Total Environment*, *764*, 142852. https://doi.org/10.1016/j.scitotenv.2020.142852.

Rockström, J., Steffen, W., Noone, K., Persson, Å., Chapin, F. S., Lambin, E., et al. (2009). Planetary boundaries: Exploring the safe operating space for humanity. *Ecology and Society*, *14*(2).

Rodriguez, R.d. G., Scanlon, B. R., King, C. W., Scarpare, F. V., Xavier, A. C., & Pruski, F. F. (2018). Biofuel-water-land nexus in the last agricultural frontier region of the Brazilian Cerrado. *Applied Energy*, *231*, 1330–1345. https://doi.org/10.1016/j.apenergy.2018.09.121.

Saladini, F., Betti, G., Ferragina, E., Bouraoui, F., Cupertino, S., Canitano, G., et al. (2018). Linking the water-energy-food nexus and sustainable development indicators for the Mediterranean region. *Ecological Indicators*, *91*, 689–697. https://doi.org/10.1016/j.ecolind.2018.04.035.

Salehi, A. A., Ghannadi-Maragheh, M., Torab-Mostaedi, M., Torkaman, R., & Asadollahzadeh, M. (2020). A review on the water-energy nexus for drinking water production from humid air. *Renewable and Sustainable Energy Reviews*, *120*, 109627. https://doi.org/10.1016/j.rser.2019.109627.

Schlör, H., Märker, C., & Venghaus, S. (2021). Developing a nexus systems thinking test–A qualitative multi- and mixed methods analysis. *Renewable and Sustainable Energy Reviews*, *138*, 110543. https://doi.org/10.1016/j.rser.2020.110543.

Schwoerer, T., Schmidt, J. I., & Holen, D. (2020). Predicting the food-energy Nexus of wild food systems: Informing energy transitions for isolated indigenous communities. *Ecological Economics*, *176*, 106712. https://doi.org/10.1016/j.ecolecon.2020.106712.

Simpson, G. B., & Jewitt, G. P. (2019). The water-energy-food nexus in the anthropocene: Moving from 'nexus thinking' to 'nexus action'. *Current Opinion in Environmental Sustainability*, *40*, 117–123. https://doi.org/10.1016/j.cosust.2019.10.007.

Steffen, W., Richardson, K., Rockström, J., Cornell, S. E., Fetzer, I., Bennett, E. M., et al. (2015). Planetary boundaries: Guiding human development on a changing planet. *Science*, *347*(6223), 1259855. https://doi.org/10.1126/science.1259855.

Vakilifard, N., Anda, M., Bahri, P. A., & Ho, G. (2018). The role of water-energy nexus in optimising water supply systems—Review of techniques and approaches. *Renewable and Sustainable Energy Reviews*, *82*, 1424–1432. https://doi.org/10.1016/j.rser.2017.05.125.

van den Heuvel, L., Blicharska, M., Masia, S., Sušnik, J., & Teutschbein, C. (2020). Ecosystem services in the Swedish water-energy-food-land-climate nexus: Anthropogenic pressures and physical interactions. *Ecosystem Services*, *44*, 101141. https://doi.org/10.1016/j.ecoser.2020.101141.

Venghaus, S., Märker, C., Dieken, S., & Siekmann, F. (2019). Linking environmental policy integration and the water-energy-land-(food-)nexus: A review of the European Union's energy, water, and agricultural policies. *Energies*, *12*(23). https://doi.org/10.3390/en12234446.

Venghaus, S., & Hake, J.-F. (2018). Nexus thinking in current EU policies–The interdependencies among food, energy and water resources. *Environmental Science and Policy*, *90*, 183–192. https://doi.org/10.1016/j.envsci.2017.12.014.

Verhoeven, H. (2015). The nexus as a political commodity: Agricultural development, water policy and elite rivalry in Egypt. *International Journal of Water Resources Development*, *31*(3), 360–374. https://doi.org/10.1080/07900627.2015.1030725.

Wang, S., Fath, B., & Chen, B. (2019). Energy–water nexus under energy mix scenarios using input–output and ecological network analyses. *Applied Energy*, *233–234*, 827–839. https://doi.org/10.1016/j.apenergy.2018.10.056.

Wang, S., Liu, Y., & Chen, B. (2018). Multiregional input–output and ecological network analyses for regional energy–water nexus within China. *Applied Energy*, *227*, 353–364. https://doi.org/10.1016/j.apenergy.2017.11.093.

Weitz, N., Strambo, C., Kemp-Benedict, E., & Nilsson, M. (2017). Closing the governance gaps in the water-energy-food nexus: Insights from integrative governance. *Global Environmental Change*, *45*, 165–173. https://doi.org/10.1016/j.gloenvcha.2017.06.006.

White, D. D., Jones, J. L., Maciejewski, R., Aggarwal, R., & Mascaro, G. (2017). Stakeholder analysis for the food-energy-water nexus in Phoenix, Arizona: Implications for nexus governance. *Sustainability (Switzerland)*, *9*(12), 2204. https://doi.org/10.3390/su9122204.

World Economic Forum. (2011). *Global risks*.

Yu, L., Xiao, Y., Zeng, X. T., Li, Y. P., & Fan, Y. R. (2020). Planning water-energy-food nexus system management under multi-level and uncertainty. *Journal of Cleaner Production*, *251*, 119658. https://doi.org/10.1016/j.jclepro.2019.119658.

Zhang, T., Tan, Q., Zhang, S., Zhang, T., & Zhang, W. (2021). A participatory methodology for characterizing and prescribing water-energy-food nexus based on improved casual loop diagrams. *Resources, Conservation and Recycling*, *164*, 105124. https://doi.org/10.1016/j.resconrec.2020.105124.

CHAPTER

The European Green Deal in the global sustainability context

Mauro Cordella[a] and Serenella Sala[b]
[a]*TECNALIA, Basque Research and Technology Alliance (BRTA), Derio, Spain*
[b]*European Commission, Joint Research Centre, Ispra, VA, Italy*

1 A 'Decade of Action' for sustainability

Far from being static, the world is an entity that is continuously changing its physical and social shape as the result of natural phenomena and anthropic interventions. Since the agricultural revolution, the activities of human beings have started to progressively shape the environmental and socio-economic state of the world and increase its complexity (Harari, 2015). Environmental impacts of human activities have become so significant in the last centuries that scholars suggest we have entered the Anthropocene, "a new human-dominated geological epoch" (Lewis & Maslin, 2015).

Changes of the world are driven by 'megatrends' that occur at different scales and paces. A number of organisations (e.g. EEA (2019), ESPAS (2019)) have been investigating on trends happening at global level, which currently include (1) growing population, urbanisation, and migration flows; (2) climate change and environmental degradation; (3) increasing scarcity of resources and global competition for the access to them; (4) accelerating technological change and connectivity; (5) power shifts in global economy and geopolitical landscapes, and associated conflicts; (6) polarising values, lifestyles, and governance approaches, as well as populism rise.

Such trends represent global challenges for achieving sustainability and call for science-based targets, internationally coordinated actions of different groups of stakeholders, as well as the availability of financial resources. In this respect, two important agreements were achieved in 2015 by the international community, during a period of economic growth and international cooperation: the Paris Agreement on Climate Change, aiming to achieve net-zero greenhouse gas (GHG) emissions by the middle of the century, and the broad agreement on 17 Sustainable Development Goals (SDGs) (see Fig. 1) to achieve by 2030, the so-called Agenda 2030 (Nature, 2020; Sachs et al., 2019).

Furthermore, the need of addressing SDGs interplays (EC-JRC, 2021a) and the role of environmental SDGs in support of the achievement of socio-economic ones is

Assessing Progress towards Sustainability. https://doi.org/10.1016/B978-0-323-85851-9.00019-5

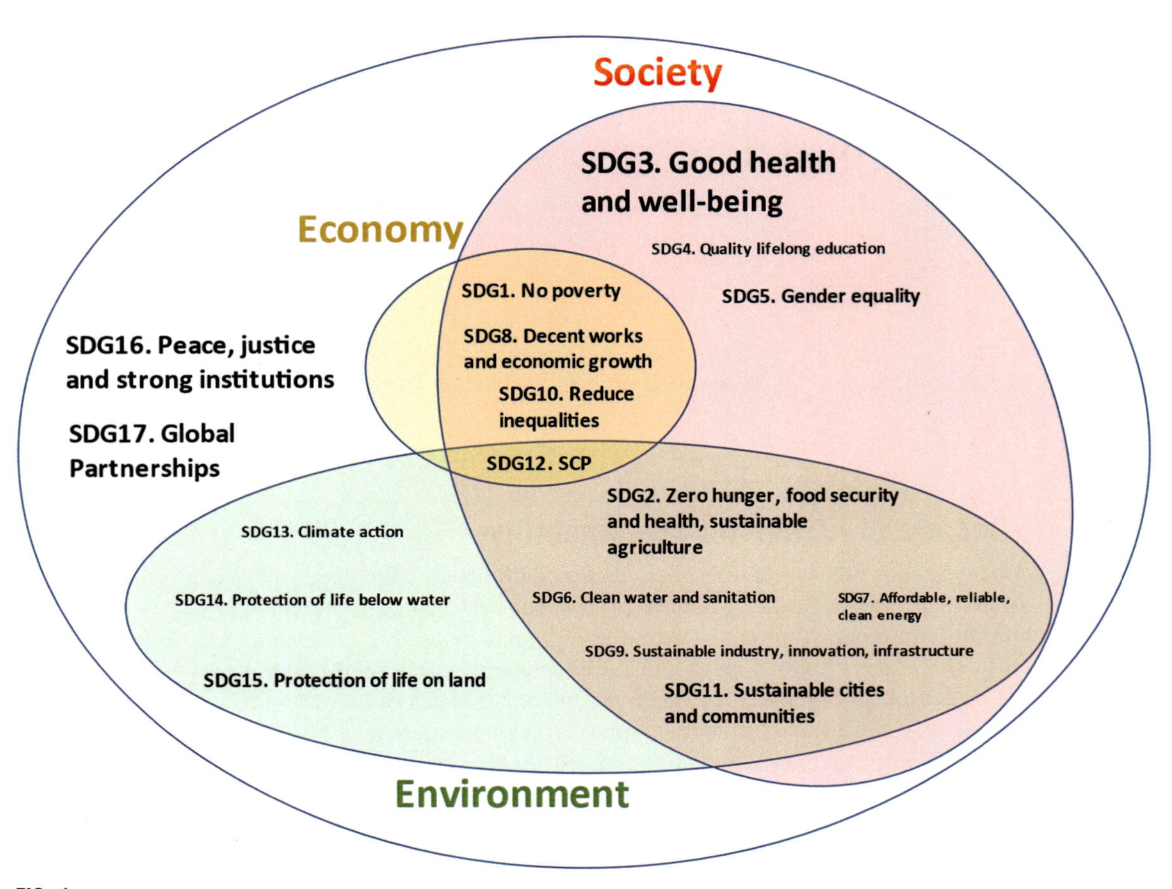

FIG. 1

Sustainability dimensions of the SDGs framework. Note: Own elaboration from UN (2020), the size of the font represents the number of indicators included in each goal (as measure of complexity).

more and more pledged (Folke, Biggs, Norström, Reyers, & Rockström, 2016). This is basically asking for the adoption of a strong sustainability approach towards SDGs achievement, and it has been mentioned as well in the prospect of sustainable solutions described by The World in 2050 initiative (2020), where socio-economic targets are expected to be achieved whilst respecting planetary boundaries.

However, the progress towards the achievement of SDGs, which depends on sustained economic growth and globalisation (Naidoo & Fisher, 2020), is slow. The world is facing a 'triple planetary crisis' of climate, nature, and pollution, which has unsustainable production and consumption at the base (UNEP, 2020a, 2021a). The situation is complicated by the health and socio-economic crisis due to Covid-19 (Heath & Boudreau, 2020), which hit in 2020–21, and which gave only a partial relaxation to the environment (Troëng, Barbier, & Rodríguez Echandi, 2020; UNEP, 2020b). Environmental issues and infectious diseases are currently perceived as top sources of global risks (World Economic Forum, 2021), so that

Covid-19 appears as one over many threats that may potentially hit global welfare and economies. In addition, Covid-19 has negatively impacted international cooperation and monitoring systems worldwide, thus complicating multilateral agreements, implementation and enforcement of environmental regulations, as well as the spread of clean innovations and investments (Nature, 2020; UNEP, 2020b).

According to Sachs et al. (2019), a common understanding of how to make SDGs operational is also lacking. Sachs et al. (2019) identified six transformation areas where to act in order to progress towards the achievement of SDGs:

(1) Education, Gender, and Inequality;
(2) Health, Well-being, and Demography;
(3) Energy Decarbonisation and Sustainable Industry;
(4) Sustainable Food, Land, Water, and Oceans;
(5) Sustainable Cities and Communities; and
(6) Digital Revolution for Sustainable Development.

The achievement of SDGs could be operationalised by decoupling SDGs from economic growth, which has been often fuelled by unsustainable activities and resulted in an unequal distribution of wealth (Nature, 2020). The way forward should be to use economic resources to achieve SDGs, and not undermine them, prioritising win-win solutions. For example, Sustainable Consumption and Production (SDG 12) embeds the characteristics to positively influence most SDGs, whilst a number of targets could contribute to reducing the likelihood of other pandemics (Naidoo & Fisher, 2020).

The state of the world highlights the urgent need of acting: if SDGs are 'the compass' and financial resources 'the fuel', science-based and internationally coordinated actions can provide 'the means' and 'the roadmap' to achieve sustainability. The year 2020 has set the start of the 'Decade of Action' to implement the Agenda 2030 and achieve the SDGs (UN Global Compact—Red Española (2020)). Initiatives such as the European Green Deal (EGD) (EC, 2019a) (see Section 2) are examples of how science-based policies (see Section 3) can promote sustainability at global level and set the boundaries for the post-pandemic economic recovery (see Section 4).

2 The European Green Deal

To overcome the most important challenges related to climate change and environmental degradation, the European Commission (EC) has defined a new growth strategy aiming at transforming the economy of the European Union (EU) into a sustainable one, i.e. a carbon-neutral and resource-efficient economy where no person nor place is left behind (EC, 2020a). The Communication from EC on the European Green Deal (EC, 2019a) provides an action plan for the transition towards an environmentally friendly, fair, prosperous, and inclusive society. This Communication envisages a modern, circular, and competitive economy, where economic growth is decoupled from resource use, and set the target for the EU to achieve carbon neutrality by 2050.

The EGD is an integral part of the strategy of the EC to implement the SDGs (SDSN, 2021; UN, 2015) and the other priorities announced in President von der Leyen's political guidelines (EC, 2019b).

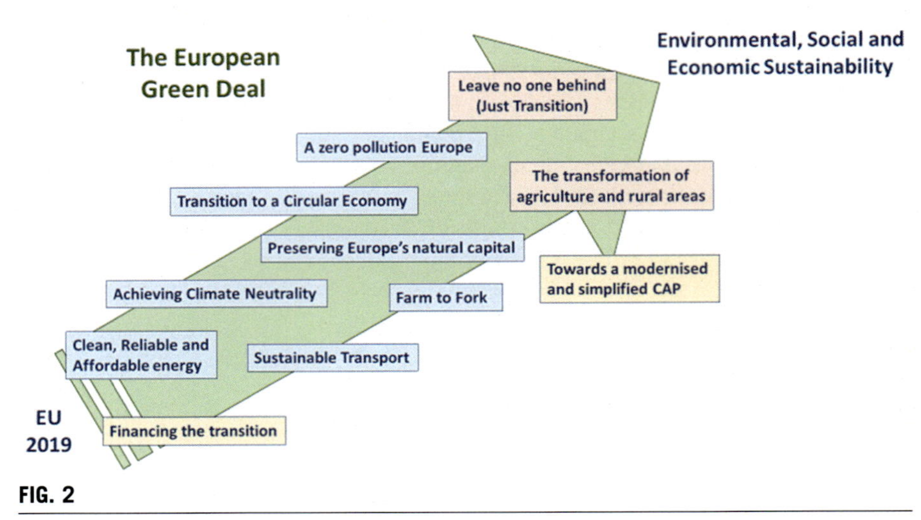

FIG. 2

Key areas of the European Green Deal.

To reach the ambitious goals of the EGD, actions are required by all sectors of the economy. For this reason, the initial roadmap of policy areas presented in the EGD (see Fig. 2) has resulted in a number of legislative documents, published in 2020–21, that complement it by tackling specific topics.

With respect to the policies linked to strong sustainability, which underpin human life and fulfil human basic needs, biodiversity conservation and food system analysis are clear priorities. Measures to protect the ecosystems are indicated in order to ensure that the biodiversity in Europe and in third countries will be on the path to recovery by 2030 (EC, 2020b). Such measures include actions for the restoration of carbon-rich habitats and climate-friendly agriculture. Biodiversity is also crucial for safeguarding EU and global food security. For the food sector (EC, 2020c), a holistic approach 'from farm to fork' is proposed that covers environmental, nutritional, and health, as well as socio-economic aspects of food, outlining the relevance of a life cycle viewpoint.

Furthermore, achieving the climate and environmental goals of the EU also requires a further push in the transition to a circular economy. The Circular Economy Action Plan (EC, 2020d) is a future-oriented agenda for achieving a circular, cleaner, and more competitive economy in Europe. The Action Plan presents a set of interrelated initiatives to establish a circular economy and reduce both resource consumption and environmental impacts. An emphasis is given to 'sustainable products' policies for the promotion of circular business models and more durable and reparable products, as well as to key sectors such as textiles, construction, electronics, and plastics. For instance, more sustainable uses of plastics and waste reduction measures are proposed for key products made of plastics (e.g. packaging, construction materials, and vehicles). In the case of buildings, the Renovation Wave initiative (EC, 2020e) instead aims at stimulating the retrofitting of the building stock taking energy efficiency and circularity principles into account.

The reduction of pollution is another crucial aspect in the EGD. Notably, the Chemicals Strategy (EC, 2020f) seeks to achieve a toxic-free environment, where chemicals are produced and used in a way that maximises their contribution to society whilst avoiding harms to the planet and to current and future generations. Relevant energy legislation will be also reviewed over time, with the goal to decarbonise the EU energy system, and with a focus on the development of clean and renewable energy from hydrogen. EU Member States will then update their national energy and climate plans in 2023, to reflect the new climate ambition. Furthermore, in 2021, the EC will adopt an action plan dedicated to achieving a zero-pollution ambition for air, water, and soil.

Besides that, the priorities related to a stronger Europe in the world, such as the New Industrial Strategy of Europe (EC, 2020g), aim at the decarbonisation and modernisation of industries, focusing in particular on energy-intensive industries such as steel and cement.

The EGD also points out the importance of finance for enabling the transition towards sustainability, in particular by qualifying investments based on criteria addressing environmental objectives (EU, 2020). In the case of investments (EC, 2020h), a transition towards climate neutrality will be enabled by the mobilisation of at least EUR 1 trillion of sustainable investments over the next decade through the EU budget. Through the Just Transition Mechanism (EC, 2020i), the EU will also provide financial support and technical assistance to help those who are most affected by the shift towards the green economy.

Other relevant fields of action are considered in the EGD such as sustainable mobility, aiming at reducing by 90% the GHG emissions in transport by 2050, and sustainable agriculture, driven by new technologies, research, and innovation, which remark how this Communication is a key reference document for shaping the future of Europe. Furthermore, digital technologies, research, and innovation are considered important enablers across the different policies, and intimately linked with other priorities besides the EGD, such as stronger Europe in the world and a Europe fit for the digital age (EC, 2019b).

The measures proposed by the EGD are expected to benefit EU companies and consumers. In particular, improved digitalisation and information flows (EC, 2020l) can enable the provision of more detailed information on products such as their origin, composition, repair, and end-of-life handling, thus complementing and supporting the other Communications in pursuing a more circular economy and other objectives set in the EGD. This can also improve the reliability of 'green claims' made by companies, therefore reducing the risk of 'greenwashing' and the proliferation of unreliable environmental labels (EC, 2020m).

Finally, it is important to observe that environmental objectives play a core role even in 'Next Generation EU', the post Covid-19 recovery plan of the EU (EC, 2020n), which demonstrates the effort of the EC to move in the direction towards sustainability, notwithstanding the dramatic and critical situation that Europe and the world are living.

3 Sustainability science in support to the European Green Deal

The EGD is proposing a set of actions to implement a transition to environmental sustainability in the EU. Several sustainability aspects are conceptually embedded. However, promoting sustainability science at full scale may require strengthening further the elements that are briefly discussed hereinafter.

Although the modern concept of sustainability had origins in the 1970s, being popularly attributed to the Club of Rome's report 'Limits to Growth' (Meadows, Meadows, Randers, & Behrens, 1972), it has been put forward much earlier, rooted in the ecology and social domains in relation to the carrying capacity of the Earth system and the inherent trade-offs between wealth generation and social justice (Purvis, Mao, & Robinson, 2019).

Over the years, the world has gained a deeper understanding of the interconnected challenges we face as human species and has recognised that sustainable development has to embrace several sustainability dimensions: from the consideration of the three fundamental pillars of environmental protection, economic growth, and social equity to the integration of institutional (O'Connor, 2006), cultural (Nurse, 2007), and technological (Vos, 2007) issues. In line with the institutional sustainability concept, partnership and peace were recognised as critical components of sustainability (UNSSC, 2021). Furthermore, as discussed earlier, there is increasing awareness about the interplay between all the sustainability pillars and related goals and targets.

Although the EGD was published in 2019, the roots of EU sustainability policy are older (e.g. EC, 2001, 2008, 2009, 2012, 2014, 2015). Environmental policies have evolved over time recognising the leading role of sustainable production and consumption to ensure an absolute decoupling of environmental impacts from socio-economic well-being.

To properly ensure a transition towards more sustainable systems, there is a growing request of evidence-based policies (e.g. in the EU (EC, 2021) and at international level (UN, 2021)) in which scientific evidence is considered and discussed during the policy-making process to tackle multifaceted and multidimensional challenges. An integrated approach to sustainability is even more necessary with the EGD, given the challenges faced and the ambitious goals set.

However, the definitions of sustainability, 'what should be sustained' (e.g. what might constitute critical natural capital), and 'to which extent', are by no means agreed on within the scientific community, with weak and strong sustainability representing two extreme economic paradigms (Neumayer, 2003). The definition is subject to value judgements (Bell & Morse, 2008; Bond et al., 2011), up to be interpreted as a shared ethical belief (Seager, Melton, & Eighmy, 2004).

Patterson, McDonald, and Hardy (2017) identified four main interpretations of the concept of sustainability: (i) ecological, (ii) economic, (iii) thermodynamic and ecological-economic, (iv) public policy and planning theory. The ecological interpretation focuses on a vision of the socio-economic system embedded in the global biophysical system; the economic emphasises the idea of social welfare, almost always measured in material terms; the thermodynamic interpretation poses ecological sustainability in the context of the entropic nature of economic-environmental interactions; the public policy and planning interpretation seeks to achieve a balance of the above-mentioned factors. Each of these interpretations implies a different scientific domain, with some issues overlapping and others diverging or overlooked.

The debate over sustainable development has led to defining a new discipline: sustainability science. Sustainable science is considered an emerging discipline that is applicative and solution-oriented, and aims to handle environmental, social, and economic issues considering cultural, historic, and institutional perspectives. The challenges of the discipline are not only related to better identifying the problems affecting sustainability but also to identify solutions enabling an actual transition towards it by adopting an integrated, comprehensive, and participatory approach. This implies the coexistence of a scientific and a social paradigm as the basis for developing an assessment of sustainability and for defining adequate solutions and interventions (Sala, Farioli, & Zamagni, 2013a, 2013b).

The need for decoupling economic growth from resource consumption and from environmental impacts is considered one of the pivotal aspects to be addressed by sustainable development. However, aiming at the decoupling may be not sufficient if the absolute pressure generated on the environment overcomes the Earth's carrying capacity (Sanyé-Mengual, Secchi, Corrado, Beylot, & Sala, 2019).

To illustrate the complexity and multidimensionality of the Earth's carrying capacity, the concept of 'planetary boundaries' has been put forward and thresholds identified for environmental pressure such as climate change and nutrient load, amongst others (Rockström et al., 2009; Steffen et al., 2015).

However, several boundaries are not yet determined (e.g. those related to chemical pollution), and their assessment poses serious challenges (Sala & Goralczyk, 2013). Another crucial challenge is related to the approach to adopt for the allocation of the safe operating spaces (assigned carrying capacity) to, e.g. countries or economic activities. In theory, each human being on Earth should receive the same allocation but, in practice, decisions regarding resource use and emissions are mostly made by national and sub-national governments, businesses, and other local actors, without coordination at global scale. Hence, to operationalise the planetary boundaries concept, there are schools of thought considering there is the need of translating boundaries into, and aligned with, targets that are relevant at these decision-making scales (Häyhä, Lucas, van Vuuren, Cornell, & Hoff, 2016). For example, one argument is for allocation based on the different levels of development of countries, which may ultimately allow certain countries to generate more pressures to reach their development goals. On the contrary, more developed countries may have to generate less pressures because their level of efficiency is supposed to be higher, or even because they have already exceeded their safe operating spaces.

Several approaches to absolute sustainability assessment, aiming at assessing the transgression of planetary boundaries, have been developed over time (see Bjørn et al. (2020) for a recent review). Besides, a calculation of transgressions for the EU has been carried out (Sala, Crenna, Secchi, & Sanyé-Mengual, 2020), highlighting the EU consumption and production patterns are leading to transgressions of several boundaries (e.g. for climate change, particulate matter, and nutrient cycle), with actions needed in the order of a factor 10 of reduction of impacts. This indicates that eco-efficiency, aimed at reducing impacts of production systems, is not enough to achieve sustainability. In fact, there is also a need to implement eco-effectiveness and handprint concepts to maximise benefits (Alvarenga et al., 2020; Hauschild, 2015), as well as act at the consumption side, as also discussed at political level in the EU (European Parliament, 2021).

Coordinated and integrated actions are needed to break the silos between territorial and product policies. Integrated policies need to address territorial and value chain perspectives, namely understanding the link between consumption patterns and resulting environmental pressures and impacts at local and global scale (UNEP, 2021b). Moreover, there is the need to improve the link between policies focusing on pressures, impacts, and their drivers. Key policy interventions of the EGD address pressures and impacts (e.g. zero pollution, biodiversity strategy), and/or their drivers (e.g. Farm to Fork, Common Agriculture Policy, clean and affordable energy, transition to a circular economy, chemical strategy), with the ambition to maximise welfare and minimise environmental and social externalities.

In this respect, holistic methods such as Life Cycle Assessment (LCA), Input–Output analysis, and general equilibrium economic models (EC-JRC, 2021b) represent powerful sustainability assessment tools whose use in the EU policy making is increasing, as also advocated by the better regulation of EC (2021). For example, the Farm to Fork strategy adopts a life cycle perspective, and LCA is used to calculate the environmental footprint of products and organisations (EC, 2013a, 2013b) and enable the implementation of the EGD.

4 Beyond the European Green Deal

4.1 Shapes of green deals

As described in the previous sections, the EGD is a macro-regional policy initiative developed in 2019, which sets an ambitious policy framework to pursue a sustainable future for Europe and the world based on scientific evidence and adherence to SDGs. However, the concept of 'green deal' is more than 10 years old (Chatzky & Siripurapu, 2021) and used beyond the EU to generally refer to means for implementing sustainability-oriented solutions.

For example, a Global Green New Deal (GGND) was proposed after the 2008–09 crisis to guide countries in the (short-term) recovery of their economies and setting the ground for sustainable development (in the medium–long term). More recently, lead economists have provided visions about what a 'green new deal' should entail, converging on the urgency of decarbonising the economy but differing on how such transition should take place (Schlosser, 2020).

A 'Green New Deal' was also proposed in the United States in 2019 (United States, 2019), consisting of a 10-year plan to depollute industries, build green infrastructure, produce 100% of energy from renewables, and improve the condition of workers (Barbier, 2019). Although the proposal did not find the political support of the former US government, a radical course change is expected to happen in 2021 with the new US presidency. The Biden's plan appears in line with the EGD and includes actions and proposals such as Paris Agreement re-joining, net-zero emission of GHGs by 2050, full sourcing of renewables-based electricity by 2035, carbon pricing, and border adjustment mechanisms (Chatzky & Siripurapu, 2021; Leonard, Pisani-Ferry, Shapiro, Tagliapietra, & Wolff, 2021). Other economies have also started implementing ambitious green policies, although giant economies such as China and India still show a huge dependency on coal and other fossil fuels (Chatzky & Siripurapu, 2021).

Green deal initiatives go even beyond the action of national governments. For example, mayors of cities worldwide announced their support for a 'Global Green New Deal' (C40Cities, 2019) committing to contributing to maintain global warming below 1.5°C by cutting GHG emissions in key sectors such as buildings, transportation, waste, energy and food, whilst at the same time ensuring adequate social protection.

4.2 Green deals in times of economic recovery

As introduced in Section 1, the progress towards sustainability is not on a good track, notwithstanding the popularity of 'green deals', and the Covid-19 pandemic is impacting hard the world, complicating further the achievement of SDGs and the necessary multilateral discussions.

A comparison can be made with the 2008–09 crisis, defined as "the worst financial and economic crisis in generations" while in the middle of oil, food, and water crises (UNEP, 2009). The GGND proposed at that time had the intention to revive economies in the short term, with a long-term sustainability view, through: (1) green investments and fiscal stimulus packages for priority areas (i.e. energy-efficient buildings, sustainable transport and renewable energy, agricultural productivity, freshwater management, and sanitation); (2) removal of harmful subsidies; (3) increase of green taxation and policies on sustainable land use, urban and water management; (4) enforcement of monitoring and accountability practices; and (5) change of international policies on trade, aid, carbon pricing, technologies, and their coordination.

However, the GGND was not able to promote a global transition towards sustainability (Barbier, 2019), mainly due to lack of long-term commitments, underpricing of fossil fuels, overuse of resources, and financing of polluting activities. In particular, adjusting the price of fossil fuels could have significantly reduced GHG emissions and air pollution deaths, whilst at the same time increasing fiscal revenues (Barbier, 2020).

This should serve as a lesson, now that the world has to recover from the Covid-19 pandemic and to put the transition towards sustainable and resilient economies on track (UNEP, 2020b): although economy and social welfare should be short-term objectives, green deals and sustainable investments are needed in the long term.

This implies financial interventions by governments and the integrated consideration of sustainability in recovery plans and multiannual financial frameworks (annual budgets). In this respect, the EU can be considered as a front-runner: its new multi-financial framework (MFF) budget for 2021–27 is above EUR 1 trillion (constant 2018 prices), with about one-third of it explicitly allocated to the protection of natural resources and the environment, to which to add the Next Generation EU fund (the EU recovery plan) (Borchardt, Barbero-Vignola, Buscaglia, Maroni, & Marelli, 2020). A text-mining analysis conducted on legislative documents publicly available during May–September 2020 (Borchardt et al., 2020) suggests that the EU recovery plan aims to deploy short-, medium-, and long-term actions addressing the 17 SDGs and covering more than half of their targets. A strong emphasis is placed on stimulating economic and employment rate growth in Member States, supporting their transition towards a circular, green, and digital economy; promoting clean and renewable energy in the EU; and tackling climate change mitigation and adaptation.

Financing the transition towards sustainability requires looking not only at the size of the green economy but also at its contribution to the overall economy. Before the Covid-19 advent, financing SDGs was USD 2.5 trillion short (Naidoo & Fisher, 2020) and only 10% of the USD 12 trillion committed in 2020 by the 50 largest economies for the pandemic recovery would contribute to build a greener society (UNEP, 2020b).[a]

However, whilst public deficit spending can have positive effects in the short term, this is not the case for financing green deals. In fact, "saddling future generations with unsustainable levels of national debt is just as dangerous as burdening them with an economy that is environmentally unsustainable." A win-win solution to finance a sustainable growth and protect human health and the environment could be the ban of harmful subsidies (e.g. for fossil fuels) and the simultaneous use of environmental taxes and market mechanisms (e.g. tradeable carbon permits) to put a cost on pollution and excessive resource use (Barbier, 2019).

[a]At the time this chapter is written, governments worldwide are still discussing on the post Covid-19 recovery measures. European and global data and analysis can be found, for example, in https:/www.greenrecoverytracker.org/ (Accessed 9 July 2021), https:/www.vivideconomics.com/casestudy/greenness-for-stimulus-index/ (Accessed 11 March 2021).

4.3 The international dimension of the European Green Deal

In the current global scenario, the EGD is a flagship initiative integrating sustainability concepts across policies. However, the EGD will also have a global impact on markets and possible geopolitical consequences (Leonard et al., 2021). Furthermore, to make a global impact and manage geopolitical risks, the EU needs for multilateral agreements, alliances, and partnerships internationally (CEPS, 2019).

For example, amongst the ambitious goals of the EU, there is the development of a low-carbon economy. Climate change is a global issue and the zero-emission objective of the EU will require reducing the dependency on fossil fuels, as well as exporting its standards and/or introducing a carbon tax coupled by a border adjustment mechanism to prevent carbon leakage. Introducing a carbon border adjustment mechanism would be a technical and political challenge that no country in the world has handled so far. Abandoning fossil fuels for renewables will also create new energy security risks and impacts related to the increased demand for raw materials, many of them critical (Leonard et al., 2021). Leonard et al. (2021) identified a series of actions to manage geopolitical risks of the EGD and foster the EU global leadership in this field, as presented in Table 1.

The EU has a great opportunity to lead a global transformation towards sustainable development (Borchardt et al., 2020), also resorting to sustainable investments, diplomacy, trade, and international cooperation, especially with neighbouring countries and powers such as China and the United States (Leonard et al., 2021).

Table 1 How to handle the geopolitical dimension of the European Green Deal.

Risk management measures	Leadership role measures
– Supporting economic diversification in neighbouring oil and gas-exporting countries, including the promotion of renewable energy and green hydrogen. – Improving the security of critical raw materials supply (e.g. through greater supply diversification, increased recycling volumes, and substitution of critical raw materials). – Setting a common carbon border adjustment mechanism with key trade partners.	– Setting global standards for the energy transition (particularly in hydrogen and green bonds) and requiring compliance with strict environmental regulations to access the EU market. – Mobilising financial resources internationally (e.g. EU budget, the EU Recovery and Resilience Fund, and EU development policy). – Promoting global platforms for climate action and the sharing of lessons learned and best practices.

Adapted from Leonard, M., Pisani-Ferry, J., Shapiro, J., Tagliapietra, S., Wolff, G. (2021). The geopolitics of the European Green Deal. *Policy contribution 04/2021, Bruegel. https://www.bruegel.org/2021/02/the-geopolitics-of-the-european-green-deal/ (Accessed 18 February 2021).*

5 Conclusions

This chapter provided an overview of the European Green Deal and how it is positioned in the global sustainability context.

The state of the world highlights the urgent need of acting to achieve sustainability. If SDGs are 'the compass' and financial resources 'the fuel', science-based and internationally coordinated actions can provide 'the means' and 'the roadmap' to achieve sustainability.

The EGD is an example of how science-based policies can promote sustainability at global level and set the boundaries for the post-pandemic economic recovery. The EGD provides an ambitious action plan for transforming the economy of the European Union into a sustainable one, i.e. a carbon-neutral, circular, and green economy where no person and no place is left behind. The initial roadmap of associated policies and measures has resulted in a number of legislative documents addressing specific areas of action.

The EGD builds on the sustainability discussion held in the last decades and the related science. This points out the need for decoupling economic growth from resource consumption and environmental impacts, as well as acting on production and consumption at system level to stay within the planetary boundaries.

The systemic nature of EGD and its policy initiatives ask for adopting and improving integrated assessment methods and approaches. In fact, EU policies are increasingly embracing multiple domains within and across sectors (e.g. the Farm to Fork strategy embraces agriculture, fisheries, food manufacturing and distribution, nutritional and health-related aspects, food safety, food labelling, food waste), which result in an even more complex evaluation of benefits and burdens of policy interventions.

In this respect, holistic sustainability assessment methods, such as life cycle sustainability assessment, and multi-disciplinary expertise should be further developed to allow embracing different dimensions of sustainability at once, or at least to complement and integrate quantitative, semi-quantitative, or qualitative approaches used to assess sustainability aspects in policy-making processes. Key research needs lie in the possibility of breaking the silos between territorial and product-related policies, namely between setting targets for specific compartments (e.g. air quality) and requirements at product and organisational level (e.g. air emission limits), and between different sustainability objectives, so that improvements for one aspect (e.g. climate change mitigation) do not come at the expense of others (the so-called 'do not significant harm' principle used for example in the EU Taxonomy). This is fundamental for consistently ensuring that policy initiatives can effectively promote the transition towards sustainability.

The EU is in a good position to be a leader in the global transition towards sustainable development, with the EGD offering a reference for shaping sustainable and resilient economies, especially now that the world has to recover from the Covid-19 crisis. However, to make a global impact and manage geopolitical risks, the EU should further promote sustainable investments and international cooperation to extend multilateral agreements, alliances, and partnerships internationally and achieve long-term commitments.

Acknowledgments

The opinions expressed in this chapter are purely those of the authors and should not be regarded as stating any official position of the European Commission. The authors are grateful to Mr. Charles Arden-Clarke (UNEP) and Ms. Katie Tuck (UNEP) for pointing at relevant sources of information in the field of sustainable economics and global environmental governance, which were consulted in the writing of this chapter.

References

Alvarenga, R. A. F., Huysveld, S., Taelman, S. E., Sfez, S., Préat, N., Cooreman-Algoed, M., et al. (2020). A framework for using the handprint concept in attributional life cycle (sustainability) assessment. *Journal of Cleaner Production*, *265*, 121743. https://doi.org/10.1016/j.jclepro.2020.121743.

Barbier, E. B. (2019). How to make the next Green New Deal work. *Nature*, *565*, 6. https://doi.org/10.1038/d41586-018-07845-5.

Barbier, E. B. (2020). Greening the post-pandemic recovery in the G20. *Environmental and Resource Economics*, *76*, 685–703. https://doi.org/10.1007/s10640-020-00437-w.

Bell, S., & Morse, S. (2008). *Sustainability indicators: Measuring the immeasurable?*. London (UK): Earthscan.

Bjørn, A., Chandrakumar, C., Boulay, A., Doka, G., Fang, K., Gondran, N., et al. (2020). Review of life-cycle based methods for absolute environmental sustainability assessment. *Environmental Research Letters*, *15*(8), 083001. https://doi.org/10.1088/1748-9326/ab89d7.

Bond, A. J., Dockerty, T., Lovett, A., Riche, A. B., Haughton, A. J., Bohan, D. A., et al. (2011). Learning how to Deal with values, frames and governance in sustainability appraisal. *Regional Studies*, *45*(8), 1157–1170. https://doi.org/10.1080/00343404.2010.485181.

Borchardt, S., Barbero-Vignola, G., Buscaglia, D., Maroni, M., & Marelli, L. (2020). *A sustainable recovery for the EU: A text mining approach to map the EU recovery plan to the sustainable development goals, EUR 30452 EN*. Luxembourg: Publications Office of the European Union, ISBN:978-92-76-25329-7. https://doi.org/10.2760/030575. 2020. JRC122301.

C40Cities. (2019). *The global green new deal. Principles of the global green new deal.* https://www.c40.org/global-green-new-deal. accessed 15 January 2021.

CEPS. (2019). *Can Europe offer a Green Deal to the world?*. https://www.ceps.eu/can-europe-offer-a-green-deal-to-the-world/. accessed 15 January 2021.

Chatzky, A., & Siripurapu, A. (2021). *Envisioning a Green New Deal: A global comparison* (last updated: February 1, 2021) https://www.cfr.org/backgrounder/envisioning-green-new-deal-global-comparison. accessed 10 February 2021.

EC. (2001). *COM(2001)264: Communication from the commission on a sustainable Europe for a better world: A European Union atrategy for sustainable development.*

EC. (2008). *COM(2008)397: Communication from the commission to the European Parliament, the council, the European economic and social committee and the committee of the regions on sustainable consumption and production and sustainable industrial policy action plan.*

EC. (2009). *COM(2009)400: Communication from the commission to the European Parliament, the council, the European economic and social committee and the committee of the regions on mainstreaming sustainable development into EU policies: 2009 review of the European Union strategy for sustainable development.*

EC. (2012). *COM(2012)60: Communication from the commission to the European Parliament, the council, the European economic and social committee and the committee of the regions on innovating for sustainable growth: A bioeconomy for Europe.*

EC. (2013a). *COM(2013)196: Communication from the commission to the European Parliament and the council: Building the single market for green products—Facilitating better information on the environmental performance of products and organisations.*

EC. (2013b). *2013/179/EU: Commission Recommendation of 9 April 2013 on the use of common methods to measure and communicate the life cycle environmental performance of products and organisations (Text with EEA relevance), Annex III, OJ L 124, 4.5.2013* (pp. 1–210).

EC. (2014). *COM(2014)445: Communication from the commission to the European Parliament, the council, the European economic and social committee and the committee of the regions on resource efficiency opportunities in the building sector.*

EC. (2015). *COM(2015)614: Communication from the commission to the European Parliament, the council, the European economic and social committee and the committee of the regions on closing the loop—An EU action plan for the circular economy.*

EC. (2019a). *COM(2019)640: Communication from the commission to the European Parliament, the European council, the council, the European economic and social committee and the committee of the regions on the European green deal.*

EC. (2019b). *A Union that strives for more: My agenda for Europe. Political guidelines for the next European Commission 2019–2024 by candidate for President of the European Commission Ursula von der Leyen.* https://ec.europa.eu/info/sites/info/files/political-guidelines-next-commission_en_0.pdf. accessed 18 February 2021.

EC. (2020a). *A European green deal. Striving to be the first climate-neutral continent.* https://ec.europa.eu/info/strategy/priorities-2019-2024/european-green-deal_en#documents. accessed 28 January 2021.

EC. (2020b). *COM(2020)380: Communication from the commission to the European Parliament, the European council, the council, the European economic and social committee and the committee of the regions on EU biodiversity strategy for 2030 bringing nature back into our lives.*

EC. (2020c). *COM(2020)381: Communication from the commission to the European Parliament, the European council, the council, the European economic and social committee and the committee of the regions on a farm to fork strategy for a fair, healthy and environmentally-friendly food system.*

EC. (2020d). *COM(2020)98: Communication from the commission to the European Parliament, the European council, the council, the European economic and social committee and the committee of the regions on a new circular economy action plan for a cleaner and more competitive Europe.*

EC. (2020e). *COM(2020)662: Communication from the commission to the European Parliament, the European council, the council, the European economic and social committee and the committee of the regions on a renovation wave for Europe—Greening our buildings, creating jobs, improving lives.*

EC. (2020f). *COM(2020)667: Communication from the commission to the European Parliament, the European council, the council, the European economic and social committee and the committee of the regions on chemicals strategy for sustainability towards a toxic-free environment.*

EC. (2020g). *COM(2020)102: Communication from the commission to the European Parliament, the European council, the council, the European economic and social committee and the committee of the regions on a new industrial strategy for Europe.*

EC. (2020h). *COM(2020)21: Communication from the commission to the European Parliament, the European council, the council, the European economic and social committee and the committee of the regions on sustainable Europe investment plan European green deal investment plan.*

EC. (2020i). *The just transition mechanism: Making sure no one is left behind.* https://ec.europa.eu/info/strategy/priorities-2019-2024/european-green-deal/actions-being-taken-eu/just-transition-mechanism_en. accessed 28 January 2021.

EC. (2020l). *COM(2020)67: Communication from the commission to the European Parliament, the European council, the council, the European economic and social committee and the committee of the regions on shaping Europe's digital future.*

EC. (2020m). *Initiative on substantiating green claims.* https://ec.europa.eu/environment/eussd/smgp/initiative_on_green_claims.htm. accessed 18 February 2021.

EC. (2020n). *Recovery plan for Europe.* https://ec.europa.eu/info/strategy/recovery-plan-europe_en. accessed 18 February 2021.

EC. (2021). *Better regulation: Guidelines and toolbox.* https://ec.europa.eu/info/law/law-making-process/planning-and-proposing-law/better-regulation-why-and-how/better-regulation-guidelines-and-toolbox_en. accessed 18 February 2021.

EC-JRC. (2021a). *KnowSDGs platform.* https://knowsdgs.jrc.ec.europa.eu/. accessed 17 February 2021.

EC-JRC. (2021b). *Input-output economics.* https://ec.europa.eu/jrc/en/research-topic/input-output-economics. accessed 17 February 2021.

EEA. (2019). *Drivers of change of relevance for Europe's environment and sustainability. EEA Report No 25/2019.* 19778449. Luxembourg: Publications Office of the European Union, ISBN:9789294802194. https://doi.org/10.2800/129404. 2020.

ESPAS. (2019). *Global trends to 2030: Challenges and choices for Europe. European strategy and policy analysis system (ESPAS) report.* https://www.iss.europa.eu/content/global-trends-2030-%E2%80%93-challenges-and-choices-europe. accessed 15 January 2021.

EU. (2020). *Regulation (EU) 2020/852 of the European Parliament and of the council of 18 June 2020 on the establishment of a framework to facilitate sustainable investment, and amending Regulation (EU) 2019/2088 (Text with EEA relevance).*

European Parliament. (2021). *MEPs call for binding 2030 targets for materials use and consumption footprint.* https://www.europarl.europa.eu/news/en/press-room/20210122IPR96214/meps-call-for-binding-2030-targets-for-materials-use-and-consumption-footprint. accessed 18 February 2021.

Folke, C., Biggs, R., Norström, A. V., Reyers, B., & Rockström, J. (2016). Social-ecological resilience and biosphere-based sustainability science. *Ecology and Society*, *21*(3), 41. https://doi.org/10.5751/ES-08748-210341.

Harari, Y. N. (2015). *Sapiens, a brief history of humankind*. London (UK): Vintage, ISBN:9780099590088.

Hauschild, M. Z. (2015). Better–but is it good enough? On the need to consider both eco-efficiency and eco-effectiveness to gauge industrial sustainability. *Procedia CIRP*, *29*, 1–7.

Häyhä, T., Lucas, P. L., van Vuuren, D. P., Cornell, S. E., & Hoff, H. (2016). From planetary boundaries to national fair shares of the global safe operating space—How can the scales be bridged? *Global Environmental Change*, *40*, 60–72. https://doi.org/10.1016/j.gloenvcha.2016.06.008.

Heath, R., & Boudreau, C. (2020). *More green deals, more global warming*. https://www.politico.com/newsletters/the-long-game/2020/09/15/more-green-deals-more-global-warming-490331. accessed 10 February 2021.

Leonard, M., Pisani-Ferry, J., Shapiro, J., Tagliapietra, S., & Wolff, G. (2021). *The geopolitics of the European Green Deal. Policy contribution 04/2021*. Bruegel. https://www.bruegel.org/2021/02/the-geopolitics-of-the-european-green-deal/. accessed 18 February 2021.

Lewis, S., & Maslin, M. (2015). Defining the anthropocene. *Nature*, *519*, 171–180. https://doi.org/10.1038/nature14258.

Meadows, D. H., Meadows, D. L., Randers, J., & Behrens, W. W. (1972). *The limits to growth*. New York (New York): Universe Books.

Naidoo, R., & Fisher, B. (2020). Reset sustainable development goals for a pandemic world. *Nature*, *583*, 198–201. https://doi.org/10.1038/d41586-020-01999-x.

Nature. (2020). Time to revise the sustainable development goals. *Nature*, *583*, 331–332. https://doi.org/10.1038/d41586-020-02002-3.

Neumayer, E. (2003). *Weak versus strong sustainability: Exploring the limits of two opposing paradigms* (2nd ed.). Cheltenham (UK)—Northampton, MA (USA): Edward Elgar Publishing, ISBN:1843764881.

Nurse, K. (2007). Culture as the fourth pillar of sustainable development. In *Vol. 11. Small states: Economic review and basic statistics* (pp. 28–40). https://doi.org/10.14217/smalst-2007-3-en.

O'Connor, M. (2006). The "four spheres" framework for sustainability. *Ecological Complexity*, *3*(4), 285–292. https://doi.org/10.1016/j.ecocom.2007.02.002.

Patterson, M., McDonald, G., & Hardy, D. (2017). Is there more in common than we think? Convergence of ecological footprinting, emergy analysis, life cycle assessment and other methods of environmental accounting. *Ecological Modelling*, *362*, 19–36. https://doi.org/10.1016/j.ecolmodel.2017.07.022.

Purvis, B., Mao, Y., & Robinson, D. (2019). Three pillars of sustainability: In search of conceptual origins. *Sustainability Science*, *14*, 681–695. https://doi.org/10.1007/s11625-018-0627-5.

Rockström, J., Steffen, W., Noone, K., Persson, Å., Chapin, F. S., Lambin, E., et al. (2009). Planetary boundaries: Exploring the safe operating space for humanity. *Ecology and Society*, *14*(2), 32. http://www.ecologyandsociety.org/vol14/iss2/art32/. accessed 18 February 2021.

Sachs, J. D., Schmidt-Traub, G., Mazzucato, M., Messner, D., Nakicenovic, N., & Rockström, R. (2019). Six transformations to achieve the sustainable development goals. *Nature Sustainability*, *2*, 805–814. https://doi.org/10.1038/s41893-019-0352-9.

Sala, S., Crenna, E., Secchi, M., & Sanyé-Mengual, E. (2020). Environmental sustainability of European production and consumption assessed against planetary boundaries. *Journal of Environmental Management*, *269*, 110686. https://doi.org/10.1016/j.jenvman.2020.110686.

Sala, S., Farioli, F., & Zamagni, A. (2013a). Progress in sustainability science: Lessons learnt from current methodologies for sustainability assessment (part I). *International Journal of Life Cycle Assessment*, *18*, 1653–1672. https://doi.org/10.1007/s11367-012-0508-6.

Sala, S., Farioli, F., & Zamagni, A. (2013b). Life cycle sustainability assessment in the context of sustainability science progress (part II). *International Journal of Life Cycle Assessment*, *18*, 1686–1697. https://doi.org/10.1007/s11367-012-0509-5.

Sala, S., & Goralczyk, M. (2013). Chemical footprint: A methodological framework for bridging life cycle assessment and planetary boundaries for chemical pollution. *Integrated Environmental Assessment and Management*, *9*(4), 623–632. https://doi.org/10.1002/ieam.1471.

Sanyé-Mengual, E., Secchi, M., Corrado, S., Beylot, A., & Sala, S. (2019). Assessing the decoupling of economic growth from environmental impacts in the European Union: A consumption-based approach. *Journal of Cleaner Production*, *236*. https://doi.org/10.1016/j.jclepro.2019.07.010, 117535.

Schlosser, K. (2020). Contrasting visions of the green new deal. *Environmental Politics*. https://doi.org/10.1080/09644016.2020.1847514.

SDSN. (2021). *Transformations for the joint implementation of agenda 2030 for sustainable development and the European green deal*. Sustainable Development Solutions Network (SDSN). https://resources.unsdsn.org/transformations-for-the-joint-implementation-of-agenda-2030-the-sustainable-development-goals-and-the-european-green-deal-a-green-and-digital-job-based-and-inclusive-recovery-from-covid-19-pandemic. accessed 18 February 2021.

Seager, T. P., Melton, J., & Eighmy, T. T. (2004). Working towards sustainable science and engineering: Introduction to the special issue on highway infrastructure. *Resources, Conservation and Recycling*, *42*(3), 205–207. https://doi.org/10.1016/j.resconrec.2004.04.001.

Steffen, W., Richardson, K., Rockström, J., Cornell, S. E., Fetzer, I., Bennett, E. M., et al. (2015). Planetary boundaries: Guiding human development on a changing planet. *Science*, *347*(6223), 1259855. https://doi.org/10.1126/science.1259855.

The World in 2050 initiative. (2020). *Innovations for sustainability. Pathways to an efficient and post-pandemic future. Report prepared by The World in 2050 initiative*. Laxenburg, Austria: International Institute for Applied Systems Analysis (IIASA). www.twi2050.org. Available at http://pure.iiasa.ac.at/id/eprint/16533/. ISBN 10: 3-7045-0157-3 10.22022/TNT/07-2020.16533.

Troëng, S., Barbier, E., & Rodríguez Echandi, C. M. (2020). *The COVID-19 pandemic is not a break for nature—Let's make sure there is one after the crisis*. https://www.weforum.org/agenda/2020/05/covid-19-coronavirus-pandemic-nature-environment-green-stimulus-biodiversity/. accessed 4 February 2021.

UN. (2015). *Transforming our world: The 2030 agenda for sustainable development*. https://sustainabledevelopment.un.org/post2015/transformingourworld. accessed 28 January 2021.

UN. (2020). *SDG indicators—Global indicator framework for the sustainable development goals and targets of the 2030 agenda for sustainable development*. https://unstats.un.org/sdgs/indicators/indicators-list/. accessed 10 February 2021.

UN. (2021). *Capacity development and evidence based policies*. https://www.un.org/development/desa/capacity-development/what-we-do/areas-of-work/evidence-based-policy/.

UN Global Compact—Red Española. (2020). *Diez tendencias que marcarán el ritmo de la Década de Acción en ODS*. https://www.pactomundial.org/2020/10/diez-tendencias-que-marcaran-el-ritmo-de-la-decada-de-accion-en-ods/. accessed 15 January 2021.

UNEP. (2009). *Global green new deal policy brief, March 2009*. https://www.unenvironment.org/resources/report/global-green-new-deal-policy-brief-march-2009. accessed 15 January 2021.

UNEP. (2020a). *The triple planetary crisis: Forging a new relationship between people and the earth*. https://www.unenvironment.org/news-and-stories/speech/triple-planetary-crisis-forging-new-relationship-between-people-and-earth. accessed 15 January 2021.

UNEP. (2020b). In E. B. Barbier (Ed.), *Building a greener recovery: Lessons from the great recession*. Geneva: United Nations Environment Programme. https://www.greengrowthknowledge.org/guidance/building-greener-recovery-lessons-great-recession. accessed 15 January 2021.

UNEP. (2021a). *Making peace with nature: A scientific blueprint to tackle the climate, biodiversity and pollution emergencies*. Nairobi https://www.unep.org/resources/making-peace-nature. accessed 12 March 2021.

UNEP. (2021b). *Catalysing science-based policy action on sustainable consumption and production—The value-chain approach & its application to food, construction and textiles*. Nairobi https://www.unep.org/resources/publication/catalysing-science-based-policy-action-sustainable-consumption-and-production. accessed 12 March 2021.

United States. (2019). *116th CONGRESS, 1st Session, H. RES. 109 Recognizing the duty of the Federal Government to create a Green New Deal*. https://www.congress.gov/bill/116th-congress/house-resolution/109/text. accessed: 18 February 2021.

UNSSC. (2021). *The 2030 agenda for sustainable development*. https://www.unssc.org/sites/unssc.org/files/2030_agenda_for_sustainable_development_kcsd_primer_en.pdf. accessed 18 February 2021.

Vos, R. O. (2007). Perspective defining sustainability: A conceptual orientation. *Journal of Chemical Technology and Biotechnology*, *82*, 334–339. https://doi.org/10.1002/jctb.1675.

World Economic Forum (Ed.). (2021). *The global risks report 2021* (16th ed.). World Economic Forum, ISBN:978-2-940631-24-7. https://www.weforum.org/reports/the-global-risks-report-2021. accessed 18 February 2021.

SECTION

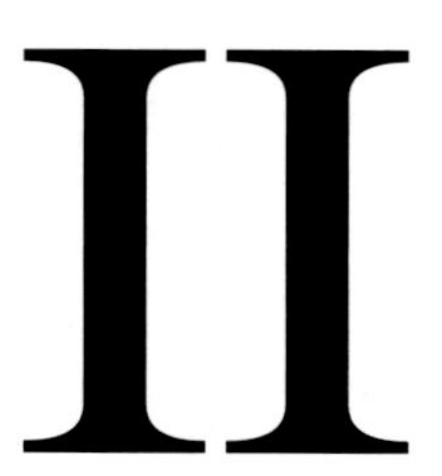

Assessment tools for sustainability-methodological issues

CHAPTER

Life Cycle Sustainability Assessment-based tools

Beatriz Rivela[a], Brandon Kuczenski[b], and Dolores Sucozhañay[c]
[a]*Inviable Life Cycle Thinking, Madrid, Spain* [b]*Institute for Social, Behavioral, and Economic Research University of California, Santa Barbara, CA, United States* [c]*Department of Space and Population, University of Cuenca, Cuenca, Ecuador*

1 What Life Cycle Thinking means: A system approach to sustainability

1.1 The need to cultivate a life cycle perspective

It is undeniable that ongoing environmental degradation limits the extent to which human society can exploit natural resources and the current situation threatens the planet's resilience and carrying capacity: contemporary society cannot be supported by over-extended systems (Griggs et al., 2013; Steffen et al., 2011). During recent decades progress has been made (Sachs, Schmidt-Traub, Mazzucato, et al., 2019), no doubt, but quoting Johan Rockström "the more we learn, the more fragile the Earth system seems to be and the faster we need to move."

The critical challenges to achieve the ambitious objectives of the Paris Agreement (UNFCCC, 2015) and the United Nations Sustainable Development Goals (UN, 2015) require not only clear scientific progress in individual disciplines but also cohesion among them. The methods now available to measure the environmental and economic impacts are greatly advanced, and experiences concerning the assessment of the social dimension of sustainability, which has a low degree of standardisation, seem promising. But there is still a significant gap between intentions and actions. This make-or-break moment requires our best effort towards enabling the use of credible assessments which are the key to monitoring progress and to support decisions. But we also need a new way of thinking to go beyond the traditional focus, understanding and including the whole environmental, social, and economic implications of our decisions to identify potential conflicts, synergies, and trade-offs.

The economy must be re-designed under the umbrella of sustainability. Circular economy (CE) is currently a popular concept promoted by the European Union, by several national governments, and by many businesses around the world (Korhonen, Nuur, Feldmann, & Birkie, 2018), which aims to redefine growth, looking beyond the current take-make-waste extractive industrial model. Despite being visionary

Assessing Progress towards Sustainability. https://doi.org/10.1016/B978-0-323-85851-9.00018-3

and provocative in its message, the research on the CE concept is emerging (Korhonen, Honkasalo, & Seppälä, 2018) and decision-making processes in CE strategies could be biased by a naive understanding of what 'closing the loop' implies (Peña et al., 2021). This is where Life Cycle Thinking (LCT) comes into play: LCT has emerged over the last decades as a promising framework, based on a systems approach, to measure the environmental, social, and economic impacts of a product, process, or service over its entire life cycle (UNEP, 2012). It provides a long-term perspective of the multiple impacts, including burden-shifting, affecting a system (Azapagic, 2017), showing the global picture to ensure a global solution. Cultivating a life cycle perspective allows to detect hidden costs and evaluate trade-offs explicitly and transparently as well as to identify the key hotspots of a product or system, for priority setting in formulating actions or strategies. Thus LCT is used by decision-makers in both the public and private sector for the development of policies and products, as well as for the procurement and provision of services (Hellweg & Canals, 2014).

LCT can successfully support the decision-making processes, structuring the sustainability intelligence to identify and develop innovative solutions, while highlighting trade-offs and avoiding the single-metric blindness that may be generated by focusing only on greenhouse gas emissions. The other key concept is Life Cycle Management (LCM), which is the systematic integration of life cycle thinking into managerial practices (Harbi, Margni, Loerincik, & Dettling, 2015). This means to manage the whole life cycle of a system—a 'system' may be a product, process, service, even an organisation—with the aim of providing societies with more sustainable goods and services.

1.2 Life cycle methodologies for sustainability assessment: Understanding tools and criteria

Life Cycle Approaches have been developed over the past three decades to assist in decision-making at all levels, from qualitative to quantitative and from conceptual to pragmatic. Concepts provide the vision, and methodologies the practical application and support decision-making by analysing data. The most effective way of supporting decisions towards sustainability is the application of life cycle concepts combined with the related tools for assessment. The enabling condition or starting point for advancing towards sustainability is the availability of transparent, credible, and easily accessible information to assess the performance of the system under analysis. The life cycle of a product—or sequence of stages 'from cradle-to-grave'—begins with the extraction of raw materials from natural resources in the ground and energy generation (Fig. 1). Materials and energy are then part of production, packaging, distribution, use, maintenance, and eventually recycling, reuse, recovery, or final disposal.

Depending on the goal, time, and resources, there are different questions that should be asked and different tools which could be used to get the needed answers. Actions towards sustainability need to be informed by reliable and trustworthy

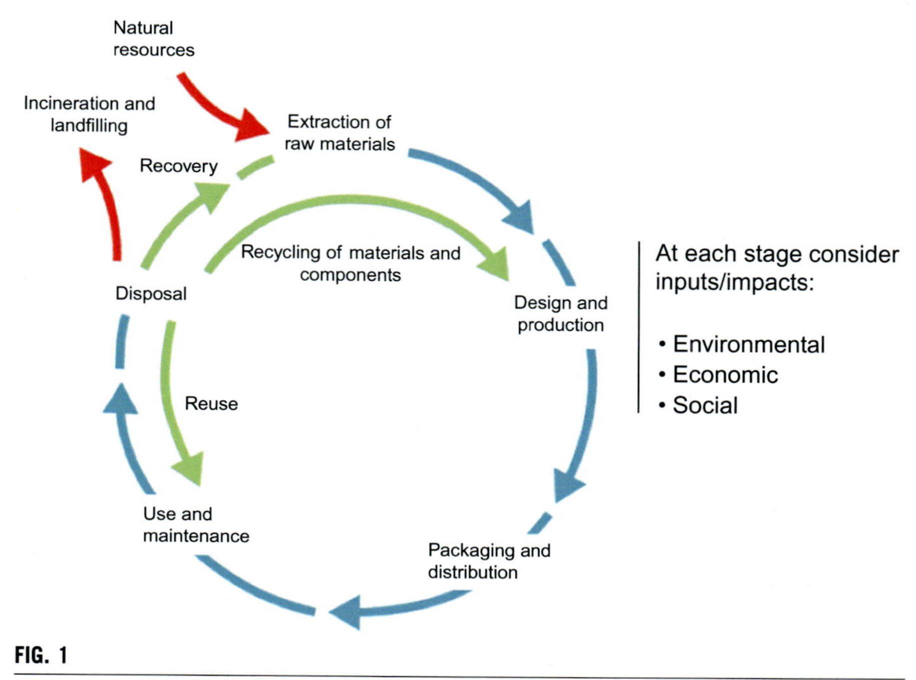

FIG. 1

Life cycle stages of a system.

science-based evidence, but they also need to be balanced with desirable speed of response and pragmatism. The misconception of Life Cycle Approaches has led, in some cases, to present their implementation as complicated, time consuming, or expensive. It is important to remark that Life Cycle Approaches are valuable and powerful approaches that, properly adopted, result in economic gains, from better supply-chain performance to internal operation efficiencies or increased institutional capacity that further enhance innovation (Life Cycle Initiative, 2016). From a first step of prioritisation to a detailed characterisation of any given sustainability issue, there is a spectrum of tools that can offer solutions to different economic, social, and environmental issues. Fig. 2 shows an overview of the main life cycle tools.

Deciding where and how to act to have the maximum impact, balancing pragmatism and science-based evidence, is of paramount importance for decision-makers. Hotspot analysis or hotspotting, the prioritisation method used by different analytical disciplines, is often a precursor to develop more detailed or granular sustainability information (Barthel et al., 2014), to identify the most significant impacts or benefits for further investigation or action. A common feature of hotspots analysis is the presentation of information and findings in accessible formats, including for non-technical audiences, who are often the key decision-makers in policy and business settings (Barthel et al., 2015).

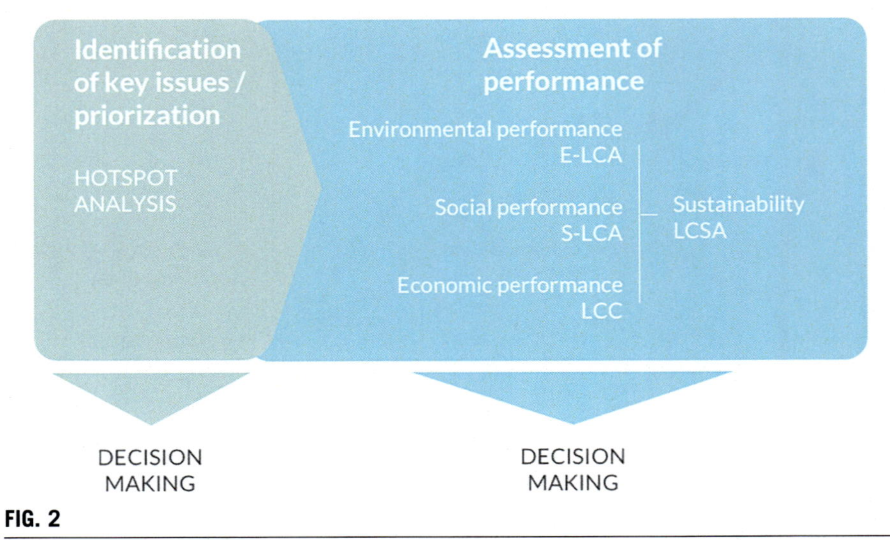

FIG. 2

Life Cycle tools: which kind of tool for which purpose?

Although there is not currently a common global approach to hotspots analysis, the Life Cycle Initiative promoted the Flagship Activity 'Global Principles and Practices for Hotspots Analysis' (2012–2017), whose aim was to develop and agree on global principles and guidance through a consensus building process. The result of this process was the publication of a common methodological framework and global guidance for sustainability hotspots analysis (Barthel et al., 2014; Barthel, Fava, James, Hardwick, & Khan, 2017), to support those wishing to commission, conduct, or use hotspots analysis studies on how to translate and apply the results of this analysis. A good example of application of the hotspot methodological framework was the development of the 'Sustainable Consumption and Production Hotspots Analysis Tool',[a] an online tool launched by UN Environment in 2019 that analyses the environmental and socio-economic performance of 171 countries over the past 25 years to provide scientific evidence of areas where improvement can be made.

2 Environmental Life Cycle Assessment

Environmental Life Cycle Assessment (LCA) is a consolidated tool or technique, based on the ISO 14040 and 14,044 standards (ISO 14040, 2006, 2006), that evaluates environmental performance throughout the life cycle of a product or service. The LCA framework conceptualises the linkages between a product's environmental

[a]The tool, intended for use by policy experts, statisticians, and the general public, can be accessed at http://scp-hat.lifecycleinitiative.org/.

interventions (Life Cycle Inventory—LCI) and the ultimate damage caused to human health, resource depletion, and ecosystem quality—information which is of critical importance to decision-makers (Fig. 3).

For decision-making in an organisation, the capabilities and applicability of LCA have been explored and a specific methodology proposed, called Organisational Life Cycle Assessment (O-LCA). This methodology uses a life cycle perspective to compile and evaluate the inputs, outputs, and potential environmental impacts of the activities associated with an organisation, and the provision of its product portfolio (UNEP, 2017). However, comparative assertions between different organisations are neither robust nor meaningful, mainly due to the lack of a consistent basis for comparison (UNEP, 2015).

2.1 Processes and flows

At the core of LCA is a conceptual model of industrial production in which different industrial activities known as 'processes' are linked together by goods, services, and other quantifiable relationships, collectively known as 'flows'. Processes and flows are inseparable from one another: a process is defined in part by the set of flows coming into and out of it, and the flows are defined by the processes which create and consume them. Each process has one or more 'reference flows' which are produced by the process (Weidema, Wenzel, Petersen, & Hansen, 2004).

Each process is understood as occurring within a specified time and space, such as 'Peru in 2020'. This is called the spatio-temporal scope of the process. Defining this information is crucial because the same industrial activity (say, steelmaking) can

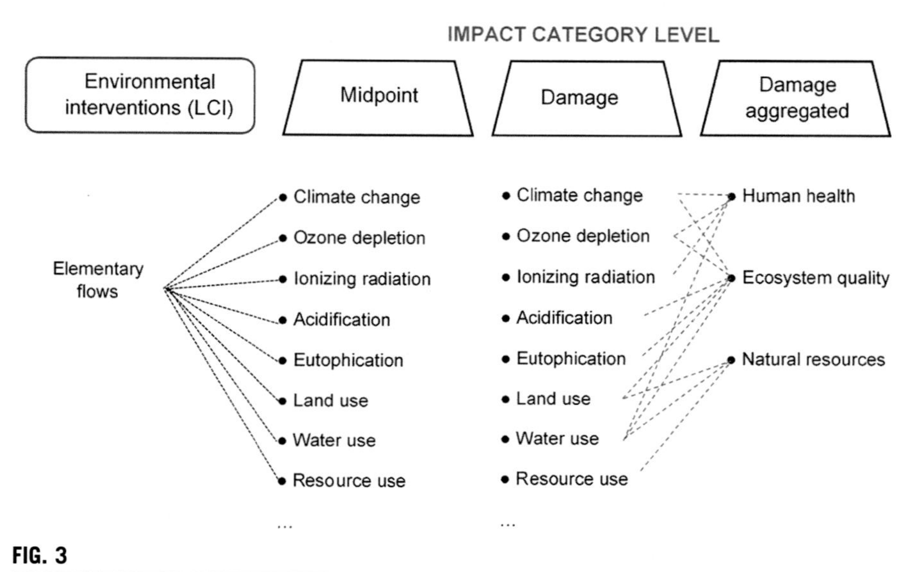

FIG. 3

The Life Cycle Impact Assessment midpoint-damage framework.

have different characteristics depending on where it is being operated, and because the way activities are carried out can also change over time (Mastrucci, Marvuglia, Benetto, & Leopold, 2020; Raschio, Smetana, Contreras, Heinz, & Mathys, 2018).

Flows, which are counterparts to processes, can be almost anything. The most straightforward flows are materials like plastic and metal, fuels like coal and gasoline [petrol], and manufactured products like computers and soccer balls. However, flows can also refer to non-materials like electricity, non-physical concepts like the occupation of space, or services like the transport of goods or the use of computing power. The set of input and output flows for a given process is called its 'inventory'. For example, the flow of 'hot coffee' could be the output of the process of operating a cafe, and the inventory for that process would likely include coffee beans, purified water, electricity for grinding the beans and heating the water, as well as a paper cup and lid, perhaps a packet of sugar, and a wooden stirring stick. Taking a very broad perspective, the cafe process could also include power to keep the lights on and run the refrigerator, service flows like the rental of commercial space, the purchase of insurance, and the payment of employee salaries for cafe operation and maintenance. Each of those things is required to operate a cafe, and so they should be captured in the inventory for that process.

2.2 A techno-economic network

When you purchase coffee in a cafe, drink it, and then dispose of the cup in a waste bin, the actions you took did not necessarily impact the environment in any way at all. After all, the cup was already made when you walked through the door, and the water may have been already hot. The central idea behind LCA, however, is that these actions could have consequences that were far away from the point of consumption. The cup was manufactured from paper; now that it is used, it cannot be used again. Electricity was generated and converted to heat; at the end of the day the garbage will be hauled and either recycled or disposed. Each of the flows in the cafe inventory has its origin or terminus in another process with its own inventory of requirements and products. In order to fully understand the impacts of your coffee consumption, it is necessary to follow all those flows back to their sources and to attribute a portion of the environmental impacts of all the involved processes to your consumption choice.

All these processes and flows can be joined together into a large network of industrial activity, a process known as life cycle inventory analysis (Vigon et al., 1993). In the parlance of industrial metabolism, activities that happen prior to the point of analysis are called 'upstream' and those that happen afterwards are 'downstream'. However, in many cases there are mutual dependencies built into the structure of the network (for instance, coal extraction is required for steelmaking, and steel is also required for coal extraction). Therefore many of the activities must be modelled together in order for the result of the analysis to be meaningful and complete. The end result is a picture of the many and varied industrial activities that are required to deliver the product of interest (Azapagic, 2017). For your coffee consumption, the quantified description of the function of the product under analysis or 'functional unit' will be the reference basis for all calculations: for example, a cup of coffee prepared from 9 g of ground coffee and 125 mL of tap water, served in a disposable cup made of paper lined with polyethylene and polystyrene lid. The definition of the functional unit is essential for modelling as it determines the materials and processes included in the assessment (Cooper, 2003).

2.3 The nature–industry boundary

After creating an organised model of the industrial economy, the next task is to describe the associated impacts of the activities to areas of ecological and social concern. This is done by imagining a physical boundary between the world of human industry, sometimes called the 'technosphere', and the natural world, and considering the set of flows that cross this boundary (Hofstetter, 1998). These 'elementary flows' have two kinds based on their directions. Inputs to the technosphere represent extraction of resources, like oil from the ground or radiation from the sun. Output flows are emissions, like exhaust from a tailpipe or debris in the ocean. Each flow has a physical extent that can in concept be measured, even though most flows go unobserved. Living things, notably humans, are also considered 'outside' the boundary, so the chemicals taken up by organisms are also considered emissions (Margni, Rossier, Crettaz, & Jolliet, 2002).

Of the flows discussed in the cafe example, none are elementary—they all exist within the economy—and as observed, none of them have direct environmental impacts. Only when a flow is found to cross that boundary does a potential impact occur. The end goal of inventory analysis is to record (or to estimate) the full set of elementary flows that occur during the life cycle of the product. Non-elementary (often 'intermediate') flows, which describe all the products and services exchanged within the economy, are thus an accounting mechanism to connect together a set of activities involving resource extraction and direct emission.

2.4 Life Cycle Impact Assessment

Different types of emissions have different kinds of implications for human and ecological health (Ridoutt et al., 2016). In order to evaluate the impacts of an activity on the areas of concern shown in Fig. 3, it is necessary to identify which flows affect each category of concern. Heat-trapping gases contribute to global warming, nitrogen and phosphorus to eutrophication, fluorocarbons to ozone depletion, and so on. Life Cycle Impact Assessment (LCIA) describes the process of characterising the potential effects of each elementary flow on an area of concern (Pennington et al., 2004; Rosenbaum et al., 2017). Often, these characterisations are given in terms of a reference substance that has a well-understood property (such as the warming effect of a kilogram of CO_2). Some substances will be more potent, and some less. The characterisation factors express how much a single unit of the intervention contributes to an impact category. The task of LCIA is to marshal the characterisations of these resources and emissions and add them up over the entire network of processes considered before, thus computing numerical 'category scores' from the results of an inventory analysis.

Each category score can be considered an indicator of the environmental burdens associated with a process. The precise meaning of those scores depends on the way the elementary flows are scored. An LCIA methodology specifies the way elementary flows are scored according to a comprehensive range of categories of concern (see Fig. 3). LCIA scores that indicate a physical equivalency, such as an equivalent amount of a reference substance, are called 'midpoint indicators'. They are generally based on a scientific model of a physical phenomenon of interest. These resulting scores can also be expressed as some valuation of the modelled effect, such as some measure of habitat loss, economic cost, or impaired human health. These 'endpoint' or 'damage indicators' are subject to considerably greater uncertainty and subjectivity. However, they may be more appropriate or easier to interpret for target audiences (Jolliet et al., 2004).

2.5 LCA in practice: Insights and approaches

LCA is commonly practiced in order to provide insight to decision-makers about the relative significance of different aspects of a product's life cycle to its overall impact (for example: making a paper cup versus growing 25 g of coffee beans) or to weigh the merits of a process change (example: selecting a new cup vendor that uses recycled paper). Performing an LCA allows a decision-maker to assign numeric scores to the different stages of a product's life cycle and compare them mathematically.

There are at least five major techniques for numerical interpretation of LCA results (Heijungs & Kleijn, 2001):

- In **Contribution analysis**, a score is divided among the different life cycle stages that make it up to identify the most important contributors.
- In **Comparative analysis** or **Scenario analysis**, a collection of models, or a single model with a number of different parameter settings, are laid side by side for comparison.
- In **Perturbation analysis**, individual parameters are varied to observe the effects on the overall score.
- In **Stochastic uncertainty analysis**, also known as Monte Carlo analysis, all uncertain parameters are varied simultaneously, and the process is repeated many times to estimate the probability distribution of the resulting scores.
- In **Discernibility analysis**, stochastic simulation and scenario analysis are combined to determine whether any significant differences exist among the different options or scenarios.

An LCA study that is performed with the objective of assigning an ecological impact value to a product or process is called an 'attributional' study, while a study whose intention is to evaluate the effects of a change is called a 'consequential' study (Baitz, 2017; Earles & Halog, 2011; Tillman, 2000). Practitioners should use different approaches depending on which of these forms applies. In an attributional study, the total sum of environmental emissions throughout the economy can be considered split up among everything produced during a time period. Every emission must be counted exactly once, and therefore the emissions from processes with multiple products must be allocated to those various products according to some measurement. Typically, mass or economic value is used as a basis for allocation. In contrast, the concept for a consequential study is some change in the status quo, after which some processes may be operated more intensively and others less so. Thus attributional studies will always include actual impacts that are incurred (Brander & Wylie, 2011), whereas consequential studies may include a mix of incurred impacts (with a positive value) and potentially avoided impacts (having a negative value). It is possible to construct a consequential study in which the effect of a change is to reduce overall impacts, leading to net negative results.

LCA studies that consider future events, such as economic forecasts or studies of emerging technologies, can be prepared using the same databases as attributional and consequential studies, but the nature of the investigation must be different (Arvidsson et al., 2018). Particularly regarding new technologies, uncertainty about the system under study can be higher. However, there are also much greater opportunities for significant improvement in the system's environmental performance. Instead of intending to be predictive, prospective LCA studies should be considered strategic—in which the objective is to understand the range of possibilities for a system that has not yet come into being. In recent years, the so-called ex ante LCA has been applied to new technologies to determine possible environmental impacts at an early stage of research and development (Tsoy, Steubing, van der Giesen, et al., 2020).

Anticipatory LCA (Wender et al., 2014) is the term given to prospective LCA performed in a deliberative process in which stakeholder perspectives are canvassed to inform modelling decisions. Anticipatory LCA performed in concert with ongoing development of a product or system can highlight areas of concern that can be addressed preventatively. Data gaps of high importance can be identified and pursued, and the findings incorporated into the study. In the case of policy, anticipatory LCA can form a basis for dialogue and cooperation between private actors and advocates for the public interest (Göswein et al., 2020).

2.6 Uncertainty and variability

An LCIA indicator score results from the combination of two highly complex synthetic models: the product model and the LCIA methodology, and is thus subject to substantial uncertainty pertaining both to the models' structures and to the numerical values themselves. LCA uncertainty is grouped into parametric uncertainty (uncertainty in data value), scenario uncertainty (uncertain description of the system under study), and model uncertainty (imprecision and inaccuracy in the model's design) (Lloyd & Ries, 2007). This third category is vast and includes questions ranging in scale from whole economic markets to 100 kW engines. It also includes uncertainty related to the practitioner's choices and preferences (Sala, Laurent, Vieira, & Van Hoof, 2020; Scrucca et al., 2020). Parametric uncertainty can be plumbed using stochastic methods such as Monte Carlo analysis, but this requires reasonable estimates about the actual uncertainty in the most important data points (Huijbregts, Gilijamse, Ragas, & Reijnders, 2003). Sometimes, parameter uncertainty is generated synthetically based on an expert assessment of the quality or representativeness of each data point (Weidema et al., 2004), but this approach has not been well validated by empirical study (Muller, Lesage, & Samson, 2016; Yang, Tao, & Suh, 2018). Model and scenario uncertainty, on the other hand, are both hard to estimate, and are often evaluated by comparing alternative models or scenarios against each other (Cherubini, Franco, Zanghelini, & Soares, 2018; Igos, Benetto, Meyer, Baustert, & Othoniel, 2019).

Uncertainty in a modelling parameter or in a result is distinguished from variability, which is the amount to which a parameter value may change during the course of activity. An example is the variation in the electricity grid mix as seasons change throughout the year. Variability can have effects in all three areas of uncertainty discussed before. It is up to the practitioner to determine and assess the variability of a system under study (AzariJafari, Yahia, & Amor, 2018; Michiels & Geeraerd, 2020).

2.7 Optional steps: Normalisation and weighting

Different LCIA indicators, especially midpoint indicators, have physical significance derived from the scientific models used to develop the impact scores. When considering a range of indicators from different categories, it is not generally possible to compare those indicators to one another because they describe different effects and are measured in different units. It is not meaningful to compare a kilogram of acid rain emissions to a square metre of habitat loss, for instance without some broader context. Endpoint indicators are meant to express ecological damage in real-world terms, but even those measurements may not be informative to a reader.

One approach to express the magnitude of an impact score in real-world terms is to compare the system under study with a much larger system, such as the total impact of an industry or a national population. This is called normalisation (Tolle, 1997), and it allows an analyst to understand which impact categories are more strongly affected by the system under study. For instance, say a consumer worked out a study of her coffee consumption. Instead of reporting the results per cup of coffee consumed, the *estimate* per cup is used to compute the *share* of the overall impact of a year of normal living for the average citizen of her country. The normalised impact scores might show her coffee consumption accounted for 0.03% of her acidification footprint, but 1.7% of her agricultural land use footprint, indicating that the coffee industry should preferentially focus on reducing its land footprint than acidification. Of course, normalised results depend on having a valid normalisation factor to represent a 'year of normal living' based on prior research.

Even knowing the relative impacts in different categories, readers often seek to understand the relative amounts of *harm* that are implied by LCIA results. Since many potential effects may be good or bad, and since decisions often result in trade-offs, decision-makers must ultimately weigh the *pros* and *cons* in the context of their own ambitions and subjective judgement. Weighting describes a technique from decision science for expressing the relative importance of different categories numerically. Multiplying the normalised scores by their appropriate weights and adding the results together allows for the computation of a single score that simplifies the comparison of different options. The selection of weighting values is irretrievably subjective and must be regarded as specific in scope to the project that made the selection. These weights could be developed through a deliberative process in order to reflect the different priorities of multiple stakeholders. Normalisation and weighting together can express LCA results in the form of a single score, but considerable work is needed to ensure that such a score is meaningful (Prado et al., 2020; Roesch, Sala, & Jungbluth, 2020).

2.8 Software and databases

Each of the two major analytic steps described before—building the network of processes and flows, and characterising their environmental impacts—is a vast and challenging undertaking, and both must be combined to compute an LCA result. This task is typically performed using specialised software (Table 1), which is available in both free and commercial forms (Ciroth, 2012).

The primary requirement is access to a Life Cycle Inventory (LCI) database, a data resource that describes the requirements, products, and emissions of a large, self-consistent set of industrial process models. Because of the global nature of the economy, such a database must span all industrial sectors and have a worldwide geographic scope. Technical requirements for preparing and maintaining such a resource are substantial (UNEP, 2011), and there are only a few large-scale LCI databases available. The most prominent premium LCI databases are ecoinvent and GaBi. Ecoinvent is the product of a partnership among a number of Swiss public agencies and private contributors, and establishes the standard for transparent,

Table 1 Brief description of the most used LCA Software Systems.

Software	Publisher	Licence	Platforms	Prominent features
OpenLCA	Green Delta	Open Source	Cross-platform (java)	Collaboration Tools; scientific computing
GaBi	Sphera	Proprietary	Windows	Extensive proprietary database; modular modelling
SimaPro	Pre Consulting	Proprietary	Windows	Scientific computing and uncertainty
Brightway2	Chris Mutel	Open Source	Cross-platform (python)	Expert LCA computation and modelling framework

scientifically reviewed, regionalised life cycle modelling. The GaBi database is maintained by Sphera Solutions GmbH, and touts many decades of close industry collaboration to develop proprietary inventory data. These premium databases are joined by a number of smaller or more specialised databases, along with an increasing range of national and regional scale data resources of both public and private provenance. Several national and international collaborations have been developed to promote the availability of LCI data, including the Global LCA Data Access network (GLAD) by the UN Environment programme, the International Reference Life Cycle Data System (ILCD) Data Network of the European Union, and the Federal LCA commons in the US (Fritter, Lawrence, Marcolin, & Pelletier, 2020).

Care must be taken when using data from different providers to ensure that the modelling approaches and background data are consistent (Suh, Leighton, Tomar, & Chen, 2016). Collaboration among authors across organisational boundaries, including reuse or exchange of product system models, is often challenging, both because of incompatibilities between software systems and because of the confidentiality of input data (Kuczenski et al., 2018). Data quality evaluation is highly dependent on the intended use or fitness for a particular purpose (Ciroth, Foster, Hildenbrand, & Zamagni, 2020; Weidema & Wesnæs, 1996), and so practitioners must spend considerable effort reviewing the characteristics of candidate datasets to find one that is appropriate.

Additional challenges are encountered when considering the impact assessment step. The Life Cycle Initiative has promoted the creation of a Global Life Cycle Impact Assessment Method (GLAM) and makes a significant effort to reach a global consensus on LCIA indicators (Jolliet et al., 2018). Similarly to LCI, there are a few widely used LCIA methodologies that are commonly used (Table 2), but it is up to each software or data provider to ensure consistency. The list of emissions resulting from an LCI computation must be matched with the list of characterisation factors for environmental impacts in order to correctly compute a result. Thus it is necessary to use standardised nomenclature or some other means for matching flows between the

Table 2 Selection of life cycle impact assessment methodologies in wide use.

Methodology	Publisher	First released	Current version	Type	Scope
LC-IMPACT	UNEP/LCI	2020	2020	Midpoint + Endpoint	Global
CML	Leiden Univ	2001	August 2016 (version 4.7)	Midpoint	Global
ReCiPe	RIVM (NL)	2008	2018	Midpoint + Endpoint	Europe/ Global
TRACI	US EPA	2003	2012 (version 2.1)	Midpoint	United States
EcoIndicator	Pre Consultants	1995	1999	Endpoint	Europe/ Global

two domains, which presents challenges in practice (Edelen et al., 2018). Regional or spatial variability of impacts is also an important concern for impacts such as water and land use, acidification, eutrophication, and others (Mutel et al., 2019). Novel impact categories such as marine debris, biodiversity or ecosystem services (Rugani et al., 2019), and social impact measures (Petti, Serreli, & Di Cesare, 2018) are increasing in abundance. However, there is much less attention in the field to systematic management of LCIA data, and no clear path for novel LCIA indicators to be integrated with existing inventory resources.

3 Life Cycle Costing

Concerning the economic domain of sustainability, Life Cycle Costing (LCC) is the tool aimed at the assessment of all costs associated with the life cycle of a product that are directly covered by one or more actors in the product life cycle (supplier, manufacturer, user or consumer, and/or end-of-life actor), with the inclusion of externalities that are anticipated to be internalised in the decision-relevant future (Swarr et al., 2011). Although LCC is older than LCA (Settanni, Notarnicola, & Tassielli, 2011), LCA has been broadly applied for environmental impact assessment over the last three decades (Laurin, 2017), but the economic impact assessment using a life cycle perspective has not been widely integrated into sustainability assessment until the last decade.

Life Cycle Costing (LCC) and Social Life Cycle Assessment (S-LCA) align with the ISO 14040 framework but differ in some respects. There are not specific standards for LCC and S-LCA, but the Life Cycle Initiative has played a key role in developing guidance documents to describe how they may be used to understand the economic costs and social impacts and risks throughout the life cycle. Fig. 4 shows the conceptual framework of LCC based on the physical product life cycle (Rebitzer & Hunkeler, 2003).

LCC studies can be classified into conventional, environmental, or societal (Hochschorner & Noring, 2011; Hunkeler, Lichtenvort, & Rebitzer, 2008):

- Conventional LCC is the compilation of all costs associated with the life cycle of a system (product, process, or activity) that are directly covered by any one or more of the actors in the system life cycle.
- Environmental LCC includes external costs that might be internalised within a foreseeable time frame.
- Societal LCC is even broader, including costs for society overall.

The relevance of LCC has been questioned and discussed by different authors (Hannouf & Assefa, 2016; Jørgensen, Herrmann, & Bjørn, 2013). Several studies have analysed the integration of LCA and LCC for single-product systems (Miah, Koh, & Stone, 2017; Moreau & Weidema, 2015) and industrial symbiosis networks (Kerdlap, Low, & Ramakrishna, 2020). More recently, (França et al., 2021) carried out a comprehensive critical review of environmental–economic studies with an

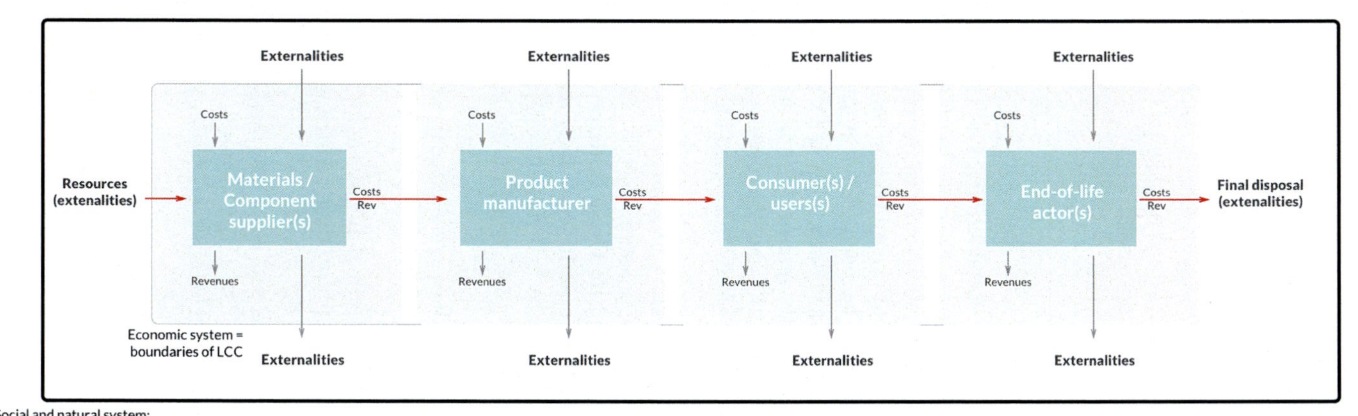

FIG. 4

The conceptual framework of LCC.

integrated use of LCA and LCC. Building design and civil construction have been identified as the areas of most advanced integration of LCA and LCC, with European countries and USA playing a leading role. Among the different strategies for the integration, the most recurrent methods are mathematical modelling and programming for optimisation, and multi-criteria decision-making. Main challenges of the integration are the differences in environmental and economic background data as well as the conversion of environmental impacts to monetary units or 'monetisation' of environmental impacts.

4 Social Life Cycle Assessment

In the context of sustainable development and the transition towards a sustainable CE, which has gained relevance as an innovative approach to achieve sustainable development, the social dimension has received less attention compared to the environmental and economic ones. However, in recent years this trend has begun to change (Padilla-Rivera, Russo-Garrido, & Merveille, 2020). Assessing social impacts is rooted in the necessity of ensuring that any project, programme, or policy preserves the well-being of the people and communities involved in light of sustainable and equitable development (Vanclay, 2003). Although moving towards the CE model may cause large social effects, the full inclusion of the social dimension within its related concepts, tools, and metrics is still to be achieved. Social Life Cycle Assessment (S-LCA), even it has not become mainstream (Zimdars, Haas, & Pfister, 2018), has been identified as a promising and suitable methodology that can be used for including social aspects of products and services with a life cycle perspective for CE actions and practices (Padilla-Rivera et al., 2020).

4.1 The S-LCA methodology

Research on the importance of integrating the social aspects in the LCA framework started in 1993 (UNEP, 2020). The greatest step towards its consolidation was the publication of the Guidelines for Social Life Cycle Assessment of Products in 2009 (UNEP, 2009), which conceptualised and standardised the S-LCA methodology (Benoît et al., 2010). From 2017 onwards, S-LCA has gained value as a real technique in sustainability science (Ramos Huarachi, Piekarski, Puglieri, & de Francisco, 2020) and eventually reached another significant milestone in 2020: the launch of the updated Guidelines for Social Life Cycle Assessment of Products and Organisations (UNEP, 2020).

The S-LCA methodology does not have a standard, but as it has been mentioned its stages are derived from the ISO 14040 framework (Popovic & Kraslawski, 2015). It entails the following four iterative stages (UNEP, 2020):

- **Goal and scope definition:** describes the intended use and the goal of the study. In this stage, the functional unit and system boundaries are defined. The selection

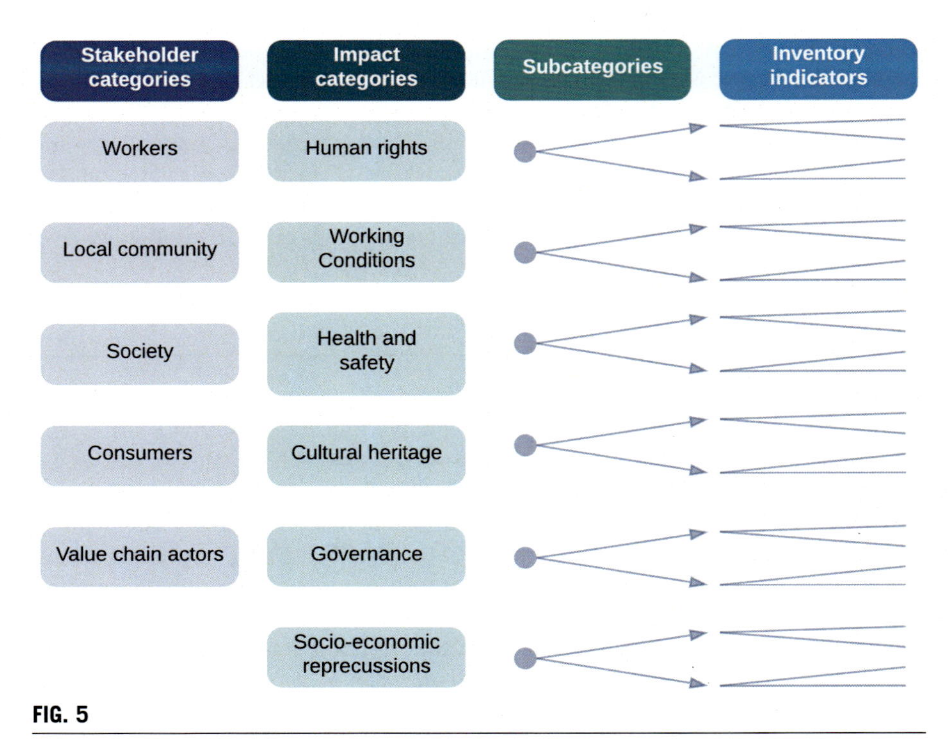

FIG. 5

Assessment system from categories to inventory data.

of stakeholder categories and impact subcategories is also performed in this stage (Fig. 5).

- **Social life cycle inventory:** entails the construction of the data inventory through the collection of significant data related to all unit processes.
- **Social life cycle impact assessment:** during this stage, data collected are processed for assessing the social impact associated with a product, service, or organisation. Two approaches are used in this stage: Type 1 (reference scale approach), which assesses social performance or social risk, and Type II (impact pathway approach), whose aim is to assess consequential social impacts (Huertas-Valdivia, Ferrari, Settembre-Blundo, & García-Muiña, 2020). The outcomes of this stage could provide insights on:
  - **Social risks**: defined as a social topic for which an adverse impact is probable.
  - **Social hotspots**: refer to a location and/or process of the life cycle where a social impact or risk is likely to occur.
  - **Social performance**: it entails the principles, practices, and outcomes of businesses' relationships with different stakeholders regarding deliberate actions and unintended business activity externalities.

- **Social footprint:** it represents the end result of the S-LCA study overall or by impact categories/subcategories.
- **Social handprint:** results of changes to business as usual, creating relative positive outcomes or impacts.

- **Interpretation:** during this stage, all inputs for drawing conclusions and recommendations are built through the revisions of previous stages, verifying, discussing, and summarising the results in the different aggregation levels (life cycle stages, impact categories or subcategories, stakeholders, or processes).

The updated S-LCA guidelines advanced on the inclusion of the following three aspects:

- Expansion of the social impact subcategories, namely: employment relationship, sexual harassment, smallholders including farmers, wealth distribution, ethical treatment of animals, poverty alleviation, education provided in the local community, health issues for children as consumers, children concerns regarding marketing practices.
- Inclusion of connections between social impact subcategories with the 17 Sustainable Development Goals (SDGs), globally accepted and adopted as part of the agenda of governments, industries, and organisations. S-LCA directly helps track the progress towards at least 14 of the 17 SDGs.
- Emphasis on the relevance of analysing positive impacts and depicts different types of positive impacts that can be assessed.

This life cycle-based methodology has specific differences compared to other methodologies of social impact assessment in terms of its object (products/services and their life cycle), scope (the entire life cycle), and the collection and reporting process in terms of social impacts and benefits across the life cycle. For example, Social Impact Assessment (SIA), which analyses, monitors, and manages both intended and unintended social consequences of planned interventions (Vanclay, 2003), lacks the life cycle perspective.

4.2 S-LCA databases

Data availability remains the biggest challenge while performing an S-LCA study. Within S-LCA, life cycle stages are associated with geographic locations, and impacts refer to potentially affected stakeholders, but usually, there are no specific data for products, services, or organisations. In response to these data scarcity situations, generic databases and modelling tools have been developed to provide information at least at the national or sectoral level, which provides a broad and widespread social scenario. Currently, the available S-LCA databases are the Social Hotspots Database (SHDB) and the Product Social Impact Life Cycle Assessment (PSILCA) database (Benoit-Norris, Cavan, & Norris, 2012; Ciroth, 2012; Huertas-Valdivia et al., 2020).

SHDB, which was developed by New Earth in the USA, presents a global input/output model based on the Global Trade Analysis Project (GTAP). SHDB is a social database containing data for 57 different sectors, for every 113 different regions corresponding to individual countries, meaning the database contains around 6441 country-specific sectors within it (Mancini et al., 2018). For its part, GreenDelta developed the PSILCA database (Ciroth & Eisfeldt, 2016); within it, the EORA Multi-Regional Input/Output (MRIO) database is used as an input–output model for depicting the interdependencies between different branches of a national economy or different regional economies. EORA provides data for 189 countries with a list of 15,909 sectors and uses monetary flows, expressed in US dollars, to link processes among different sectors and countries (Mancini et al., 2018).

4.3 Challenges and developments

S-LCA is still a rather new approach to assess social impacts along value chains of products and services and despite the significant progress achieved so far, S-LCA has to face several challenges (Petti et al., 2018; Venkatesh, 2019).

Despite the guidelines' flexibility and clarity, assessing the social dimension of sustainability remains complex. The proposed indicators tend to present a better adaptation to formal settings, presenting certain ease in obtaining information compared to informal settings. A challenge to be addressed could be the establishment of indicators for different contexts (e.g., formal–informal).

Data collection for the inventory stage, in certain contexts, remains a time-consuming and resource-intensive task. This situation is rooted in the limited availability or access to specific, compatible, high-quality, and life cycle-based data. Although the method allows the use of generic indicators in the absence of specific data, these limit the assessment to a country or sectoral level, making it difficult to provide inputs for context-specific decisions and policy-making. In this context, the challenge remains on encouraging S-LCA's use in budget constraints that require collecting primary information. Inter-institutional cooperation for constructing open S-LCA databases could be a starting point. In line with the role of stakeholders in S-LCA (UNEP, 2009), performing participatory processes is crucial, especially in unconsolidated institutions or informal sectors, to gain legitimacy and for planning and executing data collection processes.

As part of the update process, ten early pilot projects road-tested the updated guidelines in 2020 for assessing different products and services (in formal and informal environments) in different countries. This process provided valuable feedback, included in the revised Guidelines. This and other similar experiences are expected to enhance the methodology and to advance its use to different contexts (Social LC Alliance, 2020).

5 Towards Life Cycle Sustainability Assessment: The integration challenge

In line with the 'triple bottom-line' model of sustainability, one step further is to combine the evaluation of all the impacts and benefits and integrate all of them in the decision-making processes. In 2011 the Life Cycle Sustainability Assessment Framework (LCSA) (Cinelli et al., 2013; Ekener, Hansson, Larsson, & Peck, 2018) emerged as a response to the integration challenge and was presented to experts in various disciplinary fields for them to discuss and thence develop a holistic approach to effective sustainable development and sustainability decision-making. Unfortunately, the result has been scant development in the field of backup for effective decision-making (Gbededo, Liyanage, & Garza-Reyes, 2018).

6 Conclusions and looking ahead

The sustainability challenges require a systematic shift in the global economy, but the diversity of approaches and the current confusing array of issues and options—the "Wild West" of Sustainability Assessment—can create confusion among decision-makers. The Sustainable Development Goals, Resource Efficiency or Circular Economy are popular concepts currently discussed and applied both in international, regional, and national level; they are closely related to sustainable development and put the emphasis on specific aspects, making decision-reaching more feasible. This is where Life Cycle Thinking is needed, providing the global picture to ensure a global solution. The different approaches offer a "mental map" of simplification and Life Cycle Approaches provide the valuable and powerful tools that, properly adopted, support decisions towards sustainability.

Since the COVID-19 pandemic was declared, it has exposed both the fragility of the global economy and of our own lives. The call for a more resilient, equitable, circular, and low-carbon economic model has garnered support from a growing number of businesses and governments over the past few years and appears today more relevant than ever. Recent research shows an unprecedented flood of concern and commitment by companies about climate change and social inequities and reveals huge public support for putting sustainability at the heart of recovery plans. The spring of 2020 was suggestive of how much, and how quickly, we can change as a civilisation. And the biological threat provoking the crisis also made us to reflect on how our lives compromise life on this planet. Taking advantage of this unprecedented moment, basing our decisions on the best scientific knowledge available is the challenge we are facing today.

References

Arvidsson, R., Tillman, A. M., Sandén, B. A., Janssen, M., Nordelöf, A., Kushnir, D., & Molander, S. (2018). Environmental assessment of emerging technologies: Recommendations for prospective LCA. *Journal of Industrial Ecology*, *22*(6), 1286–1294. https://doi.org/10.1111/jiec.12690.

Azapagic, A. (2017). A life cycle approach to measuring sustainability. In *Encyclopedia of sustainable technologies* (pp. 71–79). Elsevier. https://doi.org/10.1016/B978-0-12-409548-9.10036-3.

AzariJafari, H., Yahia, A., & Amor, B. (2018). Assessing the individual and combined effects of uncertainty and variability sources in comparative LCA of pavements. *International Journal of Life Cycle Assessment*, *23*(9), 1888–1902. https://doi.org/10.1007/s11367-017-1400-1.

Baitz, M. (2017). *Goal and scope definition in life cycle assessment. LCA compendium—The complete world of life cycle assessment*. https://doi.org/10.1007/978-94-024-0855-3_3.

Barthel, M., Fava, J., Harnanan, C., Strothmann, P., Khan, S., & Miller, S. (2015). *Hotspots analysis: Providing the focus for action. Part of the series LCA compendium—The complete world of life cycle assessment. Life cycle management* (pp. 149–167).

Barthel, M., Fava, J., Harnanan, C., Valdivia, S., Khan, S., James, K., & Smerek, A. (2014). Hotspots analysis: Mapping of existing methodologies, tools and guidance and initial recommendations for the development of global guidance. In *UNEP/SETAC life cycle initiative-flagship project, 3*.

Barthel, M., Fava, J., James, K., Hardwick, A., & Khan, S. (2017). Hotspots analysis. An overarching methodological framework and guidance for product and sector level application. *United Nations Environment Programme*.

Benoît, C., Norris, G. A., Valdivia, S., Ciroth, A., Moberg, A., Bos, U., … Beck, T. (2010). The guidelines for social life cycle assessment of products: Just in time! *International Journal of Life Cycle Assessment*, *15*(2), 156–163. https://doi.org/10.1007/s11367-009-0147-8.

Benoit-Norris, C., Cavan, D. A., & Norris, G. (2012). Identifying social impacts in product supply chains: Overview and application of the social hotspot database. *Sustainability*, *4*(9), 1946–1965. https://doi.org/10.3390/su4091946.

Brander, M., & Wylie, C. (2011). The use of substitution in attributional life cycle assessment. *Greenhouse Gas Measurement and Management*, *1*, 161–166. https://doi.org/10.1080/20430779.2011.637670.

Cherubini, E., Franco, D., Zanghelini, G. M., & Soares, S. R. (2018). Uncertainty in LCA case study due to allocation approaches and life cycle impact assessment methods. *International Journal of Life Cycle Assessment*, *23*(10), 2055–2070. https://doi.org/10.1007/s11367-017-1432-6.

Cinelli, M., Coles, S. R., Jørgensen, A., Zamagni, A., Fernando, C., & Kirwan, K. (2013). Workshop on life cycle sustainability assessment: The state of the art and research needs—November 26, 2012, Copenhagen, Denmark. *International Journal of Life Cycle Assessment*, *18*(7), 1421–1424. https://doi.org/10.1007/s11367-013-0573-5.

Ciroth, A. (2012). Software for life cycle assessment. In *Life cycle assessment handbook: A guide for environmentally sustainable products* (pp. 143–157). John Wiley and Sons. https://doi.org/10.1002/9781118528372.ch6.

Ciroth, A., & Eisfeldt, F. (2016). *PSILCA—A product social impact life cycle assessment database*. PSILCA.

Ciroth, A., Foster, C., Hildenbrand, J., & Zamagni, A. (2020). Life cycle inventory dataset review criteria—A new proposal. *International Journal of Life Cycle Assessment*, *25* (3), 483–494. https://doi.org/10.1007/s11367-019-01712-9.

Cooper, J. S. (2003). Specifying functional units and reference flows for comparable alternatives. *International Journal of Life Cycle Assessment*, *8*, 337. https://doi.org/10.1007/BF02978507.

Earles, J. M., & Halog, A. (2011). Consequential life cycle assessment: A review. *International Journal of Life Cycle Assessment*, *16*(5), 445–453. https://doi.org/10.1007/s11367-011-0275-9.

Edelen, A., Ingwersen, W. W., Rodríguez, C., Alvarenga, R. A. F., de Almeida, A. R., & Wernet, G. (2018). Critical review of elementary flows in LCA data. *International Journal of Life Cycle Assessment*, *23*(6), 1261–1273. https://doi.org/10.1007/s11367-017-1354-3.

Ekener, E., Hansson, J., Larsson, A., & Peck, P. (2018). Developing life cycle sustainability assessment methodology by applying values-based sustainability weighting—Tested on biomass based and fossil transportation fuels. *Journal of Cleaner Production*, *181*, 337–351. https://doi.org/10.1016/j.jclepro.2018.01.211.

França, W. T., Barros, M. V., Salvador, R., de Francisco, A. C., Moreira, M. T., & Piekarski, C. M. (2021). Integrating life cycle assessment and life cycle cost: A review of environmental-economic studies. *International Journal of Life Cycle Assessment*, *26*(2), 244–274. https://doi.org/10.1007/s11367-020-01857-y.

Fritter, M., Lawrence, R., Marcolin, B., & Pelletier, N. (2020). A survey of life cycle inventory database implementations and architectures, and recommendations for new database initiatives. *International Journal of Life Cycle Assessment*, *25*(8), 1522–1531. https://doi.org/10.1007/s11367-020-01745-5.

Gbededo, M. A., Liyanage, K., & Garza-Reyes, J. A. (2018). Towards a life cycle sustainability analysis: A systematic review of approaches to sustainable manufacturing. *Journal of Cleaner Production*, *184*, 1002–1015. https://doi.org/10.1016/j.jclepro.2018.02.310.

Göswein, V., Rodrigues, C., Silvestre, J. D., Freire, F., Habert, G., & König, J. (2020). Using anticipatory life cycle assessment to enable future sustainable construction. *Journal of Industrial Ecology*, 178–192. https://doi.org/10.1111/jiec.12916.

Griggs, D., Stafford-Smith, M., Gaffney, O., Rockström, J., Öhman, M. C., Shyamsundar, P., … Noble, I. (2013). Policy: Sustainable development goals for people and planet. *Nature*, *495*(7441), 305–307. https://doi.org/10.1038/495305a.

Hannouf, M., & Assefa, G. (2016). Comments on the relevance of life cycle costing in sustainability assessment of product systems. *International Journal of Life Cycle Assessment*, *21*(7), 1059–1062. https://doi.org/10.1007/s11367-016-1136-3.

Harbi, S., Margni, M., Loerincik, Y., & Dettling, J. (2015). Life cycle management as a way to operationalize sustainability within organizations. In G. Sonnemann, & M. Margni (Eds.), *The complete world of life cycle assessment. Life cycle management* Springer. https://doi.org/10.1007/978-94-017-7221-1_3.

Heijungs, R., & Kleijn, R. (2001). Numerical approaches towards life cycle interpretation five examples. *International Journal of Life Cycle Assessment*, *6*, 141. https://doi.org/10.1007/BF02978732.

Hellweg, S., & Canals, L. M. I. (2014). Emerging approaches, challenges and opportunities in life cycle assessment. *Science*, *344*(6188), 1109–1113. https://doi.org/10.1126/science.1248361.

Hochschorner, E., & Noring, M. (2011). Practitioners' use of life cycle costing with environmental costs—A Swedish study. *International Journal of Life Cycle Assessment*, *16*(9), 897–902. https://doi.org/10.1007/s11367-011-0325-3.

Hofstetter, P. (1998). *Perspectives in life cycle impact assessment*. Springer. https://doi.org/10.1007/978-1-4615-5127-0.

Huertas-Valdivia, I., Ferrari, A. M., Settembre-Blundo, D., & García-Muiña, F. E. (2020). Social life-cycle assessment: A review by bibliometric analysis. *Sustainability (Switzerland)*, *12*(15). https://doi.org/10.3390/su12156211.

Huijbregts, M. A. J., Gilijamse, W., Ragas, A. M. J., & Reijnders, L. (2003). Evaluating uncertainty in environmental life-cycle assessment. A case study comparing two insulation options for a Dutch one-family dwelling. *Environmental Science and Technology*, *37*(11), 2600–2608. https://doi.org/10.1021/es020971+.

Hunkeler, D., Lichtenvort, K., & Rebitzer, G. (2008). *Environmental life cycle costing*. CRC Press.

Igos, E., Benetto, E., Meyer, R., Baustert, P., & Othoniel, B. (2019). How to treat uncertainties in life cycle assessment studies? *International Journal of Life Cycle Assessment*, *24*(4), 794–807. https://doi.org/10.1007/s11367-018-1477-1.

ISO 14040. (2006). *Environmental management-life cycle assessment-principles and framework*. International Organisation for Standardisation.

ISO 14044. (2006). *Environmental management-life cycle assessment-requirements and guidelines*. International Organisation for Standardisation.

Jolliet, O., Antón, A., Boulay, A.-M., Cherubini, F., Fantke, P., Levasseur, A., … Frischknecht, R. (2018). Global guidance on environmental life cycle impact assessment indicators: Impacts of climate change, fine particulate matter formation, water consumption and land use. *International Journal of Life Cycle Assessment*, *23*(11), 2189–2207. https://doi.org/10.1007/s11367-018-1443-y.

Jolliet, O., Müller-Wenk, R., Bare, J., Brent, A., Goedkoop, M., Heijungs, R., … Weidema, B. (2004). The LCIA midpoint-damage framework of the UNEP/SETAC life cycle initiative. *International Journal of Life Cycle Assessment*, *9*(6), 394–404. https://doi.org/10.1065/lca2004.09.175.

Jørgensen, A., Herrmann, I. T., & Bjørn, A. (2013). Analysis of the link between a definition of sustainability and the life cycle methodologies. *International Journal of Life Cycle Assessment*, *18*(8), 1440–1449. https://doi.org/10.1007/s11367-013-0617-x.

Kerdlap, P., Low, J. S. C., & Ramakrishna, S. (2020). Life cycle environmental and economic assessment of industrial symbiosis networks: A review of the past decade of models and computational methods through a multi-level analysis lens. *International Journal of Life Cycle Assessment*, *25*(9), 1660–1679. https://doi.org/10.1007/s11367-020-01792-y.

Korhonen, J., Honkasalo, A., & Seppälä, J. (2018). Circular economy: The concept and its limitations. *Ecological Economics*, *143*, 37–46. https://doi.org/10.1016/j.ecolecon.2017.06.041.

Korhonen, J., Nuur, C., Feldmann, A., & Birkie, S. E. (2018). Circular economy as an essentially contested concept. *Journal of Cleaner Production*, *175*, 544–552. https://doi.org/10.1016/j.jclepro.2017.12.111.

Kuczenski, B., Marvuglia, A., Astudillo, M. F., Ingwersen, W. W., Satterfield, M. B., Evers, D. P., … Laurin, L. (2018). LCA capability roadmap—Product system model description and revision. *International Journal of Life Cycle Assessment*, *23*(8), 1685–1692. https://doi.org/10.1007/s11367-018-1446-8.

Laurin, L. (2017). Overview of LCA-history, concept, and methodology. In *Encyclopedia of sustainable technologies* (pp. 217–222). Elsevier. https://doi.org/10.1016/B978-0-12-409548-9.10058-2.

Life Cycle Initiative. (2016). *E-learning module "Introduction to life cycle thinking"*. https://www.learnlifecycle.com/.

Lloyd, S. M., & Ries, R. (2007). Characterizing, propagating, and analyzing uncertainty in life-cycle assessment: A survey of quantitative approaches. *Journal of Industrial Ecology, 11*(1), 161–179. https://doi.org/10.1162/jiec.2007.1136.

Mancini, L., Eynard, U., Eisfeldt, F., Ciroth, A., Blengini, G. A., & Pennington, D. W. (2018). *Social assessment of raw materials supply chains: A life-cycle-based analysis*. https://doi.org/10.2760/470881.

Margni, M., Rossier, D., Crettaz, P., & Jolliet, O. (2002). Life cycle impact assessment of pesticides on human health and ecosystems. *Agriculture, Ecosystems & Environment, 93*(1–3), 379–392. https://doi.org/10.1016/S0167-8809(01)00336-X.

Mastrucci, A., Marvuglia, A., Benetto, E., & Leopold, U. (2020). A spatio-temporal life cycle assessment framework for building renovation scenarios at the urban scale. *Renewable and Sustainable Energy Reviews, 126*. https://doi.org/10.1016/j.rser.2020.109834.

Miah, J. H., Koh, S. C. L., & Stone, D. (2017). A hybridised framework combining integrated methods for environmental life cycle assessment and life cycle costing. *Journal of Cleaner Production, 168*(1), 846–866. https://doi.org/10.1016/j.jclepro.2017.08.187.

Michiels, F., & Geeraerd, A. (2020). How to decide and visualize whether uncertainty or variability is dominating in life cycle assessment results: A systematic review. *Environmental Modelling and Software, 133*, 10484. https://doi.org/10.1016/j.envsoft.2020.104841.

Moreau, V., & Weidema, B. P. (2015). The computational structure of environmental life cycle costing. *International Journal of Life Cycle Assessment, 20*(10), 1359–1363. https://doi.org/10.1007/s11367-015-0952-1.

Muller, S., Lesage, P., & Samson, R. (2016). Giving a scientific basis for uncertainty factors used in global life cycle inventory databases: An algorithm to update factors using new information. *International Journal of Life Cycle Assessment, 21*(8), 1185–1196. https://doi.org/10.1007/s11367-016-1098-5.

Mutel, C., Liao, X., Patouillard, L., Bare, J., Fantke, P., Frischknecht, R., … Verones, F. (2019). Overview and recommendations for regionalized life cycle impact assessment. *International Journal of Life Cycle Assessment, 24*(5), 856–865. https://doi.org/10.1007/s11367-018-1539-4.

Padilla-Rivera, A., Russo-Garrido, S., & Merveille, N. (2020). Addressing the social aspects of a circular economy: A systematic literature review. *Sustainability (Switzerland), 12*(19), 7912. https://doi.org/10.3390/SU12197912.

Peña, C., Civit, B., Gallego-Schmid, A., Druckman, A., Pires, A. C., Weidema, B., … Motta, W. (2021). Using life cycle assessment to achieve a circular economy. *International Journal of Life Cycle Assessment, 26*, 215–220. https://doi.org/10.1007/s11367-020-01856-z.

Pennington, D. W., Potting, J., Finnveden, G., Lindeijer, E., Jolliet, O., Rydberg, T., & Rebitzer, G. (2004). Life cycle assessment Part 2: Current impact assessment practice. *Environment International, 30*(5), 721–739. https://doi.org/10.1016/j.envint.2003.12.009.

Petti, L., Serreli, M., & Di Cesare, S. (2018). Systematic literature review in social life cycle assessment. *International Journal of Life Cycle Assessment, 23*(3), 422–431. https://doi.org/10.1007/s11367-016-1135-4.

Popovic, T., & Kraslawski, A. (2015). Social sustainability of complex systems. In *Vol. 36. Computer aided chemical engineering* (pp. 605–614). Elsevier B.V. https://doi.org/10.1016/B978-0-444-63472-6.00024-0.

Prado, V., Cinelli, M., Ter Haar, S. F., Ravikumar, D., Heijungs, R., Guinée, J., & Seager, T. P. (2020). Sensitivity to weighting in life cycle impact assessment (LCIA). *International Journal of Life Cycle Assessment*, *25*(12), 2393–2406. https://doi.org/10.1007/s11367-019-01718-3.

Ramos Huarachi, D. A., Piekarski, C. M., Puglieri, F. N., & de Francisco, A. C. (2020). Past and future of social life cycle assessment: Historical evolution and research trends. *Journal of Cleaner Production*, *264*. https://doi.org/10.1016/j.jclepro.2020.121506.

Raschio, G., Smetana, S., Contreras, C., Heinz, V., & Mathys, A. (2018). Spatio-temporal differentiation of life cycle assessment results for average perennial crop farm: A case study of Peruvian cocoa progression and deforestation issues. *Journal of Industrial Ecology*, *22*(6), 1378–1388. https://doi.org/10.1111/jiec.12692.

Rebitzer, G., & Hunkeler, D. (2003). Life cycle costing in LCM: Ambitions, opportunities, and limitations. *The International Journal of Life Cycle Assessment*, *8*(5), 253–256. https://doi.org/10.1007/bf02978913.

Ridoutt, B. G., Pfister, S., Manzardo, A., Bare, J., Boulay, A.-M., Cherubini, F., … Verones, F. (2016). Area of concern: A new paradigm in life cycle assessment for the development of footprint metrics. *The International Journal of Life Cycle Assessment*, *21*, 276–280. https://doi.org/10.1007/s11367-015-1011-7.

Roesch, A., Sala, S., & Jungbluth, N. (2020). Normalization and weighting: The open challenge in LCA. *International Journal of Life Cycle Assessment*, *25*(9), 1859–1865. https://doi.org/10.1007/s11367-020-01790-0.

Rosenbaum, R. K., Hauschild, M. Z., Boulay, A. M., Fantke, P., Laurent, A., Núñez, M., & Vieira, M. (2017). Life cycle impact assessment. In *Life cycle assessment: Theory and practice* (pp. 167–270). Springer International Publishing. https://doi.org/10.1007/978-3-319-56475-3_10.

Rugani, B., Maia de Souza, D., Weidema, B. P., Bare, J., Bakshi, B., Grann, B., … Verones, F. (2019). Towards integrating the ecosystem services cascade framework within the life cycle assessment (LCA) cause-effect methodology. *Science of the Total Environment*, *690*, 1284–1298. https://doi.org/10.1016/j.scitotenv.2019.07.023.

Sachs, J. D., Schmidt-Traub, G., Mazzucato, M., et al. (2019). Six transformations to achieve the sustainable development goals. *Nature Sustainability*, *2*, 805–814. https://doi.org/10.1038/s41893-019-0352-9.

Sala, S., Laurent, A., Vieira, M., & Van Hoof, G. (2020). Implications of LCA and LCIA choices on interpretation of results and on decision support. *International Journal of Life Cycle Assessment*, *25*(12), 2311–2314. https://doi.org/10.1007/s11367-020-01845-2.

Scrucca, F., Baldassarri, C., Baldinelli, G., Bonamente, E., Rinaldi, S., Rotili, A., & Barbanera, M. (2020). Uncertainty in LCA: An estimation of practitioner-related effects. *Journal of Cleaner Production*, *268*. https://doi.org/10.1016/j.jclepro.2020.122304.

Settanni, E., Notarnicola, B., & Tassielli, G. (2011). Life cycle costing (LCC). In *Vol. 6. Encyclopedia of sustainability. Measurements, indicators, and research methods for sustainability* (pp. 225–227). Berkshire Publishing.

Social LC Alliance. (2020). *The revision of the SLCA guidelines*. https://www.social-lca.org/the-revision-of-the-slca-guidelines/.

Steffen, W., Persson, A., Deutsch, L., Zalasiewicz, J., Williams, M., Richardson, K., … Svedin, U. (2011). The anthropocene: From global change to planetary stewardship. *Ambio*, *40*(7), 739–761. https://doi.org/10.1007/s13280-011-0185-x.

Suh, S., Leighton, M., Tomar, S., & Chen, C. (2016). Interoperability between ecoinvent ver. 3 and US LCI database: A case study. *International Journal of Life Cycle Assessment*, *21*(9), 1290–1298. https://doi.org/10.1007/s11367-013-0592-2.

Swarr, T. E., Hunkeler, D., Klöpffer, W., Pesonen, H. L., Ciroth, A., Brent, A. C., & Pagan, R. (2011). Environmental life-cycle costing: A code of practice. *International Journal of Life Cycle Assessment*, *16*(5), 389–391. https://doi.org/10.1007/s11367-011-0287-5.

Tillman, A.-M. (2000). Significance of decision-making for LCA methodology. *Environmental Impact Assessment Review*, *20*(1), 113–123. https://doi.org/10.1016/S0195-9255(99)00035-9.

Tolle, D. A. (1997). Regional scaling and normalization in LCIA. *The International Journal of Life Cycle Assessment*, *2*, 197. https://doi.org/10.1007/bf02978416.

Tsoy, N., Steubing, B., van der Giesen, C., et al. (2020). Upscaling methods used in ex ante life cycle assessment of emerging technologies: A review. *International Journal of Life Cycle Assessment*, *25*, 1680–1692. https://doi.org/10.1007/s11367-020-01796-8.

UN. (2015). *Transforming our world: The 2030 agenda for sustainable development*. UN General Assembly.

UNEP. (2009). In C. Benoît Norris, & B. Mazijn (Eds.), *Guidelines for social life cycle assessment of products* United Nations Environment Programme (UNEP).

UNEP. (2011). *Global guidance principles for life cycle assessment databases—A basis for greener processes and products*. United Nations Environment Programme.

UNEP. (2012). *Greening the economy through life cycle thinking ten years of the UNEP/SETAC life cycle initiative*.

UNEP. (2015). *Guidance on organizational life cycle assessment*. United Nations Environment Programme.

UNEP. (2017). *Road testing organizational life cycle assessment around the world: Applications, experiences and lessons learned*. United Nations Environment Programme.

UNEP. (2020). In C. B. Norris, M. Traverso, S. Neugebauer, E. Ekener, T. Schaubroeck, S. R. Garrido, … G. Arcese (Eds.), *Guidelines for social life cycle assessment of products and organizations 2020* United Nations Environment Programme (UNEP). https://bit.ly/2Y3mLr9.

UNFCCC. (2015). *Paris agreement to the United Nations framework convention on climate change*.

Vanclay, F. (2003). International principles for social impact assessment. *Impact Assessment and Project Appraisal*, *21*(1), 5–11.

Venkatesh, G. (2019). Critique of selected peer-reviewed publications on applied social life cycle assessment: Focus on cases from developing countries. *Clean Technologies and Environmental Policy*, *21*(2), 413–430. https://doi.org/10.1007/s10098-018-1644-x.

Vigon, B. W., Tolle, D. A., Cornaby, B. W., Latham, H. C., Harrison, C. L., Boguski, T. L., … Sellers, J. D. (1993). *Life-cycle assessment: Inventory guidelines and principles*. EPA/600/R-92/245.

Weidema, B., Wenzel, H., Petersen, C., & Hansen, K. (2004). *The product, functional unit, and reference flows in LCA*.

Weidema, B. P., & Wesnæs, M. S. (1996). Data quality management for life cycle inventories—An example of using data quality indicators. *Journal of Cleaner Production*, *4*(3–4), 167–174. https://doi.org/10.1016/S0959-6526(96)00043-1.

Wender, B. A., Foley, R. W., Hottle, T. A., Sadowski, J., Prado-Lopez, V., Eisenberg, D. A., … Seager, T. P. (2014). Anticipatory life-cycle assessment for responsible research and innovation. *Journal of Responsible Innovation*, *1*, 200–207. https://doi.org/10.1080/23299460.2014.920121.

Yang, Y., Tao, M., & Suh, S. (2018). Geographic variability of agriculture requires sector-specific uncertainty characterization. *International Journal of Life Cycle Assessment*, *23*(8), 1581–1589. https://doi.org/10.1007/s11367-017-1388-6.

Zimdars, C., Haas, A., & Pfister, S. (2018). Enhancing comprehensive measurement of social impacts in S-LCA by including environmental and economic aspects. *International Journal of Life Cycle Assessment*, *23*(1), 133–146. https://doi.org/10.1007/s11367-017-1305-z.

CHAPTER

Footprint tools

7

Yvonne Lewis[a] and Brett Cohen[a,b]

[a]*The Green House, Cape Town, South Africa* [b]*Department of Chemical Engineering, University of Cape Town, Cape Town, South Africa*

1 Introduction

There is no universal definition of a footprint indicator or tool, rather since its introduction in the 1990s the footprint concept has been applied to a wide range of approaches that aim to assess and communicate the extent of human activities that influence environmental sustainability in some way (Fang et al., 2016). The 'footprint' metaphor is effective and compelling, describing the mark humans leave on nature. It also has negative inferences that underscore the damage already done by humans on the planet (Box 1) and help communicate the sustainable development imperative to policy makers, corporate decision makers, and the general public.

Many footprints, including carbon, water, nitrogen, and others introduced in this chapter, address single environmental issues. Others, such as the Product Environmental Footprint (PEF), attempt to assess sustainability as a whole by covering multiple stressors on the environment. One commonality is that life cycle thinking often underpins the approaches. Increasingly the term 'footprint family' has been used to collectively describe the breadth of tools that identify themselves as 'footprints' (Fang et al., 2016). Although attempts have been made to develop a universal definition for footprints and standardise the concepts towards defining a single methodological framework and approach (Fang & Heijungs, 2015; Fang, Heijungs, & De Snoo, 2014; Ridoutt & Pfister, 2013), the scientific community has started to embrace the diversity of tools and approaches (even for footprints with the same name but completely different underlying logic) (Fang & Heijungs, 2015). The result is a classification scheme that maps members of the footprint family according to a number of key distinguishing features (Fang et al., 2016).

In this review of footprint tools, the classification scheme of Fang et al. (2016) is used in describing the various footprint tools in terms of:

- Scale: The scale of footprint analysis ranges from Global, National, Regional, Organisations, Industries, Products, all the way to Individual consumers.

Assessing Progress towards Sustainability. https://doi.org/10.1016/B978-0-323-85851-9.00014-6

Box 1

"A footprint is a mark one never meant to leave: a revealing clue in a garden, a blemish on an otherwise sparkling floor. It evokes both the weight of whoever left it and that being's ominous absence. Verbs popular among environmentalists include trample, tread, oppress and dominate. Heavy feet imply antagonism and the lack of intimacy between humans and the non-human world."

(Martin, 2014)

- Perspective: Two perspectives are common to footprint studies, namely a consumption-based approach and a production-based approach.
- Domain: Most footprints deal with environmental issues, but increasingly socio-economic footprint approaches are being developed (UNEP, 2020; Weidema, 2018).
- Group: Footprints typically deal with resources (raw materials extracted from nature or used to meet demand) or emissions (outputs to the environment as a result of activities). Resource-based footprints include land, water and material footprints, whereas examples of emission footprints are carbon, nitrogen and chemical footprints.
- Category: The category relates to whether or not the footprints sum the resource use or emissions generated in an inventory approach, or whether the inventory is further translated into a contribution to environmental impacts, drawing on life cycle assessment methodologies. For many footprints both inventory and impact versions have been developed.

Fig. 1 maps key environmental footprints according to group (resource vs. emission) and category (inventory oriented vs. impact oriented) based on (Fang et al., 2016; Fang & Heijungs, 2015). While most footprints can be classified as either emission or resource footprints, the ecological footprint is unique in that it includes measures of both emissions and resource consumption. The vertical arrows in the figure demonstrate the transition over time of footprint concepts from inventory oriented to impact oriented as attempts are made to overcome the simplifications inherent in inventory-oriented approaches.

In developing this review, bibliometric data was collated from the Web of Science to identify the footprint tools that have been most widely developed and applied. The number of journal articles and other publications per year per from 2001 to June 2021 is presented in Fig. 2, which also shows the increase in the use of footprint tools as reflected by academic publications.

2 Ecological footprint

The Ecological Footprint (EF) is recognised as the first of the footprint tools (Wackernagel & Rees, 1996). It was developed in the 1990s as one of the first measures of sustainability and gained popularity due to the fact that the footprint concept

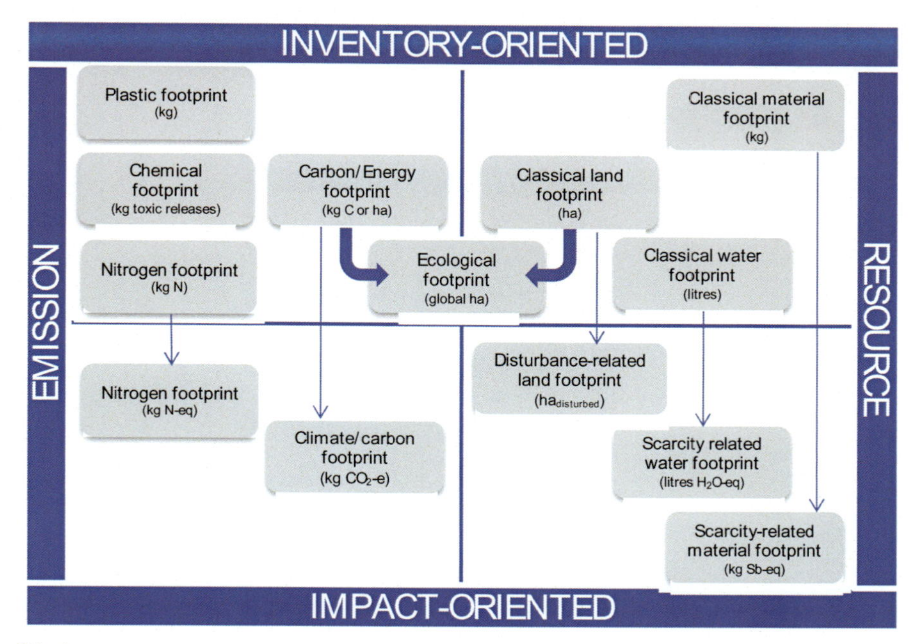

FIG. 1

Mapping of environmental footprints according to group (resource vs emission) and category (inventory oriented vs impact oriented).

was relatable, tangible, and meaningful. The footprint terminology was also well suited given that the ecological footprint is expressed in terms of hectares of land area. The methodology attempts to quantify all the productive land areas that are required to support a given population or produce a particular product. This includes cropland, grazing land, fishing grounds, and forests that provide natural resources as well as occupied or built-up land and the land area required to absorb carbon emissions that arise due to energy use (Lin et al., 2018). The resulting Ecological Footprint represents the demands on nature. On the supply side, the methodology allows comparison with the available biocapacity. This is a measure of a region's available land areas expressed in the same units of global hectares. By comparing the two values, it provides a measure of the degree to which society can remain within the regenerative capacity of the earth.

The global footprint network (GFN) and the EF approach have been taken up by policy makers at national and sub-national government level as well as non-governmental organisations (Matuštík & Kočí, 2021). Further characteristics are summarised in Table 1. The GFN publishes National Footprint Accounts, which presents global and country EF and biocapacities to highlight those countries with biological deficits or reserves (Lin et al., 2018). See Box 2. The global EF and biocapacity are expressed in different ways that add to their mainstream

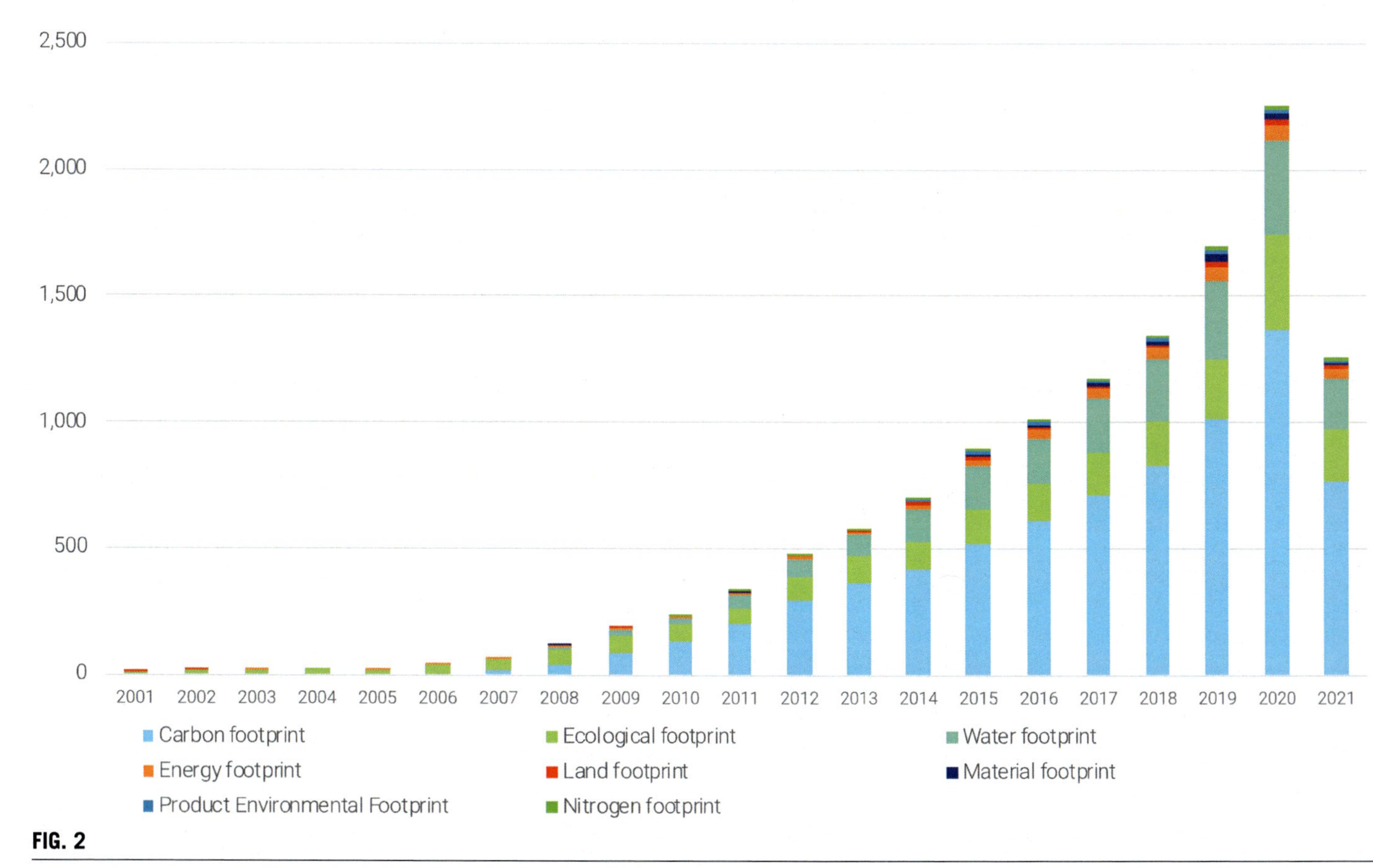

FIG. 2

Number of publications per year for key footprint tools identified on Web of Science.

Table 1 Ecological footprint characteristics.

Characteristic	Classification of ecological footprints[a]					
Scale	**Global**	**National**	**Regional**	**Organisation**	**Product**	**Individual**
Perspective	Consumption based			Production based		
Domain	Environment			Socio-economic		
Group	Resources			Emissions		
Category	Inventory			Impact		

Box 2 Ecological footprint calculators and tools

The Global Footprint Network has a personal ecological footprint calculator www.footprintcalculator.org.

The National Footprint Accounts: Ecological footprint and biocapacity data for all countries can be explored and downloaded from the data portal (data.footprintnetwork.org). Additional data can be purchased under licence through the footprint network. This includes country-specific consumption land-use matrices (https:/www.footprintnetwork.org/licenses/).

accessibility: the first relates to how many planets would be required to support current consumption levels and the second is identifying the day in the year when biocapacity equals the EF, namely earth overshoot day. In this way, the ecological footprint provides an intuitive and graphical tool for communicating a range of impacts, characteristics that have contributed to its widespread utilisation (Matuštík & Kočí, 2021).

3 Carbon footprint

The carbon footprint represents a measure of the impact of activities on the earth's climate (Table 2). A number of different measures of carbon footprint are identified (Alvarez et al., 2016). The simplest of these is an inventory of tonnes or kilogrammes of emissions of carbon dioxide (CO_2) and other greenhouse gases (GHGs), including methane, nitrous oxide, and others. Another expression of carbon footprint is as an impact indicator, where the emissions of gases other than CO_2 are expressed in carbon dioxide equivalents (CO_2-e) most commonly using 100-year global warming potentials, a measure of warming equivalency relative to CO_2. When all greenhouse gases are included, the term climate footprint is often used. A land-based carbon footprint has also been proposed as part of the ecological footprint methodology described before, being a measure of the forest land needed to sequester anthropogenic CO_2 emissions (Mancini et al., 2016).

Table 2 Carbon footprint characteristics.

Characteristic	Classification of carbon footprints[a]					
Scale	**Global**	**National**	**Regional**	**Organisation**	**Product**	**Individual**
Perspective	Consumption based			Production based		
Domain	Environment			Socio-economic		
Group	Resources			Emissions		
Category	Inventory			Impact		

[a]*The shaded fields in the table indicate the characteristics of carbon footprints.*

Carbon footprints have been successfully used in the communication of the issue of climate change, thanks in part to the ease of interpretation and the ability for the information to be readily communicated to a wide audience (Alvarez et al., 2016). However, the use of carbon footprint as a stand-alone measure of environmental performance has the danger of leading to undesirable trade-offs and problem shifting. For example, efforts to reduce carbon footprints through substitution of fossil diesel with biodiesel can have an unintended consequence of destroying forest or agricultural land to provide the requisite crops for production of biofuel (Alvarez et al., 2016). There are also a host of methodological choices which are made in the calculation of carbon footprints which need to be clearly communicated and need to be taken into account when attempting to compare results. These include the system boundaries and data used in the analysis (life cycle, input–output data, or point source emissions), which IPCC Assessment Report is used to obtain the Global Warming Potentials to convert emissions of other gases into CO_2 equivalents and how biogenic carbon is accounted for. Further standardisation problems link to different standards being available for different scales (see Table 3).

At the territorial level, production-based GHG accounting is often criticised as it cannot capture carbon leakage (Ottelin et al., 2019). The alternative, consumption-based accounting, can be used to overcome these issues, although methodologies for consumption-based accounting are equally challenged by methodological issues, with little consensus on how such quantification is to be approached (Heinonen et al., 2020).

4 Water footprint

The Water Footprint was introduced in 2002 to represent an 'indicator of water use behind all the goods and services consumed by one individual or the individuals of a country' (Hoekstra, 2017). Although the water footprint, like the ecological footprint, initially focused on national scales, the scope was later broadened to cover other scales of analysis, including companies, products, and individual footprints (Hoekstra, 2017). The Water Footprint Network (WFN) (waterfootprint.org) led

Table 3 Carbon footprint standards and tools.

Standard/tool	Comment	Web link
International Standards Organisation (ISO) 14067:2018	Guides the calculation of product carbon footprints	https://www.iso.org/home.html
International Standards Organisation (ISO) 14064-1:2018	Guides the calculation organisational GHG inventories	https://www.iso.org/home.html
Greenhouse Gas Protocol	A range of easy-to-use standards, based in part on the ISO standards, for calculating organisational, product, and value chain carbon footprints	https://ghgprotocol.org
Online calculators	Personal and household carbon footprint each requiring differing levels of inputs and based on different assumptions	For example https://www.nature.org/en-us/get-involved/how-to-help/carbon-footprint-calculator/, https://footprint.wwf.org.uk/#/, https://www3.epa.gov/carbon-footprint-calculator/
Life cycle analysis	Allows for the calculation of carbon footprint of products and services as one of the impact categories	Various LCA tools available in the public domain
CCaLC	Carbon footprinting of industrial activities and supply chains	http://www.ccalc.org.uk

the development of a Global Water Footprint Assessment Standard (Hoekstra et al., 2011). In parallel to the standard developed by the WFN, the LCA community has developed similar methodologies for life cycle water accounting, resulting in the development of the ISO 14046:2014 water footprint standard (Box 3) (Table 4).

A water footprint is measured as the volume of water appropriated for a certain purpose, making it unavailable for another purpose (Hoekstra, 2017). Water footprints are divided into three categories. The 'green water footprint' measures rainwater that is evapo-transpired or incorporated by plants; the 'blue water footprint' measures the surface- or ground-water withdrawn from a water body; while the 'grey water footprint' measures a hypothetical amount of water necessary for assimilation of discharged pollution (Aldaya, 2012). The quantification of a water footprint does, however, need to be put in a broader context, and a life cycle assessment approach is used to guide the analysis. Given that water-use impacts are largely related to local and temporal water availability, LCA methodologies usually compare the volumetric water footprint to local scarcity factors, giving rise to the term 'water scarcity footprint' (Boulay et al., 2018).

Box 3 Water footprint standards, calculators, and tools

The Global Water Footprint Assessment Standard was developed by the Water Footprint Network to support assessment of water footprints (Hoekstra et al., 2011). The Water Footprint is also defined by an ISO standard: ISO 14046:2014.

The life cycle initiative established the WULCA working group focusing on water-use assessment and water footprinting from a life cycle perspective (https:/wulca-waterlca.org).

The water footprint network (waterfootprint.org) provides a simplified and extended **personal water footprint online calculator**.

The water footprint network also provides a **national water footprint explorer**, which is the water footprint of national consumption (i.e. the total amount of fresh water used to produce the goods and services consumed by the country's inhabitants). There is also a **product gallery**, which summarises the water footprint of common food products.

The water footprint assessment tool (www.waterfootprintassessmenttool.org) is a free online web application that provides the blue, green, and grey water footprint, an assessment of water pollution and blue water scarcity both in tabular and map form. Currently the tool can be used for countries, water basins, and the world, with raw materials, products, and organisations to follow.

The underlying data for these tools is available on the **waterstat water footprint database** (https:/waterfootprint.org/en/resources/waterstat/)

Table 4 Water footprint characteristics.

Characteristic	Classification of water footprints[a]					
Scale	**Global**	**National**	**Regional**	**Organisation**	**Product**	**Individual**
Perspective	Consumption based			Production based		
Domain	Environment			Socio-economic		
Group	Resources			Emissions		
Category	Inventory			Impact		

[a]*The shaded fields in the table indicate the characteristics of water footprints.*

5 Nitrogen and phosphorous footprints

Levels of reactive nitrogen in the atmosphere have grown dramatically as a result of human-based activities, including agriculture, transport, and industrial processes. Once this nitrogen has entered the environment, it moves through the atmosphere, forests, grasslands, and waters, resulting in a host of environmental changes, including smog, acid rain, forest dieback, coastal 'dead zones', biodiversity loss, stratospheric ozone depletion, and an enhanced greenhouse effect. These all impact negatively on both ecosystems and humans (Galloway et al., 2003).

The nitrogen footprint was thus developed to measure the contribution of people products, organisations, and countries to reactive nitrogen levels in the environment

Box 4 Nitrogen footprint calculators and tools

A personal **Nitrogen footprint calculator** is available for selected countries (www.n-print.org). The tool can be scaled for use by communities, organisations, or other countries, while an institution-level nitrogen footprint tool is under development.

(Einarsson & Cederberg, 2019; Galloway et al., 2014; Leach et al., 2012). The nitrogen footprint has been defined as the "total amount of Nr [reactive nitrogen, all other forms than N_2 released to the environment as a result of [...] consumption" (Leach et al., 2012).

The first tool for calculation of nitrogen footprint, the N-calculator (Box 4), was developed by Leach et al. (2012). This calculator allows an individual to assess their contribution to nitrogen releases into the environment from food consumption and production, housing (utility usage), transportation, and goods and services. N-Calculators are available for the United States and the Netherlands (Galloway et al., 2014; Leach et al., 2012). Various specific applications have been explored, for example those that focus mainly on diet (Galloway et al., 2014), nitrogen use efficiency (Hutchings et al., 2020), or nitrogen embedded in international trade (Oita et al., 2016).

One of the critiques of the nitrogen footprint is that it aggregates emissions in different locations, at different times, and in different forms. This makes understanding the environmental relevance of such releases challenging, an issue which is being worked on by the LCA community (Einarsson & Cederberg, 2019). Other challenges include data availability, how to discriminate nitrogen species, system boundaries, and the transformation of Nr through nitrification (Einarsson & Cederberg, 2019).

A similar concept exists for measuring phosphorous footprint. Although nitrogen and phosphorus use is largely linked, they are typically calculated in separate footprints (Oita et al., 2020). As for nitrogen, phosphorous is introduced into the environment through agriculture and industrial processes, with a build-up of excessive phosphorous resulting in multiple negative environmental impacts, notably on water bodies causing algal blooms. While a range of nitrogen footprint tools have been developed, fewer tools have been developed for calculation of phosphorous footprints (Metson et al., 2020). Those that are available have been developed by Metson et al. (2020) and Li, Wiedmann, and Hadjikakou (2019).

6 Product environmental footprint (PEF)

Life cycle approaches to determining the impact of a product (or service) are based on generic or sector-specific guidelines, and are potentially subject to different outcomes being calculated by different parties depending on choices made about the goal, scope, and assessment approaches. The Product Environmental Footprint (PEF) approach was developed by the European Union in recognition of the need

Table 5 Product environmental footprint characteristics.

Characteristic	Classification of product environmental footprints[a]					
Scale	**Global**	**National**	**Regional**	**Organisation**	**Product**	**Individual**
Perspective	Consumption based			Production based		
Domain	Environment			Socio-economic		
Group	Resources			Emissions		
Category	Inventory			Impact		

[a]*The shaded fields in the table indicate the characteristics of product environmental footprints.*

for a universally comparable life cycle standard for individual products that was more specific than generic Life Cycle Assessments or sector-related standards. Characteristics of product environmental footprints are summarised in Table 5.

A PEF is a measure of the environmental performance of a good or service on a life cycle basis, from extraction of raw materials, through production and use, to final waste management. It provides a basis for understanding the environmental impacts of goods and services, across their life cycles (EC-JRC, 2012). A total of 16 impact categories can be used in the calculation of the PEF profile, if relevant to a product group. These impact categories are climate change, ozone depletion, human toxicity (cancer and non-cancer), particulate matter, ionising radiation, photochemical ozone formation, acidification, eutrophication (terrestrial, freshwater, and marine), freshwater ecotoxicity, land use, water use, and resource use (minerals and metals, and fossil fuels) (European Commission, 2017).

PEFs are being developed for a wide range of products from bottled water to dog food to computer cables. Related guidelines have also been developed for determining the environmental impacts of organisations, known as Organisational Environmental Footprints (OEFs).

The intention of the PEF approach is thus to increase the reproducibility and comparability of impacts calculated by different parties for individual products, through decreasing the flexibility in calculation approaches (European Commission, 2017; Bach et al., 2018). The audience for PEFs includes producers to identify opportunities for changes in production to reduce key impacts, consumers and customers wishing to compare products against alternatives and benchmarks, and policy makers seeking to implement policies to reduce environmental impacts in specific areas.

Product Environmental Footprint Category Rules (PEFCRs) have been developed which describe a set of rules, requirements, and guidelines for calculating PEFs. Through following the Category Rules, it can be ensured that a common way of measuring life cycle environmental performance is applied (EC-JRC, 2012; European Commission, 2017). A comprehensive description of how PEFs are calculated for individual product categories is presented by the EC-JRC (2012).

Given that PEFs are based on a life cycle approach, the tools used in Life Cycle Assessment (LCA) are relevant here. Life Cycle Assessment software package developers have produced an environmental footprint database for use in developing PEFs (e.g. https://simapro.com/2019/using-the-environmental-footprint-database-in-simapro/). The data requirements for developing PEFs are aligned with those required for conducting LCAs, including water, energy, and materials inputs at all stages in the value chain.

In parallel with the development of PEFs, the EU has also developed rules for the calculation of Organisational Environmental Footprints, to guide the measurement of the environmental performance of organisations from a life cycle perspective (Rimano et al., 2019).

7 Other footprint tools

The previous sections have presented the footprint tools that have been most widely developed and applied (see Fig. 2). This section presents a short description of a number of the less widely used footprint tools.

7.1 Energy footprint

A significant contributor to global warming is the increase in CO_2 emissions in the atmosphere from the energy sector. The energy footprint represents a subset of the ecological footprint and was originally defined to be the equivalent amount of forest land required to absorb CO_2 emissions from fossil fuel combustion and electricity generation for a product, organisation sub-national area, country, region, or the planet, over a particular time period, calculated using a global average forest sequestration potential. An alternative definition is the sum of areas required to sequestrate CO_2 emissions from consumption of non-food and non-feed energy (Fang, Heijungs, & de Snoo, 2013). A further definition extends the coverage to include land coverage for hydropower production, forests for fuelwood, and cropland for fuel crops (Global Footprint Network, 2010). Subsequent to the various definitions being presented, little academic literature is found where energy footprint is calculated in this way. The term is now more commonly used to describe the electricity required, in kWh for example, for the production of a product or delivery of a service, often on a life cycle basis.

Given the limited application of energy footprint as measured in terms of land area, no tools or software packages exist that have been custom designed for its calculation. However, calculation of CO_2 sequestration potential of forests is well documented, with the Intergovernmental Panel on Climate Change's guidelines providing a widely used reference for this purpose (IPCC, 2006). In calculating the energy footprint, it would first be necessary to determine the CO_2 emissions from energy consumption, and the land area required to absorb this CO_2 could be calculated using the IPCC guidelines.

If, however, an energy footprint measured in kWh is the required metric, a plethora of tools exist to calculate the energy usage for a product or process. For calculating emissions on a life cycle basis, the use of one of the commercially available software packages can be beneficial, given the ability to draw on in-built background data sets.

7.2 Land footprint

Productive land is a scarce global resource. One of the challenges associated with assessing the impacts on productive land of human-based activities is that global trade means that the location of consumption is often remote from the location of production. The land footprint thus attempts to bridge the gap between production and consumption (Matuštík & Kočí, 2021; O'Brien, Schütz, & Bringezu, 2015). Various calculation approaches that are commonly used for calculating other footprints have also been used to determine land footprints, including Multi-Regional Input–Output analysis (MRIO), material flow analysis, and life cycle assessment (Perminova et al., 2016). Different assessments also use different units of measure. This leads to challenges in comparing results from different studies. While the land footprint can provide insights into, for example, how consumption in developed countries drives the land use in developing countries or limits of bioeconomy, the land-use effect on the environment is not always clear and linear, since soil and climate characteristics, land-use type, and intensity of use vary between locations (Matuštík & Kočí, 2021).

7.3 Material footprint

The material footprint provides a measure of the use of materials from the perspective of consumption. This is achieved through allocating globally extracted and used raw materials to domestic final demand (Wiedmann et al., 2015). Four material categories are considered: metal ores, non-metallic minerals, fossil fuels, and biomass (crops, crop residues, wood, wild fish catch, etc.). As with the land footprint, the material footprint demonstrates the disjunct between production and consumption, and the shifting of impacts between developed and developing countries, through international trade and consumption (Giljum, Bruckner, & Martinez, 2015). The material footprint is typically reported as total mass of materials, rather than weighting materials according to scarcity, and is calculated using Multi-Regional Input–Output (MRIO) approaches (Giljum et al., 2015). To overcome this limitation, a resource depletion footprint, where material footprint is weighted by depletion potentials, has been proposed by (Fang & Heijungs, 2014).

7.4 Chemical and ozone footprint

The chemical footprint presents a measure of chemical substances released into the environment, which in turn can lead to impacts including those of ecotoxicity and human toxicity. An extensive list of chemical substances is considered, including pesticides and heavy metals (Sala & Goralczyk, 2013). An ozone footprint provides a measure of emission of gases that fall under the Montreal Protocol for Ozone Depleting Substances. Given that N_2O, which is a major ozone-depleting gas, is not included in the Montreal protocol, the nitrogen footprint is complementary to the ozone footprint in assessing the planetary boundary for stratospheric ozone depletion (Meyer & Newman, 2018).

7.5 Biodiversity footprint

Human activities, including fragmentation of habitats, overpopulation, chemical pollution, invasive species, and over-exploitation of resources in hunting and fishing, threaten species on the planet, leading to species becoming extinct at a far greater rate than the background rate (Ceballos et al., 2015). The biodiversity footprint is used to provide a measure of the impact of human activities on biodiversity in a similar fashion to a carbon or energy footprint. There is no single agreed methodology for determining a biodiversity footprint, and a range of biodiversity footprinting tools is available, with the choice of tool being determined by the purpose of the assessment, sector, and data availability (Goedicke, Sabag, & Keijzer, 2020).

The Millennium Ecosystem Assessment (MEA) identifies five drivers of biodiversity loss, namely land occupation and transformation, pollution, climate change, species overexploitation, and invasive species. Studies to determine biodiversity impact have typically focused on a single indicator, with the most common being impacts on number or composition of species, and have focused on different scales of analyses. However, comprehensive assessments of biodiversity impact should consider all of the aspects of biodiversity and impacts at different scales (Marquardt et al., 2019).

When examining biodiversity footprint from a life cycle perspective, it is highlighted that existing LCA tools have historically considered impacts of products and services on biodiversity in terms of three of the five drivers, namely land occupation and transformation, pollution, and climate change. More recent work has sought to develop a product biodiversity footprint which also includes species overexploitation and invasive species. The Product Biodiversity Footprint thus seeks to bridge the gap between LCA and ecology (Asselin et al., 2020).

Similar to the other footprinting methodologies presented, the biodiversity footprint can be used by policy and decision makers and planners to guide implementation of measures to reduce supply chain, product or site level biodiversity impacts; for guiding investors, for ranking and benchmarking, for voluntary disclosure, and for risk assessments (Goedicke et al., 2020).

7.6 Waste absorption footprint

Waste products from human activity are either disposed of to land or incinerated, and liquid wastes are disposed of to water bodies. The ecological footprint considers the land area required to absorb carbon dioxide emissions related to energy use but does not consider the land (or water) area required to assimilate other wastes. The Waste

Absorption Footprint was therefore introduced to address this shortcoming and provide a measure of the impact of human demand on the biosphere's capacity to assimilate wastes. Waste emissions are translated into absorptive land and water areas, and the results are expressed in units of average hectares, through scaling different land-use types in proportion to their relative absorptive capacity. The approach covers a variety of wastes from human activity, rather than being restricted to any one kind of waste (Jiao et al., 2013). The waste absorption footprint is considered a subset or extension of the ecological footprint concept (Jiao et al., 2015). If nitrogen and phosphorous releases to water are included in the waste absorption footprint, then there would be an overlap with nitrogen and phosphorous footprints if quantified (Gavrilescu, Teodosiu, & David, 2020).

7.7 Plastic footprint

Plastic pollution is increasingly being recognised as a significant environmental challenge, the long-term impacts of which are only starting to be understood (Ali et al., 2021; Kurniawan et al., 2021). Without an understanding of the multiple pathways that plastics take in the environment, from estimating losses, how plastics breakdown into macro- and micro-plastics, and the effects on ecosystems and humans, means that an inventory-oriented plastic footprint is a necessary default (Boucher et al., 2019, 2020). A number of plastic footprint tools and approaches have been developed to quantify the scale of the impact and raise awareness (Boucher et al., 2019). The Plastic Leaks Project is a methodology that quantifies plastic releases into the environment across of a product or business across regions and life cycle stages. It covers both micro- and macro-plastics (Peano et al., 2020). The marine plastic footprint methodology is also advanced but focuses specifically on developing an inventory of plastic flows into the marine environment (Boucher et al., 2020). Other projects and working groups have been established to develop the life cycle impact assessment methodology for plastics in the environment.

8 Conclusions

Our understanding of sustainability, and indeed the metrics used to measure our progress towards sustainability, is constantly evolving, expanding, and developing. This chapter has presented an overview of the application of environmental footprinting as a framework for calculating and communicating single or multiple impacts associated with the provision of goods and services; the activities of individuals, companies and institutions, regions and countries; and for society as a whole. Moving forward, some scientific researchers support the development of integrated and harmonised tools to overcome some of the challenges associated with single-issue footprints, whereas others lean towards classification as a means to retain the value of simpler easy-to-understand footprints as part of a footprint family. The latter approach is particularly important when new issues that impact on sustainability arise that require urgent action.

References

Aldaya, M. M. (2012). *The water footprint assessment manual*. https://doi.org/10.4324/9781849775526.

Ali, S. S., et al. (2021). Degradation of conventional plastic wastes in the environment: A review on current status of knowledge and future perspectives of disposal. *Science of the Total Environment*, *771*, 144719. https://doi.org/10.1016/j.scitotenv.2020.144719.

Alvarez, S., et al. (2016). Strengths-weaknesses-opportunities-threats analysis of carbon footprint indicator and derived recommendations. *Journal of Cleaner Production*, *121*. https://doi.org/10.1016/j.jclepro.2016.02.028.

Asselin, A., et al. (2020). Product biodiversity footprint—A novel approach to compare the impact of products on biodiversity combining life cycle assessment and ecology. *Journal of Cleaner Production*, *248*, 119262. https://doi.org/10.1016/j.jclepro.2019.119262.

Bach, V., et al. (2018). Product environmental footprint (PEF) pilot phase-comparability over flexibility? *Sustainability (Switzerland)*, *10*(8), 2898. https://doi.org/10.3390/su10082898.

Boucher, J., et al. (2019). *Review of plastic footprint methodologies: Laying the foundation for the development of a standardised plastic footprint measurement tool*. Gland, Switzerland: IUCN. https://doi.org/10.2305/iucn.ch.2019.10.en.

Boucher, J., et al. (2020). *The marine plastic footprint: Towards a science-based metric for measuring marine plastic leakage and increasing the materiality and circularity of plastic*. Gland, Switzerland: IUCN. https://doi.org/10.2305/iucn.ch.2020.01.en.

Boulay, A. M., et al. (2018). The WULCA consensus characterization model for water scarcity footprints: Assessing impacts of water consumption based on available water remaining (AWARE). *International Journal of Life Cycle Assessment*, *23*(2), 368–378. https://doi.org/10.1007/s11367-017-1333-8.

Ceballos, G., et al. (2015). Accelerated modern human-induced species losses: Entering the sixth mass extinction. *Science Advances*, *1*(5), e1400253. https://doi.org/10.1126/sciadv.1400253.

EC-JRC. (2012). *Product environmental footprint (PEF) guide*. European Commission Joint Research Centre. Available at https://ec.europa.eu/environment/eussd/pdf/footprint/PEFmethodologyfinaldraft.pdf. (Accessed 16 March 2021).

Einarsson, R., & Cederberg, C. (2019). Is the nitrogen footprint fit for purpose? An assessment of models and proposed uses. *Journal of Environmental Management*, *240*, 198–208. https://doi.org/10.1016/j.jenvman.2019.03.083.

European Commission. (2017). *PEFCR guidance document,—Guidance for the development of product environmental footprint category rules (PEFCRs), version 6.3*.

Fang, K., & Heijungs, R. (2014). Moving from the material footprint to a resource depletion footprint. *Integrated Environmental Assessment and Management*, *10*, 596–598. https://doi.org/10.1002/ieam.1564.

Fang, K., & Heijungs, R. (2015). Investigating the inventory and characterization aspects of footprinting methods: Lessons for the classification and integration of footprints. *Journal of Cleaner Production*, *108*, 1028–1036. https://doi.org/10.1016/j.jclepro.2015.06.086.

Fang, K., Heijungs, R., & de Snoo, G. (2013). The footprint family: Comparison and interaction of the ecological, energy, carbon and water footprints. *Revue de Métallurgie*, *110*(1), 77–86. https://doi.org/10.1051/metal/2013051.

Fang, K., Heijungs, R., & De Snoo, G. R. (2014). Theoretical exploration for the combination of the ecological, energy, carbon, and water footprints: Overview of a footprint family. *Ecological Indicators*, *36*, 508–518. https://doi.org/10.1016/j.ecolind.2013.08.017.

Fang, K., et al. (2016). The footprint's fingerprint: On the classification of the footprint family. *Current Opinion in Environmental Sustainability*, *23*, 54–62. https://doi.org/10.1016/j.cosust.2016.12.002.

Galloway, J. N., et al. (2003). The nitrogen cascade. *Bioscience*, *53*, 341–356. https://doi.org/10.1641/0006-3568(2003)053[0341:TNC]2.0.CO;2.

Galloway, J. N., et al. (2014). Nitrogen footprints: Past, present and future. *Environmental Research Letters*, *9*(11), 115003. https://doi.org/10.1088/1748-9326/9/11/115003.

Gavrilescu, D., Teodosiu, C., & David, M. (2020). Environmental assessment of wastewater discharges at river basin level by means of waste absorption footprint. *Sustainable Production and Consumption*, *21*, 33–46. https://doi.org/10.1016/j.spc.2019.10.006.

Giljum, S., Bruckner, M., & Martinez, A. (2015). Material footprint assessment in a global input-output framework. *Journal of Industrial Ecology*, *19*(5). https://doi.org/10.1111/jiec.12214.

Global Footprint Network. (2010). *Ecological footprint atlas*.

Goedicke, R., Sabag, O., & Keijzer, M. (2020). *A compass for navigating the world of biodiversity footprinting tools: An introduction for companies and policy makers*. Amsterdam: IUCN. Available at https://www.iucn.nl/files/publicaties/a_compass_for_navigating_biodiversity_footprint_tools_-_final_1.pdf.

Heinonen, J., et al. (2020). Spatial consumption-based carbon footprint assessments—A review of recent developments in the field. *Journal of Cleaner Production*, *256*, 120335. https://doi.org/10.1016/j.jclepro.2020.120335.

Hoekstra, A. Y. (2017). Water footprint assessment: Evolvement of a new research field. *Water Resources Management*, *31*(10), 3061–3081. https://doi.org/10.1007/s11269-017-1618-5.

Hoekstra, A. Y., et al. (2011). *The water footprint assessment manual: Setting the global standard*. London, UK: Earthscan.

Hutchings, N. J., et al. (2020). Measures to increase the nitrogen use efficiency of European agricultural production. *Global Food Security*, *26*, 100381. https://doi.org/10.1016/j.gfs.2020.100381.

IPCC. (2006). *2006 IPCC guidelines for national greenhouse gas inventories*. Available at: https://www.ipcc-nggip.iges.or.jp/public/2006gl/.

Jiao, W., et al. (2013). The waste absorption footprint (WAF): A methodological note on footprint calculations. *Ecological Indicators*, *34*, 356–360. https://doi.org/10.1016/j.ecolind.2013.05.024.

Jiao, W., et al. (2015). Evaluating environmental sustainability with the waste absorption footprint (WAF): An application in the Taihu Lake Basin, China. *Ecological Indicators*, *49*, 39–45. https://doi.org/10.1016/j.ecolind.2014.09.032.

Kurniawan, S. B., et al. (2021). Current state of marine plastic pollution and its technology for more eminent evidence: A review. *Journal of Cleaner Production*, *278*, 123537. https://doi.org/10.1016/j.jclepro.2020.123537.

Leach, A. M., et al. (2012). A nitrogen footprint model to help consumers understand their role in nitrogen losses to the environment. *Environmental Development*, *1*(1), 40–66. https://doi.org/10.1016/j.envdev.2011.12.005.

Li, M., Wiedmann, T., & Hadjikakou, M. (2019). Towards meaningful consumption-based planetary boundary indicators: The phosphorus exceedance footprint. *Global Environmental Change*, *54*, 227–238. https://doi.org/10.1016/j.gloenvcha.2018.12.005.

Lin, D., et al. (2018). Ecological footprint accounting for countries: Updates and results of the national footprint accounts, 2012-2018. *Resources*, *7*(3), 2012–2018. https://doi.org/10.3390/resources7030058.

Mancini, M. S., et al. (2016). Ecological footprint: Refining the carbon footprint calculation. *Ecological Indicators*, *61*, 390–403. https://doi.org/10.1016/j.ecolind.2015.09.040.

Marquardt, S. G., et al. (2019). Consumption-based biodiversity footprints—Do different indicators yield different results? *Ecological Indicators*, *103*, 461–470. https://doi.org/10.1016/j.ecolind.2019.04.022.

Martin, L. J. (2014). *Is a footprint the right metaphor for ecological impact ?*. Scientific American.

Matuštík, J., & Kočí, V. (2021). What is a footprint? A conceptual analysis of environmental footprint indicators. *Journal of Cleaner Production*, *285*, 124833. https://doi.org/10.1016/j.jclepro.2020.124833.

Metson, G. S., et al. (2020). The U.S. consumer phosphorus footprint: Where do nitrogen and phosphorus diverge? *Environmental Research Letters*, *15*(10), 105022. https://doi.org/10.1088/1748-9326/aba781.

Meyer, K., & Newman, P. (2018). The planetary accounting framework: A novel, quota-based approach to understanding the impacts of any scale of human activity in the context of the planetary boundaries. *Sustainable Earth*, *1*(1), 4. https://doi.org/10.1186/s42055-018-0004-3.

O'Brien, M., Schütz, H., & Bringezu, S. (2015). The land footprint of the EU bioeconomy: Monitoring tools, gaps and needs. *Land Use Policy*, *47*(September 2015), 235–246. https://doi.org/10.1016/j.landusepol.2015.04.012.

Oita, A., et al. (2016). Substantial nitrogen pollution embedded in international trade. *Nature Geoscience*, *9*(2), 111–115. https://doi.org/10.1038/ngeo2635.

Oita, A., et al. (2020). Trends in the food nitrogen and phosphorus footprints for Asia's giants: China, India, and Japan. *Resources, Conservation and Recycling*, *157*, 104752. https://doi.org/10.1016/j.resconrec.2020.104752.

Ottelin, J., et al. (2019). What can we learn from consumption-based carbon footprints at different spatial scales? Review of policy implications. *Environmental Research Letters*, *14*, 093001. https://doi.org/10.1088/1748-9326/ab2212.

Peano, L., et al. (2020). *Plastic leak project methodological guidelines*. https://doi.org/10.46883/onc.3404.

Perminova, T., et al. (2016). Methods for land use impact assessment: A review. *Environmental Impact Assessment Review*, *60*, 64–74. https://doi.org/10.1016/j.eiar.2016.02.002.

Ridoutt, B. G., & Pfister, S. (2013). Towards an integrated family of footprint indicators. *Journal of Industrial Ecology*, *17*(3), 337–339.

Rimano, M., et al. (2019). Life cycle approaches for the environmental impact assessment of organizations: Defining the state of the art. *Administrative Sciences*, *9*(4), 94. https://doi.org/10.3390/admsci9040094.

Sala, S., & Goralczyk, M. (2013). Chemical footprint: a methodological framework for bridging life cycle assessment and planetary boundaries for chemical pollution. *Integrated Environmental Assessment and Management*, *9*(4), 623–632. https://doi.org/10.1002/ieam.1471.

UNEP. (2020). In C. Benoît Norris, et al. (Eds.), *Guidelines for social life cycle assessment of products and organizations 2020* United National Environment Programme (UNEP). Available at: http://www.unep.fr/shared/publications/pdf/DTIx1164xPA-guidelines_sLCA.pdf.

Wackernagel, M., & Rees, W. (1996). *Our ecological footprint: Reducing human impact on the earth*. Gabriola Island, BC, Canada: New Society Publishers.

Weidema, B. P. (2018). The social footprint—A practical approach to comprehensive and consistent social LCA. *The International Journal of Life Cycle Assessment*, *23*(3), 700–709. https://doi.org/10.1007/s11367-016-1172-z.

Wiedmann, T. O., et al. (2015). The material footprint of nations. *Proceedings of the National Academy of Sciences of the United States of America*, *112*(20), 6271–6276. https://doi.org/10.1073/pnas.1220362110.

CHAPTER

The combined use of life cycle assessment and data envelopment analysis to analyse the environmental efficiency of multi-unit systems

Jara Laso[a], Jorge Cristóbal[a], María Margallo[a], Rubén Aldaco[a], and Ian Vázquez-Rowe[b]

[a]*Department of Chemical and Biomolecular Engineering, University of Cantabria, Santander, Spain* [b]*Peruvian LCA and Industrial Ecology Network (PELCAN), Department of Engineering, Pontificia Universidad Católica del Perú, San Miguel, Lima, Peru*

1 The importance of eco-efficiency in sustainable development

The use of composite indexes, which merge or combine different assessment methods, has been demanded to reduce the complexity in the interpretation of sustainability results, but at the same it is difficult to draw any detailed conclusions from aggregated scores (Kalbar, Birkved, Nygaard, & Hauschild, 2017). For instance, in recent years Diaz-Sarachaga, Jato-Espino, and Castro-Fresno (2018) developed the Sustainable Development Goals Index (SDGs Index), which comprises 99 indicators for measuring the degree of sustainability in more than 150 countries. Thereafter, Jabbari, Shafiepour Motlagh, Ashrafi, and Abdoli (2020) proposed the use of the Development Index (DEVI), which has a high correlation with the Human Development Index.

Nevertheless, the aggregation of indicators is not always possible since conflicting objectives and criteria can occur in the decision-making process (Kumar et al., 2017). For these cases, multiple criteria decision-making methods (MCDMs) help to find optimal results in complex scenarios including various indicators to meet technical and market-related constraints, whilst maximising profit margins and boosting competitiveness (Guillén-Gosálbez, You, Galán-Martín, Pozo, & Grossmann, 2019). In fact, the evaluation of sustainability is often treated as a complex multi-criteria

Assessing Progress towards Sustainability. http://doi.org/10.1016/B978-0-323-85851-9.00008-0

problem due to its multidimensional nature and the trade-off between scientific objectivity and policy relevant values (Opon & Henry, 2020). MCDMs show two branches: multiple criteria optimisation and multi-criteria decision analysis (MCDA). MCDA deals with multiple criteria problems that have a small number of alternatives often in an environment of uncertainty, whereas multiple criteria optimisation is directed at problems formulated within a mathematical programming framework, but with several objectives instead of just one. MCDMs have evolved considerably, especially during the past four decades. Whilst in the 1970s multiple objective optimisation was focused on theories of linear programming, from the 1990s the discipline progressed to fuzzy multiple objective programming, multiple criteria heuristics, evolutionary algorithms in multiple criteria optimisation, or the integration of data envelopment analysis (DEA) (Ehrgott & Gandibleux, 2003).

DEA is included within MCDMs but is used in specific cases, for performance evaluation or benchmarking, where no subjective inputs are required. In fact, DEA supports decisions in efficiency (or inefficiency) alternatives (Nakayama, Arakawa, & Yun, 2003). Measuring the efficiency of an industry has an enormous interest for organisations to improve its productivity. However, solving this complex problem has required several attempts. In the late 1950s, it was not possible to find the answer to combine multiple inputs into any satisfactory measure of efficiency. In the late 1970s, the development of the DEA methodology allowed measuring the relative efficiency of multi-input multi-output units (Cook & Seiford, 2009).

DEA has evolved through time increasing the application to sustainability research. DEA methods used in this field can be classified into several groups, highlighting: (i) traditional DEA models including Cooper and Rhodes (CCR) (Charnes, Cooper, & Rhodes, 1978) and the Banker, Charnes, and Cooper (BCC) models; (ii) slack-based models (SBM) and intertemporal DEA models, especially the DEA-based Malmquist productivity index; (iii) models handling special types of data such as fuzzy, ordinal, qualitative, negative data; (iv) two-stage network DEA; and (v) two-stage contextual factor evaluation. However, researchers have developed a new combined methodology in which the optimisation of multiple units attained through DEA can be merged with life cycle methodologies, namely life cycle assessment (LCA), in order to maximise certain advantages of the two methods, allowing a more holistic environmental evaluation of production systems that enables the quantification of environmental benchmarks and cleaner production thresholds. In this context, the upcoming sections of this chapter will focus on providing a review on current practices of the joint application of LCA + DEA, as well as point towards the methodological challenges and opportunities for the future (Zhou, Yang, Chen, & Zhu, 2018).

2 The LCA + DEA method

The combination of LCA and DEA methodologies has been employed over the past years to assess the eco-efficiency of a wide spectrum of production systems (Rebolledo-Leiva, Angulo-Meza, Iriarte, González-Araya, & Vásquez-Ibarra, 2019). On the one hand, LCA is a standardised commonly used tool that allows the evaluation of the environmental performance of products, processes, and/or services (ISO, 2006). It provides a holistic view of the potential environmental impacts since it considers the whole life cycle of the system under study (Aldaco et al., 2019). On the other hand, DEA is a well-established methodology for non-parametrically quantifying the relative efficiency of multiple homogeneous units usually referred to as Decision-Making Units (DMUs) (Charnes et al., 1978). The joint application of LCA and DEA makes it possible to benefit from the potential advantages of each methodology, as well as to minimise the weaknesses and limitations of both methodologies. Fig. 1 displays the main methodological aspects to be considered in an LCA + DEA study and most widely applied LCA + DEA methods.

Considering that in most cases, all the information obtained from an LCA will be used in the DEA computational phase, it is necessary to correctly define each methodological aspect. Hence, the goal and scope stage of an LCA study must define a functional unit to which all the data collected and the results estimated are referred, as well as the system boundaries, since the overall results depend on their definition (Laso, Hoehn, et al., 2018; Laso, Vázquez-Rowe, Margallo, Irabien, & Aldaco, 2018).

The life cycle inventory (LCI) is the most time-consuming step in LCA studies (Laso, Margallo, et al., 2018; Laso, Vázquez-Rowe, et al., 2018). This is mainly due to two reasons: data availability is sometimes limited and the necessity to collect high quality data (Iribarren, Vázquez-Rowe, Moreira, & Feijoo, 2010). The latter involves gathering data from a high number of similar facilities to ensure sample representativeness for a particular system. In a conventional LCA study, the common way to handle inputs/outputs from a high number of facilities is to estimate an LCI from the average input and output values. However, the high degree of variability of the data could be a barrier. Another option is to perform individual LCA studies for each LCI, but results interpretation may not be straightforward (Vázquez-Rowe, Iribarren, Moreira, & Feijoo, 2010). To deal with this problem, the LCA + DEA method provides additional information to complement LCA, such as efficiency scores and percentages of potential improvements, which allows going a step further by delivering potential benchmarks of best performing units (Vázquez-Rowe & Iribarren, 2015).

Required inputs and outputs during the life cycle of a product, process, or service are converted into environmental impacts in the life cycle impact assessment (LCIA) step by means of LCIA methods, such as CML (Guinée et al., 2002) or ReCiPe (Goedkoop et al., 2009). Consequently, the selection of the environmental indicators is related to the LCIA method. Depending on the case study assessed and the technical aspects of the DEA model, environmental impacts can be used as inputs or outputs in the DEA matrix. Hence, operational efficiency can be linked to environmental impacts (Lozano, Iribarren, Moreira, & Feijoo, 2009).

To perform a DEA analysis, some methodological aspects to be defined are (i) construction of the DEA matrix, (ii) selection of the model orientation, (iii) selection of the returns to scale, and (iv) selection of the DEA model. The definition of these variables will depend on the characteristics of the case study.

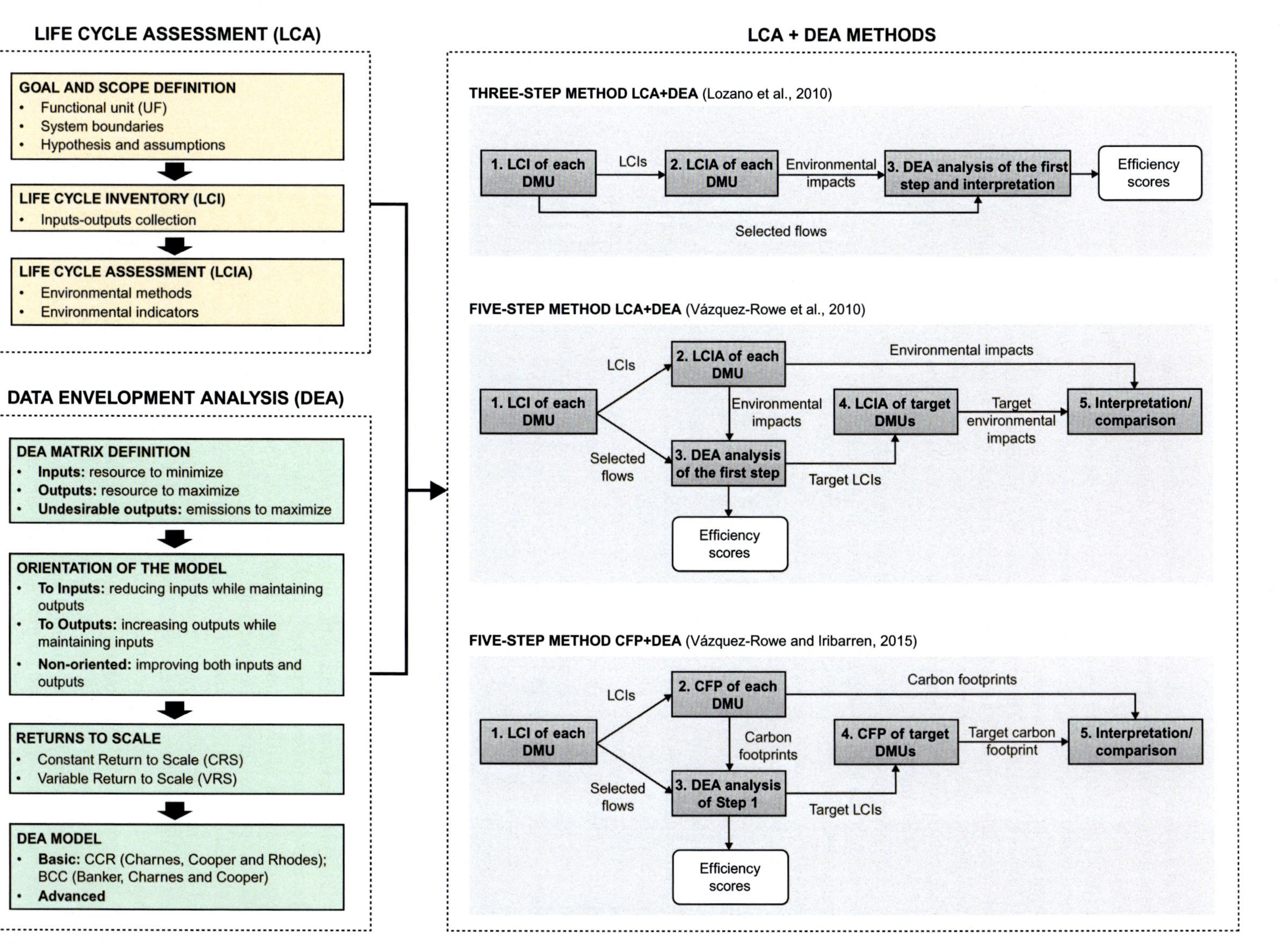

FIG. 1

Graphical representation of commonly used life cycle assessment (LCA) and data envelopment analysis (DEA) combined approaches. *DMU*, decision-making unit.

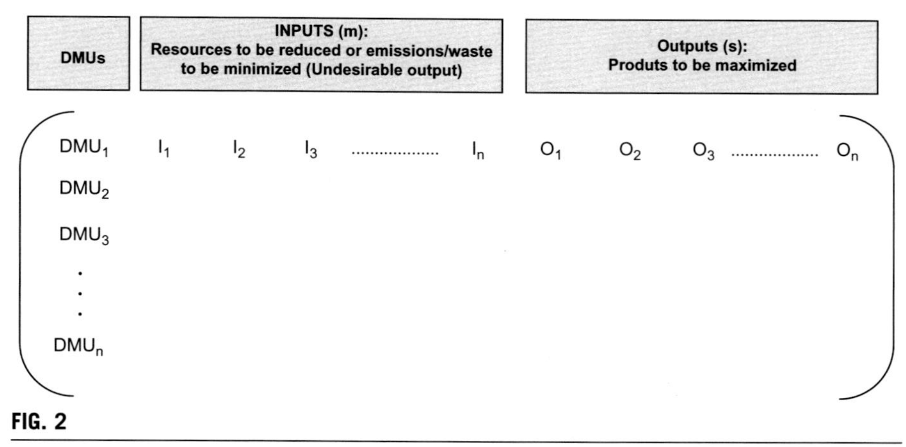

FIG. 2

Structure of the DEA matrix.

2.1 DEA matrix construction

The development of the LCI and the computation of the LCIA for each DMU are essential for the definition of the DEA matrix, which will be composed of as many rows as the number of facilities assessed, and as many columns as inputs and outputs considered (Charnes et al., 1978) (Fig. 2). The definition of the inputs and outputs should be carefully selected to ensure that they represent the system under study as much as possible. In general, inputs are resources that must be reduced, whilst outputs are products to be maximised. However, in some cases the outcomes of a system are emissions that should be minimised. This type of variable is called 'undesirable output' or 'bad output' and requires different treatment in DEA models (Rebolledo-Leiva, Angulo-Meza, Iriarte, & González-Araya, 2017). For more details on the dealing with undesirable outputs, see Section 4.1.

Additionally, for the DEA matrix the minimum sample size of DMUs (n) should be defined based on the number of inputs (m) and outputs (s). It is important to highlight that many inputs and outputs compared to the number of DMUs may diminish the discriminatory power of DEA (Cook, Tone, & Zhu, 2014). Thus Golany and Roll (1989) suggested that the number of DMUs should be at least twice the number of inputs and outputs combined. For Cooper, Seiford, and Tone (2007) the suggested number of DMUs to be analysed should be at least three times the combined number of inputs and outputs: $n \geq \max \{m \times s, 3 \times (m + s)\}$. The selection of input and output items is crucial for successful application of DEA.

2.2 Model orientation

DEA models can be oriented to inputs or to outputs. The objective of models oriented to inputs is to determine how many inputs may be reduced to operate efficiently to maintain the same output levels. For instance, Iribarren et al. (2010) used an input-oriented approach in the application of LCA + DEA to milk production, where the consumption of diesel, electricity, water, silage plastic, maize silage, concentrate, grass silage, and alfalfa were inputs to be reduced; the emissions of CH_4, NH_3, and N_2O were considered undesirable output to be also minimised, and milk production was the output to be maintained. In contrast, an oriented to output approach aims to determine how much output levels could be increased operating efficiently but maintaining input levels. For example, Gonzalez-Garcia, Manteiga, Moreira, and Feijoo (2018) used an output-oriented approach to assess the sustainability of 26 representative Spanish cities using monetary variables as outputs, which should be maximised. Although mathematically viable, some authors have highlighted the risks of using an output-oriented approach in production systems that may be constrained by planetary boundaries or finite resources (e.g. fishing stocks) (Laso, Hoehn, et al., 2018). Finally, a non-oriented approach seeks the combined improvement of both inputs and outputs. However, as far as the authors were able to ascertain, this perspective has not been applied to date in LCA + DEA studies.

2.3 Returns to scale

Three common assumptions are usually made to perform a DEA analysis: convexity, scalability, and free disposability (also named strong disposability) of inputs and outputs. When the three assumptions are made, it is supposed that inputs increase proportionally to outputs, so a constant return to scale (CRS) approach is applied. Nevertheless, if convexity and free disposability, but not scalability are assumed, a variable return to scale (VRS) approach is used (Laso, Vázquez-Rowe, et al., 2018). For instance, Laso, Vázquez-Rowe, et al. (2018) assessed the eco-efficiency of the Cantabrian purse-seining fishing fleet comparing both CRS and VRS approaches. They observed that the amount of efficient DMUs was higher when a VRS approach was used. Consequently, the average efficiency increased. This may be because the constraint set for CRS is less restrictive than the VRS approach, obtaining lower efficiency scores and more efficient DMUs for a VRS approach.

2.4 DEA model

Finally, a wide array of DEA models have been developed and are available in the scientific literature. The most established basic models are the CCR (Charnes et al., 1978) and the BCC models (Banker, Charnes, & Cooper, 1984), which assume CRS and VRS returns to scale, respectively. CCR is built on the assumption of CRS characteristics, estimating the efficiency as the maximum of a ratio of weighted outputs to weighted inputs subject to the condition that the similar ratios for every DMU be less than or equal to unity (Charnes et al., 1978). Banker et al. (1984) added a convexity constraint in the CCR model, assuming VRS for the BCC model. Regarding more advanced models, these may assume different types of returns to scale (Vásquez-Ibarra, Rebolledo-Leiva, Angulo-Meza, González-Araya, & Iriarte, 2020). Table 1 collects the DEA models used in the literature and the productive

Table 1 Summary of DEA models used in different productive sectors on the literature.

DEA model	Sector	References
Charnes Cooper Rhodes (CCR)	Agri-food	Ribal, Sanjuán, Clemente, and Fenollosa (2009), Elhami, Akram, and Khanali (2016), Feijoo, Sanmartin, and Moreira (2017), Khoshnevisan et al. (2015), Kouchaki-Penchah, Nabavi-Pelesaraei, O'Dwyer, and Sharifia (2017), Mohammadi et al. (2013), Mohseni, Borghei, and Khanali (2018), Nabavi-Pelesaraei, Rafiee, Mohtasebi, Hosseinzadeh-Bandbafha, and Chau (2017), Payandeh, Kheiralipour, Karimi, and Khoshnevisan (2017), Paramesh, Arunachalam, Nikkhah, Das, and Ghnimi (2018), and Gamboa, Bojacá, Schrevens, and Maertens (2020)
	Energy	Ewertowska, Galán-Martín, Guillén-Gosálbez, Gavaldá, and Jiménez (2015), Onat et al. (2017), Ren et al. (2014), and Belucio, Rodrigues, Antunes, Freire, and Dias (2021)
	Waste management	Cristóbal, Limleamthong, Manfredi, and Guillén-Gosálbez (2016) and Lorenzo-Toja et al. (2018)
	Construction	Tatari and Kucukvar (2012)
Banker–Charnes–Cooper (BCC)	Agri-food	Kouchaki-Penchah et al. (2017), Nabavi-Pelesaraei et al. (2017), Payandeh et al. (2017), Ullah, Perret, Gheewala, and Soni (2016), Rebolledo-Leiva et al. (2017), Jradi, Chameeva, Delhomme, and Jaegler (2018), Mohseni et al. (2018), Paramesh et al. (2018), Rebolledo-Leiva et al. (2019), and Elhami et al. (2016)
	Energy	Ewertowska, Pozo, Gavaldà, Jiménez, and Guillén-Gosálbez (2017)
	Several sectors at regional level	Rybaczewska-Błażejowska and Gierulski (2018) and Rybaczewska-Błażejowska and Masternak-Janus (2018)
Slack-Based Measure (SBM)	Agri-food	Iribarren et al. (2010, 2011), Vázquez-Rowe, Villanueva-Rey, Iribarren, Teresa Moreira, and Feijoo (2012), Mohammadi et al. (2015),

Continued

Table 1 Summary of DEA models used in different productive sectors on the literature. *Continued*

DEA model	Sector	References
		Grados, Heuts, Vetters, and Schrevens (2017), Lijó et al. (2017), and Masuda (2016), Cecchini, Venanzi, Pierri, and Chiorri (2018), Laso, Hoehn, et al. (2018), Rebolledo-Leiva et al. (2019), Grados and Schrevens (2019), and Cortés, Feijoo, Fernández, and Moreira (2021)
	Energy	Iribarren, Martín-Gamboa, and Dufour (2013), Martín-Gamboa et al. (2016, 2018)
	Waste management	Lorenzo-Toja et al. (2015)
	Construction	Iribarren et al. (2015)
	Fisheries	Lozano, Iribarren, Moreira, and Feijoo (2010), Vázquez-Rowe et al. (2010, 2011), Iribarren and Vázquez-Rowe (2013), Vázquez-Rowe and Tyedmers (2013), Avadí, Vázquez-Rowe, and Fréon (2014), Ramos et al. (2014), Feijoo et al. (2017), González-García et al. (2015), and Laso, Vázquez-Rowe et al. (2018)
	Chemicals	Gavião, Meza, Lima, Sant'Anna, and Soares de Mello (2017)
	Several sectors at city level	Gonzalez-Garcia et al. (2018)
	Services	Álvarez-Rodríguez, Martín-Gamboa, and Iribarren (2019)
Dynamic Network Slacks-Based Measure (DNSBM)	Services	Álvarez-Rodríguez, Martín-Gamboa, and Iribarren (2020)
Additive (ADD)	Electronics	Barba-Gutiérrez, Adenso-Díaz, and Lozano (2009)
Directional Distance Function (DDF)	Agri-food	Beltrán-Esteve, Reig-Martínez, and Estruch-Guitart (2017)
Russell Measure (RM)	Electronics	Lozano, Adenso-Díaz, and Barba-Gutiérrez (2011)
Enhance Russell Measure (ERM)	Fisheries	Lozano et al. (2009)
Range Adjusted Measure (RAM)	Agri-food	Soteriades et al. (2016)

Table 1 Summary of DEA models used in different productive sectors on the literature. *Continued*

DEA model	Sector	References
Epsilon-Based Measure (EBM)	Fisheries	Feijoo et al. (2017)
Multiple Objective Ratio Optimisation with Dominance (MORO-D)	Agri-food	Angulo-Meza, González-Araya, Iriarte, Rebolledo-Leiva, and Soares de Mello (2019)
Slacks-Based Inefficiency (SBI)	Agri-food	Gutiérrez, Aguilera, Lozano, and Guzmán (2017)
Super-Efficiency (SEF)	Agri-food	Iribarren et al. (2010)

sector in which they were applied. It can be observed that the SBM model is the most commonly used model in LCA+DEA studies. This may occur because the SBM model deals directly with the excess of inputs and the deficit of outputs (Tonne, 2001). The use of the SBM model is followed by the CCR and BCC models. As mentioned before, based on Table 1, the agri-food industry is the most assessed sector in the literature. In general, CCR, BCC, and SBM are those most applied models, but there are other advanced methods employed to estimate the eco-efficiency of systems with technological and managerial differences. For instance, the Directional Distance Function (DDF) model was used to estimate the eco-efficiency of nearly 200 Spanish citrus farms using environmental impacts as undesirable outputs considering economic behaviour and environmental performance (Beltrán-Esteve et al., 2017). Lozano et al. (2009) assessed the eco-efficiency of mussel cultivation in rafts using the Enhance Russell Measure (ERM) method whose aim is to minimise the ratio of average inputs reduction to average output increase in relative terms. Similarly, Soteriades et al. (2016) assessed dairy farm sustainability at the animal- and farm-levels by applying the Range Adjusted Measure (RAM) model and using five environmental impacts as inputs and three outputs. Other advanced DEA models, such as Multiple Objective Ratio Optimisation with Dominance (MORO-D), Slacks-Based Inefficiency (SBI), or Super-Efficiency (SEF), have also been applied to agri-food systems to assess the eco-efficiency of organic blueberry orchards (Angulo-Meza et al., 2019), conventional and organic rain-fed cereals in Spain (Gutiérrez et al., 2017), and milk production in Galicia (Spain) (Iribarren et al., 2010), respectively. More information about the sectors to which the LCA+DEA method has been applied is given in Section 3.

2.5 LCA+DEA approaches

The way in which both methodologies are combined depends on the systems under study and how both operational and environmental variables are integrated into the DEA model. Fig. 1 displays the three LCA+DEA methods most commonly applied in the literature. The three-step method proposed by Lozano et al. (2010) uses LCI data and environmental impacts from the LCIA inputs and production levels as output in the DEA analysis. In contrast, the five-step method proposed by Vázquez-Rowe et al. (2010) performs an environmental characterisation before the use of the DEA model and, thereafter, it uses the potential reduction environmental impacts obtained from the DEA analysis to determine a new prospective environmental characterisation for each DMU. Finally, the method makes a comparison between the potential environmental impacts of the improved DMUs and the current DMUs. Considering that carbon footprint is the most recurrently used indicator in an LCA scope, Vázquez-Rowe and Iribarren (2015) proposed a modified five-step structure combining carbon footprinting (CFP) and DEA. In this case, the environmental characterisation was based on the global warming potential category exclusively, but from a methodological point of view does not differ from the one above mentioned. In addition, there is an intermediate method, based on the three-step method, developed by Rebolledo-Leiva et al. (2017), that proposes four steps to use the CFP indicator as an undesirable output in DEA analysis. Undesirable outputs, which are discussed in Section 4.1, must be computationally minimized in the DEA matrix.

3 A brief history of the joint use of LCA+DEA

The use of LCA as a tool to assess the environmental profile of a given product or system dates back to the late 1960s and early 1970s, when the Midwest Research Institute (MRI) carried out a study about resource requirements, emissions, and waste flows of different beverage containers of Coca-Cola Company (Hunt, Franklin, & Hunt, 1996). In this case, the MRI employed the term Resource and Environmental Profile Analysis (REPA). However, it was not until the first decade of the 21st century when the interest in LCA studies increased. In parallel, the use of the DEA as a tool to measure the efficiency emerged during a study of E. Rhodes, A. Charnes, and W.W. Cooper to analyse schools, which resulted in the formulation of the CCR DEA model (Charnes et al., 1978).

The application of LCA together with DEA is relatively new since the first study that combined both methodologies was published in 2009. Barba-Gutiérrez et al. (2009) studied the eco-efficiency of different electrical and electronic facilities using the output of the LCA analysis, in terms of ecopoints, associated with each product for the DEA efficiency analysis. Therefore, environmental impacts were used as inputs and retail price as output to the DEA model. The study used an input-oriented and an ADD advanced model, considering a VRS approach due to the heterogeneity of the facilities analysed. Authors highlighted that, although the proposed approach was applied to a specific case study, it could be extended to any other system chosen. Later, in the same year, the first three-step LCA+DEA method proposed by Lozano et al. (2009) was applied to the fishery sector, in particular to mussel cultivation in rafts. The definition of this method allowed setting targets and quantifying potential improvements. In 2010, Vázquez-Rowe et al. (2010) went a step further and presented the five-step LCA+DEA method to assess the eco-efficiency of Galician (North of Spain) trawling vessels during 2008. These two LCA+DEA methods are the most used in the literature but, as mentioned before, a modified five-step procedure (Vázquez-Rowe & Iribarren, 2015) or a four-step method (Rebolledo-Leiva et al., 2017) is also currently employed.

The number of published articles that apply LCA+DEA has considerably increased since 2015 (see Table 1). One of the reasons is the versatility and flexibility of the methodology that made it possible to analyse a high variety of sectors. Recently, however, this methodology has focused on the agri-food sector, following the trend of previous years (Cortés et al., 2021; Gamboa et al., 2020; Soteriades et al., 2016; Zhong et al., 2020).

For instance, the LCA+DEA method has been used to estimate the eco-efficiency of dairy farms (Cecchini et al., 2018; Cortés et al., 2021; Iribarren et al., 2010; Iribarren, Hospido, Moreira, & Feijoo, 2011; Soteriades et al., 2016), orange production (Ribal et al., 2009), organic blueberry orchards (Angulo-Meza et al., 2019; Rebolledo-Leiva et al., 2017), and raspberry orchards (Rebolledo-Leiva et al., 2019). The eco-efficiency of fishing vessels has been also analysed using LCA+DEA, especially the Spanish fleet. Vázquez-Rowe, Iribarren, Hospido, Moreira, and Feijoo (2011) assessed deep-sea, offshore, and coastal fleets from Galicia; Ramos et al. (2014) analysed the Basque trawling fleet and Laso, Vázquez-Rowe, et al. (2018) studied the Cantabrian purse seining fleet. Another important linked to fishing was conducted by Avadí et al. (2014) regarding the Peruvian anchoveta fleet. In the energy sector, Ewertowska, Galán-Martín, Guillén-Gosálbez, Gavaldá, and Jiménez (2016) assessed the electricity mix of the top European economies, whereas Ewertowska et al. (2017) analysed 11 technologies for electricity generation. Renewable energies were studied by Iribarren et al. (2013) using the 5-step LCA+DEA method to benchmark the operational and environmental performance of 25 wind farms in Spain. The treatment of residues, such as wastewater (Lorenzo-Toja et al., 2015, 2018) or food waste (Cristóbal et al., 2016), has also been evaluated using LCA+DEA and, to a lesser extent, some studies were focused on the construction (Iribarren et al., 2015; Tatari & Kucukvar, 2012) and chemical (Gavião et al., 2017) sectors.

4 Recent methodological advances in LCA+DEA

The combined use of LCA and DEA presents multiple challenges during its conceptualisation, development, and implementation phases. This section addresses the main methodological challenges faced by LCA+DEA practitioners, some of them already mentioned before, and presents the recent advances proposed in the literature to solve them.

4.1 Dealing with undesirable outputs

One of the most challenging methodological issues when applying LCA+DEA is dealing with undesirable outputs, as mentioned in Section 2.1. When measuring performance or efficiency with DEA, it is required that the isotonicity property is maintained for the functions relating inputs and outputs (i.e. an increase in any input should result in some output increase and not a decrease in any output) (Bowlin, 1998). However, as mentioned before, in LCA it is common to have LCI output flows of emissions and pollutants, as well as LCIA results of impact categories such as global warming potential, that negatively affect the environment (i.e. outputs to be minimised). The so-called undesirable outputs have been largely discussed in the literature and different approaches have been reported for their treatment (Pishgar-Komleh, Zylowski, Rozakis, & Kozyra, 2020).

One of the approaches, even if not recommended, is ignoring or not considering undesirable outputs. This approach might be reasonable in LCA+DEA studies that use exclusively operational LCI data as inputs for the DEA analysis (i.e. in the five-step methodology) as done by Vázquez-Rowe et al. (2011). However, if a study ignores arbitrarily undesirable outputs that include results from the LCIA in the DEA analysis, as may happen mostly in the three-step methodology, it is recognised that the efficiency of the DMU scores may be underestimated and, thus, misleading results may be provided (He, Zhang, Lei, Fu, & Xu, 2013).

The most common approach used in the literature is to treat undesirable outputs as inputs. This approach was introduced as an 'ecological extension of DEA' by Dyckhoff and Allen (2001) based on the logic of using the original data and that the desired minimisation is directly considered. In many cases it has been used to include results from the LCIA. For instance, Soteriades et al. (2016) evaluated five impact categories as inputs. On a similar note, Iribarren et al. (2011) included LCI data directly, accounting for wastewater and direct livestock emissions in dairy farms. LCIA results and LCI data can also be combined as inputs as done by Belucio et al. (2021). This approach should be applied with care either for output- or input-oriented models, since it can lead to incomplete conclusions (i.e. certain ecological inefficiencies in the factors and in the emissions might remain undetected) or misleading conclusions (e.g. the natural interpretation of the potential reduction factors may have no natural interpretation), respectively (Pishgar-Komleh et al., 2020).

Another approach to deal with undesirable outputs is based on data transformation. This is done by taking reciprocals as proposed by Golany and Roll (1989), and afterwards introduced by Knox Lovell, Pastor, and Turner (1995) to compare the macroeconomic performance of 19 OECD countries. As a variation of taking reciprocals, undesirable outputs can also be transformed by translation (i.e. by adding a sufficiently large scalar to the additive inverse). For example, Rebolledo-Leiva et al. (2017) used a multiplicative inverse transformation to CFP as an undesirable output. It is also common that authors use a transformation in the scale based on a normalisation procedure as shown by Ewertowska et al. (2017) in order to give parameter values between 0 and 1. Normalisation should be applied with caution since it can lead to loss of information.

A further approach considers treating the undesirable outputs as regular outputs by applying ratio functions and models. This approach is conditioned by the assumption on the disposability (either weak or strong (i.e. free)) selected for the case study that should be based on evidence. It is important to highlight that most efficiency analysis applications assume of strong disposability of inputs and outputs (see Mehdiloo and Podinovski (2019) for further discussion on this topic).

Recently, new models in the literature are coming forward to deal with undesirable outputs. An approach to include both desirable and undesirable outputs in SBM models was introduced by Tone (2004). Based on that, Zhou, Ang, and Wang (2012) introduced a non-radial SBM model combined with a ratio approach that incorporates the undesirable outputs. As an example, Zhong et al. (2020) applied an SBM model to calculate the eco-efficiency value for maize seed production in China, including five environmental impacts as undesirable output.

It is important to highlight that depending on the case study under assessment, different techniques might be more appropriate than others (Halkos & Petrou, 2019), and that a combination of different approaches is also possible, e.g. data transformation and treating undesirable outputs as inputs (Ewertowska et al., 2017). Pishgar-Komleh et al. (2020) summarised the six most common methods for treating undesirable outputs and compared their results to select the most appropriate LCA+DEA model in a case study for winter wheat farms in Poland.

4.2 Handling uncertainty

It is recognised that uncertainty is inherent to LCA, since there is always a discrepancy between the real values and the ones used in the study, thus impacting on the reliability of LCA results. This uncertainty can be either epistemic that stems from limited data (commonly referred to as parameter, stochastic, or data uncertainty), and from limited knowledge whilst modelling the process (called model uncertainty), or can also stem from appropriateness of modelling and methodological choices such as functional unit, time horizon, geographical scale (also called scenario uncertainty) (Bamber et al., 2020). In LCA, epistemic uncertainty effects can be addressed by uncertainty analysis (i.e. a systematic technique that quantifies the uncertainty in LCI results due to variability and inaccuracy of data and model), and scenario uncertainty can be addressed by sensitivity analysis (systematic technique that assesses the effect of methodological choice and data on results) (Ling-Chin, Heidrich, & Roskilly, 2016).

When combining LCA+DEA, the main challenge lies in the nature of the DEA analysis where inputs and outputs are based on known and deterministic data. Recent LCA+DEA publications still consider managing uncertainty as a future methodological development of their research to be done (Soteriades et al., 2016; Torregrossa, Marvuglia, & Leopold, 2018). Only a few studies addressed the issue of dealing with uncertainty. Thus Egilmez, Gumus, Kucukvar, and Tatari (2016) developed a fuzzy LCA+DEA to deal with EIO-LCI uncertainty propagation (through fuzzy crisp efficiency scores that capture the uncertainty in inputs (i.e. selected environmental impacts) and outputs (i.e. total production)) and included a sensitivity analysis of environmental impact indicators (through the super-efficiency DEA model). Lorenzo-Toja et al. (2018) used an adapted version of ANOVA for repeated measurements and the Friedman's test to understand whether individual DMU (i.e. wastewater treatment plants) scores showed no significant differences through time in a window analysis. Ewertowska et al. (2017) applied stochastic modelling through Monte Carlo analysis to capture uncertainty in LCI entries and damage factors of the LCIA models. After solving the stochastic DEA model for 500 scenarios on technologies considered for electricity generation, the authors highlighted the importance to analyse the dispersion of efficiencies and targets to assess the robustness of nominal DEA results. Finally, Paramesh et al. (2018) conducted an uncertainty analysis considering six different impact assessment methodologies (that present different characterisation methods for global warming potential and ozone layer depletion impact categories) in the LCA performed after the energy flows of arecanut production were optimised using DEA.

Ignoring uncertainty lowers the credibility of LCA results. For this reason, uncertainty must be addressed to give a confidence level of LCA results, since these should not be a simple deterministic point. There is still room for more advances capturing uncertainty within LCA+DEA studies such as addressing uncertainty in allocation methods or characterisation models, as concluded by Vásquez-Ibarra et al. (2020).

4.3 Ranking efficient units: Super-efficiency analysis

LCA+DEA-based results usually lead to a number of DMUs being deemed 'efficient' that, in some cases and depending on modelling decisions, can be very high (i.e. low discrimination between efficient and inefficient DMUs). Therefore, one of the main challenges within these studies is ranking efficient DMUs when discrimination is scarce. In fact, depending on the scope of the study, a low discrimination between the DMUs can be insufficient for decision-making processes such as, for example, proposing environmental benchmarks. When those DMUs need to be ranked, in an objective way and without further information from the decision-maker (i.e. one of the main strengths of DEA), different methods are reported such as the super-efficiency model, the cross-evaluation procedure, and multiple objective linear programmes (Cooper et al., 2007). The most widely used method in LCA+DEA is super-efficiency, as introduced by Per and Christian (1993). It is based on the idea of excluding the DMU from the comparison set under evaluation and, thus, since it is not compared against itself, efficient DMUs can receive better scores than one for input-oriented or lower scores than one for output-oriented. Iribarren et al. (2010) applied a super-efficiency analysis perspective (input-oriented slack-based with CRS) to mussel cultivation in order to rank best performing entities (i.e. 18 out of 67 DMUs) and, thus, propose climate change benchmarks. In the same line (Iribarren et al., 2011) applied the same model to dairy farms to rank the 31 that were deemed as efficient out of 72 DMUs included. Vázquez-Rowe et al. (2012) applied the same super-efficiency model to vineyards to enhance the discrimination amongst efficient DMUs (i.e. 24 out of 40) after the DEA analysis to propose benchmarks for environmental policy. Mohammadi et al. (2015) ranked rice paddy fields in Iran to enhance the efficiency discrimination (i.e. 26 out of 82), and the new rank obtained of the best performing units was used as a reference value for environmental decision-making. Cristóbal et al. (2016) used the input-oriented SBM model proposed by Tonne (2001) to rank four waste management options that were deemed as efficient out of a group of six options. Paramesh et al. (2018) performed the super-efficiency analysis to the production of arecanut to rank the 28 farmers deemed efficient, out of 70 included, to select best management practices to enhance energy use efficiency.

However, Seiford and Zhu (1999) reported infeasibility problems for the super-efficiency method under certain conditions. Hence, other authors proposed different approaches to rank efficient units. Egilmez et al. (2016) used a super-efficiency sensitivity algorithm (proposed by (Zhu, 2001)) that assumed synchronous proportional variation in all inputs and outputs for the DMUs under assessment, whilst data for the remaining DMUs were assumed as fixed. This approach was applied to a fuzzy EIO-LCA+DEA for sustainability benchmarking in which 33 food manufacturing sectors in the United States were ranked. Tavana, Izadikhah, Farzipoor Saen, and Zare (2021) used a modified integer non-radial enhanced Russell model to rank 41 types of construction floor covering systems. In contrast to the super-efficiency analysis introduced by Per and Christian (1993), Tavana et al. (2021) used the approach developed by Izadikhah, Saen, and Ahmadi (2017) that does not remove the DMU under consideration from the reference set of the model.

The selection of an efficiency score (i.e. cut-off criterion) amongst the super-efficient values for further decision-making still seems to be, most of the time, an arbitrary choice. Iribarren et al. (2010, 2011) set a cut-off criterion of 1.05 (to select four out of 18 efficient DMUs, and 13 out of 31 efficient DMUs, respectively), and Mohammadi et al. (2015) applied a value of 1.2 (to select nine out of 26 efficient DMUs). Paramesh et al. (2018) also applied a cut-off criterion (according to Iribarren et al. (2010)) of 1.2 (to select nine out of 28 efficient DMUs). Tavana et al. (2021) selected the four best DMUs for a later comparison out of 41 without setting a specific cut-off criterion. In contrast, some authors such as Vázquez-Rowe et al. (2012), presented results of super-efficient vineyards grouped by production level instead of declaring a cut-off criterion.

5 Advantages and limitations of applying the LCA+DEA method: Future outlook

After a decade of relatively high proliferation of the LCA+DEA method in the scientific literature, it is fair to say that this combined methodology has aided in the identification of potential benchmarks to reduce the environmental impacts of complex systems (Laso, Vázquez-Rowe, et al. (2018)). In fact, the relative comparison between multiple DMUs has been deemed as an important asset that in many cases has allowed delivering an individual environmental diagnosis of units within broader systems. More specifically, an inefficient DMU does not only receive a diagnosis that flags it as performing worse than its better-performing peers, but it also receives a mathematical computation relating to the input or output minimisations needed to operate at the same level of efficiency as the above-mentioned best-performing peers (i.e. those located in the production frontier).

Despite the utility of this mathematical benchmark, many LCA+DEA studies have presented these revised values as potential reductions that could make the production system attain succulent improvements in environmental impacts. However, a vast majority of the LCA+DEA literature retrieved do not delve into the actual implementation of these reductions. In fact, many of these studies have been carried out for agricultural and fishing studies, in which the inherent variability and complexity of the natural environment, which is in direct contact with the technosphere, generates considerable uncertainty. For instance, vineyards may be close to each other in a specific geographical area (e.g. valley, appellation), but certain soil (e.g. lower organic matter) or climatic (e.g. shading) conditions may impede a reduction in inefficiencies for the least productive plots. Similarly, the DEA benchmark computation may not consider that, e.g. fishing vessels are competing for a limited stock and, therefore, converging all DMUs (i.e. vessels) to higher levels of efficiency is biophysically unfeasible. Consequently, it should be expected that a higher interconnection between LCA+DEA studies and innovation strategies should be sought in the near future to upgrade the importance of this method in decision-making and policy support.

Window analysis in DEA systems has shown to be an interesting model to be used in LCA+DEA methods (Torregrossa et al., 2018; Vázquez-Rowe & Tyedmers, 2013), since it allows monitoring the relative efficiency of a DMU throughout a certain period of time. This allows to extract conclusions by comparing a single DMU to itself in different time periods or to compare DMUs between them in a given period of time. Unfortunately, however, LCIs and, to a certain extent, LCIAs lack this temporal perspective in most of the mainstream LCA databases and software available. Consequently, we argue that there is still a wide array of opportunities to couple window analysis DEA (and probably other DEA models) with dynamic LCA systems (Pigné et al., 2020), through the temporalisation of background and foreground systems in LCIs. This would allow LCA+DEA practitioners not only to model changes in primary data obtained from their foreground systems, but to adapt changing technology and operations through time to the DEA window analysis matrix. Whilst more complex to model, we argue that this perspective could allow reducing uncertainties when calculating benchmarks and environmental impact reduction potential.

Uncertainty, as discussed in Section 4.2, is also an important conundrum to be solved in LCA+DEA methods. Although some interesting advancements have been performed in certain studies, this remains an important limitation to understand, as, for instance, the robustness of efficiency distances between DMUs. Whilst we expect new studies to delve further into new statistical analyses of LCA+DEA methods, we advocate standardisation processes for this joint method. Considering the growth of the method in the past decade, it is plausible to assume that an increasing number of studies will continue to appear in the literature. Guidance or standardisation reports would help in providing advice on how uncertainty, the use of LCIA methods within the framework, the selection of DEA models, and other grey areas should be addressed in LCA+DEA.

Acknowledgments

"Prof. Vázquez-Rowe wishes to thank the *Dirección de Fomento de la Investigación* at PUCP for funding for this study." This would be very much appreciated.

References

Aldaco, R., Butnar, I., Margallo, M., Laso, J., Rumayor, M., Dominguez-Ramos, A., ... Dodds, P. E. (2019). Bringing value to the chemical industry from capture, storage and use of CO_2: A dynamic LCA of formic acid production. *Science of the Total Environment*, *663*, 738–753. https://doi.org/10.1016/j.scitotenv.2019.01.395.

Álvarez-Rodríguez, C., Martín-Gamboa, M., & Iribarren, D. (2019). Combined use of Data Envelopment Analysis and Life Cycle Assessment for operational and environmental benchmarking in the service sector: A case study of grocery stores. *Science of the Total Environment*, *667*, 799–808. https://doi.org/10.1016/j.scitotenv.2019.02.433.

Álvarez-Rodríguez, C., Martín-Gamboa, M., & Iribarren, D. (2020). Sustainability-oriented efficiency of retail supply chains: A combination of Life Cycle Assessment and dynamic network Data Envelopment Analysis. *Science of the Total Environment*, *705*, 135977. https://doi.org/10.1016/j.scitotenv.2019.135977.

Angulo-Meza, L., González-Araya, M., Iriarte, A., Rebolledo-Leiva, R., & Soares de Mello, J. C. (2019). A multiobjective DEA model to assess the eco-efficiency of agricultural practices within the CF+DEA method. *Computers and Electronics in Agriculture*, *161*, 151–161. https://doi.org/10.1016/j.compag.2018.05.037.

Avadí, A., Vázquez-Rowe, I., & Fréon, P. (2014). Eco-efficiency assessment of the Peruvian anchoveta steel and wooden fleets using the LCA+DEA framework. *Journal of Cleaner Production*, *70*, 118–131. https://doi.org/10.1016/j.jclepro.2014.01.047.

Bamber, N., Turner, I., Arulnathan, V., Li, Y., Zargar Ershadi, S., Smart, A., et al. (2020). Comparing sources and analysis of uncertainty in consequential and attributional life cycle assessment: Review of current practice and recommendations. *International Journal of Life Cycle Assessment*, *25*(1), 168–180. https://doi.org/10.1007/s11367-019-01663-1.

Banker, D., Charnes, A., & Cooper, W. W. (1984). Some models for estimating technical and scale inefficiencies in data envelopment analysis. *Management Science*, *30*(9), 1078–1092. https://doi.org/10.1287/mnsc.30.9.1078.

Barba-Gutiérrez, Y., Adenso-Díaz, B., & Lozano, S. (2009). Eco-efficiency of electric and electronic appliances: A data envelopment analysis (DEA). *Environmental Modeling and Assessment*, *14*(4), 439–447. https://doi.org/10.1007/s10666-007-9134-2.

Beltrán-Esteve, M., Reig-Martínez, E., & Estruch-Guitart, V. (2017). Assessing eco-efficiency: A metafrontier directional distance function approach using life cycle analysis. *Environmental Impact Assessment Review*, *63*, 116–127. https://doi.org/10.1016/j.eiar.2017.01.001.

Belucio, M., Rodrigues, C., Antunes, C. H., Freire, F., & Dias, L. C. (2021). Eco-efficiency in early design decisions: A multimethodology approach. *Journal of Cleaner Production*, *283*. https://doi.org/10.1016/j.jclepro.2020.124630.

Bowlin, W. F. (1998). Measuring performance: An introduction to data envelopment analysis (DEA). *The Journal of Cost Analysis*, *15*, 3–27. https://doi.org/10.1080/08823871.1998.10462318.

Cecchini, L., Venanzi, S., Pierri, A., & Chiorri, M. (2018). Environmental efficiency analysis and estimation of CO2 abatement costs in dairy cattle farms in Umbria (Italy): A

SBM-DEA model with undesirable output. *Journal of Cleaner Production*, *197*, 895–907. https://doi.org/10.1016/j.jclepro.2018.06.165.

Charnes, A., Cooper, W. W., & Rhodes, E. (1978). Measuring the efficiency of decision making units. *European Journal of Operational Research*, *2*(6), 429–444. https://doi.org/10.1016/0377-2217(78)90138-8.

Cook, W. D., & Seiford, L. M. (2009). Data envelopment analysis (DEA)—Thirty years on. *European Journal of Operational Research*, *192*(1), 1–17. https://doi.org/10.1016/j.ejor.2008.01.032.

Cook, W. D., Tone, K., & Zhu, J. (2014). Data envelopment analysis: Prior to choosing a model. *Omega (United Kingdom)*, *44*, 1–4. https://doi.org/10.1016/j.omega.2013.09.004.

Cooper, W. W., Seiford, L. M., & Tone. (2007). *Data envelopment analysis: A comprehensive text with models and applications*.

Cortés, A., Feijoo, G., Fernández, M., & Moreira, M. T. (2021). Pursuing the route to eco-efficiency in dairy production: The case of Galician area. *Journal of Cleaner Production*, *285*. https://doi.org/10.1016/j.jclepro.2020.124861.

Cristóbal, J., Limleamthong, P., Manfredi, S., & Guillén-Gosálbez, G. (2016). Methodology for combined use of data envelopment analysis and life cycle assessment applied to food waste management. *Journal of Cleaner Production*, *135*, 158–168. https://doi.org/10.1016/j.jclepro.2016.06.085.

Diaz-Sarachaga, J. M., Jato-Espino, D., & Castro-Fresno, D. (2018). Is the sustainable development goals (SDG) index an adequate framework to measure the progress of the 2030 agenda? *Sustainable Development*, *26*(6), 663–671. https://doi.org/10.1002/sd.1735.

Dyckhoff, H., & Allen, K. (2001). Measuring ecological efficiency with data envelopment analysis (DEA). *European Journal of Operational Research*, *132*(2), 312–325. https://doi.org/10.1016/S0377-2217(00)00154-5.

Egilmez, G., Gumus, S., Kucukvar, M., & Tatari, O. (2016). A fuzzy data envelopment analysis framework for dealing with uncertainty impacts of input-output life cycle assessment models on eco-efficiency assessment. *Journal of Cleaner Production*, *129*, 622–636. https://doi.org/10.1016/j.jclepro.2016.03.111.

Ehrgott, M., & Gandibleux, X. (2003). *Multiple criteria optimization. State of the art annotated bibliographic surveys*. New York: Kluwer Academic Publishers.

Elhami, B., Akram, A., & Khanali, M. (2016). Optimization of energy consumption and environmental impacts of chickpea production using data envelopment analysis (DEA) and multi objective genetic algorithm (MOGA) approaches. *Information Processing in Agriculture*, *3*, 190–205. https://doi.org/10.1016/j.inpa.2016.07.002.

Ewertowska, A., Galán-Martín, A., Guillén-Gosálbez, G., Gavaldá, J., & Jiménez, L. (2016). Assessment of the environmental efficiency of the electricity mix of the top European economies via data envelopment analysis. *Journal of Cleaner Production*, *116*, 13–22. https://doi.org/10.1016/j.jclepro.2015.11.100.

Ewertowska, A., Pozo, C., Gavaldà, J., Jiménez, L., & Guillén-Gosálbez, G. (2017). Combined use of life cycle assessment, data envelopment analysis and Monte Carlo simulation for quantifying environmental efficiencies under uncertainty. *Journal of Cleaner Production*, *166*, 771–783. https://doi.org/10.1016/j.jclepro.2017.07.215.

Feijoo, G., Sanmartin, S., & Moreira, M. T. (2017). Implementation of linear programming and life cycle approach in an excel application to determine ecoefficiency. In A. Espuña, M. Graells, & L. Puigjaner (Eds.), *27th European Symposium on Computer Aided Process Engineering, Computer Aided Chemical Engineering*. Elsevier, pp. 2731–2736. https://doi.org/10.1016/B978-0-444-63965-3.50457-8.

Gamboa, C., Bojacá, C. R., Schrevens, E., & Maertens, M. (2020). Sustainability of small-holder quinoa production in the Peruvian Andes. *Journal of Cleaner Production*, *264*. https://doi.org/10.1016/j.jclepro.2020.121657.

Gavião, L. O., Meza, L. A., Lima, G. B. A., Sant'Anna, A. P., & Soares de Mello, J. C. C. B. (2017). Improving discrimination in efficiency analysis of bioethanol processes. *Journal of Cleaner Production*, *168*, 1525–1532. https://doi.org/10.1016/j.jclepro.2017.06.020.

Goedkoop, M. J., Heijungs, R., Huijbregts, M., De Schryver, A., Struijs, J., & Van Zelm, R. (2009). *ReCiPe 2008—A life cycle impact assessment method which comprises harmonised category indicators at the midpoint and the endpoint level; first edition report I: Characterization* (1st ed.). 6 January 2009 http://www.lcia-recipe.net/.

Golany, B., & Roll, Y. (1989). An application procedure for DEA. *Omega*, *17*(3), 237–250. https://doi.org/10.1016/0305-0483(89)90029-7.

Gonzalez-Garcia, S., Manteiga, R., Moreira, M. T., & Feijoo, G. (2018). Assessing the sustainability of Spanish cities considering environmental and socio-economic indicators. *Journal of Cleaner Production*, *178*, 599–610. https://doi.org/10.1016/j.jclepro.2018.01.056.

González-García, S., Villanueva-Rey, P., Belo, S., Vázquez-Rowe, I., Moreira, M. T., Feijoo, G., & Arroja, L. (2015). Cross-vessel eco-efficiency analysis. A case study for purse seining fishing from North Portugal targeting European pilchard. *International Journal of Life Cycle Assessment*, *20*, 1019–1032. https://doi.org/10.1007/s11367-015-0887-6.

Grados, D., Heuts, R., Vetters, E., & Schrevens, E. (2017). A model-based comprehensive analysis of technical sustainability of potato production systems in the Mantaro Valley, Central Highlands, Peru. Acta Hortic. 1154. ISHS 2017. In *Proc. V int. symposium on applications of modelling as an innovative technology in the horticultural supply chain*, doi:10.17660/ActaHortic.2017.1154.20.

Grados, D., & Schrevens, E. (2019). Multidimensional analysis of environmental impacts from potato agricultural production in the Peruvian Central Andes. *Science of the Total Environment*, *663*, 927–934. https://doi.org/10.1016/j.scitotenv.2019.01.414.

Guillén-Gosálbez, G., You, F., Galán-Martín, Á., Pozo, C., & Grossmann, I. E. (2019). Process systems engineering thinking and tools applied to sustainability problems: Current landscape and future opportunities. *Current Opinion in Chemical Engineering*, *26*, 170–179. https://doi.org/10.1016/j.coche.2019.11.002.

Guinée, J., Gorrée, M., Heijungs, R., Huppes, G., Kleijn, R., de Koning, A., et al. (2002). *Handbook on life cycle assessment*. Dordrecht: Kluwer Academic Publishers.

Gutiérrez, E., Aguilera, E., Lozano, S., & Guzmán, G. I. (2017). A two-stage DEA approach for quantifying and analysing the inefficiency of conventional and organic rain-fed cereals in Spain. *Journal of Cleaner Production*, *149*, 335–348. https://doi.org/10.1016/j.jclepro.2017.02.104.

Halkos, G., & Petrou, K. N. (2019). Treating undesirable outputs in DEA: A critical review. *Economic Analysis and Policy*, *62*, 97–104. https://doi.org/10.1016/j.eap.2019.01.005.

He, F., Zhang, Q., Lei, J., Fu, W., & Xu, X. (2013). Energy efficiency and productivity change of China's iron and steel industry: Accounting for undesirable outputs. *Energy Policy*, *54*, 204–213. https://doi.org/10.1016/j.enpol.2012.11.020.

Hunt, R. G., Franklin, W. E., & Hunt, R. G. (1996). LCA—How it came about. *The International Journal of Life Cycle Assessment*, *1*, 4–7. https://doi.org/10.1007/BF02978624.

Iribarren, D., Hospido, A., Moreira, M. T., & Feijoo, G. (2011). Benchmarking environmental and operational parameters through eco-efficiency criteria for dairy farms. *Science of the Total Environment*, *409*(10), 1786–1798. https://doi.org/10.1016/j.scitotenv.2011.02.013.

Iribarren, D., Martín-Gamboa, M., & Dufour, J. (2013). Environmental benchmarking of wind farms according to their operational performance. *Energy*, *61*, 589–597. https://doi.org/10.1016/j.energy.2013.09.005.

Iribarren, D., Marvuglia, A., Hild, P., Guiton, M., Popovici, E., & Benetto, E. (2015). Life cycle assessment and data envelopment analysis approach for the selection of building components according to their environmental impact efficiency: A case study for external walls. *Journal of Cleaner Production*, *87*(1), 707–716. https://doi.org/10.1016/j.jclepro.2014.10.073.

Iribarren, D., & Vázquez-Rowe, I. (2013). Is labor a suitable input in LCA + DEA studies? Insights on the combined use of economic, environmental and social parameters. *Social Sciences*, *2*, 114–130. https://doi.org/10.3390/socsci2030114.

Iribarren, D., Vázquez-Rowe, I., Moreira, M. T., & Feijoo, G. (2010). Further potentials in the joint implementation of life cycle assessment and data envelopment analysis. *Science of the Total Environment*, *408*(22), 5265–5272. https://doi.org/10.1016/j.scitotenv.2010.07.078.

ISO. (2006). *14040: Environmental management—Life cycle assessment—Principles and framework*. International Organization for Standardization.

Izadikhah, M., Saen, R. F., & Ahmadi, K. (2017). How to assess sustainability of suppliers in the presence of dual-role factor and volume discounts? A data envelopment analysis approach. *Asia Pacific J. Oper. Res.*, *34*, 1–25.

Jabbari, M., Shafiepour Motlagh, M., Ashrafi, K., & Abdoli, G. (2020). Differentiating countries based on the sustainable development proximities using the SDG indicators. *Environment, Development and Sustainability*, *22*(7), 6405–6423. https://doi.org/10.1007/s10668-019-00489-z.

Jradi, S., Chameeva, T. B., Delhomme, B., & Jaegler, A. (2018). Tracking carbon footprint in French vineyards: A DEA performance assessment. *Journal of Cleaner Production*, *192*, 43–54. https://doi.org/10.1016/j.jclepro.2018.04.216.

Kalbar, P. P., Birkved, M., Nygaard, S. E., & Hauschild, M. (2017). Weighting and aggregation in life cycle assessment: Do present aggregated single scores provide correct decision support? *Journal of Industrial Ecology*, *21*(6), 1591–1600. https://doi.org/10.1111/jiec.12520.

Khoshnevisan, B., Bolandnazar, E., Shamshirband, S., Shariati, H. M., Anuar, N. B., & Kiah, M. L. M. (2015). Decreasing environmental impacts of cropping systems using life cycle assessment (LCA) and multi-objective genetic algorithm. *Journal of Cleaner Production*, *86*, 67–77. https://doi.org/10.1016/j.jclepro.2014.08.062.

Knox Lovell, C. A., Pastor, J. T., & Turner, J. A. (1995). Measuring macroeconomic performance in the OECD: A comparison of European and non-European countries. *European Journal of Operational Research*, *87*(3), 507–518. https://doi.org/10.1016/0377-2217(95)00226-X.

Kouchaki-Penchah, H., Nabavi-Pelesaraei, A., O'Dwyer, J., & Sharifia, M. (2017). Environmental management of tea production using joint of life cycle assessment and data envelopment analysis approaches. *Environmental Progress & Sustainable Energy*, 1–7.

Kumar, A., Sah, B., Singh, A. R., Deng, Y., He, X., Kumar, P., et al. (2017). A review of multi criteria decision making (MCDM) towards sustainable renewable energy development. *Renewable and Sustainable Energy Reviews*, *69*, 596–609. https://doi.org/10.1016/j.rser.2016.11.191.

Laso, J., Hoehn, D., Margallo, M., García-Herrero, I., Batlle-Bayer, L., Bala, A., et al. (2018). Assessing energy and environmental efficiency of the Spanish agri-food system using the LCA/DEA methodology. *Energies*, *11*(12). https://doi.org/10.3390/en11123395.

Laso, J., Margallo, M., García-Herrero, I., Fullana, P., Bala, A., Gazulla, C., … Rubén, R. (2018). Combined application of Life Cycle Assessment and linear programming to evaluate food waste-to-food strategies: Seeking for answers in the nexus approach. *Waste Management*, *80*, 186–197. https://doi.org/10.1016/j.wasman.2018.09.009.

Laso, J., Vázquez-Rowe, I., Margallo, M., Irabien, Á., & Aldaco, R. (2018). Revisiting the LCA + DEA method in fishing fleets. How should we be measuring efficiency? *Marine Policy*, *91*, 34–40. https://doi.org/10.1016/j.marpol.2018.01.030.

Lijó, L., Lorenzo-Toja, Y., González-García, S., Bacenetti, J., Negri, M., & Moreira, M. T. (2017). Eco-efficiency assessment of farm-scaled biogas plants. *Bioresource Technology*, *237*, 146–155. https://doi.org/10.1016/j.biortech.2017.01.055.

Ling-Chin, J., Heidrich, O., & Roskilly, A. P. (2016). Life cycle assessment (LCA)—From analysing methodology development to introducing an LCA framework for marine photovoltaic (PV) systems. *Renewable and Sustainable Energy Reviews*, *59*, 352–378. https://doi.org/10.1016/j.rser.2015.12.058.

Lorenzo-Toja, Y., Vázquez-Rowe, I., Chenel, S., Marín-Navarro, D., Moreira, M. T., & Feijoo, G. (2015). Eco-efficiency analysis of Spanish WWTPs using the LCA+DEA method. *Water Research*, *68*, 651–666. https://doi.org/10.1016/j.watres.2014.10.040.

Lorenzo-Toja, Y., Vázquez-Rowe, I., Marín-Navarro, D., Crujeiras, R. M., Moreira, M. T., & Feijoo, G. (2018). Dynamic environmental efficiency assessment for wastewater treatment plants. *International Journal of Life Cycle Assessment*, *23*(2), 357–367. https://doi.org/10.1007/s11367-017-1316-9.

Lozano, S., Adenso-Díaz, B., & Barba-Gutiérrez, Y. (2011). Russell non-radial eco-efficiency measure and scale elasticity of a sample of electric/electronic products. *Journal of the Franklin Institute*, *348*, 1605–1614. https://doi.org/10.1016/j.jfranklin.2011.02.005.

Lozano, S., Iribarren, D., Moreira, M. T., & Feijoo, G. (2009). The link between operational efficiency and environmental impacts. A joint application of life cycle assessment and data envelopment analysis. *Science of the Total Environment*, *407*(5), 1744–1754. https://doi.org/10.1016/j.scitotenv.2008.10.062.

Lozano, S., Iribarren, D., Moreira, M. T., & Feijoo, G. (2010). Environmental impact efficiency in mussel cultivation. *Resources, Conservation and Recycling*, *54*(12), 1269–1277. https://doi.org/10.1016/j.resconrec.2010.04.004.

Masuda, K. (2016). Measuring eco-efficiency of wheat production in Japan: A combined application of life cycle assessment and data envelopment analysis. *Journal of Cleaner Production*, *126*, 373–381. https://doi.org/10.1016/j.jclepro.2016.03.090.

Mehdiloo, M., & Podinovski, V. V. (2019). Selective strong and weak disposability in efficiency analysis. *European Journal of Operational Research*, *276*(3), 1154–1169. https://doi.org/10.1016/j.ejor.2019.01.064.

Mohammadi, A., Rafiee, S., Jafari, A., Dalgaard, T., Knudsen, M. T., Keyhani, A., … Hermansen, J. E. (2013). Potential greenhouse gas emission reductions in soybean farming: A combined use of life cycle assessment and data envelopment analysis. *Journal of Cleaner Production*, *54*, 89–100. https://doi.org/10.1016/j.jclepro.2013.05.019.

Mohammadi, A., Rafiee, S., Jafari, A., Keyhani, A., Dalgaard, T., Knudsen, M. T., et al. (2015). Joint life cycle assessment and data envelopment analysis for the benchmarking of environmental impacts in rice paddy production. *Journal of Cleaner Production*, *106*, 521–532. Elsevier Ltd https://doi.org/10.1016/j.jclepro.2014.05.008.

Mohseni, P., Borghei, A. M., & Khanali, M. (2018). Coupled life cycle assessment and data envelopment analysis for mitigation of environmental impacts and enhancement of energy efficiency in grape production. *Journal of Cleaner Production*, *197*, 937–947. https://doi.org/10.1016/j.jclepro.2018.06.243.

Nabavi-Pelesaraei, A., Rafiee, S., Mohtasebi, S. S., Hosseinzadeh-Bandbafha, H., & Chau, K. (2017). Energy consumption enhancement and environmental life cycle assessment in paddy production using optimization techniques. *Journal of Cleaner Production*, *162*, 571–586. https://doi.org/10.1016/j.jclepro.2017.06.071.

Nakayama, H., Arakawa, M., & Yun, Y. B. (2003). Multiple criteria optimization: State of the art annotated bibliographic surveys. *International Series in Operations Research & Management Science*, *52*. https://doi.org/10.1007/0-306-48107-3_7.

Onat, N. C., Noori, M., Kucukvar, M., Zhao, Y., Tatari, O., & Chester, M. (2017). Exploring the suitability of electric vehicles in the United States. *Energy*, *121*, 631–642. https://doi.org/10.1016/j.energy.2017.01.035.

Opon, J., & Henry, M. (2020). A multicriteria analytical framework for sustainability evaluation under methodological uncertainties. *Environmental Impact Assessment Review*, *83*. https://doi.org/10.1016/j.eiar.2020.106403.

Paramesh, V., Arunachalam, V., Nikkhah, A., Das, B., & Ghnimi, S. (2018). Optimization of energy consumption and environmental impacts of arecanut production through coupled data envelopment analysis and life cycle assessment. *Journal of Cleaner Production*, *203*, 674–684. https://doi.org/10.1016/j.jclepro.2018.08.263.

Payandeh, Z., Kheiralipour, K., Karimi, M., & Khoshnevisan, B. (2017). Joint data envelopment analysis and life cycle assessment for environmental impact reduction in broiler production systems. *Energy*, *127*, 768–774. https://doi.org/10.1016/j.energy.2017.03.112.

Per, A., & Christian, P. N. (1993). A procedure for ranking efficient units in data envelopment analysis. *Management Science*, 1261–1264. https://doi.org/10.1287/mnsc.39.10.1261.

Pigné, Y., Gutiérrez, T. N., Gibon, T., Schaubroeck, T., Popovici, E., Shimako, A. H., et al. (2020). A tool to operationalize dynamic LCA, including time differentiation on the complete background database. *International Journal of Life Cycle Assessment*, *25*(2), 267–279. https://doi.org/10.1007/s11367-019-01696-6.

Pishgar-Komleh, S. H., Zylowski, T., Rozakis, S., & Kozyra, J. (2020). Efficiency under different methods for incorporating undesirable outputs in an LCA+DEA framework: A case study of winter wheat production in Poland. *Journal of Environmental Management*, *260*. https://doi.org/10.1016/j.jenvman.2020.110138.

Ramos, S., Vázquez-Rowe, I., Artetxe, I., Moreira, M. T., Feijoo, G., & Zufia, J. (2014). Operational efficiency and environmental impact fluctuations of the basque trawling fleet using LCA+DEA methodology. *Turkish Journal of Fisheries and Aquatic Sciences*, *14*(1), 77–90. https://doi.org/10.4194/1303-2712-v14_1_10.

Rebolledo-Leiva, R., Angulo-Meza, L., Iriarte, A., & González-Araya, M. C. (2017). Joint carbon footprint assessment and data envelopment analysis for the reduction of greenhouse gas emissions in agriculture production. *Science of the Total Environment*, *593–594*, 36–46. https://doi.org/10.1016/j.scitotenv.2017.03.147.

Rebolledo-Leiva, R., Angulo-Meza, L., Iriarte, A., González-Araya, M. C., & Vásquez-Ibarra, L. (2019). Comparing two CF+DEA methods for assessing eco-efficiency from theoretical and practical points of view. *Science of the Total Environment*, *659*, 1266–1282. https://doi.org/10.1016/j.scitotenv.2018.12.296.

Ren, J., Tan, S., Dong, L., Mazzi, A., Scipioni, A., & Sovacool, B. K. (2014). Determining the life cycle energy efficiency of six biofuel systems in China: A data envelopment analysis. *Bioresource Technology*, *162*, 1–7. https://doi.org/10.1016/j.biortech.2014.03.105.

Ribal, J., Sanjuán, N., Clemente, G., & Fenollosa, M. L. (2009). Eco-efficiency measurement in agricultural production. A case study on citrus fruits production. *Economia Agraria y Recursos Naturales*, *9*, 125–148.

Rybaczewska-Błażejowska, M., & Gierulski, W. (2018). Eco-efficiency evaluation of agricultural production in the EU-28. *Sustainability*, *10*, 1–21. https://doi.org/10.3390/su10124544.

Rybaczewska-Błażejowska, M., & Masternak-Janus, A. (2018). Eco-efficiency assessment of polish regions: Joint application of life cycle assessment and data envelopment analysis. *Journal of Cleaner Production*, *172*, 1180–1192. https://doi.org/10.1016/j.jclepro.2017.10.204.

Seiford, L. M., & Zhu, J. (1999). Infeasibility of super-efficiency data envelopment analysis models. *INFOR Journal*, *37*(2), 174–187. https://doi.org/10.1080/03155986.1999.11732379.

Soteriades, A. D., Faverdin, P., Moreau, S., Charroin, T., Blanchard, M., & Stott, A. W. (2016). An approach to holistically assess (dairy) farm eco-efficiency by combining life cycle analysis with data envelopment analysis models and methodologies. *Animal*, *10*(11), 1899–1910. https://doi.org/10.1017/S1751731116000707.

Tatari, O., & Kucukvar, M. (2012). Eco-efficiency of construction materials: Data envelopment analysis. *Journal of Construction Engineering and Management*, *138*(6), 733–741. https://doi.org/10.1061/(ASCE)CO.1943-7862.0000484.

Tavana, M., Izadikhah, M., Farzipoor Saen, R., & Zare, R. (2021). An integrated data envelopment analysis and life cycle assessment method for performance measurement in green construction management. *Environmental Science and Pollution Research*, *28*(1), 664–682. https://doi.org/10.1007/s11356-020-10353-7.

Tone, K. (2004). Dealing with undesirable outputs in DEA: A slacks-based measure (SBM) approach. *National Graduate Institute For Policy Studies (GRIPS) Research Report Series I-2003-0005*.

Tonne, K. (2001). A slacks-based measure of efficiency in data envelopment analysis. *European Journal of Operational Research*, *130*, 407–509. https://doi.org/10.1016/S0377-2217(99)00407-5.

Torregrossa, D., Marvuglia, A., & Leopold, U. (2018). A novel methodology based on LCA + DEA to detect eco-efficiency shifts in wastewater treatment plants. *Ecological Indicators*, *94*, 7–15. https://doi.org/10.1016/j.ecolind.2018.06.031.

Ullah, A., Perret, S. R., Gheewala, S. H., & Soni, P. (2016). Eco-efficiency of cotton-cropping systems in Pakistan: An integrated approach of life cycle assessment and data envelopment analysis. *Journal of Cleaner Production*, *134*, 623–632. https://doi.org/10.1016/j.jclepro.2015.10.112.

Vásquez-Ibarra, L., Rebolledo-Leiva, R., Angulo-Meza, L., González-Araya, M. C., & Iriarte, A. (2020). The joint use of life cycle assessment and data envelopment analysis methodologies for eco-efficiency assessment: A critical review, taxonomy and future research. *Science of the Total Environment*, *738*. https://doi.org/10.1016/j.scitotenv.2020.139538.

Vázquez-Rowe, I., & Iribarren, D. (2015). Review of life-cycle approaches coupled with data envelopment analysis: Launching the CFP + DEA method for energy policy making. *Scientific World Journal*, *2015*. https://doi.org/10.1155/2015/813921.

Vázquez-Rowe, I., Iribarren, D., Hospido, A., Moreira, M. T., & Feijoo, G. (2011). Computation of operational and environmental benchmarks within selected galician fishing fleets.

Journal of Industrial Ecology, *15*(5), 776–795. https://doi.org/10.1111/j.1530-9290.2011.00360.x.

Vázquez-Rowe, I., Iribarren, D., Moreira, M. T., & Feijoo, G. (2010). Combined application of life cycle assessment and data envelopment analysis as a methodological approach for the assessment of fisheries. *International Journal of Life Cycle Assessment*, *15*(3), 272–283. https://doi.org/10.1007/s11367-010-0154-9.

Vázquez-Rowe, I., & Tyedmers, P. (2013). Identifying the importance of the\skipper effect \within sources of measured inefficiency in fisheries through data envelopment analysis (DEA). *Marine Policy*, *38*, 387–396. https://doi.org/10.1016/j.marpol.2012.06.018.

Vázquez-Rowe, I., Villanueva-Rey, P., Iribarren, D., Teresa Moreira, M., & Feijoo, G. (2012). Joint life cycle assessment and data envelopment analysis of grape production for vinification in the Rías Baixas appellation (NW Spain). *Journal of Cleaner Production*, *27*, 92–102. https://doi.org/10.1016/j.jclepro.2011.12.039.

Zhong, F., Jiang, D., Zhao, Q., Guo, A., Ullah, A., Yang, X., et al. (2020). Eco-efficiency of oasis seed maize production in an arid region, Northwest China. *Journal of Cleaner Production*, *268*. https://doi.org/10.1016/j.jclepro.2020.122220.

Zhou, H., Yang, Y., Chen, Y., & Zhu, J. (2018). Data envelopment analysis application in sustainability: The origins, development and future directions. *European Journal of Operational Research*, *264*(1), 1–16. https://doi.org/10.1016/j.ejor.2017.06.023.

Zhou, P., Ang, B. W., & Wang, H. (2012). Energy and CO_2 emission performance in electricity generation: A non-radial directional distance function approach. *European Journal of Operational Research*, *221*(3), 625–635. https://doi.org/10.1016/j.ejor.2012.04.022.

Zhu, J. (2001). Super-efficiency and DEA sensitivity analysis. *European Journal of Operational Research*, *129*(2), 443–455. https://doi.org/10.1016/S0377-2217(99)00433-6.

CHAPTER

Territorial Life Cycle Assessment

Eléonore Loiseau[a,b], Thibault Salou[a,b], and Philippe Roux[a,b]
[a]*ITAP, Univ Montpellier, INRAE, Institut Agro, Montpellier, France* [b]*Elsa, Research Group for Environmental Lifecycle and Sustainability Assessment, Montpellier, France*

1 Introduction

Life cycle assessment (LCA) has been initially designed to assess the environmental performance of a product or a service, according to a life cycle perspective (Rebitzer et al., 2004). Although product oriented, proposals have been made to broaden the object of analysis of LCA towards meso-scale and macro-scale objects (Guinée et al., 2011). More and more LCA applications are being carried out on a broader scale including LCA of organisations, lifestyles, or even countries (Hellweg & Mila i Canals, 2014). Amongst several tools and methods (e.g. material and energy flow analysis, the ecological footprint, emergy and exergy analysis), Loiseau, Junqua, Roux, and Bellon-Maurel (2012) identified LCA as a promising approach to assess the environmental performances of territories due to key features. First, by taking a lifecycle and multi-criteria perspective, LCA allows to identify burden shifting between geographical areas and impact categories (Finnveden et al., 2009). Second, LCA relies on the use of a functional unit (FU) to compare alternatives on a same reference, thus making it possible to assess the environmental performance of very different systems but providing the same service.

Territories can be defined as the interface between a geographical space and a group of stakeholders who use, manage, and develop it (Moine, 2006). They cover three dimensions, i.e. (i) the material dimension of a geographic area defined by the physical properties that can be considered as opportunities or constraints for the development of human systems; (ii) the organisational dimension defined by social and institutional actors structured within activities, organisations, or jurisdictions that embody the strategies of territorial development; and (iii) the identity dimension defined by the way social and institutional stakeholders think and implement a project for their territory (Laganier, Villalba, & Zuindeau, 2002). Closely linked to environmental issues and to local stakeholders, territories have a central role to play in the ecological transition of societies as a driving force for innovation. This concerns territories located in rural areas as well as large urban metropolitan areas, and includes

Assessing Progress towards Sustainability. http://doi.org/10.1016/B978-0-323-85851-9.00011-0

all types of sub-national units such as cities, city-region areas, group of municipalities, and all levels of the Nomenclature of Territorial Units for Statistics defined for the countries of the European Union (Regulation (EC), 2003). Many initiatives led by local authorities have emerged around the world to foster the transition towards sustainability such as the Cities Climate Leadership Group,[a] the World Cities Network,[b] the Energy Cities,[c] the Regions of Climate Action,[d] or the Local Governments for Sustainability.[e] To support these initiatives, quantitative assessment methods are needed to provide objective information on their environmental performance.

First guidelines to account for energy consumption and greenhouse gas (GHG) emissions at community scale have been proposed by the World Resource Institute (WRI) and the World Bank Committee on Sustainable Development (WBCSD) (WRI, 2004) and by the British standard (PAS 2070, 2014). However, broader perspectives including multi-criteria approaches are currently lacking for assessing the environmental impacts of land planning policies defined at sub-national scales such as those required by the Environmental Impact Assessment (EIA) or the Strategic Environmental Assessment (SEA) European directives (European Commission, 2009). Consequently, several authors proposed to use LCA in territorial planning to adopt a life cycle and multi-criteria approach (Albertí, Balaguera, Brodhag, & Fullana-i-Palmer, 2017; Beloin-Saint-Pierre et al., 2017; Bidstrup, 2015; Björklund, 2012; Larrey-Lassalle et al., 2017; Loiseau, Roux, Junqua, Maurel, & Bellon-Maurel, 2013; Mazzi, Toniolo, Catto, De Lorenzi, & Scipioni, 2017). Given these elements, broadening the object of analysis in LCA points the way towards a new LCA-based approach, that can be called 'territorial LCA' (Loiseau et al., 2013).

This chapter describes the main principles of territorial LCA approaches (Loiseau et al., 2018) according to the four LCA phases defined in the ISO Standards (ISO 14040, 2006; ISO 14044, 2006) and by discussing different methodological choices. It will then give an extensive overview of the applications of territorial LCAs in the peer-reviewed scientific literature. This review allows identifying the main areas of use of territorial LCA approaches and inventorying the main methodological limitations (e.g. data collection, considering spatial and temporal variability or territorial organisation). To address them, proposals have been made to use other tools resulting from works on urban metabolism using material flow analysis (MFA), but also tools such as geographic information system (GIS) or economic modelling. These couplings are discussed. Finally, this chapter concludes on the main research areas to be explored in the future.

[a]http://www.c40.org/.
[b]http://www.worldcitiesnetwork.org.
[c]http://www.energy-cities.eu/.
[d]http://regions20.org.
[e]http://www.iclei.org.

2 Main principles of territorial LCA

According to (Loiseau et al., 2018), two main types of territorial LCA approaches can be defined (see Fig. 1): (i) type A, which allows the contextualisation of the LCA of a sector or a specific supply chain with a strong local anchorage (e.g. waste management), and (ii) type B, which adopts a cross-sectoral approach up to the consideration of all the human activities of a given territory (all production and consumption activities). These two types of approaches are described according to the four LCA phases defined in the ISO Standards (ISO 14040, 2006; ISO 14044, 2006).

2.1 Goal and scope definition

Territorial LCA approaches have two main goals, i.e. (i) providing a comprehensive environmental baseline, and (ii) comparing the environmental performance of managing and land planning scenarios. The main outputs of the approach will provide valuable and objective information for decision-makers and local stakeholders to elaborate policy, manage territorial services, or to communicate to citizens on territorial environmental performances. The shift from a product-oriented approach to a territorial-based approach raises two challenges in terms of goal and scope definition (Loiseau et al., 2013). The first deals with the system boundaries and the second with the functional unit (FU) definition.

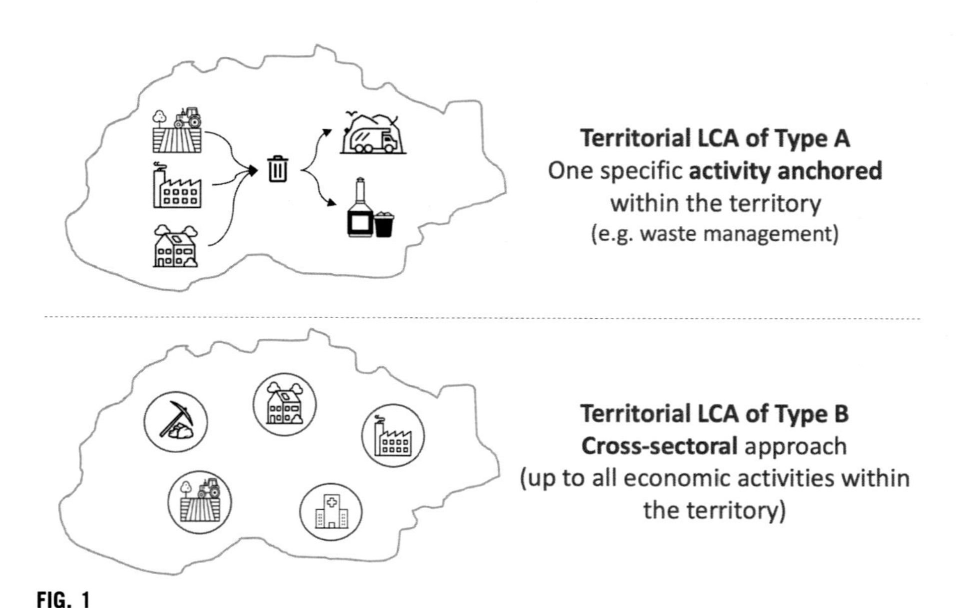

FIG. 1

Illustration of the two main types of territorial LCA approaches.

System boundaries and territorial responsibility

In a territorial LCA approach, the issue of the system boundaries under study can be broken down into three main points, i.e. (i) the choice of the geographical boundaries, (ii) the scope of the activities under study, and (iii) the definition of the territorial responsibility principle (see Fig. 2).

One straightforward choice for the geographical boundaries is to consider the administrative ones as they are objective and they can better support the data gathering process (Loiseau et al., 2013; Mirabella, Allacker, & Sala, 2019). Yet, Albertí, Brodhag, and Fullana-i-Palmer (2019) showed that other choices may seem more justified to stakeholders, especially as the territory boundaries may be ambiguous. The choice of administrative boundaries can be arbitrary, and it is also possible to delimit a territory by taking into account an entire population catchment area or by taking into account a density of services that goes beyond administrative boundaries.

The second point concerns the territorial activities considered in the evaluation. In the case of type B territorial LCAs, it is a question of having a cross-sectoral approach. However, it is still necessary to clearly define the activities to be considered. This may involve looking at activities related to the lifestyles of the populations in place (inhabitants, tourists, workers, beneficiaries of services, etc.), or to the economy as a whole, or related to the provision of public services.

Then, the last point deals with the territorial responsibility in terms of environmental impacts. Different perspectives can be adopted to allocate these impacts. On

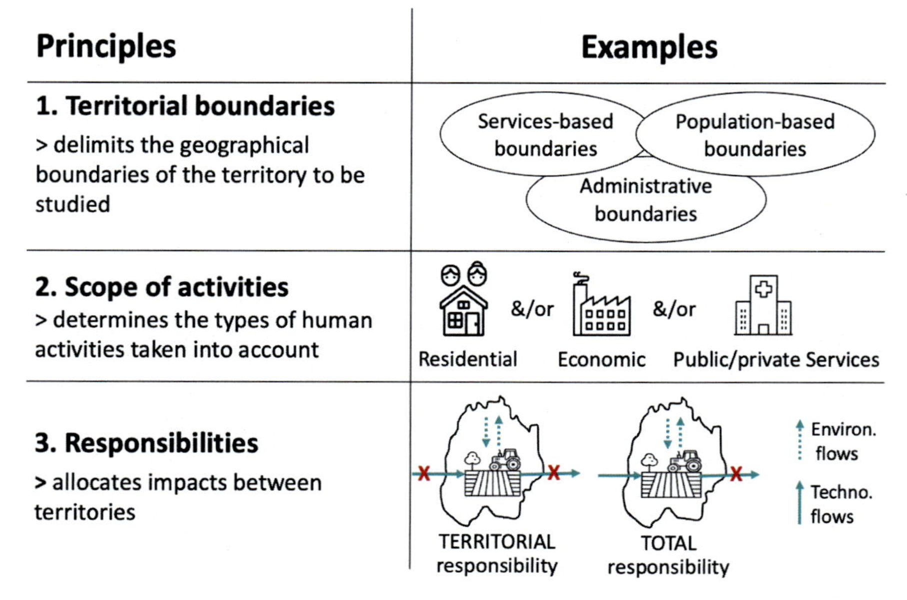

FIG. 2

The three principles to consider when defining system boundaries in territorial LCA.

one hand, a production-based perspective focuses on the direct emissions and resources consumption generated by the activities within a defined territory (Ferng, 2003). This is the principle of territorial responsibility, generally adopted for national accounting in the case of greenhouse gas emissions. On the other hand, a consumption-based approach includes all the upstream and downstream impacts related to the territorial activities (Muradian, O'Connor, & Martinez-Alier, 2002). Eder and Narodoslawsky (1999) defined variants of territorial responsibility based on these two approaches. In territorial LCA, Loiseau et al. (2013) opted for the principle of total territorial responsibility based on the consideration of all impacts that take place within the territory including the indirect impacts incurred by the imports of goods and services. According to this principle, exports are not included. From a practical point of view, it seems impossible to know all the downstream life cycles related to exported products (e.g. an exported agricultural raw material may have multiple potential uses in the rest of the world that cannot be determined). At a global scale, this principle of responsibility leads to a double counting issue. However, on a territorial scale, this provides an exhaustive view of the direct and induced impacts of a territory's activities and enables pollution transfers to be clearly identified.

Albertí, Brodhag, and Fullana-i-Palmer (2019) investigated other possibilities such as responsibility based on value creation or on the type of impacts considered. The latter suggests attributing global impacts (e.g. climate change) where the consumption takes place, whilst local impacts (e.g. eutrophication, water use) would be allocated where the impacts actually occur. However, these proposals require a lot of data and processing to be implemented in practice.

Ultimately, system boundary definition relies on a range of methodological choices that must be transparent and be made in concertation with the stakeholders of the territory in line with the objectives of the study.

Functional unit definition

Territories are by definition multifunctional systems, which deliver many services such as the provision of food, housing, work, or transportation, or the access to health, cultural, or recreational services (Banski et al., 2011; Pérez-Soba et al., 2008). It is therefore necessary to consider this diversity of functions in quantifying the environmental performance of these systems to carry out robust comparative assessments. Yet, dealing with multifunctionality in LCA remains an important issue, intensified in the case of territorial systems.

In territorial LCA of type A, a main function is defined for the specific territorial activity under study considering the local context, and system expansion or allocation is performed to handle multifunctionality (Loiseau et al., 2018).

In territorial LCA of type B, two approaches are proposed to deal with multifunctionality. On one hand, a unique FU needs to be defined, based for instance on the number of residents, and normalisation is proposed to consider other socio-economic features specific to the territory under study. Mirabella et al. (2019) proposed to define a population equivalent to take into account the share of people that takes advantages of the territorial services (e.g. inhabitants, tourists, commuters, etc.). Albertí, Roca, Brodhag, and Fullana-i-Palmer (2019) introduced the City Prosperity

Index (CPI) developed by the United Nations Human Settlements Programme (UN-Habitat) (Moreno & Murguía, 2015) as technical performance within the FU. It considers all the dimensions of the Human Development Index and other ones (e.g. economic productivity, or governance and legislation). On the other hand, Loiseau et al. (2013) proposed to adapt the concept of FU for territorial LCA. It is no longer a question of quantifying an environmental intensity (e.g. kg CO_2 eq/FU), but of quantifying an eco-efficiency. Territorial eco-efficiency is defined as a ratio between a basket of services provided by a territory and the environmental impacts generated by this same system (Seppälä et al., 2005). The starting point is the definition of a reference flow as the combination of the studied territory and a given land planning scenario (i.e. baseline, prospective, or retrospective scenarios), and the quantification of the services provided is an output of the approach.

2.2 Life cycle inventory

Territorial life cycle inventory (LCI) can be divided into two stages (Loiseau et al., 2013; Loiseau, Roux, Junqua, Maurel, & Bellon-Maurel, 2014; Roibás, Loiseau, & Hospido, 2017). The first stage consists of collecting primary data on the system under study, i.e. the direct flows of products, services, materials, or energy circulating in the territory. The second step is to compute secondary data on background processes (upstream or downstream of life cycles). Primary data can be derived from local data, surveys, statistics, or from national or regional data after a downscaling process. Secondary data mainly come from existing LCA databases. Depending on the size of the sector or the number of processes to model and the geographical scale of analysis, two main types of LCA modelling can be used, i.e. (i) process-based LCA approach (bottom–up approach) or (ii) environmental input–output tables (EIO) LCA (top–down approach) (see Fig. 3) (Peters, 2010).

Process-based LCA approach computes LCIs for products or services. These process inventories can be modified to be more representative of the studied territory (e.g. by adapting the electricity mix used) (Patouillard et al., 2018). If the processes do not exist in LCA databases (such as Ecoinvent), it is necessary to create them or to use similar processes as proxies (e.g. the impacts of producing 1 kg of pears in the orchard do not exist, it is possible to use those of apples) (Mila i Canals et al., 2011). EIO-LCA relies on IO tables with environmental extensions to assess the environmental impacts of economic sectors as a whole. With as inputs monetary data on the amounts of goods and services produced or consumed, it is possible to estimate the impacts of territorial industrial sectors or household consumption.

In any case, data collection remains a crucial and time-consuming step, and strategies can be deployed to make it more effective depending on the available data and the objectives of the study (Roibás, Loiseau, & Hospido, 2018a, 2018b).

2.3 Life cycle impact assessment

This phase is based on existing life cycle impact assessment (LCIA) methods. Methods based on midpoint and endpoint indicators such as ReCiPe (Huijbregts et al., 2016) or Impact World+ (Bulle et al., 2019) have the advantage of aggregating

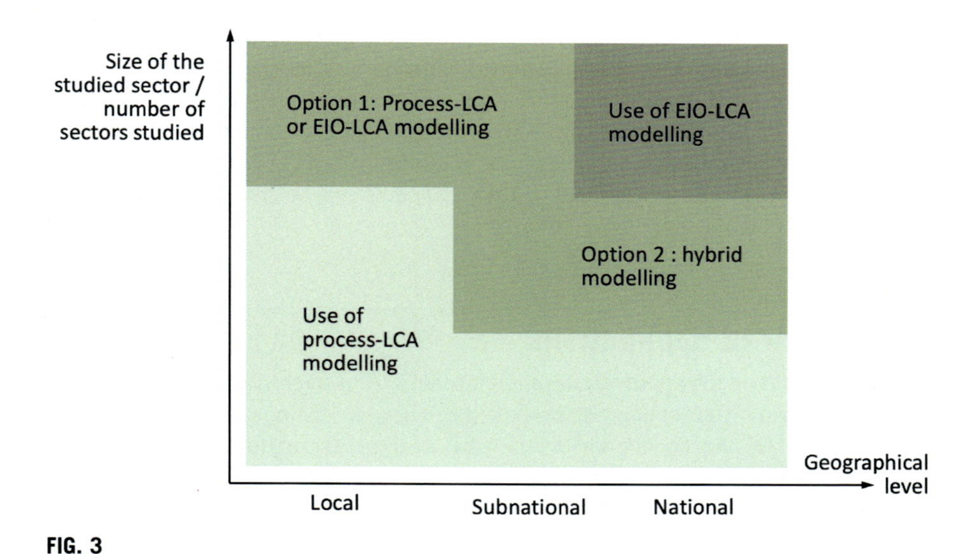

FIG. 3

Choice of LCA modelling approaches depending on the size of the studied territorial system.

the midpoint contributions in damage over the three protection areas (i.e. human health, ecosystem quality, and resources). It also allows proposing a hierarchy of environmental issues based on biophysical models. Besides, significant methodological advances have been made in recent years to take into account spatial variability in regionalised LCIA methods (Mutel et al., 2019). It is therefore strongly recommended to use these methods for territorial approaches, at least for the local and regional impacts induced by the environmental flows located in the studied area.

2.4 Interpretation

An important point in the interpretation of the results consists in correctly identifying the pollution transfers between the studied territory and the rest of the world. In the impact methods commonly used to evaluate projects and developments on territories, only direct impacts are usually taken into account, considering that these are the ones of interest to local stakeholders (Bidstrup, 2015). To address this issue, it is possible to make a distinction between 'on-site' and 'off-site' impacts in LCA (Larrey-Lassalle et al., 2017; Loiseau et al., 2013). 'On-site' impacts are those generated by direct pollutant emissions on the territory or by local resource consumption. 'Off-site' impacts are induced by territorial activities and are taking place all around the world. In the case of a wastewater treatment plant (WWTP), 'on-site' impacts are induced by direct emissions of GHGs into the air and of nitrates into water bodies, and 'off-site' impacts are due to the electricity production used for the WWTP operation such as resource depletion or ionising radiations.

Moreover, it is possible to distinguish between global impacts, regardless of where environmental flows take place, the impacts will be the same (e.g. climate change impacts) and regional/local impacts where the intensity of the impact depends on the conditions of the places where resources are consumed or pollutants are emitted (e.g. eutrophication) (Potting & Hauschild, 2006). In this case, more in-depth analyses on local/regional 'on-site' impacts can be carried out to improve the robustness of the results.

3 Overview of applications

A state of the art on the applications of territorial LCA in scientific literature has been performed. Two different approaches have been used to distinguish between Type A and Type B. As the conceptual framework of type B territorial LCA is recent (Loiseau et al., 2012), there are few applications, and most of them have been identified and reported in this section. Conversely, type A applications, which consist of a territorial contextualisation of LCAs carried out for a specific sector of activity, are more numerous. In this case, the overview is rather a question of analysing, on the basis of a corpus of publications, the main areas of application, and the specific questions addressed to LCA.

3.1 Type A: Bibliometric analysis

Although relevant, the term 'territory' is little used in English language. Queries on the database of peer-reviewed literature using the keywords 'LCA' and 'territory' therefore yield few results. To inventory existing work, it is necessary to use other keywords. Two families of keywords were chosen, namely 'urban/city' and 'region', to identify the corpus of papers relating to type A territorial LCA. The searches were carried out on the ISI Web of Science database without geographical or temporal limits. In addition, the corpus of papers were analysed using VOSviewer, a freely available text mining software to generate bibliometric maps based on co-occurrence matrix using keywords (van Eck & Waltman, 2010).

Results for the search on 'LCA' and 'urban/city'

There were 1778 results corresponding to this query. After processing the corpus on VOSviewer to identify the main keywords used, the map displayed in Fig. 4 was produced. Four clusters have been determined and make it possible to highlight the main fields of application of this type of LCA. The first group in red is related to the building sector and focuses in particular on the issues of energy consumption and CO_2 emissions. This is also the case for the second group, in yellow, which focuses on mobility. The third group, in green, relates to water management with issues related to impacts, especially eutrophication and acidification, and also to the FU. Finally, the last group, in blue, is concerned with waste management, with keywords dealing with the collection and treatment stages (landfill, incineration) and energy recovery.

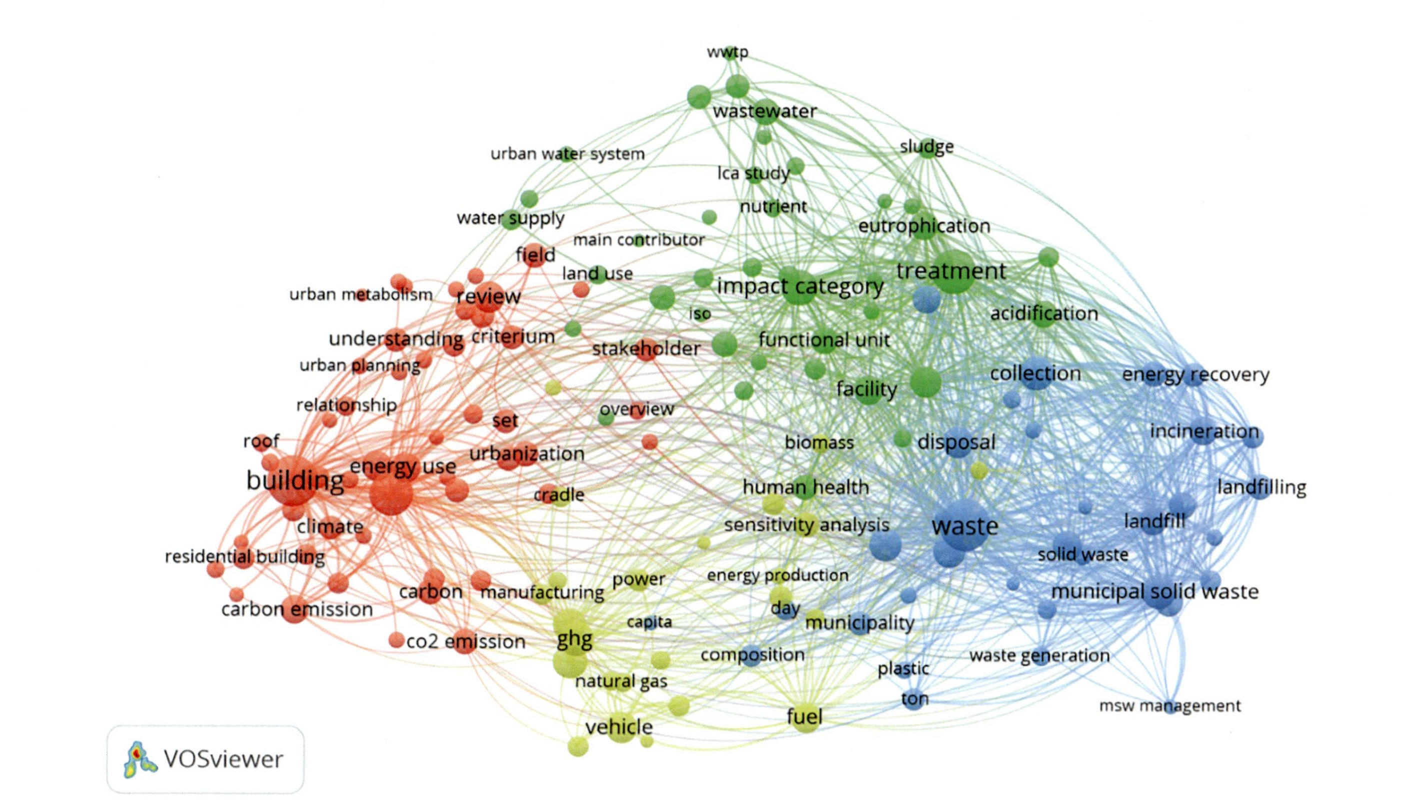

FIG. 4

Co-occurrence of keywords used in the field of 'urban/city' LCA (created with VOSviewer).

This meta-analysis provides a quick overview of the sectors studied and the issues addressed in type A LCA in an urban context. The results are consistent with other review works carried out on the implementation of LCA in urban systems (Mirabella et al., 2019; Petit-Boix et al., 2017). There is a strong emphasis on waste, water, and building sectors according to (Petit-Boix et al., 2017) and mono-criteria LCAs (e.g. carbon footprint analysis) were common in 20% of the studies, highlighting the need to integrate more impact categories to prevent trade-offs in studies.

Results for the search on 'LCA' and 'region'

There were 2675 results corresponding to this query. After processing the corpus on VOSviewer to identify the main keywords used, the map displayed in Fig. 5 was produced. This time, five clusters have been formed. Four groups relate to sectors of activity, i.e. agriculture (in red), bioenergy (in green), mobility (in purple), and waste management (in blue) showing a rural context prevalence. The last group focuses on the life cycle impact assessment (LCIA) phase. This result is not surprising as the regionalisation of LCA impacts is a research field since the late 1990s (Potting & Hauschild, 2006), and numerous works have been published in order to develop regionalised LCIA methods for LCA practitioners such as the latest LC-Impact (Verones et al., 2020) and Impact World+ (Bulle et al., 2019) methods. Moreover, this explains why the term 'territorial' was preferred to 'regional' when referring to LCAs conducted on meso-scale systems to avoid confusion.

Mobility and waste management have also emerged in urban LCAs as key sectors of activity. However, on closer scrutiny, the issues addressed differed. In the case of mobility, the emphasis seems to be here on the electric vehicles, whereas the issue of biogas stands out in waste management.

In the field of agriculture, studies seem to focus on both crop and livestock production with a focus on yields and productivity. Moreover, environmental concerns in terms of fertilisation (i.e. use of fertiliser, nitrogen emissions, and eutrophication), phytosanitary treatments, and land use are also highlighted. Finally, the main issues addressed in bioenergy sector relate to GHGs, energy consumption, and land use.

This meta-analysis provides an overview of the sectors of activity studied in type A territorial LCAs and the questions addressed. Although bibliometric research has identified more papers (results that can be explained by the bias due to the papers on the regionalisation of LCA impacts) than for urban LCAs, to the authors' knowledge there are currently no reviews of type A territorial LCA in a regional context. More detailed analysis would need to be carried out to clarify the aims, methods, and limitations of such studies.

3.2 Type B: Comparative study of the peer-reviewed papers

A bibliographical search identified 16 scientific papers dealing with territorial LCA of type B with the implementation on a real or theoretical case study (see Table 1). The works of Roibás et al. (2017) and (Roibás et al., 2018a) have been grouped together because the latter proposes methodological adaptations of the former,

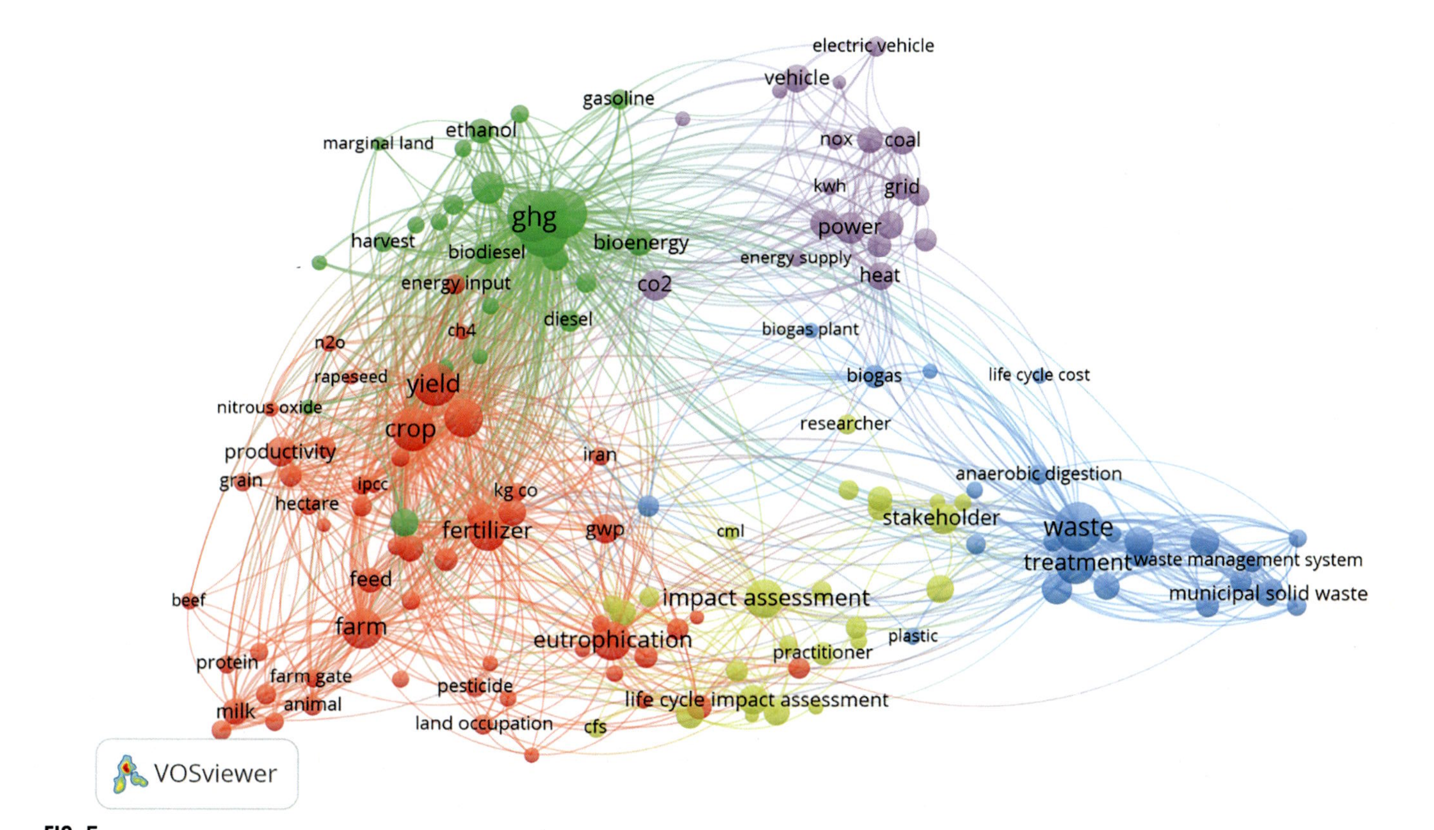

FIG. 5

Co-occurrence of keywords used in the field of 'region' LCA (created with VOSviewer).

Table 1 Main methodological characteristics of type B territorial LCA approaches applied to case studies in the peer-reviewed scientific literature.

References	Aim	Audience	Boundaries	FU	LCI	LCIA
Goldstein, Birkved, Quitzau, and Hauschild (2013)	Identification of the main impact sources, inter-city comparisons, and tracking temporal shifts in a city's sustainability resulting from policy interventions	Policy makers	**1.** Administrative (cities) **2.** Focus on main metabolic flows of residents and industries (e.g. food, construction materials, energy, water, other materials, waste and wastewater management) **3.** Total responsibility	No FU but results are normalised to the per capita level (used as a conceptual average citizen)	Primary data for the foreground system based on metabolic flow computation and LCA database for the background system (Ecoinvent)	Global characterisation factors (ReCiPe LCIA method)
Loiseau et al. (2014)	Provision of a consistent territorial environmental baseline to identify main impact contributors and environmental issues	Local decision-makers and managers	**1.** Administrative (a group of municipalities) **2.** All production and consumption activities **3.** Total responsibility	No FU but a basket of land-use functions (environmental, economic, environmental) used to quantify territorial eco-efficiency	Primary data for the foreground system based on local statistics and surveys and downscaling of regional/national data, and secondary data based on both Ecoinvent and EEIO databases (hybrid modelling)	Global characterisation factors (ReCiPe and Impact World+ LCIA method)

Nitschelm, Aubin, Corson, Viaud, and Walter (2016)	Land planning purpose (activity locations)	Farmers, natural resources managers	**1.** Geographical or defined by agricultural land coverage **2.** Only farm activities **3.** Total responsibility (farm gate)	Choice of one main function amongst basket of agricultural territory functions (environmental, economic, and societal)	Definition of farm activity typologies and pedo-climatic conditions to compute spatially differentiated foreground LCIs. Secondary data based on LCA database	Definition of biophysical environment typologies to quantify local on-site impacts and CML 2001 for other impacts
Avadí, Nitschelm, Corson, and Vertès (2016)	Diagnosis and assessment of the effects of innovative agricultural practices on a region's impacts	Farmers, local stakeholders	**1.** Agricultural area **2.** Only farm activities (crops and livestocks) **3.** Total responsibility (farm gate)	Multiple FUs: ha, kg fat-and-protein-corrected milk, kg live weight (LW)	Proposal of a general approach to build farm typologies and regional extrapolation of farm-level LCAs to compute representative LCIs of farming regions	Global characterisation factors (CML 2001 LCIA method)
Mazzi et al. (2017)	Prioritisation of the feasible measures and help for the adoption of the criteria of public commitment to sustainability	Stakeholders and administrators	**1.** Administrative (a municipality) **2.** Focus on water, electricity, natural gas, and waste management services consumed by territorial activities	The execution and provision of human, commercial, and administrative activities by the municipality during 1 year	Primary sources collected within the municipality and secondary data based on LCA databases (Ecoinvent)	Global characterisation factors (ReCiPe LCIA method)

Continued

Table 1 Main methodological characteristics of type B territorial LCA approaches applied to case studies in the peer-reviewed scientific literature. *Continued*

References	Aim	Audience	Boundaries	FU	LCI	LCIA
			(inhabitants, tourists, economic activities) and the municipality **3.** Total responsibility			
Roibás et al. (2017, 2018a)	Environmental diagnosis and identification of the main contributing activities to carbon footprint	Decision-makers	**1.** Administrative (a region) **2.** All production and consumption activities **3.** Total responsibility	No FU but an average inhabitant and tourist are used as reporting units for consumption activities	Different approaches are compared to compute data based only on EEIO databases or using a hybrid modelling. Particular attention is paid to avoid double counting between the activities within the territory	Only carbon footprint
García-Guaita, González-García, Villanueva-Rey, Moreira,	Evaluation of the city's environmental profile and identification of	Decision-makers	**1.** Administrative (a city) **2.** Focus on resident consumption related to six	An average conceptual citizen (inhabitant equivalent)	Downscaling of regional and national data for primary sources and literature papers for the	Global characterisation factors (ReCiPe LCIA method)

and Feijoo (2018)	the key responsible flows		main items (water, food, transport, energy, waste, and wastewater management) **3.** Total responsibility		background system	
Roibás et al. (2018b)	Providing normalisation factors at sub-national scale (a Spanish region)	LCA practitioners and decision-makers	**1.** Administrative (a region) **2.** All production and consumption activities **3.** Total responsibility	No FU but normalisation factors are reported in absolute terms, per capita and per worker.	Primary data for the foreground system based on local statistics and surveys and downscaling of regional/national data, and secondary data based on both Ecoinvent and EEIO databases (hybrid modelling)	Global characterisation factors (ReCiPe LCIA method)
González-García and Dias (2019)	Identification of environmental hotspots, evaluation of decision-making strategies, and education of city's residents about sustainable lifestyles	Decision-makers, residents	**1.** Administrative (a city) **2.** Focus on resident consumption with six main groups of flows (construction materials, fossil fuels, energy, food, waste, and other flows such	Consumption of materials and the generation of wastes per inhabitant per year	Primary data based on local, and downscaling of regional/national data and use of LCA database for the background system (Ecoinvent, LCA food or literature)	Global characterisation factors (ReCiPe LCIA method)

Continued

Table 1 Main methodological characteristics of type B territorial LCA approaches applied to case studies in the peer-reviewed scientific literature. *Continued*

References	Aim	Audience	Boundaries	FU	LCI	LCIA
			as paper, glass, cardboard, plastic, wood) **3.** Total responsibility			
Qi, Zhang, Jiang, Hou, and Li (2019)	Assessment of the environmental protection performance of cities, identification of the main contributors of the environmental impact to provide direction for future environmental protection decision-making	Decision-makers	**1.** Administrative **2.** Focus on all production and consumption activities **3.** Total responsibility	FU: per annual gross domestic product (GDP) of a studied and quantification of a set of land-use functions	Primary data from local statistics and secondary data based on LCA database (USLCI)	Global characterisation factors (ReCiPe LCIA method)
Cremer, Müller, Berger, and Finkbeiner (2020)	City's environmental performance for understanding, communication, developing roadmaps, adapting tender requirements and subsidies	Citizens/ employees, local and the national government, private city service providers, and other commercial sectors	**1.** Administrative (a city) **2.** Four levels of assessment: (i) public service provision by city-owned companies, (ii) general public service provision, and (iii) activities beyond public service provision	No FU but a reporting unit as in OLCAs with (1) the reporting organisation and (2) the reporting flow (based on product portfolio)	Based on the reporting flow for the primary data and no indication for background data	All types of LCIA methods can be used

			3. Total responsibility for level 1 and 2 (issues with level 3)			
Ding, Bourrelly, and Achten (2020)	Facilitating decision making in the design of land-use planning.	Decision-makers	**1.** Administrative (a region) **2.** Focus on farm activities **3.** Total responsibility (farm gate)	FU: numbers of ha	Development of an in- and off-territorial Emission Factor (tEF) to consider spatial variability in agricultural practices	Global characterisation factors (Impact World+ method)
Zhang, Su, Zhu, and Li (2020)	Identification of environmental hotspots, provision of directions and references for management, prioritisation and justification of the future development of a region	Urban administrators, decision-makers	**1.** Administrative (district area) **2.** All buildings and infrastructures for all types of production and consumption activities and all traffic activities **3.** Total responsibility	The incremental construction and operation of all the buildings and infrastructures in the planning area in the target year	Primary data based on local data and statistics and secondary data based on LCA database (no more information)	BEPAS method (Zhang, Wu, Yang, & Zhu, 2006) and policy distance to target information for weighting
Borghino et al. (2021)	Supporting informed decision making by assessing scenarios of change	Local actors	**1.** PDO (Protected Designation of Origin) of a French Cheese (Brie de Meaux) **2.** Focus on farm activities	FU: unit area (ha of agricultural land) and biomass produced (kg of products) and total impacts for	Primary data based on local interviews and expertise and secondary based on Agribalyse and Ecoinvent except	ILCD for water, CML for abiotic resource depletion, acidification potential, eutrophication

Continued

Table 1 Main methodological characteristics of type B territorial LCA approaches applied to case studies in the peer-reviewed scientific literature. *Continued*

References	Aim	Audience	Boundaries	FU	LCI	LCIA
			3. Including life cycles of agricultural products from raw material extraction to storage (for grain crops) or to the first stage of processing even if they occurred outside of the geographical boundaries	the farming activities	for direct on-farm emissions which are computed using MEANS-InOut software	potential, and land competition, and IPCC for climate change
González-García, Caamaño, Moreira, and Feijoo (2021)	Performing a sustainability diagnosis of a large city	Policy makers, citizens, local government	**1.** Administrative (a city) **2.** All consumption and production activities with a focus on seven main items (tap water, food, energy, transport, manufactures, waste, wastewater) **3.** Total responsibility	Inhabitant equivalent unit based on the amount of permanent population and the floating population (visitors, workers, students)	Based on local and regional data for the foreground system and LCA database (Ecoinvent) and scientific papers for the background system	Global characterisation factors (ReCiPe LCIA method)

leading to 15 case studies. The reference to territorial LCA is not necessarily explicit in all the papers. This is particularly the case for papers with an urban application; however, the city is also a territory, and all the identified papers deal with the implementation of LCA for a cross-sectoral evaluation of a territory. Table 1 compares the methodological choices made to carry out the type B territorial LCAs according to the goal and scope definition, the LCI and the LCIA phase.

In the goal and scope phase, there are many similarities regarding the objectives of the studies and the targeted audience. Most studies carry out an environmental diagnosis for decision-makers in order to identify the main hotspots and contributors to the impacts with the aim of prioritising the measures to be implemented and defining sustainable roadmaps. Some studies also aim to educate local stakeholders (González-García & Dias, 2019), to take into account environmental criteria to guide public tenders and subsidies (Cremer et al., 2020), to determine the best place to locate an activity (Nitschelm et al., 2016), or to compare scenarios (Borghino et al., 2021). An original study proposes to use the territorial LCA approach to develop normalisation factors at a sub-national scale (Roibás et al., 2018a).

On the other hand, there are great differences in the definition of system boundaries and FUs. Although almost all the studies take into account the administrative criterion for delimiting geographical boundaries and adopt the principle of total responsibility, the scope of activities and flows taken into account vary greatly from one study to another. Three studies focus solely on agricultural activities and propose methodological developments to better take into account the diversity of agricultural practices or integrate environmental conditions in the assessment (Avadí et al., 2016; Borghino et al., 2021; Ding et al., 2020; Nitschelm et al., 2016). One study only includes road traffic, buildings, and infrastructure related to all urban activities (residents and economic activities) (Zhang et al., 2020). Two studies focus mainly on the environmental performance of public services including the municipality management (Cremer et al., 2020; Mazzi et al., 2017). It should also be noted that these two studies focus on two quite distinct evaluation procedures, respectively, Organisational LCA and Environmental Management System (EMAS) which explain the focus on the municipality organisation level. Four studies investigate the combination between urban metabolism and LCA (García-Guaita et al., 2018; Goldstein et al., 2013; González-García et al., 2021; González-García & Dias, 2019). They take into account a limited number of flows, mainly related to the inhabitant consumption, with sometimes the consideration of other activities such as services or local industries. Finally, four studies consider all the production and consumption activities located in a given territory (Loiseau et al., 2014; Qi et al., 2019; Roibás et al., 2017, 2018b).

The system boundary choices have a direct influence on the FU definition and can explain the differences between the case studies. If the focus is on a rural territory, then FUs taking into account agriculture multifunctionality will be defined such as the surface area to support biodiversity and recreational activities, or the production of food and revenues (Nitschelm et al., 2016). In the case of urban LCAs, FUs based on inhabitant equivalents have been proposed with more or less degree of refinement. For instance, González-García et al. (2021) include information on the time spent by each type of population in the calculation of the inhabitant equivalent. Finally, two studies that focus on all territorial production and consumption activities define a basket of land-use functions to assess territorial performance. At the end, all studies agree that this question of FU definition remains tricky.

The LCI phase is largely based on local data for the foreground system, with regional and national downscaling where necessary (González-García & Dias, 2019; Loiseau et al., 2014). For secondary data, LCA databases such as Ecoinvent are frequently used. Only Loiseau et al. (2014), Roibás et al. (2017) and Roibás et al. (2018b) use EIO-LCA to model some sector activity (manufacturing industries and household consumption). Roibás et al. (2017) also propose an approach to take into account double counting within territorialised value chains. For example, in their case study, there are coal power plants, and the electricity produced is used both for production and consumption activities on the territory, and for exports. It is required not to count the impacts of this energy twice in territorial activities. To do this, it is necessary to first identify and quantify these intra-territorial flows, and then allocate the impacts between the upstream and downstream value chains. In this study, the choice was made to allocate the impacts to the most downstream activities in these chains. Furthermore, Roibás et al. (2018a) discuss different options to simplify the data collection by using only EEIO databases. To ensure the robustness of the results, it is recommended to use, if available, databases at a sub-national level taking into account local characteristics.

Finally, all studies rely on global LCIA methods for impact assessment, although they recommend considering spatial variability.

4 Combinations with other tools

The analysis of the literature pointed to a significant amount of case studies in the field of type A territorial LCAs, and that type B applications are beginning to develop. However, there are still methodological barriers that need to be overcome to facilitate territorial LCA implementation, particularly for data collection and the inclusion of spatial variability. Several studies have mobilised other tools to address these issues, such as metabolism studies or GIS tools. In addition, the interest of economic modelling to integrate socio-economic interactions has been underlined (Borghino et al., 2021; Loiseau et al., 2018; O'Keeffe, Majer, Bezama, & Thrän, 2016). These combinations are discussed in this section.

4.1 Territorial metabolism studies

The concept of metabolism has been applied to cities to define the "the sum of the technical and socioeconomic processes that occur, resulting in growth, production of energy, and elimination of waste" (Kennedy, Cuddihy, & Engel-Yan, 2007). Using methods implemented in territorial metabolism studies such as material flow analyses (MFA) facilitates data collection in territorial LCAs. MFA establishes the material and energy balances of a system according to the law of mass conservation (Pincetl, 2012).

In the type B territorial LCAs described in Table 1, four of the studies used MFA. Goldstein et al. (2013) inventoried significant metabolic flows, and the appropriate up- and downstream processes from the direct metabolism were found in line with a process-based LCA approach as in García-Guaita et al. (2018), González-García et al. (2021) and González-García and Dias (2019). This simplified analysis gives a first overview of the environmental impacts of city, highlighting in most cases the significant transfer of pollution to the rest of the world. However, data gaps can lead to the exclusion of numerous flows and the black box modelling adopted do not allow the link between the impacts and their drivers. Other modelling approaches could be implemented such as the network system modelling to disaggregate city into different components and to describe the links between these components. Beloin-Saint-Pierre et al. (2017) pointed out that this structural information is critical when performing environmental assessment at urban or territorial scales.

4.2 GIS tools

GIS tools could be applied to any LCA phase to increase the representativeness of the territorial system by the inclusion of spatial information. These tools were used, for example, to model the system under study and define its main functions, taking into account the potentials and constraints offered by the territory in terms of biophysical characteristics, and the economic activities and infrastructures in place (Loiseau et al., 2018). GIS tools can also be used for spatialised LCI and LCIA (Nitschelm et al., 2016). Although regionalised impact methods are becoming more common with new LCIA methods such as LC-Impact (Verones et al., 2020) or Impact World+ (Bulle et al., 2019), no type B territorial LCA studies listed in Table 1 include spatial differentiation in impact assessment. This can be due to a lack of integration of regionalised data into existing LCA software and databases (Reap, Roman, Duncan, & Bras, 2008). The LCIA models developed are inherently consistent but remain disconnected from LCIs and do not rely on spatialised inventory data (Bare, 2010).

To address this issue, Ding et al. (2020) proposed to combine GIS tools with territorial LCA to facilitate the inventory stage and to reflect territorial characteristics in terms of input inventory for agricultural areas. Given a land-use planning map, in which GIS can read the locations and land occupation information (e.g. area, types of cultivation land), total emissions of territory can be computed through the calculation of the emission per unit area of each land-use type.

Nitschelm et al. (2016) developed a methodological framework to define and locate both activity and environment typologies within an agricultural territory.

Environmental typologies allow to take into account local soil and climatic conditions to calculate direct emissions, and the vulnerability of the area concerned to compute the impacts. In addition, they discussed the best way to represent environmental impacts on a map to support decision-making. This framework needs to be applied on real case studies to test its feasibility and interest for local stakeholders.

4.3 Economic modelling

Many studies in the literature combine LCA with economic modelling to provide exhaustive and quantitative information to support stakeholders in decision-making processes (Beaussier, Caurla, Bellon-Maurel, & Loiseau, 2019). Economic models used are based on the equilibrium theory which assumes that markets can reach an equilibrium for supply and demand, under the assumption of pure and perfect competition. Market prices are thus determined endogenously (Arrow Kenneth & Debreu, 1954). Two types of equilibrium models can be distinguished: (i) computable general equilibrium model (CGEM) and (ii) partial equilibrium model (PEM). CGEM aims at representing the whole economic sectors and allows for the provision of socio-economic indicators, whilst PEM includes a detailed representation of one or few connected sectors. It is therefore reasonable to consider that PEM may be more relevant for type A territorial LCA, whilst CGEM might be more suitable to address type B. Combined with an LCA approach, both PEM and CGEM have the potential to address important territorial LCA issues such as consequential effects, temporal dynamics, and quantification of socio-economic indicators.

Both type A and B of territorial LCA aimed at comparing the environmental performance of managing and land planning scenarios. Scenario design, system boundaries definition, and inventory process are thus critical steps for territorial LCA to capture all environmental impacts linked to changes in land planning. As connections between markets for different goods and services are strong, it is therefore likely that changes in territorial planning lead to effects of a consequential nature, i.e. indirect and substitution effects via the modification of price ratios of goods and services. Both PEM and CGEM are relevant tools to capture these effects in the economy and design scenarios. This information challenges territorial LCA to define system boundaries and territorial responsibility, as the effects of the original disturbance may affect sectors or activities outside the territory boundaries of the reference scenario. Changes in price ratio also lead to substitution effects between production factors for a given production technology. This can be difficult and resource consuming for process-based LCA, as LCIs must be adapted for each scenario assessed, to ensure consistency between LCA model and the economic model. Another point to consider is the temporal dynamics. Indeed, territorial planning is supposed to be medium- to long-term horizon and thus considering evolution of the technological and socio-economic contexts is a key element for territorial LCA. Due to their ability to consider both technological and macroeconomic context change, PEM and CGEM have been identified as promising tools to combine with

LCA to address this issue. Finally, a key feature of these models is that supply and demand are explicitly modelled and commodity prices are endogenously determined. It allows the calculation of additional indicators such as consumer and producer surpluses, two economic indicators of well-being that can be used in eco-efficiency ratios.

However, there are some limitations that reduce the operability of these models. First, most of these models are developed at macro-level. Adaptations are thus required to fit the scale of territorial LCA, as defined in this chapter. These adaptations are time consuming and require data that may be difficult to obtain. Moreover, using these economic modelling tools requires econometric skills to draw out their full potential, which can only be achieved in multidisciplinary projects. Finally, most of coupling PEM or CGEM with LCA identified in the literature consists in low coupling, i.e. outputs from the economic model are used to feed the LCA model. It is likely that moving towards high coupling, i.e. considering feedback loops between environmental targets and economic sphere, is advisable to fully assess environmental impacts of territorial planning scenarios and be consistent with strong sustainability objectives (Loiseau et al., 2016).

5 Conclusions and perspectives

Territories are a key element in the necessary ecological transition of our societies. To support stakeholders in defining sustainable trajectories, quantitative assessment tools are needed. Implementing an LCA at the scale of a territory allows to identify pollution transfers, and to quantify the environmental impacts of trajectories regarding a bouquet of services rendered, through the quantification of an eco-efficiency. Dealing with territorial multifunctionality is one of the main adaptations of type B territorial LCAs. This is all the more necessary as the ongoing transition towards a more circular economy makes it unavoidable by connecting local activities to each other. Conventional LCA approach, which consists in studying activities one by one through a multiplicity of FUs and a set of allocation or substitution rules to model the flow loop, is no longer adapted to address these challenges. The case studies described in this chapter show the methodological developments provided to operationalise and deepen territorial LCA approaches. Next step relies on the consideration of absolute eco-efficiency (Hauschild, 2015) to design territorial trajectories in line with global and regional planetary boundaries (Bjørn et al., 2020). The inclusion of a safe operating space into territorial LCA approaches will broaden its utility for decision support. The question is no longer whether planning scenario A is better for the environment (more eco-efficient) than scenario B. Instead, the question to be answered is as follows: is the planning scenario A (including all territorial activities and all exchanges with the rest of the world) enough eco-efficient (i.e. compatible with a safe operating space for humanity?). This is a research challenge for the coming decade.

Acknowledgments

The authors are members of the ELSA research group (Environmental Life Cycle and Sustainability Assessment, http://www.elsa-lca.org/) and thank all ELSA members for their advice.

References

Albertí, J., Balaguera, A., Brodhag, C., & Fullana-i-Palmer, P. (2017). Towards life cycle sustainability assesent of cities. A review of background knowledge. *Science of the Total Environment*, *609*, 1049–1063. https://doi.org/10.1016/j.scitotenv.2017.07.179.

Albertí, J., Brodhag, C., & Fullana-i-Palmer, P. (2019). First steps in life cycle assessments of cities with a sustainability perspective: A proposal for goal, function, functional unit, and reference flow. *Science of the Total Environment*, *646*, 1516–1527. https://doi.org/10.1016/j.scitotenv.2018.07.377.

Albertí, J., Roca, M., Brodhag, C., & Fullana-i-Palmer, P. (2019). Allocation and system boundary in life cycle assessments of cities. *Habitat International*, *83*, 41–54. https://doi.org/10.1016/j.habitatint.2018.11.003.

Arrow Kenneth, J., & Debreu, G. (1954). Existence of an equilibrium for a competitive economy. *Econometrica*, *265*. https://doi.org/10.2307/1907353.

Avadí, A., Nitschelm, L., Corson, M., & Vertès, F. (2016). Data strategy for environmental assessment of agricultural regions via LCA: Case study of a French catchment. *International Journal of Life Cycle Assessment*, *21*(4), 476–491. https://doi.org/10.1007/s11367-016-1036-6.

Banski, J., Bednarek, M., Danes, M., Feliu, E., Fons Esteve, J., Garcia, G., et al. (2011). *EU-LUPA European Land Use Patterns, report, EPSON 2013 Programme*. Luxembourg: European Commission.

Bare, J. C. (2010). Life cycle impact assessment research developments and needs. *Clean Technologies and Environmental Policy*, *12*(4), 341–351. https://doi.org/10.1007/s10098-009-0265-9.

Beaussier, T., Caurla, S., Bellon-Maurel, V., & Loiseau, E. (2019). Coupling economic models and environmental assessment methods to support regional policies: A critical review. *Journal of Cleaner Production*, *216*, 408–421. https://doi.org/10.1016/j.jclepro.2019.01.020.

Beloin-Saint-Pierre, D., Rugani, B., Lasvaux, S., Mailhac, A., Popovici, E., Sibiude, G., et al. (2017). A review of urban metabolism studies to identify key methodological choices for future harmonization and implementation. *Journal of Cleaner Production*, *163*, S223–S240. https://doi.org/10.1016/j.jclepro.2016.09.014.

Bidstrup, M. (2015). Life cycle thinking in impact assessment-current practice and LCA gains. *Environmental Impact Assessment Review*, *54*, 72–79. https://doi.org/10.1016/j.eiar.2015.05.003.

Björklund, A. (2012). Life cycle assessment as an analytical tool in strategic environmental assessment. Lessons learned from a case study on municipal energy planning in Sweden. *Environmental Impact Assessment Review*, *32*(1), 82–87. https://doi.org/10.1016/j.eiar.2011.04.001.

Bjørn, A., Sim, S., King, H., Patouillard, L., Margni, M., Hauschild, M. Z., et al. (2020). Life cycle assessment applying planetary and regional boundaries to the process level: A model case study. *International Journal of Life Cycle Assessment*, *25*(11), 2241–2254. https://doi.org/10.1007/s11367-020-01823-8.

Borghino, N., Corson, M., Nitschelm, L., Wilfart, A., Fleuet, J., Moraine, M., et al. (2021). Contribution of LCA to decision making: A scenario analysis in territorial agricultural production systems. *Journal of Environmental Management*, *287*, 112288. https://doi.org/10.1016/j.jenvman.2021.112288.

Bulle, C., Margni, M., Patouillard, L., Boulay, A. M., Bourgault, G., De Bruille, V., et al. (2019). IMPACT World+: A globally regionalized life cycle impact assessment method. *International Journal of Life Cycle Assessment*, *24*(9), 1653–1674. https://doi.org/10.1007/s11367-019-01583-0.

Cremer, A., Müller, K., Berger, M., & Finkbeiner, M. (2020). A framework for environmental decision support in cities incorporating organizational LCA. *International Journal of Life Cycle Assessment*, *25*(11), 2204–2216. https://doi.org/10.1007/s11367-020-01822-9.

Ding, T., Bourrelly, S., & Achten, W. M. J. (2020). Operationalising territorial life cycle inventory through the development of territorial emission factor for European agricultural land use. *Journal of Cleaner Production*, *263*. https://doi.org/10.1016/j.jclepro.2020.121565.

Eder, P., & Narodoslawsky, M. (1999). What environmental pressures are a region's industries responsible for? A method of analysis with descriptive indices and input-output models. *Ecological Economics*, *29*(3), 359–374. https://doi.org/10.1016/S0921-8009(98)00092-5.

European Commission. (2009). *Report from the commission to the council, the European Parliament, the European economic and social committee and the Committee of the Regions on the application and the effectiveness of the directive on strategic environmental assessment (directive 2001)*.

Ferng, J. J. (2003). Allocating the responsibility of CO2 over-emissions from the perspectives of benefit principle and ecological deficit. *Ecological Economics*, *46*(1), 121–141. https://doi.org/10.1016/S0921-8009(03)00104-6.

Finnveden, G., Hauschild, M. Z., Ekvall, T., Guinée, J., Heijungs, R., Hellweg, S., et al. (2009). Recent developments in life cycle assessment. *Journal of Environmental Management*, *91*(1), 1–21. https://doi.org/10.1016/j.jenvman.2009.06.018.

García-Guaita, F., González-García, S., Villanueva-Rey, P., Moreira, M. T., & Feijoo, G. (2018). Integrating urban metabolism, material flow analysis and life cycle assessment in the environmental evaluation of Santiago de Compostela. *Sustainable Cities and Society*, *40*, 569–580. https://doi.org/10.1016/j.scs.2018.04.027.

Goldstein, B., Birkved, M., Quitzau, M. B., & Hauschild, M. (2013). Quantification of urban metabolism through coupling with the life cycle assessment framework: Concept development and case study. *Environmental Research Letters*, *8*(3). https://doi.org/10.1088/1748-9326/8/3/035024.

González-García, S., Caamaño, M. R., Moreira, M. T., & Feijoo, G. (2021). Environmental profile of the municipality of Madrid through the methodologies of Urban Metabolism and Life Cycle Analysis. *Sustainable Cities and Society*, *64*. https://doi.org/10.1016/j.scs.2020.102546.

González-García, S., & Dias, A. C. (2019). Integrating lifecycle assessment and urban metabolism at city level: Comparison between Spanish cities. *Journal of Industrial Ecology*, *23*(5), 1062–1076. https://doi.org/10.1111/jiec.12844.

Guinée, J. B., Heijungs, R., Huppes, G., Zamagni, A., Masoni, P., Buonamici, R., et al. (2011). Life cycle assessment: Past, present, and future. *Environmental Science and Technology*, *45*(1), 90–96. https://doi.org/10.1021/es101316v.

Hauschild, M. Z. (2015). Better—But is it good enough? On the need to consider both eco-efficiency and eco-effectiveness to gauge industrial sustainability. In *Vol. 29. Procedia CIRP* (pp. 1–7). Elsevier B.V. https://doi.org/10.1016/j.procir.2015.02.126.

Hellweg, S., & Mila i Canals, L. (2014). Emerging approaches, challenges and opportunities in life cycle assessment. *Science*, *344*(6188), 1109–1113. https://doi.org/10.1126/science.1248361.

Huijbregts, M. A. J., Steinmann, Z. J. N., Elshout, P. M. F., Stam, G., Verones, F., Vieira, M. D. M., et al. (2016). *ReCiPe2016: A harmonized life cycle impact assessment method at mid-point and endpoint level.*

ISO 14040. (2006). *ISO 14040: Environmental management—Life cycle assessment—Principles and framework.* ISO.

ISO 14044. (2006). *ISO 14044. Environmental management—Life cycle assessment—Requirements and guidelines.* ISO.

Kennedy, C., Cuddihy, J., & Engel-Yan, J. (2007). The changing metabolism of cities. *Journal of Industrial Ecology*, *11*(2), 43–59. https://doi.org/10.1162/jie.2007.1107.

Laganier, R., Villalba, B., & Zuindeau, B. (2002). Le développement durable face au territoire : éléments pour une recherche pluridisciplinaire. *Développement durable et territoires*. https://doi.org/10.4000/developpementdurable.774.

Larrey-Lassalle, P., Catel, L., Roux, P., Rosenbaum, R. K., Lopez-Ferber, M., Junqua, G., et al. (2017). An innovative implementation of LCA within the EIA procedure: Lessons learned from two Wastewater Treatment Plant case studies. *Environmental Impact Assessment Review*, *63*, 95–106. https://doi.org/10.1016/j.eiar.2016.12.004.

Loiseau, E., Aissani, L., Le Féon, S., Laurent, F., Cerceau, J., Sala, S., et al. (2018). Territorial Life Cycle Assessment (LCA): What exactly is it about? A proposal towards using a common terminology and a research agenda. *Journal of Cleaner Production*, *176*, 474–485. https://doi.org/10.1016/j.jclepro.2017.12.169.

Loiseau, E., Junqua, G., Roux, P., & Bellon-Maurel, V. (2012). Environmental assessment of a territory: An overview of existing tools and methods. *Journal of Environmental Management*, *112*, 213–225. https://doi.org/10.1016/j.jenvman.2012.07.024.

Loiseau, E., Roux, P., Junqua, G., Maurel, P., & Bellon-Maurel, V. (2013). Adapting the LCA framework to environmental assessment in land planning. *International Journal of Life Cycle Assessment*, *18*(8), 1533–1548. https://doi.org/10.1007/s11367-013-0588-y.

Loiseau, E., Roux, P., Junqua, G., Maurel, P., & Bellon-Maurel, V. (2014). Implementation of an adapted LCA framework to environmental assessment of a territory: Important learning points from a French Mediterranean case study. *Journal of Cleaner Production*, *80*, 17–29. https://doi.org/10.1016/j.jclepro.2014.05.059.

Loiseau, E., Saikku, L., Antikainen, R., Droste, N., Hansjürgens, B., Pitkänen, K., et al. (2016). Green economy and related concepts: An overview. *Journal of Cleaner Production*, *139*, 361–371. https://doi.org/10.1016/j.jclepro.2016.08.024.

Mazzi, A., Toniolo, S., Catto, S., De Lorenzi, V., & Scipioni, A. (2017). The combination of an Environmental Management System and Life Cycle Assessment at the territorial level. *Environmental Impact Assessment Review*, *63*, 59–71. https://doi.org/10.1016/j.eiar.2016.11.004.

Mila i Canals, L., Azapagic, A., Doka, G., Jefferies, D., King, H., Mutel, C., et al. (2011). Approaches for addressing life cycle assessment data gaps for bio-based products. *Journal of Industrial Ecology*, *15*(5), 707–725. https://doi.org/10.1111/j.1530-9290.2011.00369.x.

Mirabella, N., Allacker, K., & Sala, S. (2019). Current trends and limitations of life cycle assessment applied to the urban scale: Critical analysis and review of selected literature. *International Journal of Life Cycle Assessment*, *24*(7), 1174–1193. https://doi.org/10.1007/s11367-018-1467-3.

Moine, A. (2006). Le territoire comme un système complexe : un concept opératoire pour l'aménagement et la géographie. *L'Espace géographique*, 115. https://doi.org/10.3917/eg.352.0115.

Moreno, E. L., & Murguía, R. O. (2015). *The city prosperity initiative: 2015 global city report*. UN-Habitat programme.

Muradian, R., O'Connor, M., & Martinez-Alier, J. (2002). Embodied pollution in trade: Estimating the "environmental load displacement" of industrialised countries. *Ecological Economics*, *41*(1), 51–67. https://doi.org/10.1016/S0921-8009(01)00281-6.

Mutel, C., Liao, X., Patouillard, L., Bare, J., Fantke, P., Frischknecht, R., et al. (2019). Overview and recommendations for regionalized life cycle impact assessment. *International Journal of Life Cycle Assessment*, *24*(5), 856–865. https://doi.org/10.1007/s11367-018-1539-4.

Nitschelm, L., Aubin, J., Corson, M. S., Viaud, V., & Walter, C. (2016). Spatial differentiation in Life Cycle Assessment LCA applied to an agricultural territory: Current practices and method development. *Journal of Cleaner Production*, *112*, 2472–2484. https://doi.org/10.1016/j.jclepro.2015.09.138.

O'Keeffe, S., Majer, S., Bezama, A., & Thrän, D. (2016). When considering no man is an island—Assessing bioenergy systems in a regional and LCA context: A review. *International Journal of Life Cycle Assessment*, *21*(6), 885–902. https://doi.org/10.1007/s11367-016-1057-1.

PAS 2070. (2014). PAS 2070: 2013—Specification for the assessment of greenhouse gas emissions of a city. In *PAS 2070*.

Patouillard, L., Bulle, C., Querleu, C., Maxime, D., Osset, P., & Margni, M. (2018). Critical review and practical recommendations to integrate the spatial dimension into life cycle assessment. *Journal of Cleaner Production*, *177*, 398–412. https://doi.org/10.1016/j.jclepro.2017.12.192.

Pérez-Soba, M., Petit, S., Jones, L., Bertrand, N., Briquel, V., Omodei-Zorini, L., et al. (2008). Land use functions—A multifunctionality approach to assess the impact of land use changes on land use sustainability. In *Sustainability impact assessment of land use changes* (pp. 375–404). Berlin Heidelberg: Springer. https://doi.org/10.1007/978-3-540-78648-1_19.

Peters, G. P. (2010). Carbon footprints and embodied carbon at multiple scales. *Current Opinion in Environmental Sustainability*, *2*(4), 245–250. https://doi.org/10.1016/j.cosust.2010.05.004.

Petit-Boix, A., Llorach-Massana, P., Sanjuan-Delmás, D., Sierra-Pérez, J., Vinyes, E., Gabarrell, X., et al. (2017). Application of life cycle thinking towards sustainable cities: A review. *Journal of Cleaner Production*, *166*, 939–951. https://doi.org/10.1016/j.jclepro.2017.08.030.

Pincetl, S. (2012). A living city: Using urban metabolism analysis to view cities as life forms. In *Metropolitan sustainability: Understanding and improving the urban environment* (pp. 3–25). Elsevier Ltd. https://doi.org/10.1533/9780857096463.1.3.

Potting, J., & Hauschild, M. Z. (2006). Spatial differentiation in life cycle impact assessment: A decade of method development to increase the environmental realism of LCIA. *International Journal of Life Cycle Assessment*, *11*(1), 11–13. https://doi.org/10.1065/lca2006.04.005.

Qi, Y., Zhang, Y., Jiang, H., Hou, H., & Li, J. (2019). Life cycle assessment in urban territories: A case study of Dalian city, China. *International Journal of Life Cycle Assessment*, *24*(7), 1194–1208. https://doi.org/10.1007/s11367-018-1465-5.

Reap, J., Roman, F., Duncan, S., & Bras, B. (2008). A survey of unresolved problems in life cycle assessment. Part 1: Goal and scope and inventory analysis. *International Journal of Life Cycle Assessment*, *13*(4), 290–300. https://doi.org/10.1007/s11367-008-0008-x.

Rebitzer, G., Ekvall, T., Frischknecht, R., Hunkeler, D., Norris, G., Rydberg, T., et al. (2004). Life cycle assessment part 1: Framework, goal and scope definition, inventory analysis, and applications. *Environment International*, *30*(5), 701–720. https://doi.org/10.1016/j.envint.2003.11.005.

Regulation (EC). (2003). *Regulation (EC) No 1059/2003 of the European Parliament and of the Council of 26 May 2003 on the establishment of a common classification of territorial units for statistics (NUTS)*.

Roibás, L., Loiseau, E., & Hospido, A. (2017). Determination of the carbon footprint of all Galician production and consumption activities: Lessons learnt and guidelines for policymakers. *Journal of Environmental Management*, *198*, 289–299. https://doi.org/10.1016/j.jenvman.2017.04.071.

Roibás, L., Loiseau, E., & Hospido, A. (2018a). A simplified approach to determine the carbon footprint of a region: Key learning points from a Galician study. *Journal of Environmental Management*, *217*, 832–844. https://doi.org/10.1016/j.jenvman.2018.04.039.

Roibás, L., Loiseau, E., & Hospido, A. (2018b). On the feasibility and interest of applying territorial Life Cycle Assessment to determine subnational normalisation factors. *Science of the Total Environment*, *626*, 1086–1099. https://doi.org/10.1016/j.scitotenv.2018.01.126.

Seppälä, J., Melanen, M., Mäenpää, I., Koskela, S., Tenhunen, J., & Hiltunen, M. R. (2005). How can the eco-efficiency of a region be measured and monitored? *Journal of Industrial Ecology*, *9*(4), 117–130. https://doi.org/10.1162/108819805775247972.

van Eck, N. J., & Waltman, L. (2010). Software survey: VOSviewer, a computer program for bibliometric mapping. *Scientometrics*, *84*(2), 523–538. https://doi.org/10.1007/s11192-009-0146-3.

Verones, F., Hellweg, S., Antón, A., Azevedo, L. B., Chaudhary, A., Cosme, N., et al. (2020). LC-IMPACT: A regionalized life cycle damage assessment method. *Journal of Industrial Ecology*, *24*(6), 1201–1219. https://doi.org/10.1111/jiec.13018.

WRI. (2004). *The greenhouse gas protocol—A corporate accounting and reporting standard*.

Zhang, B., Su, S., Zhu, Y., & Li, X. (2020). An LCA-based environmental impact assessment model for regulatory planning. *Environmental Impact Assessment Review*, *83*. https://doi.org/10.1016/j.eiar.2020.106406.

Zhang, Z., Wu, X., Yang, X., & Zhu, Y. (2006). BEPAS—A life cycle building environmental performance assessment model. *Building and Environment*, *41*. https://doi.org/10.1016/j.buildenv.2005.02.028.

CHAPTER

Environmental impact and risk assessment

10

Brindusa Sluser, Oana Plavan, and Carmen Teodosiu

Department of Environmental Engineering and Management, "Gheorghe Asachi" Technical University of Iasi, Iasi, Romania

1 EIA concepts, goals, and target audience

The consequences of various projects, plans, programmes, or policies can generate negative effects on environmental quality and are assessed by means of the environmental impact assessment (EIA) (Gilbuena et al., 2013). EIA represents an environmental policy tool and a major component of sustainable development to estimate the environmental effects of activities at the earliest stage possible (Glasson, Therivel, & Chadwick, 1994; Leopold, Clarke, Hanshaw, & Balsley, 1971; Morris & Therivel, 1995). EIA was introduced in 1969 by the USA National Environmental Policy Act (NEPA), and, following that, it became widely used and received a significant boost in Europe as well (Glasson & Therivel, 2019). Therefore EIA has become one of the important instruments of environmental management used in environmental policy and planning worldwide; however, only introducing EIA into legislation does not ensure its successful implementation and sometimes proved to be detrimental, given the lack of knowledge, precision, uncertainty, and performance (Bond, Fischer, & Fothergill, 2017; Khosravi, Jha-Thakur, & Fischer, 2019).

The EIA procedure relies on tools for either impact identification (scoping) or impact prediction (impact assessment) that have a common base with risk assessment (RA), both being highly interdisciplinary. The key concepts applied within EIA and RA are remarkably similar and are based on the same objectives to achieve efficient environmental management (Zeleňáková et al., 2020). On the other hand, life cycle assessment (LCA) is a robust methodology that assesses the environmental impacts of products or processes, and it can be particularly applied in EIA, environmental comparison of processes, and abatement alternatives (Corominas et al., 2020; Neamtu, Sluser, Plavan, & Teodosiu, 2021; Teodosiu, Barjoveanu, Sluser, Popa, & Trofin, 2016). Thus LCA is an analytical tool to assess the environmental impacts of the whole production chain, while EIA is a procedure to support the decision-makers (ISO 14040, 2006; Tukker, 2000).

Assessing Progress towards Sustainability. http://doi.org/10.1016/B978-0-323-85851-9.00004-3

2 Environmental impact assessment methodology

The EIA methodology was developed in the late 1960s and during the last three decades has been widely applied, yet there is still room for improvement of the ways to mitigate undesirable long-term negative environmental effects (Kuitunen, Jalava, & Hirvonen, 2008; Saffari, Ataei, Sereshki, & Naderi, 2019). **EIA** has various **definitions** in the scientific literature. For instance, Glasson and Therivel (2019) defined it as '*the need to identify and predict the impacts on the environment and on man's health and well-being of legislative proposals, policies, programmes, projects and operational procedures, and to interpret and communicate information about the impacts*'. Similarly, the United Nations Environment Programme (UNEP) described it as '*a systematic process to identify, predict and evaluate the environmental effects of proposed actions and projects*'. In EIAs, a major focus is on preventing, mitigating, and balancing the negative effects of proposed projects and therefore EIA has been largely accepted as a major decision-making support tool for adequate project planning (Khosravi et al., 2019; Rocha, Ramos, & Fonseca, 2019; Roos, Cilliers, Retief, Alberts, & Bond, 2020).

EIA is currently used for various projects, such as tourism (Canteiro, Córdova-Tapia, & Brazeiro, 2018), transport (Soria-Lara et al., 2020), coastline projects (Robu et al., 2015), mining activities (Aryafar, Yousefi, & Doulati Ardejani, 2013; Ştefănescu, Robu, & Ozunu, 2013; Zhang, Lu, Xi, & Chun, 2015), cement plants (Saffari et al., 2019), climate change (Jiricka-Pürrer et al., 2018), flood mitigation measures (Gilbuena et al., 2013), landfill (Sajjadi, Aliakbari, Matlabi, Biglari, & Rasouli, 2017), and water management (Barjoveanu, Comandaru, Rodriguez-Garcia, Hospido, & Teodosiu, 2014; Shakib-Manesh, Hirvonen, Jalava, Ålander, & Kuitunen, 2014; Teodosiu, Barjoveanu, Robu, & Ene, 2012), and it can be applied for all environmental issues related to unsustainable implementation of various activities (Saffari et al., 2019).

EIA covers several types of evaluation instruments (Fig. 1), applies them to various activities, and it is reliable at local, national, and international scales. It delivers environmental impact statements, such as an environmental report, environmental impact report, or audit compliances, which can help decision-makers to improve and better motivate their decisions (McCabe & Sadler, 2002). Thus, for instance, in the case of EIA of new projects, the result is the environmental impact report, based on which the Environmental Protection Agency (EPA) will make a decision. In the case of industrial activities and processes, the result of EIA is the audit compliance. Strategic EIA refers to the evaluation of the negative effects in case of plans, programmes, or policies so that the consequences that might appear are diminished from the first stage of implementation.

2.1 EIA stages

The EIA process involves various steps (Fig. 2), starting at the project screening and scoping that aim to identify, in the earliest possible stages, the impacts of the project under evaluation (i.e. stages 1–8). Further, the description of the project/development and environmental assessment includes geological, morphological, hydrological

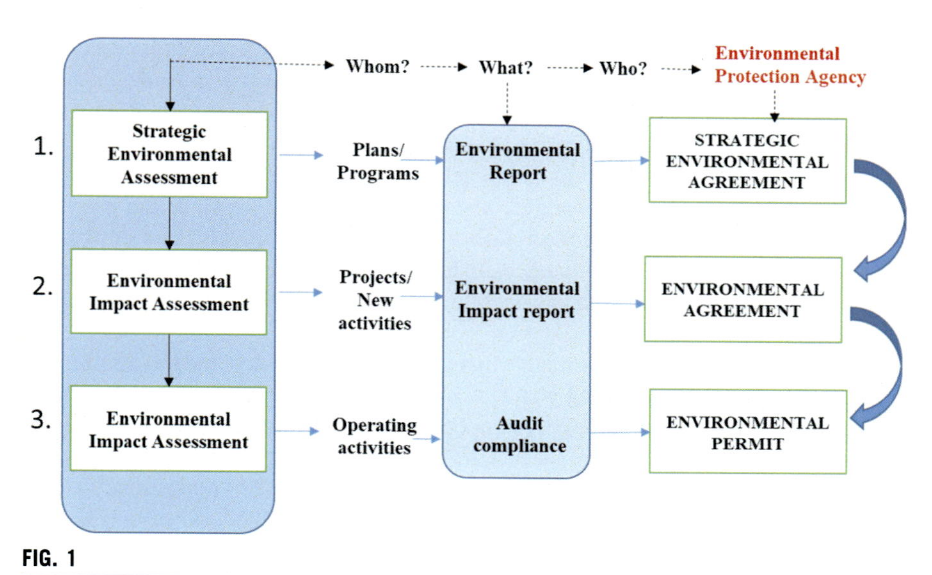

FIG. 1

Procedures for environmental impact assessment.

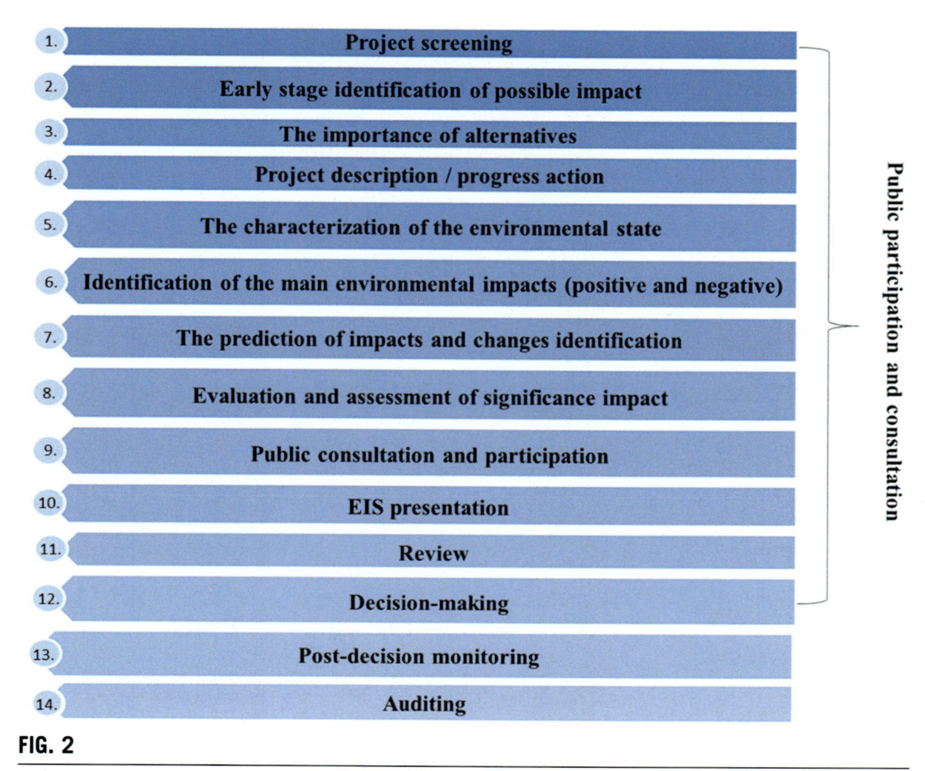

FIG. 2

The main stages in environmental impact assessment procedure (Glasson et al., 1994; Morris & Therivel, 1995).

data about the location, environmental quality (air, soil, and water), processes, and current/future environmental conditions considering at least two alternatives of the initial proposed project/development/plan. The identification of the main impacts is conducted by environmental experts able to define and quantify the significance of those impacts, and the prediction of impacts is based on the estimation of the magnitude of the environmental changes. The significance evaluation of the estimated impacts allows prioritisation of the main adverse impacts. The mitigation step mainly includes measures to avoid, minimise, and compensate for the consequences on environmental quality and human health, while public participation and consultation are mandatory.

The success of any environmental management plan that considers monitoring actions depends on the auditing step (i.e. stages 12–14 at Fig. 2) in which the prediction and mitigation measures are proposed (Glasson et al., 1994). The most significant aspect required for any successful EIA implementation proved to be public participation (i.e. stages 9–11 at Fig. 2). Thus, to assure the transparency of the decision-making process, public participation allows the collection of opinions on environmental issues (e.g. health, family wellness, and living environment) and assures the success of a project implementation (Hasan, Nahiduzzaman, & Aldosary, 2018). Thus Khosravi et al. (2019) described the EIA main stages as, firstly, framing/screening the project's goals; secondly, defining the field of evaluation and public participation; and thirdly, analysis of alternatives, evaluation of effects and impact minimisation, drafting the report on the EIA process, and the decision-making process. Therefore EIA is an anticipatory, participatory, and flexible environmental management tool (Khosravi et al., 2019; Muralikrishna & Manickam, 2017a).

The target groups, main actors, and personnel categories involved in any EIA procedure and their interactions are presented in Fig. 3. All kinds of EIAs include the

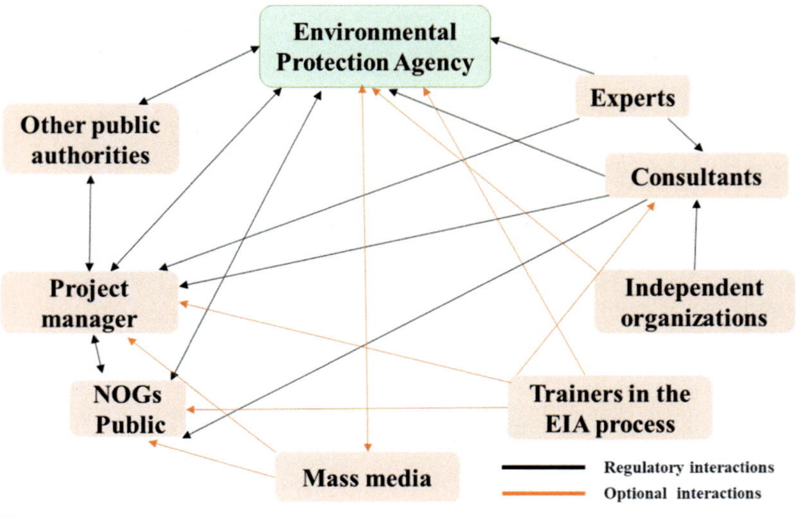

FIG. 3

Stakeholders and target groups in the EIA process.

stakeholders' involvement as an integral part, and anyone with an interest can take part, from government ministers and statutory bodies to the public and local communities. The bidirectional arrows show the communication and interactions from both involved parties like a public consultation, while the unidirectional arrows show the communication of such claims or other requirements from one part to the other part involved in EIA decision-making process. According to Yao, He, and Bao (2020), the procedure of public participation provides a platform where multiple stakeholders can express their opinions and participate in decision-making, and EIA sometimes risks having more of a socio-political nature than a technical one.

2.2 EIA tools

EIA involves the use of both qualitative and quantitative methods (Table 1) to provide support to authorities to decide based on scientific facts. Thus EIA-related tools were developed targeting faster, objective, and flexible methods, addressing qualitative or quantitative evaluation.

According to Gilbuena et al. (2013) and Josimovic, Petric, and Milijic (2014), the most common methods for EIA quantification are shown in Fig. 4. Quantitative methods have gained more attention since the earlier applications of EIA, so, together with the checklists, the most frequently employed quantitative tools were the Leopold and the Rapid Impact Assessment matrices, as well as the Index of Global Pollution (I_{GP}) (Cojocaru et al., 2017; Glasson et al., 1994; Neamtu et al., 2021; Robu et al., 2007). At present, the focus has moved to

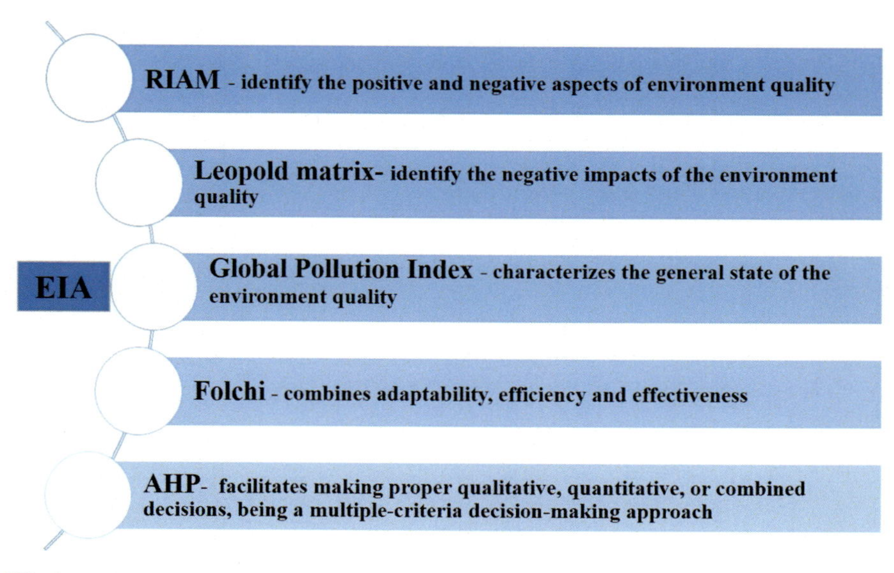

FIG. 4

The most applied methods for environmental impact assessment.

Table 1 Tools used in environmental impact assessment.

EIA tools	Definition	References
Qualitative		
Checklists for impact assessment	Preliminary evaluation based on an existing list of questions that can be applied by non-environmental experts as well. It is considered a very subjective method.	Glasson et al. (1994)
Environmental risk assessment (ERA)	The main goal is to identify, analyse, and evaluate the risks that could arise from certain anthropic or industrial activities. The method completes the environmental impact assessment procedure as volume of information, quality, and objectivity, being an important tool for the decision-making process (see Section 3).	Robu, Căliman, Beţianu, and Gavrilescu (2007) Zeleňáková et al. (2020)
Quantitative		
Leopold Matrix	It is one of the first methods of EIA that is still used, and it evaluates the harmful effects of development projects. It was developed within the NEPA US requirements in 1970 (see 'Leopold's matrix' section).	Leopold et al. (1971) Scheid (1998) Morris and Therivel (1995) Pastakia (1998) Robu et al. (2015) Mohebali, Maghsoudy, Doulati Ardejani, and Shafaei (2019) Neamtu et al. (2021)
Battelle Columbus method (BEES)	It has been identified as one of the quantitative methods designed to evaluate the impacts of water resource developments, water quality management plans, etc. The environmental impact assessment is divided into four categories: ecology, pollution, aesthetics, and human interest.	Batelle (1972) Scheid (1998) Goyal and Deshpande (2001) Wagh and Gujar (2014) Al-Nasrawi, Kareem, and Saleh (2020)
Importance Scale Matrix (ISM)	It involves comparison of the parameters considered in impact assessment, using the importance scale (slightly, moderately, strongly, dominantly, etc.). The parameters may refer to the quality indicators of environmental components (air, soil, water). The importance matrix is developed based on the opinion of the expert group. It is considered to have minimum subjectivity with more accurate and reliable results.	Batelle (1972) Scheid (1998) Goyal and Deshpande (2001)

Rapid Impact Assessment Matrix (RIAM)	Its main goal is to perform a series of operations to compare various variables, and the evaluation steps included in this method allow both qualitative and quantitative estimations (see 'Rapid impact assessment matrix' section).	Morris and Therivel (1995) Kuitunen et al. (2008) Neamtu et al. (2021)
Index of Global Pollution (IGP)	This methodology allows the global estimation of the affected ecosystem and environmental quality by anthropogenic activities, considering both the ideal and the realistic scenarios of environmental quality (see 'Global pollution index' section).	Cojocaru, Cocârţâ, Istrate, and Creţescu (2017) Zaharia and Surpateanu (2018)
Analytical Hierarchy Process (AHP)	It is the most important multi-criteria decision-making approach that facilitates proper qualitative, quantitative, or combined decision-making. This method has the capacity to control and reduce the inconsistency of specialist judgements (see 'Analytical hierarchy process method' section).	Pourghasemi, Pradhan, and Gokceoglu (2012) Aminbakhsh, Gunduz, and Sonmez (2013) Mohebali et al. (2019)
Folchi	This method combines adaptability, comprehensiveness, repeatability, efficiency, and effectiveness, and it is one of the most objective approaches in EIA (see 'Folchi method' section).	Phillips (2013) Saffari et al. (2019)
Life Cycle Assessment (LCA)	It was developed to identify and evaluate the environmental impacts generated by products and services for the entire cycle, from cradle to grave. It is considered a complex, time-consuming, and expensive process. A large volume of input data is required (see Section 5).	ISO 14040 (2006) Azapagic and Perdan (2010) Corominas et al. (2020)
Integrated approaches/ integrated index (EIRA)	The integrated method was developed to quantify simultaneously the environmental impacts and the associated risks, based on the same input data used in EIA and based on an algorithm. In its first step, it quantifies the impacts as a function of the measured concentration of certain pollutants, and then the associated risk to each impact is quantified as a direct functionality. The method was further adapted and improved to be applied to various situations, considering four environmental components (surface water, ground water, air, and soil), more, or just one (see Section 4).	Ştefănescu et al. (2013) Robu et al. (2015) Teodosiu, Robu, Cojocariu, and Barjoveanu (2015)

multi-criteria analysis (a set of systematic procedures for designing, evaluating, and selecting decision alternatives based on conflicting and incommensurate criteria), Folchi approach, Analytical Hierarchy Process (AHP) (Al-Nasrawi et al., 2020; Malczewski, 2018; Mohebali, Maghsoudy, & Ardejani, 2020; Saffari et al., 2019).

The **working principles** applied when quantifying environmental impacts with the **RIAM**, **Leopold's Matrix**, or $\mathbf{I_{GP}}$ take into consideration the measured concentrations (MCs) of certain quality indicators (i), as compared to the maximum allowed concentration (MAC) and alert level (AL), as stated by the environmental standards (Fig. 5). On this basis, an evaluation of the types of pollution and impacts can be made. Thus if the MC is under the AL (section A, Fig. 5), then there is an insignificant impact, and, even so, prevention measures are still required so that pollution will not increase. If the MC is between AL and MAC (section B, Fig. 4), then there is likely significant pollution/impact, and pollution control measures are mandatory, while, if the MC is higher than MAC (section C, Fig. 5), then there is significant pollution/impact, and remediation actions are to be taken, considering the applicable legislation.

In the case of the RIAM method, which uses two types of criteria—group A and group B (Table 2)—the values for criteria A and B were assigned based on the MCs for each pollutant, as an average of multi-annual measurements and compared to the maximum limits established by environmental standards. In the case of criterion A_1 (Fig. 5), value 4 was assigned for MCs higher than the maximum allowed, value 3 was assigned for MCs between alert/threshold and maximum limit, value 2 was assigned in the case of measurements under threshold, and value 1 could be assigned if the MC of a certain pollutant is nearly zero. Criterion A_2 has the value -3 in the case of measurements higher than maximum limits, while value -1 was assigned in the case of MCs under the threshold. Considering criterion B, value 1 was assigned when the MC was under threshold, while value 3 was assigned when the MC was higher than the maximum limits according to environmental standards (Fig. 5).

In the case of the I_{GP}, the evaluation grade (NB) is assigned based on the MC of certain pollutants, compared to MACs, so that, if the MC is much higher than the maximum allowed, then the evaluation grade that can be assigned is 1 (highly polluted environment). If the MC is lower than 10% of MAC, then there is insignificant pollution, and the evaluation grade that can be assigned is 10. In the case of Leopold's Matrix, the assigning values for magnitude (M) and importance (I) are vice versa as compared to the I_{GP}, as shown in Fig. 5. The RIAM digitised method, based on the described working principle, proved that the impact quantification became more objective and not influenced by the decision-makers' experience, comprising a complex database of quality

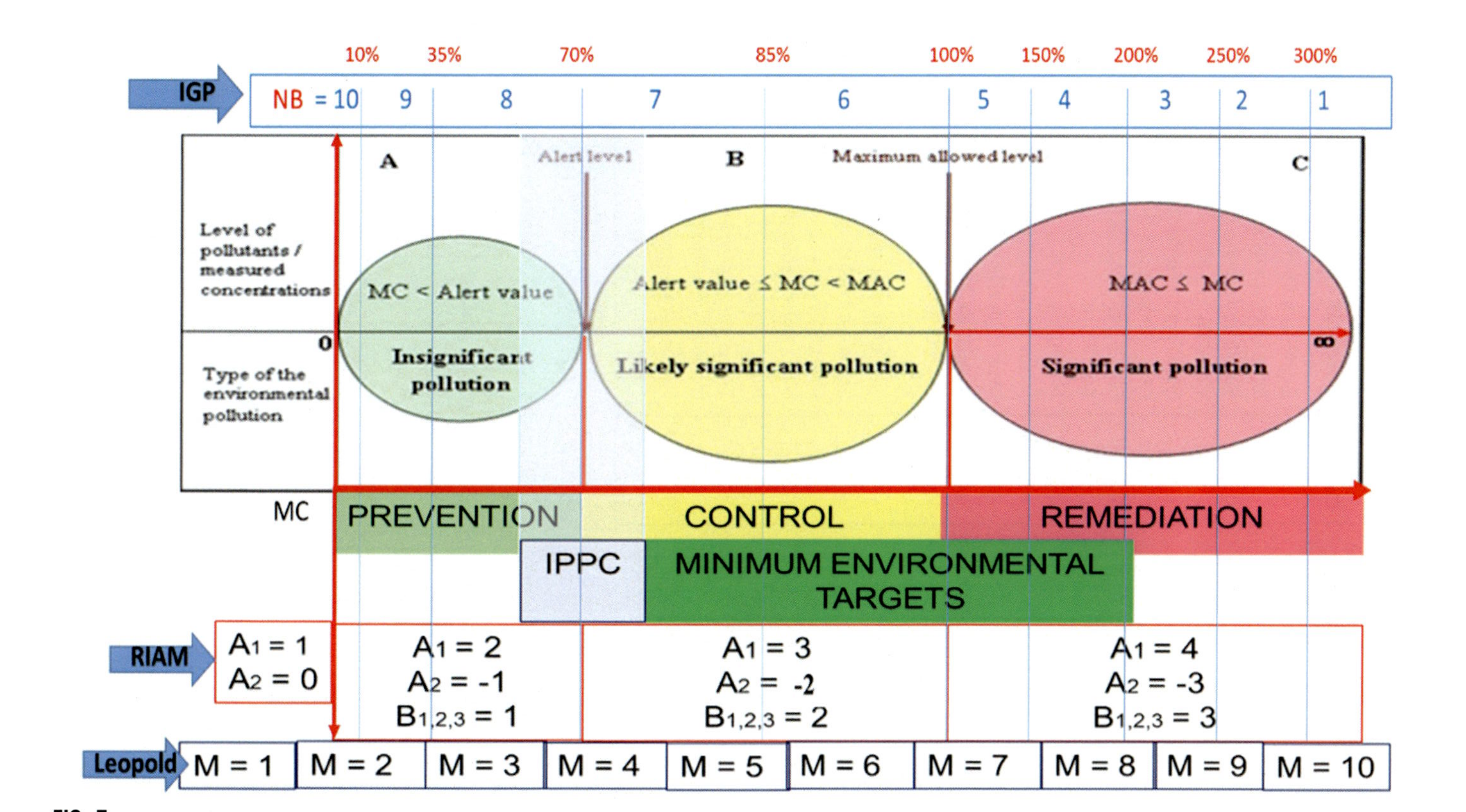

FIG. 5

Working principles considered in impact assessment evaluation for RIAM, I_{GP}, and Leopold Matrix (Sluşer, Şchiopu, Bălan, & Pruteanu, 2017). *MC*, measured concentration; *MAC*, maximum allowed concentration; *IPPC*, integrated pollution prevention control requirements; *NB*, evaluation grades in the case of the global pollution index method; *M*, the magnitude in the case of the Leopold Matrix; A_1, A_2, B_1, B_2, B_3, criteria used in the case of the RIAM method.

Table 2 Definition of criteria employed in the RIAM (Kuitunen et al., 2008; Mohebali et al., 2019; Robu et al., 2015).

Criterion	Scale	Description
A_1—The significance/importance of the condition/effect	4	Significant for the national/international interests
	3	Significant for the regional/national interests
	2	Significant only for the zones found near the local zone
	1	Significant only for the local conditions
	0	No significance
A_2—The magnitude of the change/effect	+3	Major significant benefit
	+2	Meaningful benefit for the status quo
	+1	Benefit for the status quo
	0	Lack of change/status quo/no influence
	−1	Negative change of status quo
	−2	Significant disadvantages or negative changes
	−3	Major disadvantages or changes
B_1—Permanence **B_2**—Reversibility **B_3**—Cumulativeness	1	No changes
	2	Temporary/reversible/non-cumulative/unique
	3	Permanent/irreversible/cumulative/synergetic

indicators; thus, the experience of evaluators did not influence the results (Suditu & Robu, 2012).

In comparison to the above-described methods, the **Folchi Matrix** and the **AHP** refer to the efforts of scientists to assess industrial activities in an integrated manner, considering the impacts and associated risks induced in the environment, to make the quantitative assessment more accessible and easier to be applied, such as digitising the conventional methods (Mohebali et al., 2019; Suditu & Robu, 2012) or applying qualitative modelling for integrated assessment (Xu, Haase, Su, Wang, & Pauleit, 2020) and a multi-criteria assessment decision support system (MCA-DSS) (Schetke, Haase, & Kötter, 2012). Usually, interaction matrices are the most common because they are easy to apply, inexpensive, and allow fast implementation. For instance, Mohebali et al. (2019) carried out an EIA by applying four different methods: the Leopold Matrix, modified Folchi Matrix, AHP, and RIAM. The results were integrated by using the combined environmental impact assessment (C-EIA) algorithm to provide a unique solution to assess the environmental impacts. Other methods, such as the I_{GP} or integrated approaches for integrated environmental impact and RA (Robu et al., 2015), were employed to assess the impact of various pollution sources that generate inorganic or organic priority pollutants.

Rapid impact assessment matrix

Pastakia (1998) used the RIAM for the first time, which provided transparency and significantly reduced the execution time for an EIA. The selected environmental components were, therefore, much easier and faster to track, due to the simple shape of this matrix. The RIAM has the possibility to perform a series of operations to compare various variables, and the evaluation steps included in this method allow quantitative estimations. This method was developed by Pastakia (1998), applied by Kuitunen et al. (2008) and Sandham, Hoffmann, and Retief (2008), digitised by Suditu and Robu (2012), and takes into consideration two types of criteria (Table 2): criteria of group A (A_1 and A_2), which can individually influence the results, and criteria of group B (B_1, B_2, B_3), which cannot individually influence the environmental score (ES).

ESs are calculated as presented in Eqs. (1)–(3), based on the values assigned to each group of criteria, and can be between (−108) and 108, so the lower the value is, the more significant the impact is (Mohebali et al., 2019; Mondal, Rashmi, & Dasgupta, 2010; Robu et al., 2015; Suthar & Sajwan, 2014).

$$A_1 * A_2 = AT \tag{1}$$

$$B_1 + B_2 + B_3 = Bt \tag{2}$$

$$AT * Bt = ES \tag{3}$$

The RIAM method was applied for various cases, both project development and operating industrial activities. For instance, Robu et al. (2015) followed the RIAM methodology when evaluating the environmental impact generated in the southern part of the Romanian Black Sea coast, and the results revealed that the surface water is mostly affected by heavy metal pollution, negatively influencing the biota as well. Kuitunen et al. (2008) used the RIAM to compare the results of EIA and strategic environmental assessment (SEA) and reported that there are many similarities in the initial approach, but it is suitable to evaluate the impact in both situations helping to prioritise the environmental issues. Ijäs, Kuitunen, and Jalava (2010) evaluated in the context of impact significance assessment the applicability of the RIAM, and Mondal et al. (2010) used the RIAM for impact quantification of Varanasi's municipal solid waste. Shakib-Manesh et al. (2014) studied the usability of the RIAM as an instrument to prioritise proposals for small-scale water restoration projects in relation to their potential to improve the environment, while Suthar and Sajwan (2014) used the RIAM for selection of a new waste disposal site for disposing urban solid wastes in Dehradun city (India). Their results showed that this method is a reliable tool to identify the suitability of the site in accordance with the ecological, physical, biological, social/cultural, and economic issues of the envisaged project.

Leopold's matrix

Leopold's Matrix (Leopold et al., 1971) was one of the first methods to assess the harmful effects of development projects. According to Mohebali et al. (2019), this method is still used due to its simplicity and efficiency, and it examines the impacts of individual project activities on the environment. So, Sajjadi et al. (2017) evaluated the environmental impacts of the municipal solid waste disposal site, and Josimovic et al. (2014) assessed the impact of a wind farm. By using Leopold's Matrix (Table 3), each action needs to consider the possibility to impact certain environmental components (Ashofteh, Bozorg-Haddad, & Loáiciga, 2017; Josimovic et al., 2014; Mohebali et al., 2020), and the impact (**Imp**) is described according to its **magnitude (M)** and the **importance (I)/severity (S) of the effect**. The **magnitude** of an interaction represents its amplitude and is described numerically on an evaluation scale ranked from 1 to 10, where 1 represents the natural state of the environmental component, and 10 represents the highly polluted environment. The importance/severity is also assigned on a scale from 1 to 10, where 10 represents the most important/severe situation, and it correlates the type of pollutant (e.g. toxic, carcinogenic, or priority pollutants) with its MC. In practice, in the case of regular pollutants, the importance is the values assigned for magnitude plus 1, and in the case of priority, toxic, or carcinogenic pollutants, the importance is the values assigned for the magnitude plus 2 units. Some studies use the following equations to determine the magnitude of the impact in two working scenarios, different than the scenarios presented in Fig. 5. Thus the first scenario considers the impact quantification based on the ratio between the MC (c_{det}) and the MAC in Eq. (4), while the second one considers the MC (c_{det}) as compared to the AL or other reference (c_{ref}) in Eq. (5) (Al-Nasrawi et al., 2020).

Table 3 Example of Leopold's Matrix (Ştefănescu et al., 2013).

Environmental component	**Quality indicator (i)**	**Process P1**			**Process P2**			**... Process Pn**			**Average on row**
Magnitude and severity	*i*	M	S	**Imp**	M	S	**Imp**	M	S	**Imp**	
Air											
Ground water											
Soil											
…..											
Average on column											

$$M = \frac{C_{det}}{MAC} \quad (4)$$

$$M = \frac{C_{det}}{C_{ref}} \quad (5)$$

In Table 3, an example is given of the employed EIA based on the mentioned working principles. The impact induced on one environmental component, by all assessed processes, is represented by the average of impacts obtained on rows, whereas the average of impacts obtained on columns emphasises the impact induced by one process into all environmental components.

Global pollution index

The $\boldsymbol{I_{GP}}$ considers both the ideal and the realistic scenarios of environmental quality and involves going through a step-by-step synergistic assessment based on quality indicators that reflect the general state of one or more environmental components, followed by their correlations through graphical methods (Zaharia & Surpateanu, 2018). The first step is to establish the environmental components under evaluation, sampling, and analysing. In this way, it is suggested to snapshot the quality of each environmental component on a credit worthiness scale by marking them such as to express the proximity and the distance from an ideal state. The evaluation scale varies from 1 to 10 (Fig. 5), the latter corresponding to the natural state of the environment (MC is nearly 0), whereas the former to an irreversible degradation of the studied ecosystems. For each quality indicator assessed, an individual score is assigned from 1 to 10, based on the concentration of the targeted pollutants (see Fig. 5). The next step is the calculation of the evaluation grades for each environmental component, as an arithmetic mean of the grades received by each quality indicator analysed. These values are used for a graphical representation. If three environmental components were analysed, the geometric figure is a triangle; if four environmental components were analysed, the geometric figure is a square, etc. (Fig. 6). In this diagram, the ideal state is represented by a regular geometric figure, with equal rays having a value of 10 units. The real state is expressed by connecting the dots that resulted from evaluation—an irregular geometric figure having a smaller area, enclosed within the regular geometric figure that corresponds to the ideal state. In the end, the I_{GP} is calculated as defined by Eq. (6):

$$I_{GP} = \frac{S_i}{S_r} \quad (6)$$

where S_i is the area of the ideal state of the environment, and S_r is the area that represents the real state (evaluated situation).

An easier method was proposed to calculate the I_{GP}, solely based on the arithmetic mean of evaluation scores directly correlated with the global state of the ecosystem (Cojocaru et al., 2017). The higher the I_{GP} is, the higher the impact on environmental quality. This method has several benefits, the most important being that it provides a general picture of the health status of the environment (i.e. its

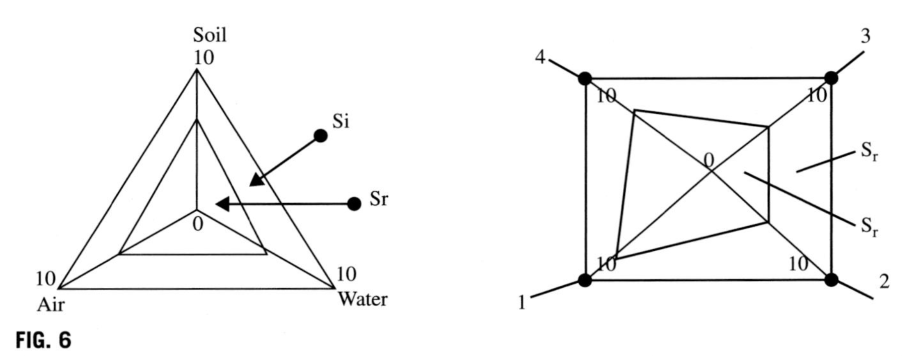

FIG. 6

Geometric representation of global pollution index calculation using Rojanschi's method: (A) three environmental components are assessed; (B) four environmental components are assessed.

quality at a moment in time) and allows the comparison of various environmental areas, based on the same quality indicators. On the other hand, its main disadvantage is the subjectivity whenever assigning the evaluation grades, which, in turn, depends on the evaluators' experience and exigency.

Folchi method

The method was introduced by Folchi in 2003 to deal with complex EIA issues given the desired mixture of adaptability, comprehensiveness, repeatability, efficiency, and effectiveness (Saffari et al., 2019). According to Aryafar et al. (2013), the Folchi method is simpler and more comprehensible than Leopold's method, being one of the most significant quantitative methods used in EIA but not suitable at a large scale. It comprises seven stages (Fig. 7) and was initially applied to mining activities (Samimi Namin et al., 2011). Firstly, the existing environmental context is characterised from the points of view of the geology, geotechnics, hydrology, weather, and economy, and then (second stage) all the impacting factors that may modify the pre-existing environmental condition in the specific activity are identified. In the third stage, the range of magnitude of the variation caused by each impacting factor is defined (between 1 and 10), and the environmental components under evaluation are selected referring to those pre-existing conditions that could be modified because of the mining activities. In the next two steps (fifth and sixth), the correlation between the environmental components and the specific magnitude of each impacting factor is realised. Finally, the weighted sum of the environmental impacts on each environmental component is calculated as a global impact.

Mirmohammadi, Gholamnejad, Fattahpour, Seyedsadri, and Ghorbani (2009) developed the Folchi method to be applied to mining activities, while Saffari et al. (2019) assessed the environmental impact of a cement plant considering two approaches: the Folchi method and the Fuzzy Delphi method; thus, the subjectivity of the weights estimated in the Folchi method is highly reduced. Within the sustainability context, according to Mohebali et al. (2020), the stakeholders claim that the

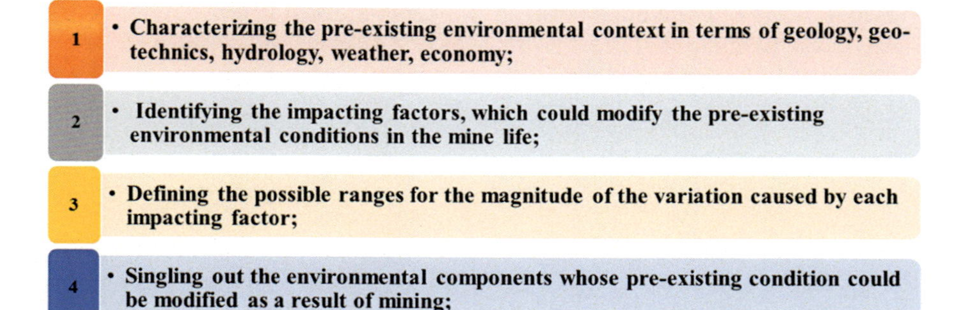

FIG. 7

Stages in the Folchi method (Phillips, 2013; Samimi Namin, Shahriar, & Bascetin, 2011).

environmental benefits of the proposed projects should be considered too. Basically, the main sustainability components are both the minimisation of the environmental impact of an industrial project and the maximisation of the social and economic development (benefits).

Analytical hierarchy process method

The AHP method was developed by Saaty (1980, 1990) and due to its simplicity, efficiency, and safety it was used by scientists who dealt with decision-making processes (Mohebali et al., 2019). Since EIA is a complex multi-dimensional process, involving multiple criteria and multiple actors, Ramanathan (2001) proposed the AHP for EIA as a multi-criteria technique. It was considered that the AHP is the most important multiple-criteria decision-making approach that facilitates making adequate qualitative, quantitative, or combined decisions (Mohebali et al., 2019; Pourghasemi et al., 2012). Pourghasemi et al. (2012) considered a fuzzy logic-based approach and AHP to produce landslide susceptibility maps at the Haraz watershed in Iran. Through this method, combined quantitative and qualitative tools were used, and it involved different groups of stakeholders and opinions expressed by many experts. As procedure, after having determined the assessment criteria and alternatives in the AHP, the next step is to conduct a comparison of paired criteria and to build a pairwise comparison matrix. The ratio of row score to column score in each cell is the pairwise comparison (Mohebali et al., 2019). To apply the method, it is

Table 4 Scale of preference between two parameters in AHP (Pourghasemi et al., 2012).

Scales	Degree of preference	Explanation
1	Equally	Two activities contribute equally to the objective
3	Moderately	Experience and judgement slightly to moderately favour one activity over another
5	Strongly	Experience and judgement strongly or essentially favour one activity over another
7	Very strongly	An activity is strongly favoured over another, and its dominance is showed in practice
9	Extremely	The evidence of favouring one activity over another is of the highest degree possible of an affirmation
2, 4, 6, 8	Intermediate values	Used to represent compromises between the preferences in weights 1, 3, 5, 7, and 9
Reciprocals	Opposites	Used for inverse comparison

necessary that each factor be evaluated and rated against every other factor by assigning a relative dominant value between 1 and 9 (Table 4) (Kayastha, Dhital, & De Smedt, 2013; Pourghasemi et al., 2012).

This method is widely used and involves several stages (Pourghasemi et al., 2012): (1) break a complex unstructured problem down into its component factors, which are the parameters chosen in the study; (2) arrange these factors in a hierarchical order; (3) assign numerical values according to their subjective relevance to determine the relative importance of each factor; and (4) synthesise the rating to determine the priorities to be assigned to these factors. According to Kayastha et al. (2013), the main disadvantage is the subjectivity of ranking the environmental factors, which can vary from one expert to another.

3 Risk assessment methodology

Risk assessment (RA) completes the volume of information, quality, and objectivity in the EIA procedure and has become an important tool for the decision-making process (Robu et al., 2007; Zelenakova, Zvijakova, & Singovszka, 2017). According to the 'Risk Assessment and Mapping Guidelines' (European Commission, SEC, 2010), most EU member states have developed RA methodologies that increase the capacity to respond to the identified risks through prevention, preparation, and response measures. The European Commission has initiated a process to create a methodological framework for RA to enable the development of common European strategies and policies based on comparable results at the EU level. RA includes the

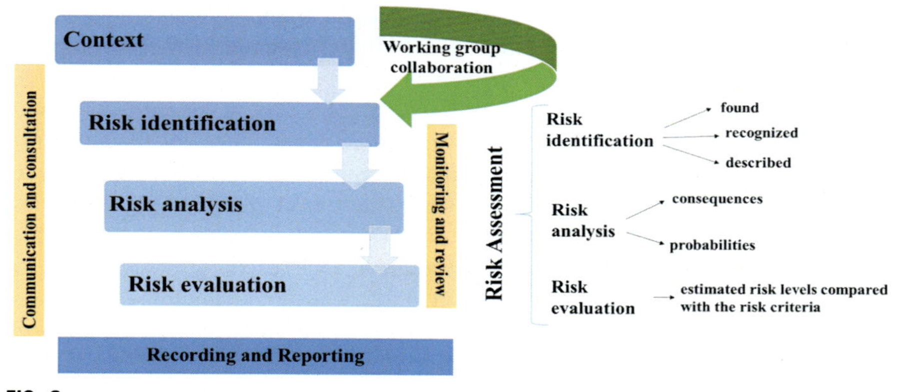

FIG. 8

The risk management based on ISO 31000:2018 (Fuentes-Bargues, Bastante-Ceca, Ferrer-Gisbert, & González-Cruz, 2020).

basic elements of the risk management process, defined by ISO 31000:2018 (Fig. 8): (a) communication and consultation; (b) the context; (c) RA (referring to risk identification, risk analysis, and evaluation); (d) recording and reporting; and (e) monitoring and review (Fuentes-Bargues et al., 2020).

The scientific community recommends that the environmental risk assessment (**ERA**) should be a quantitative approach for comparing the risks for environmental degradation and human health due to natural and anthropogenic contamination (Cothern, 1996).

The definitions of ERA may vary substantially from one source to another. For instance, ERA evaluates the environmental status to estimate the consequences of certain hazards or other urban–industrial actions, and it breaks into human health RA, ecological RA, and applied industrial RA (Muralikrishna & Manickam, 2017b). Other definitions commonly used for risk and hazard are as follows: '*risk is the likelihood that a harmful consequence will occur as the results of an action or condition*' (Muralikrishna & Manickam, 2017b) or '*risk is the combination of the possibility of occurrence of an event and likelihood and strength of undesirable effects of that event*' (Topuz, Talinli, & Aydin, 2011). Hazard is commonly defined as '*the potential to cause harm*' (Kaikkonen, Venesjärvi, Nygård, & Kuikka, 2018; Manuilova, 2003). Chartres, Bero, and Norris (2019) defined hazard as '*any natural or man-made substance, chemical, physical or biological agent, that is capable of causing an adverse health outcome in certain circumstances and the risk is an estimation of the effect of an adverse health outcome when exposed to a hazard*' (Eq. 7).

$$R = P \times Q \tag{7}$$

where R is the level of risk, P the probability of exposure, and Q the severity of the consequences.

According to Fig. 8, in the 1st stage **risk identification**, the working groups and end-users identify the context and the basis of RA (Papathoma-Köhle, Promper, & Glade, 2016). Defining the risk criteria constitutes an important stage of RA that will eventually show whether a risk is acceptable/tolerable. Several experimental studies in the risk management methodology revealed that risks can be calculated by estimating the theoretical level of human exposure and the potential severity of health effects (Melko & Ievins, 2012). According to Chartres et al. (2019) and Manuilova (2003), **risk assessment** is a multi-stage process, which includes hazard identification, hazard characterisation, exposure assessment, and risk characterisation (Fig. 9).

In the case of **hazard identification**, three questions should be addressed: (1) *what may go wrong?*; (2) *what are the consequences?*; and (3) *what are the probabilities of the consequences?* (Liu & Ramirez, 2017)? According to Muralikrishna and Manickam (2017c), increased attention should be paid to ***risk management***, seen as the decision process that emphasises three activities: risk estimation, risk evaluation, and risk control. At the same time, SEA and sustainability evaluation can be completed by RA, ensuring a framework to evaluate economic, social, and physical outcomes (including environmental impacts) of proposed activities or projects.

ERA can be developed by various methods, such as diagrams, checklists, matrices, or combined methods (Chartres et al., 2019; Robu et al., 2007; Ştefănescu et al., 2013). The best approach for ERA remains the combination of both qualitative and quantitative methods, since qualitative methods require a sound level of knowledge and experience, while the use of quantitative methods requires a significant level of reliable information. For most studies, experts are willing to develop complete evaluation methods that are more objective to address the uncertainty and complexity of environmental risks.

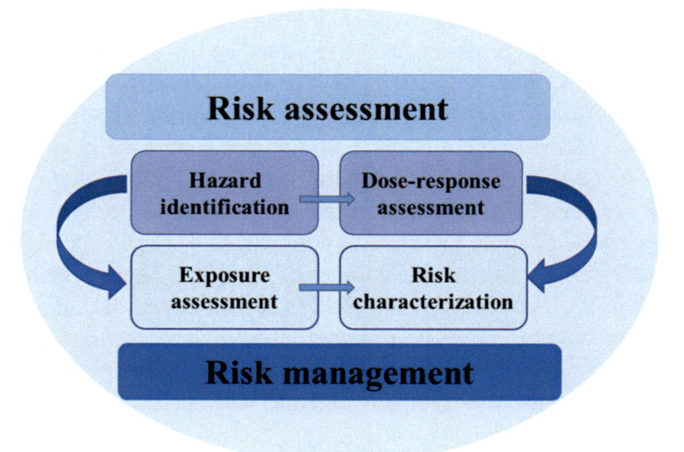

FIG. 9

Risk assessment approach (Manuilova, 2003).

Methods like **fault tree analysis**, **event tree analysis**, bow-tie method, and other mathematical methods are considered efficient tools in environmental risk source identification (Ding et al., 2020). For instance, Ding et al. (2020) highlight that AHP-based methods are easy and flexible for application to identify the important risk sources. Aminbakhsh et al. (2013) mentioned the following methods used for RA: influence diagram method; Monte Carlo simulation to assess concurrent change in multiple independent variables; decision analysis bearing decision matrices and decision trees along with multi-criteria decision-making techniques, such as the simple multi-attribute rating technique; AHP; and universal matrix risk analysis (UMRA) used in both environmental assessment methods (ERA and EIA). Melko and Ievins (2012) identify the most common RA methods and techniques, which are used for RA: methods that are qualitative, quantitative, and semi-quantitative or hybrid. Qualitative methods are commonly used; these methods are simple and easy to use, and they do not need private knowledge.

Fuentes-Bargues et al. (2020) propose **five risk techniques** that could be used: SWIFT, cause and effect analysis, scenario analysis, failure mode and effects analysis (FMEA), and consequence/likelihood matrix. On the other hand, according to Zelenakova et al. (2017), UMRA is one of the proposed methodologies of **risk analysis** not only to enhance the transparency and sensitivity of the evaluation process but also to cope with the requirements of the EIA system at the European level. The basic principle of the methodology is the calculation of the risk index that would allow prioritisation of measures needed to be undertaken.

4 Integrated approach of environmental impact and risk assessments

Since EIA and RA have the same final goal—to support the decision-making process—the trend is to integrate RA into EIA (Robu et al., 2007). It is widely accepted (Al-Nasrawi et al., 2020; Goyal & Deshpande, 2001; Khosravi et al., 2019; Kuitunen et al., 2008) that EIA and RA can be applied together over the whole project life cycle, starting from the initial stage of impact prediction and evaluation, during its implementation, and all the way to its post-closure stage to support an appropriate decision-making process (Ştefănescu et al., 2013). The aim for integrating the RA principles within those of the EIA comes from the necessity of providing objective information about the environmental quality as well as about the risks to which environmental components are exposed. Both type of assessments, highly interdisciplinary, are based on remarkably similar concepts and, generally, address the same environmental issues (Zeleňáková et al., 2020). RA is based on the principle that envisaged activities are connected to stressful factors for the environment, and it perfectly integrates as information, quality, and goals into EIA, becoming an important tool for decision-making and planning (Fuentes-Bargues et al., 2020). Moreover, the international policies emphasise that environmental impact and risk assessments

(EIRA) should analyse the negative effects on environmental quality by quantifying the environmental risk as a function of environmental impact (Robu et al., 2007).

4.1 EIRA assumptions and methodology

The **first step** for performing EIRA is to define the system, i.e. what are the economic activities or projects to be evaluated and what are the likely environmental components exposed to? The **second step** is to assign importance (from 0 to 1) to each environmental component considered and, by using the matrix calculation, to obtain the weight scores and the importance units (IU) (Tables 5 and 6) that will be further used in impact quantification (EI). The **third step** is to quantify the environmental quality (Q) as the ratio between MAC for a certain quality indicator (i) and its MC for any specific environmental component (air, water, soil, etc.). The **fourth step** is to quantify the impact on environmental quality (EI), considering its IU and its calculated quality (Q), and then the environmental risk (ER) is quantified as the probability (p) of impact occurrence (Eqs. 8–11). The probability of impact occurrence is calculated following the same algorithm by applying the matrix for calculating the IU for each environmental component.

$$Q = \frac{MAC_i}{MC_i} \tag{8}$$

Table 5 Example of importance units (IU) calculation (Ştefănescu et al., 2013).

Environmental component	***Assigned value*** ↓	Groundwater (l)	Surface water (m)	Air (n)	Soil (p)
Maximum value is →	**1.00**	*1.00*	*1.25*	*1.67*	*2.50*
Groundwater (l)	**1.00**	1.00=l/l	1.25=l/m	1.67=l/n	2.50=l/p
Surface water (m)	**0.80**	0.80=m/l	1.00=m/m	1.33=m/n	2.00=m/p
Air (n)	**0.60**	0.60=n/l	0.75=n/m	1.00=n/n	1.50=n/p
Soil (p)	**0.40**	0.40=p/l	0.50=p/m	0.67=p/n	1.00=p/p

The bold numbers are assigned by the evaluators, while the italics numbers are obtained calculating the matrix.

Table 6 Importance units (IU) for each environmental component.

Component	**Weights (W)**	**Importance units (IU)**
Groundwater	0.36 = 1/(1.00 + 0.80 + 0.60 + 0.40)	160
Surface water	0.27 = 1/(1.25 + 1.00 + 0.75 + 0.50)	190
Air	0.21 = 1/(1.67 + 1.33 + 1.00 + 0.67)	260
Soil	0.14 = 1/(2.50 + 2.00 + 1.50 + 1.00)	390
Total (check key)	*1.00*	*1000*

The italic numbers refer to the checking key and show if the calculus was done properly. If the sum is 1, then the matrix is correctly solved.

$$EI_i = \frac{IU}{Q} \tag{9}$$

where

EI_i—environmental impact considering the quality indicator i.
IU—the importance units assigned to the environmental component (air, water, soil).
Q_i—quality of the environmental component considering the quality indicator i.

$$ER_j = EI_j \times p_j \tag{10}$$

where

ER_j—risk to the environmental component j.
EI_j—impact on the environmental component j.
p_j—probability of impact occurrence on the environmental component j.

$$EI_j = \frac{\sum_{i=1}^{n} EI_i}{n} \tag{11}$$

The results show that if the EI and the ER have low values, then the impact is insignificant, and the associated risk is negligible. The higher the obtained values are, the higher the impact and risk. The method is flexible and can be extended or retracted to more environmental components or quality indicators, and it has the main advantage that it is a facile instrument that offers information in a transparent manner about impact and associated risk simultaneously.

4.2 EIRA index

The EIRA index addresses the legal drawback and evaluates the environmental impacts based on the pollutant loads, which means that the impact magnitude is represented by considering the product between the pollutant concentration and pollutant loads (e.g. wastewaters), as presented in Eq. (12) (Teodosiu et al., 2015):

$$EI = \frac{C_{det} \times Q_{det}}{C_{max} \times Q_{max}} \times IU \tag{12}$$

where Q_{det} is the average wastewater flow discharged by the pollution source (measured value), m^3/s; Q_{max} the maximum contracted wastewater flow, according to the environmental permit, m^3/s; C_{det} the average pollutant concentration in the effluent, mg/L; C_{max} the MAC, mg/L; and IU the importance unit (dimensionless).

This new approach associates the impact magnitude or extent caused by discharging treated wastewaters with the quality of the receiving water body, without necessarily establishing a causality relationship between these aspects. The IU are assigned by the evaluator, considering the river quality classes (Teodosiu et al., 2015), with the ranges between 0.2 and 1, where value 1 indicates the 'maximum importance', assigned for poor water quality ('very polluted') and value 0.2 represents the 'minimum importance', assigned for unpolluted water ('very good water quality'). The probability units (p_j) of impact occurrence are calculated using historic data series as the frequency of discharging wastewater with concentrations higher than 70% of MAC (Eq. 13) (Tartakovsky, 2013; Topuz et al., 2011).

$$P = \frac{n}{m} \quad (13)$$

where n is the number of 70% MAC reached over the data series (number of 'pollution events') and m the total number of measurements of the data series.

The global EI and ER values are obtained based on an original method, as previously described. The new EIRA index was applied by Teodosiu et al. (2012) for a scenario when a point pollution source discharges wastewater effluent with the reference at 70% MAC ($C_{ref} = 0.7 \cdot C_{max}$) and with a flow equal to the maximum allowed by the environmental permit ($Q_{ref} = Q_{max}$). For example, if the pollution sources discharge the wastewater effluent into a receiving river, classified in the second quality class, the IU that corresponds to this are $IU_{ref} = 0.4$, indicating if the polluting source may or may not deviate the water quality status. Furthermore, for the reference risk, the probability of the environmental impact occurrence was assumed to be $p_{ref} = 100\%$ (Eqs. 14, 15).

$$\mathrm{EI}_{ref} = \frac{C_{ref} \times Q_{ref}}{C_{max} \times Q_{max}} \times UI_{ref} = \frac{0.7 \times C_{max} \times Q_{max}}{C_{max} \times Q_{max}} \times 0.4 = 0.28 \quad (14)$$

$$\mathrm{ER}_{ref} = \mathrm{EI}_{ref} \times p_{ref} = 0.28 \times 1 = 0.28 \quad (15)$$

The EIRA methodology could be successfully used as an instrument for identifying hotspots. Firstly, it provides a common scale to estimate impacts and risks at a river basin level, and it can be used to prioritise among the impacts and risks. Secondly, the environmental impacts and associated risks are calculated based on the MC or considering the pollutant loads and the receiving river water quality (Teodosiu et al., 2016).

5 Environmental impact assessment and life cycle assessment

LCA has been intensively applied for the assessment of environmental impacts of products, processes, or services (Barjoveanu et al., 2014; ISO 14040, 2006).

LCA includes a planning phase (goal and scope definition), an input/output analysis phase (life cycle inventory), an EIA phase (life cycle impact assessment), and an interpretation phase (interpretation of results). But even being standardised, LCA still lacks comprehensive approaches to evaluate the local environmental impacts associated, for instance, with water cycle management (Teodosiu et al., 2016). In this sense, Teodosiu et al. (2012) underlined the idea of using LCA for water system performance evaluation and changed the traditional approach towards water resources from merely a transport medium for other impact generators to a product with economic value, thus contributing to the integration of LCA as a comprehensive instrument in water resources management. This approach enabled the evaluation of water service system performance considering their development perspectives in terms of improving technologies and spatial (urban/rural) expansion (Barjoveanu et al., 2014).

Besides, LCA seems to be an useful tool to get a better picture of the performance of the environmental assessments. Even so, there are few publications that have proposed an integrative methodology of LCA and EIA (Larrey-Lassalle et al., 2017; Wang, Chan, & Li, 2015), but none of them have compared the results obtained with the existing EIA conclusions without LCA (Larrey-Lassalle et al., 2017). For instance, Wang et al. (2015) adopted the concept of LCA and introduced a comprehensive method that integrates the fuzzy analysis for environmental performance assessment. The proposed approach was used as a screening tool to reduce the number of eco-design options and to identify key improvement areas (Wang et al., 2015).

6 Conclusions

EIA is required in cases of new proposed economic activities or new development projects, as well as in the cases of economic or industrial activities, as it identifies at an early stage the possible impacts on environmental quality and ecosystems, while RA is intended to identify and mostly evaluate the hazards to human health, ecosystems, and environment. Once the NEPA was approved by the US Government in 1969, and any public or private project had to be under impact assessment procedure, the first method applied was the simple interaction matrix, the so-called Leopold's Matrix. From this point further, the European regulations were adapted to implement and assure the same principles of impact assessment. Thus several methods have been developed to perform EIA, such as Leopold's Matrix, RIAM, I_{GP}, integrated index, and LCA, which are the most applied in practice. Later, the trend was to integrate the RA into EIA so that the decision-making process would be highly improved and reasoned. The integrated method and the index proposed to quantify the environmental impacts and associated risk are based on measured or estimated concentrations of certain pollutants considered to characterise the environmental quality (as quality indicators), using as reference the MAC or AL (70% MAC), according to environmental standards. Currently, scientists have integrated different environmental management tools so that a better picture is given, and the decision-making process is much improved.

This chapter has presented the main impact assessment procedures, their goals and target audience, as well as the main methodologies usually applied by environmental experts for impact identification and quantification. At the same time, it was underlined that RA has, in its first stages, the same main goals as impact assessment, focusing next on hazard identification and risk estimation on human health and ecosystems. Thus the current trend is to develop methods that are more objective and offer precise information about the likely impacts and risks, with the final purpose to help decision-makers when approving new environmental policies or development strategies that could impact environmental quality or human health.

Acknowledgments

The authors would like to acknowledge the support of the "Environmental Education – OERs for Rural Citizens (EnvEdu – OERs)" project, code 19-COP-0038, with financial support from the Education, Scholarships, Apprenticeships and Youth Entrepreneurship Programme (ESAYEP), European Economic Area Financial Mechanism 2014–2021.

References

Al-Nasrawi, F. A., Kareem, S. L., & Saleh, L. A. (2020). Using the Leopold matrix procedure to assess the environmental impact of pollution from drinking water projects in Karbala city, Iraq. *IOP Conference Series: Materials Science and Engineering*, *671*. https://doi.org/10.1088/1757-899X/671/1/012078.

Aminbakhsh, S., Gunduz, M., & Sonmez, R. (2013). Safety risk assessment using analytic hierarchy process (AHP) during planning and budgeting of construction projects. *Journal of Safety Research*, *46*, 99–105. https://doi.org/10.1016/j.jsr.2013.05.003.

Aryafar, A., Yousefi, S., & Doulati Ardejani, F. (2013). The weight of interaction of mining activities: Groundwater in environmental impact assessment using fuzzy analytical hierarchy process (FAHP). *Environment and Earth Science*, *68*(8). https://doi.org/10.1007/s12665-012-1910-x.

Ashofteh, P. S., Bozorg-Haddad, O., & Loáiciga, H. A. (2017). Multi-criteria environmental impact assessment of alternative irrigation networks with an adopted matrix-based method. *Water Resources Management*, *31*. https://doi.org/10.1007/s11269-016-1554-9.

Azapagic, A., & Perdan, S. (2010). *Assessing environmental sustainability: Life cycle thinking and life cycle assessment*. https://doi.org/10.1002/9780470972847.ch3.

Barjoveanu, G., Comandaru, I. M., Rodriguez-Garcia, G., Hospido, A., & Teodosiu, C. (2014). Evaluation of water services system through LCA. A case study for Iasi City, Romania. *International Journal of Life Cycle Assessment*, *19*, 449–462. https://doi.org/10.1007/s11367-013-0635-8.

Battelle. (1972). *Final report on environmental evaluation system for water resource planning*. Columbus, OH: Battelle, Columbus Laboratories.

Bond, A., Fischer, T. B., & Fothergill, J. (2017). Progressing quality control in environmental impact assessment beyond legislative compliance: An evaluation of the IEMA EIA Quality Mark certification scheme. *Environmental Impact Assessment Review*, *63*, 160–171. https://doi.org/10.1016/j.eiar.2016.12.001.

Canteiro, M., Córdova-Tapia, F., & Brazeiro, A. (2018). Tourism impact assessment: A tool to evaluate the environmental impacts of touristic activities in Natural Protected Areas. *Tourism Management Perspectives*, *28*, 220–227. https://doi.org/10.1016/j.tmp.2018.09.007.

Chartres, N., Bero, L. A., & Norris, S. L. (2019). A review of methods used for hazard identification and risk assessment of environmental hazards. *Environment International*, *123*, 231–239. https://doi.org/10.1016/j.envint.2018.11.060.

Cojocaru, C., Cocârţă, D. M., Istrate, I. A., & Creţescu, I. (2017). Graphical methodology of global pollution index for the environmental impact assessment using two environmental components. *Sustainability*, *9*. https://doi.org/10.3390/su9040593.

Corominas, L., Byrne, D. M., Guest, J. S., Hospido, A., Roux, P., Shaw, A., et al. (2020). The application of life cycle assessment (LCA) to wastewater treatment: A best practice guide and critical review. *Water Research*, *184*. https://doi.org/10.1016/j.watres.2020.116058.

Cothern, C. R. (1996). Introduction and overview of difficulties encountered in developing comparactive rankings of environmental problems. In *Comparative environmental risk assessment*. USA: Lewis Publishers, ISBN:0-87371-605-1.

Ding, G., Xin, L., Guo, Q., Wei, Y., Li, M., & Liu, X. (2020). Environmental risk assessment approaches for industry park and their applications. *Resources, Conservation and Recycling*, *159*, 104844. https://doi.org/10.1016/j.resconrec.2020.104844.

European Commission, SEC. (2010). *Risk assessment and mapping guidelines for disaster management*. https://ec.europa.eu/transparency/regdoc/rep/2/2010/EN/SEC-2010-1626-F1-EN-MAIN-PART-1.PDF.

Fuentes-Bargues, J. L., Bastante-Ceca, M. J., Ferrer-Gisbert, P. S., & González-Cruz, M. C. (2020). Study of major-accident risk assessment techniques in the environmental impact assessment process. *Applied Sciences*, *12*, 1–16. https://doi.org/10.3390/su12145770.

Gilbuena, R., Kawamura, A., Medina, R., Amaguchi, H., Nakagawa, N., & Du Bui, D. (2013). Environmental impact assessment of structural flood mitigation measures by a rapid impact assessment matrix (RIAM) technique: A case study in Metro Manila, Philippines. *Science of the Total Environment*, *456–457*, 137–147. https://doi.org/10.1016/j.scitotenv.2013.03.063.

Glasson, J., & Therivel, R. (2019). *Introduction to environmental impact assessment* (5th ed.). London: Routledge. https://doi.org/10.4324/9780429470738.

Glasson, J., Therivel, R., & Chadwick, A. (1994). *Introduction to environmental impact assessment*. Taylor & Francis Group.

Goyal, S. K., & Deshpande, V. A. (2001). Comparison of weight assignment procedures in evaluation of environmental impacts. *Environmental Impact Assessment Review*, *21*, 553–563. https://doi.org/10.1016/S0195-9255(01)00086-5.

Hasan, M. A., Nahiduzzaman, K. M., & Aldosary, A. S. (2018). Public participation in EIA: A comparative study of the projects run by government and non-governmental organizations. *Environmental Impact Assessment Review*, *72*, 12–24. https://doi.org/10.1016/j.eiar.2018.05.001.

Ijäs, A., Kuitunen, M. T., & Jalava, K. (2010). Developing the RIAM method (rapid impact assessment matrix) in the context of impact significance assessment. *Environmental Impact Assessment Review*, *30*, 82–89. https://doi.org/10.1016/j.eiar.2009.05.009.

ISO 14040. (2006). *Environmental management. Life cycle assessment. Principles and framework*. Switzerland: International Organization for Standardization.

ISO 31000. (2018). *Risk management—Guidelines*. Switzerland: International Organization for Standardization.

Jiricka-Pürrer, A., Czachs, C., Formayer, H., Wachter, T. F., Margelik, E., Leitner, M., et al. (2018). Climate change adaptation and EIA in Austria and Germany—Current consideration and potential future entry points. *Environmental Impact Assessment Review*, *71*, 26–40. https://doi.org/10.1016/j.eiar.2018.04.002.

Josimovic, B., Petric, J., & Milijic, S. (2014). The use of the Leopold matrix in carrying out the EIA for wind farms in Serbia. *Energy and Environment Research*, *4*. https://doi.org/10.5539/eer.v4n1p43.

Kaikkonen, L., Venesjärvi, R., Nygård, H., & Kuikka, S. (2018). Assessing the impacts of seabed mineral extraction in the deep sea and coastal marine environments: Current methods and recommendations for environmental risk assessment. *Marine Pollution Bulletin*, *135*, 1183–1197. https://doi.org/10.1016/j.marpolbul.2018.08.055.

Kayastha, P., Dhital, M. R., & De Smedt, F. (2013). Application of the analytical hierarchy process (AHP) for landslide susceptibility mapping: A case study from the Tinau watershed, west Nepal. *Computational Geosciences*, *52*, 398–408. https://doi.org/10.1016/j.cageo.2012.11.003.

Khosravi, F., Jha-Thakur, U., & Fischer, T. B. (2019). Enhancing EIA systems in developing countries: A focus on capacity development in the case of Iran. *Science of the Total Environment*, *670*, 425–432. https://doi.org/10.1016/j.scitotenv.2019.03.195.

Kuitunen, M., Jalava, K., & Hirvonen, K. (2008). Testing the usability of the Rapid Impact Assessment Matrix (RIAM) method for comparison of EIA and SEA results. *Environmental Impact Assessment Review*, *28*, 312–320. https://doi.org/10.1016/j.eiar.2007.06.004.

Larrey-Lassalle, P., Catel, L., Roux, P., Rosenbaum, R. K., Lopez-Ferber, M., et al. (2017). An innovative implementation of LCA within the EIA procedure: lessons learned from two wastewater treatment plant case studies. *Environmental Impact Assessment Review*, *63*, 95–106. https://doi.org/10.1016/j.eiar.2016.12.004.

Leopold, L. B., Clarke, F. E., Hanshaw, B. B., & Balsley, J. E. (1971). *Procedure for evaluating environmental impact*. US Geological Survey Circular.

Liu, W., & Ramirez, A. (2017). State of the art review of the environmental assessment and risks of underground geo-energy resources exploitation. *Renewable and Sustainable Energy Reviews*, *76*, 628–644. https://doi.org/10.1016/j.rser.2017.03.087.

Malczewski, J. (2018). *Multicriteria analysis in earth systems and environmental sciences* (pp. 197–217). https://doi.org/10.1016/B978-0-12-409548-9.09698-6.

Manuilova, A. (2003). *Methods and tools for assessment of environmental risk* (p. 21). Akzo Nobel.

McCabe, M., & Sadler, B. (2002). *Environmental impact assessment. Training resource manual* (2nd ed., pp. 1–600).

Melko, A., & Ievins, J. (2012). Methods of the environmental risk analysis and assessment, the modified method of the risk index. *Safety of Technogenic Environment*, *2*, 50–56.

Mirmohammadi, M., Gholamnejad, J., Fattahpour, V., Seyedsadri, P., & Ghorbani, Y. (2009). Designing of an environmental assessment algorithm for surface mining projects. *Journal of Environmental Management*, *90*, 2422–2435. https://doi.org/10.1016/j.jenvman.2008.12.007.

Mohebali, S., Maghsoudy, S., & Ardejani, F. D. (2020). Application of data envelopment analysis in environmental impact assessment of a coal washing plant: A new sustainable approach. *Environmental Impact Assessment Review*, *83*, 106389. https://doi.org/10.1016/j.eiar.2020.106389.

Mohebali, S., Maghsoudy, S., Doulati Ardejani, F., & Shafaei, F. (2019). Developing a coupled environmental impact assessment (C-EIA) method with sustainable development approach for environmental analysis in coal industries. *Environment, Development and Sustainability*, *22*, 6799–6830. https://doi.org/10.1007/s10668-019-00513-2.

Mondal, M. K., Rashmi, & Dasgupta, B. V. (2010). EIA of municipal solid waste disposal site in Varanasi using RIAM analysis. *Resources, Conservation and Recycling*, *54*, 541–546. https://doi.org/10.1016/j.resconrec.2009.10.011.

Morris, P., & Therivel, R. (1995). *Methods of environmental impact assessment*. Oxford: UBC Press.

Muralikrishna, I. V., & Manickam, V. (2017a). Environmental impact assessment. In *Environmental management*. https://doi.org/10.1016/B978-0-12-811989-1.00006-3.

Muralikrishna, I. V., & Manickam, V. (2017b). Environmental risk assessment. In *Environmental management*. https://doi.org/10.1016/B978-0-12-811989-1.00008-7.

Muralikrishna, I. V., & Manickam, V. (2017c). Energy management and audit. In *Environmental management*. https://doi.org/10.1016/b978-0-12-811989-1.00009-9.

Neamtu, R., Sluser, B., Plavan, O., & Teodosiu, C. (2021). Environmental monitoring and impact assessment of Prut river cross-border pollution. *Environmental Monitoring and Assessment*, *193*, 340. https://doi.org/10.1007/s10661-021-09110-1.

Papathoma-Köhle, M., Promper, C., & Glade, T. (2016). A common methodology for risk assessment and mapping of climate change related hazards-implications for climate change adaptation policies. *Climate*, *4*(1), 8. https://doi.org/10.3390/cli4010008.

Pastakia, C. M. R. (1998). The rapid impact assessment matrix (RIAM)—A new tool for environmental impact assessment. *Environmental Impact Assessment Review*, 8–18.

Phillips, J. (2013). The application of a mathematical model of sustainability to the results of a semi-quantitative Environmental Impact Assessment of two iron ore opencast mines in Iran. *Applied Mathematical Modelling*, *37*, 7839–7854. https://doi.org/10.1016/j.apm.2013.03.029.

Pourghasemi, H. R., Pradhan, B., & Gokceoglu, C. (2012). Application of fuzzy logic and analytical hierarchy process (AHP) to landslide susceptibility mapping at Haraz watershed, Iran. *Natural Hazards*, *63*, 965–996. https://doi.org/10.1007/s11069-012-0217-2.

Ramanathan, R. (2001). A note on the use of the analytic hierarchy process for environmental impact assessment. *Journal of Environmental Management*, *63*, 27–35. https://doi.org/10.1006/jema.2001.0455.

Robu, B., Jitar, O., Teodosiu, C., Strungaru, S. A., Nicoara, M., & Plavan, G. (2015). Environmental impact and risk assessment of the main pollution sources from the Romanian black sea coast. *Environmental Engineering and Management Journal*, *14*, 331–340. https://doi.org/10.30638/eemj.2015.033.

Robu, B. M., Căliman, F. A., Beţianu, C., & Gavrilescu, M. (2007). Methods and procedures for environmental risk assessment. *Environmental Engineering and Management Journal*, *6*, 573–592. https://doi.org/10.30638/eemj.2007.074.

Rocha, C. F., Ramos, T. B., & Fonseca, A. (2019). Manufacturing pre-decisions: A comparative analysis of environmental impact statement (EIS) reviews in Brazil and Portugal. *Sustainability*, *11*. https://doi.org/10.3390/SU11123235.

Roos, C., Cilliers, D. P., Retief, F. P., Alberts, R. C., & Bond, A. J. (2020). Regulators' perceptions of environmental impact assessment (EIA) benefits in a sustainable development context. *Environmental Impact Assessment Review*, *81*, 106360. https://doi.org/10.1016/j.eiar.2019.106360.

Saaty, T. L. (1980). *The analytical hierarchy process, planning, priority*. Resource Allocation RWS Publications USA.

Saaty, T. L. (1990). How to make a decision: The analytic hierarchy process. *European Journal of Operational Research*, *48*, 9–26. https://doi.org/10.1016/0377-2217(90)90057-I.

Saffari, A., Ataei, M., Sereshki, F., & Naderi, M. (2019). Environmental impact assessment (EIA) by using the Fuzzy Delphi Folchi (FDF) method (case study: Shahrood cement plant, Iran). *Environment, Development and Sustainability*, *21*, 817–860. https://doi.org/10.1007/s10668-017-0063-1.

Sajjadi, S. A., Aliakbari, Z., Matlabi, M., Biglari, H., & Rasouli, S. S. (2017). Environmental impact assessment of Gonabad municipal waste landfill site using Leopold Matrix. *Electronic Physician*, *9*(2), 3714–3719. https://doi.org/10.19082/3714.

Samimi Namin, F., Shahriar, K., & Bascetin, A. (2011). Environmental impact assessment of mining activities. A new approach for mining methods selection. *Gospodarka Surowcami Mineralnymi*, *27*, 113–143.

Sandham, L. A., Hoffmann, A. R., & Retief, F. P. (2008). Reflections on the quality of mining EIA reports in South Africa. *Journal of the Southern African Institute of Mining and Metallurgy*, *108*, 701–706.

Scheid, F. (1998). *Theory and problems of numerical analysis. Schaum's outline series* (pp. 348–349). Singapore: McGrawHill.

Schetke, S., Haase, D., & Kötter, T. (2012). Towards sustainable settlement growth: A new multi-criteria assessment for implementing environmental targets into strategic urban planning. *Environmental Impact Assessment Review*, *32*, 195–210. https://doi.org/10.1016/j.eiar.2011.08.008.

Shakib-Manesh, T. E., Hirvonen, K. O., Jalava, K. J., Ålander, T., & Kuitunen, M. T. (2014). Ranking of small scale proposals for water system repair using the Rapid Impact Assessment Matrix (RIAM). *Environmental Impact Assessment Review*, *49*, 49–56. https://doi.org/10.1016/j.eiar.2014.06.001.

Sluşer, B. M., Şchiopu, A. M., Bălan, C., & Pruteanu, M. (2017). Postclosure influence of emissions resulted from municipal waste dump sites: A case study of the north-east region of Romania. *Environmental Engineering and Management Journal*, *16*, 1017–1026. https://doi.org/10.30638/eemj.2017.103.

Soria-Lara, J. A., Batista, L., Le Pira, M., Arranz-López, A., Arce-Ruiz, R. M., Inturri, G., et al. (2020). Revealing EIA process-related barriers in transport projects: The cases of Italy, Portugal, and Spain. *Environmental Impact Assessment Review*, *83*, 106402. https://doi.org/10.1016/j.eiar.2020.106402.

Ştefănescu, L., Robu, B. M., & Ozunu, A. (2013). Integrated approach of environmental impact and risk assessment of Rosia Montana Mining Area, Romania. *Environmental Science and Pollution Research*, *20*, 7719–7727. https://doi.org/10.1007/s11356-013-1528-x.

Suditu, G. D., & Robu, B. M. (2012). Digitization of the environmental impact quantification process. *Environmental Engineering and Management Journal*, *11*, 841–848. https://doi.org/10.30638/eemj.2012.107.

Suthar, S., & Sajwan, A. (2014). Rapid impact assessment matrix (RIAM) analysis as decision tool to select new site for municipal solid waste disposal: A case study of Dehradun city, India. *Sustainable Cities and Society*, *13*, 12–19. https://doi.org/10.1016/j.scs.2014.03.007.

Tartakovsky, D. M. (2013). Assessment and management of risk in subsurface hydrology: A review and perspective. *Advances in Water Resources*, *51*, 247–260. https://doi.org/10.1016/j.advwatres.2012.04.007.

Teodosiu, C., Barjoveanu, G., Robu, B., & Ene, S. A. (2012). Sustainability in the water use cycle: Challenges in the Romanian context. *Environmental Engineering and Management Journal*, *11*(11), 1987–2000. https://doi.org/10.30638/eemj.2012.248.

Teodosiu, C., Barjoveanu, G., Sluser, B. R., Popa, S. A. E., & Trofin, O. (2016). Environmental assessment of municipal wastewater discharges: A comparative study of evaluation methods. *International Journal of Life Cycle Assessment*, *21*, 395–411. https://doi.org/10.1007/s11367-016-1029-5.

Teodosiu, C., Robu, B., Cojocariu, C., & Barjoveanu, G. (2015). Environmental impact and risk quantification based on selected water quality indicators. *Natural Hazards*, *75*, 89–105. https://doi.org/10.1007/s11069-013-0637-7.

Topuz, E., Talinli, I., & Aydin, E. (2011). Integration of environmental and human health risk assessment for industries using hazardous materials: A quantitative multi criteria approach for environmental decision makers. *Environment International*, *37*, 393–403. https://doi.org/10.1016/j.envint.2010.10.013.

Tukker, A. (2000). Life cycle assessment as a tool in environmental impact assessment. *Environmental Impact Assessment Review*, *20*(4), 435–456. https://doi.org/10.1016/s0195-9255(99)00045-1.

Wagh, C. H., & Gujar, M. G. (2014). The environmental impact assessment by using the Battelle method. *International Journal of Science and Research*, *3*, 82–86.

Wang, X., Chan, H. K., & Li, D. (2015). A case study of an integrated fuzzy methodology for green product development. *European Journal of Operational Research*, *241*(1), 212–223. https://doi.org/10.1016/j.ejor.2014.08.007.

Xu, C., Haase, D., Su, M., Wang, Y., & Pauleit, S. (2020). Assessment of landscape changes under different urban dynamics based on a multiple-scenario modeling approach. *Environment and Planning B-Urban Analytics and City Science*, *47*, 1361–1379. https://doi.org/10.1177/2399808320910161.

Yao, X., He, J., & Bao, C. (2020). Public participation modes in China's environmental impact assessment process: An analytical framework based on participation extent and conflict level. *Environmental Impact Assessment Review*, *84*, 106400. https://doi.org/10.1016/j.eiar.2020.106400.

Zaharia, C., & Surpateanu, M. (2018). Environmental impact assessment using the method of global pollution index applied for a heat and power cogeneration plant. *Environmental Engineering and Management Journal*, *5*, 1141–1152. https://doi.org/10.30638/eemj.2006.092.

Zeleňáková, M., Labant, S., Zvijáková, L., Weiss, E., Čepelová, H., Weiss, R., et al. (2020). Methodology for environmental assessment of proposed activity using risk analysis. *Environmental Impact Assessment Review*, *80*, 106333. https://doi.org/10.1016/j.eiar.2019.106333.

Zelenakova, M., Zvijakova, L., & Singovszka, E. (2017). Universal matrix of risk analysis method for flood mitigation measures in Vyšná Hutka, Slovakia. *Fresenius Environmental Bulletin*, *26*, 1216–1224.

Zhang, Y., Lu, W., Xi, Y., & Chun, Q. (2015). The impacts of mining exploitation on the environment in the Changchun–Jilin–Tumen economic area, Northeast China. *Natural Hazards*, *76*, 1019–1038. https://doi.org/10.1007/s11069-014-1533-5.

CHAPTER

11 Multi-criteria decision-making

Claudia Labianca[a,b], Sabino De Gisi[b], and Michele Notarnicola[b]

[a]*Department of Civil and Environmental Engineering, Hong Kong Polytechnic University, Hung Hom, Kowloon, Hong Kong, China* [b]*Department of Civil, Environmental, Land, Building Engineering and Chemistry (DICATECh), Polytechnic University of Bari, Bari, Italy*

1 Introduction

Environmental decisions are often difficult and incorporate multi-disciplinary knowledge about natural, chemo-physical, and social sciences, politics, economics, and ethics. Decision-makers rely on experimental investigations, computational models, and tools to assess environmental impacts to human health and ecological risks associated with anthropic and industrial stressors. Firstly, it is essential to be aware that multi-disciplinary problems are not always easy to be addressed. Secondly, for such a complex problem it is appropriate to include stakeholders of the area using a participatory approach.

Public involvement is a fundamental principle of the Environmental Impact Assessment (EIA) process. The main advantages of public participation in environmental issues are educating people and making them more aware of environmental issues, improving plans and policies on the basis of stakeholders' knowledge and experience, and more transparency and support in decision-making. In 2011 the international organisations launched the Life Cycle Sustainability Assessment Framework for experts from different disciplinary fields to develop an integrated approach to combine impacts and benefits but in particular to support effective sustainable development and sustainability decision-making (Gbededo, Liyanage, & Garza-Reyes, 2018) in the final vision of the Sustainable Development Goals (Halisçelik & Soytas, 2019). The entire decision-making process must be clear and traceable, taking into account the necessities of the zone (De Feo & De Gisi, 2010a). As such, the integration of heterogeneous and sometimes uncertain information requires a systematic and easily understandable framework to organise and process technical information though an expert judgement.

Multi-Criteria Decision-Making (MCDM) provides a scientifically sound decision methodology to combine different types of inputs with stakeholder views and cost/benefit information. In complex decision-making problems, there is generally

Assessing Progress towards Sustainability. https://doi.org/10.1016/B978-0-323-85851-9.00003-1

no obvious solution, and as such it is necessary to use the preferences of decision-makers to differentiate between solutions (Langemeyer, Gómez-Baggethun, Haase, Scheuer, & Elmqvist, 2016). MCDM is also referred to as MCDA (Multi-Criteria Decision-Making Analysis) or MCA (Multi-criteria analysis). The application of MCDM has grown and developed significantly over the past decades. Fig. 1A shows the increasing number of documents published in the decade 2020–2010; the subject areas covered are numerous although the main ones are computer science, engineering, mathematics, environmental science, and decision science (Fig. 1B).

MCDM was applied to improve the selection of regulations in the remediation of contaminated sites, the reduction of pollutants entering the ecosystem, and the management optimisation of environmental resources. There are several classifications of MCDM problems and methods, although an important distinction is based on whether the solutions to the problem are explicitly (direct method) or implicitly defined (indirect method). In the first case, there are a finite number of alternatives and each alternative is 'quantified' with respect to specific, specially defined criteria. The goal of MCDM is to identify the optimal solution but also to sort/classify the alternatives. In the second case, the alternatives are not explicitly known. The number of alternatives can be counted or not, but in general it is exponentially large. An alternative (solution) can be found by solving a mathematical model.

Some of the most technically sound MCDM methods in literature are Analytical Hierarchy Process (AHP) (Saaty, 1990), Elimination et Choix Traduisant la Realité (ELECTRE) (Çalı & Balaman, 2019), technique for order preference by similarity to an ideal solution (TOPSIS) (Ozturk & Batuk, 2011), Decision Support sYstem for the REqualification of contaminated sites (DESYRE) (Cappuyns, 2016), preference ranking organisation method for enrichment evaluation (PROMETHEE) (Lolli et al., 2016), and others. The above-mentioned methodologies are widely used in several fields such as engineering, environmental management, and public administration (Ahmadi, Behzadian, Ardeshir, & Kapelan, 2017; Özkan, Sarıçiçek, & Özceylan, 2020) to provide a reliable and sustainable road map to decision-makers. These methods are usually implemented in specialised decision-making software, in many cases available on the market. In Section 3 several applications of these methods will be shown.

Choosing between alternatives is a natural decision-making approach as old as human history. However, 'traditional' decision-making usually involves evaluating pros and cons in an intuitive or holistic manner. MCDM framework aims at solving decision problems impartially and more technically in a structured way, resulting in more transparent and consistent decisions.

This chapter will show the main steps and advantages of developing MCDM methods applied to the resolution of complex environmental problems.

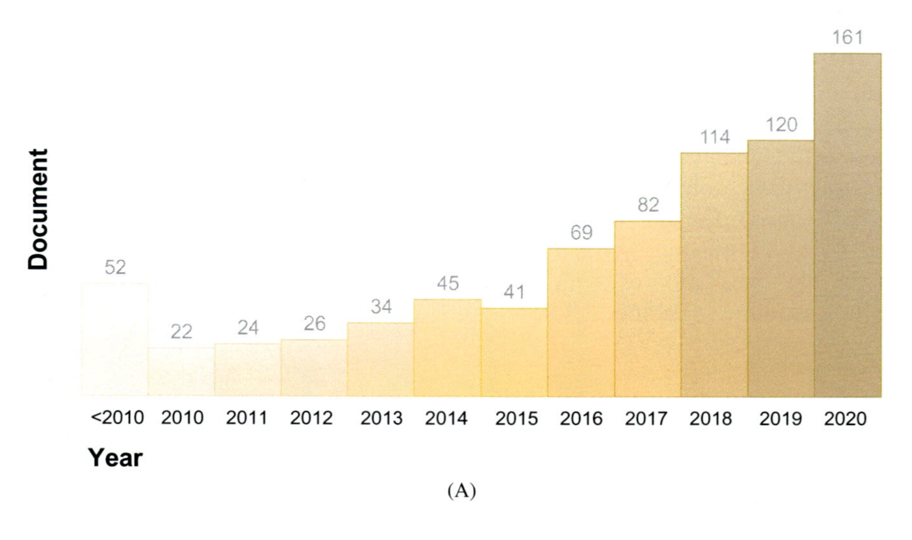

(A)

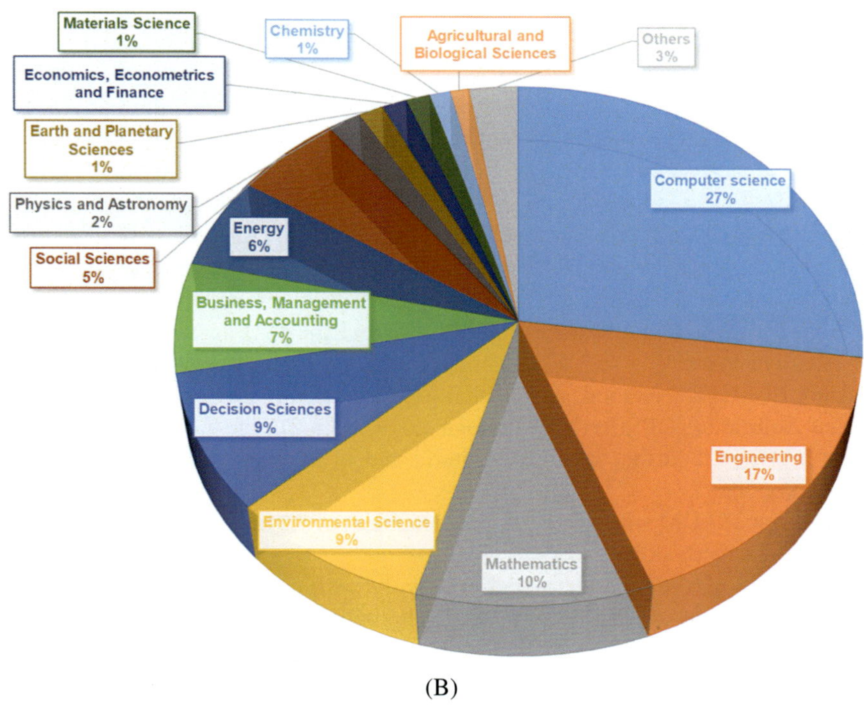

(B)

FIG. 1

Number of documents published in 2010–2020 in the environmental-MCDM field (A) and subject areas (B).

Source: Scopus.

2 Multi-criteria decision-making methods and framework

Multi-criteria analyses have been utilised (a) for evaluating multi-attribute discrete options (Multi-Attribute Decision-Making, MADM), also referred to as Multiple Criteria Evaluation Problems or Discrete Multiple Criteria Problems, and (b) for selecting from continuous sets of options (Multi-Objective Decision-Making, MODM), also referred to as Multiple Criteria Design Problem or Continuous Multiple Criteria Problem.

For the formulation of a multi-criteria decisional process, the function $F(x)$ is defined as follows (Paolucci, 2000):

$$\max F(x) = [f_1(x), \ldots, f_k(x)]^T \quad (1)$$

$$x \in X \subseteq R^n \quad (2)$$

with

x vector of decisional variables,
f_i ith objective,
X set of admissible alternatives.

In brief, MADM is associated with problems with predetermined alternatives. It is used to rank, classify, or select alternatives concerning criteria categorised under outranking, interactive, multi-attribute utility theory (MAUT), and multi-attribute value theory (MAVT).[a] On the other hand, MODM is associated with problems in which the alternatives have been non-predetermined and it is used for vector optimisation in mathematical programming (Kazimieras Zavadskas, Antucheviciene, & Kar, 2019).

In the context of MADM methods, a variety of resolution methods are available: Methods for Cardinal Preference of Attribute over Linear Assignment method, Simple Additive Weighting (SAW) method, Hierarchical Additive Weighting method (HAW), ELECTRE method, and Technique for Order of Preference by Similarity to Ideal Solution (TOPSIS) (Vinogradova, 2019) (Fig. 2). The most commonly used is the SAW method that combines criteria and their weights into a single value (Podvezko, 2011).

Amongst MADM approaches, the development of a MCDM can be based on the following six methodological steps (Fig. 3).

The first step (Problem Identification) involves structuring and framing the decision problem being addressed. All possible alternatives need to be identified through a preliminary screening by a Technical Working Group (TWG) made of unbiased experts or members of the Verification Bodies and the European Commission (EU Technical Working Groups, 2021). For instance, in the context of the remediation of a contaminated site, Labianca, De Gisi, Todaro, and Notarnicola (2020)

[a]MAVT presupposes that the decision criteria and the outcomes are certain, whilst MAUT handles also uncertainty. Unlike MAVT, MAUT does not require independence on the criteria (Lindell, 2017).

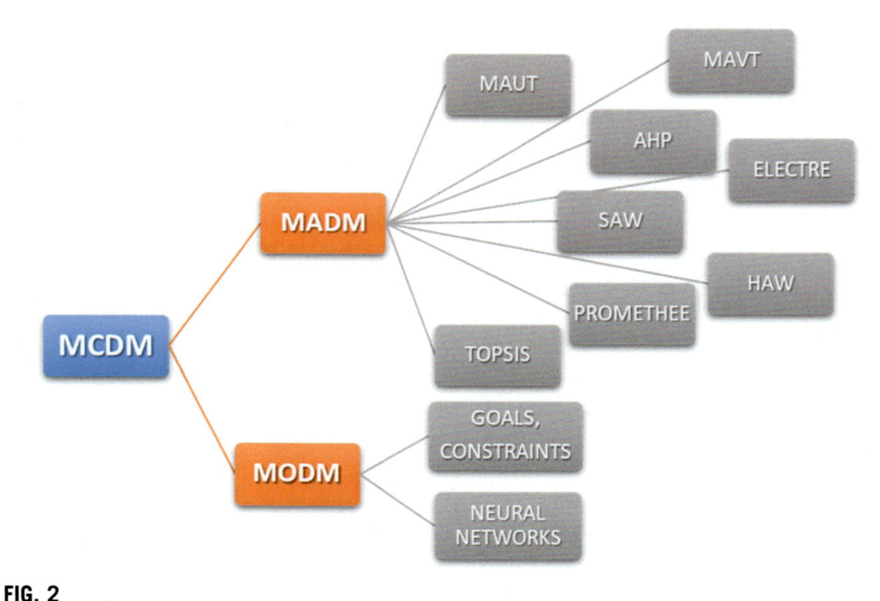

FIG. 2

MCDM classification methods.

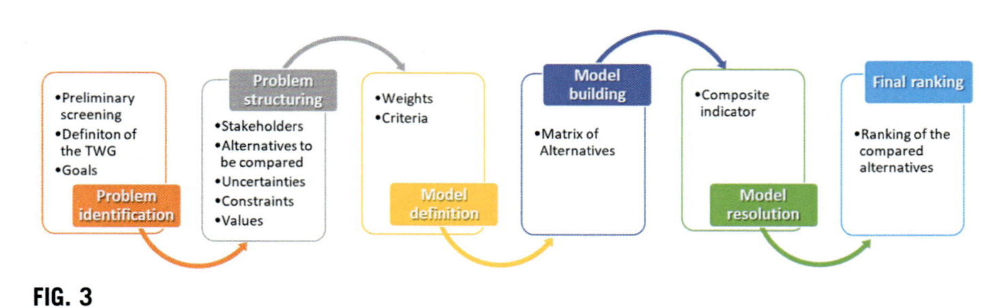

FIG. 3

MCDM general framework (MADM approach).

listed a total of 50 possible remediation technologies that involved different physical, chemical, and biological treatments in order to reach the required environmental goals. In this context, the main stakeholders involved need to be clearly defined. Freeman (1984) described a stakeholder as an individual or a group of individuals able to influence the goal of an organisation or can be influenced themselves by that goal. Later on, according to Macharis, Milan, and Verlinde (2014), stakeholders consist of any group of people, organised or not, who share a common interest in a particular issue. Therefore a stakeholder should be defined according to his/her stake in the issue, as this determines if that stakeholder can affect or will be affected by the ultimate decision. Some examples of key stakeholders are creditors, directors, employees, government agencies, owners (shareholders), suppliers, unions, and the local communities (Post, Preston, & Sauter-Sachs, 2002).

Secondly (Problem Structuring), all evaluation criteria need to be defined to assess and compare technical, environmental, social, and economic aspects of the alternatives. The criteria should be valid, reliable, and relevant to the decision-makers and other stakeholders without major overlaps (double-counting) or redundancy (irrelevant criteria). The selection factors represent the relevant criteria to compare all the alternatives. Each criterion/attribute is related to an objective in the given decision context and can be defined through technical and scientific reports. Numerical attributes express the absolute measure of a technical parameter (e.g. material properties, costs, etc.). However, the values of numerical attributes are often imprecise and depend on the measuring techniques (Jahan, Edwards, & Bahraminasab, 2016). Sometimes attributes can be non-numerical; in fact, some properties are often described by linguistic terms or ordinal data (e.g. poor, adequate, good), and some can only be explained in text, images, or interviews (Ashby, Brechet, Cebon, & Salvo, 2004).

The third step (Model Definition) is crucial because it entails the definition of the weights to be attributed to each evaluation criteria, based on the knowledge of experts and stakeholders. Each evaluation criterion (C_i) is given a weight (w_i), according to the relation with the other criteria. Then, the corresponding vector of weights (W) is defined by the TWG as criteria priorities.

Various expression modes are available for comparing different criteria as direct methods: direct rating (Bottomley, Doyle, & Green, 2000), swing weighting (Von Winterfeldt & Edwards, 1993), the simple multi-attribute rating technique (SMART) (Edwards, 1977), pairwise comparison methods such as the AHP (Saaty, 1977); and as indirect methods: decomposition methods such as discrete choice experiments (DCE) (de Bekker-Grob, Ryan, & Gerard, 2012), bisection and difference methods (Von Winterfeldt & Edwards, 1993) or the measuring attractiveness by a categorical-based evaluation technique (MACBETH) (Ryan, Gerard, & Currie, 2012), conjoint analysis (Green, Krieger, & Wind, 2001), and the potentially all pairwise rankings of all possible alternatives (PAPRIKA) methodology (Hansen & Ombler, 2008). Each mode has its advantages and disadvantages, with an important influence on the final result (Table 1).

For instance, the Paired Comparison Technique (PCT) can be used for a pairwise confrontation amongst different criteria (De Feo & De Gisi, 2014). The PCT matrix is constructed by using scores that represent specialists' judgements when comparing each criterion in relation to the others (e.g. using three numerical judgements: 1 means 'more important than'; 0.5 means 'equal importance to'; 0 means 'less important than'). The weight of each criterion (w_i) is calculated according to the following formula (Eq. 3):

$$w_i = \sum_{j=1}^{n} S_j \quad (3)$$

where S_j represents the scores of technical judgements.

Table 1 Main direct and indirect methods.

Direct methods	**Description**	**Reference**
Direct rating	Ratings of the alternatives' performance on criterion are allocated in natural scales appropriate for the criterion and converted to a common one (e.g. SMART technique is based on a linear additive model).	Bottomley et al. (2000)
AHP	Either criteria or option performance is compared in pairs with a point scale that can be graphic, verbal, or numeric. The comparisons are entered into a matrix.	Saaty (1977)
Swing weighting	It is used to determine trade-off weights by comparing overall value gain in one criterion for change from worst to best performance against the corresponding change in other criteria.	Von Winterfeldt and Edwards (1993)
MACBETH	It is a scoring method based on the additive value model. Questions compare two options at a time (on each criterion or amongst criteria), asking the responder for only a qualitative preference differences judgement using seven semantic categories.	Ryan et al. (2012)
Indirect methods	**Description**	**Reference**
Conjoint analysis	It is composed of choice sets that are identified from the ranking of the alternatives. The criteria weights are identified using this regression technique for all alternatives.	Green et al. (2001)
DCE	It involves presenting respondents with a series of hypothetical scenarios (choice sets) described using the criteria of interest. Respondents are then asked to choose their preferred intervention.	de Bekker-Grob et al. (2012)
Bisection and difference methods	In the bisection method, the responder is asked to identify the value point on the attribute scale which is halfway between the two endpoints on the scale. In the difference method the decision-maker must consider different increments on the objectively measured scale and relate these to the difference in values.	Von Winterfeldt and Edwards (1993)
PAPRIKA	It asks questions based on choosing between two hypothetical alternatives defined on only two criteria/attributes at a time. Based on the answers, it adapts and chooses next questions to ask. It may be recognised as a type of adaptive conjoint analysis.	Hansen and Ombler (2008)

Criteria are scored and weighted usually automatically by MCDM software (see Section 3). In particular, for the direct methods and the PAPRIKA (indirect) method, each alternative's scores on the criteria are multiplied by the weights; then the weighted scores are summed across the criteria in order to calculate each alternative's total score. For the conjoint analysis or DCE method, the regression technique estimates each alternative's value/utility or its probability of being chosen by the decision-makers.

The fourth step (Model Building) involves the construction and normalisation of the decision matrix, called Matrix of Alternatives (MOA), which can consist of the alternatives to be compared in the rows and the evaluation criteria in the columns. The normalisation consists in making all the indexes comparable on the same scale. It is shown that normalisation can be applied in different ways; however, it might also affect the final ranking. The main objectives of the normalisation process are (i) capability in removing scales, (ii) symmetry in normalisation for cost and benefit criteria, (iii) transforming other types of criteria to the benefit criteria, (iv) ranking reversals, (v) handling negative values (Jahan et al., 2016). When using a matrix-based structure, a proper normalisation method can enhance the effectiveness of the final resolution.

For instance, the SAW method (Afshari, Mojahed, & Yusuff, 2010) calculates an evaluation score for each alternative by multiplying each element of the matrix with the weight of the corresponding column assigned by the decision-maker. The advantage of this technique is that it is a proportional linear transformation of the raw data, which implies that the relative order of magnitude of the standardised scores remains the same.

As a fifth step (Model Resolution), the composite indicator, parameter for choosing the best alternatives, needs to be constructed. The composite indicator can be influenced by previous choices. As a consequence, the selection method to calculate the composite indicator should consider the dimensionality of the phenomena under study, include reliable sub-indicators, and quantify the uncertainty of previous steps (Langhans, Reichert, & Schuwirth, 2014; OECD-JRC, 2008).

As such, the composite indicator has to be validated and calibrated (e.g. by means of case studies). The goal is to evaluate its robustness through techniques able to quantify the uncertainties generated by the methodological assumptions. In addition, a sensitivity analysis can reveal the reliability of the composite indicator results. The JRC (Joint Research Centre of the European Commission) has released important guidelines about the role of the uncertainty (UA) and sensitivity analysis (SA). More specifically, the UA describes the effect of uncertainty in the composite indicator outcome, without recognising which assumptions are predominantly responsible (Saltelli et al., 2019). The SA focuses on the contribution of each individual variable uncertainty to the output variance (OECD-JRC, 2008; Paruolo, Saisana, & Saltelli, 2013). In this regard, Table 2 collects some examples of composite indicators recently developed in different environmental sectors highlighting the approach adopted for their definition.

Table 2 Composite indicator construction.

Index construction	Investigation field	Goal	Size of the case study	References
Linear method of aggregation	Solid waste	Development of an indicator system for the selection of the best site for composting plant.	4 potential sites for the localisation of composting plants.	De Feo and De Gisi (2010b)
Information aggregated by using a Goal Programming Synthetic Indicator (GPSI)	Tourism	Development of an indicator system easy to implement the sustainability of tourism activities considering social, economic, and environmental areas and 85 indicators.	181 selected cultural tourist destinations in Andalusia (Spain).	Lozano-Oyola, Blancas, González, and Caballero (2012)
Composite index constructed with a weighted arithmetic mean function (ratio or percentage differences from the mean)	Air, water	Assessment of the pollution status of the case study (the middle section of the Lower Seyhan River Basin, Turkey).	–	Golge, Yenilmez, and Aksoy (2013)
Ratio or percentage differences from the mean	SEA (Strategic Environmental Assessment)	Development of an indicator system to foster sustainability in strategic planning in China.	–	Wang, Lam, Harder, Ma, and Yu (2013)
Linear method of aggregation	Water	Development of an indicator system for the selection of the best combination of coagulant/flocculant and dosage for physical–chemical treatment of municipal wastewater.	–	De Feo, De Gisi, and Galasso (2008)
Re-scaled values	Air	Evaluation of the air pollution index in Turkey for the period 1990–2011 by using global warming potential, acidification potential, tropospheric ozone forming potential, and particulate formation potential.	–	Köne and Büke (2014)

Continued

Table 2 Composite indicator construction. *Continued*

Index construction	Investigation field	Goal	Size of the case study	References
Linear method of aggregation	Wastewater	Assessment of the sustainability of small wastewater treatment plants embracing economic, environmental, and social areas.	7 WWTP technologies for secondary treatment.	Molinos-Senante, Gómez, Garrido-Baserba, Caballero, and Sala-Garrido (2014)
Linear method of aggregation	Wastewater	Prioritisation of critical WWTPs.	44 WWTPs.	De Gisi, Sabia, Casella, and Farina (2015)
Construction of the composite index with an algorithm	Wastewater	Development of a composite index to measure energy efficiency in a wastewater treatment plant.	44 WWTPs.	Mauricio-Iglesias, Longo, and Hospido (2020)

Adapted from Sabia, G., De Gisi, S., Farina, R. (2016). Implementing a composite indicator approach for prioritizing activated sludge-based wastewater treatment plants at large spatial scale. Ecological Indicators, 71, *1–18.*

Lastly, the sixth step (Final Ranking) involves the final ranking of the compared alternatives in order to select the best solution. The MCDM results can be presented in tables or in graphs for decision-makers to review.

It is worthwhile highlighting that MCDM methodology is intended to serve as a tool to support communities in reaching a decision, which must be theirs and not the tool's decision. Where appropriate, the MCDM results can be used to communicate and to explain the final decision to stakeholders. The six-step approach can be generalised for the case of the MODM methodology, although there are several differences with MADM techniques. For instance, in MODM more than one goal (objective) is defined and needs to be optimised. In MADM the goal is one, but several are the criteria (attributes) (Ervural, Evren, & Delen, 2018).

3 Tools and software

There are several web-based software that allow developing indices and ranking alternatives through multiple combinations of normalisation methods. Additionally, aggregation methods can be applied to combine different normalised indicators in the overall score. In general, the MCDM tools aim at allowing the users (i) to load easily a dataset directly from a spreadsheet with alternatives, (ii) to build multiple indices and to choose a normalisation method with an aggregation function, and (iii) to visualise and compare final scores and rankings. As previously said, Uncertainty (UA) and Sensitivity Analyses (SA) are required for reliable and robust outcomes. UAs and SAs provide multiple outputs and rankings of the alternatives, which should be discussed properly and effectively (Burgass, Halpern, Nicholson, & Milner-Gulland, 2017). As a result, several software packages have been developed with various graphical interfaces to support the interpretation phase. A study by Mustajoki and Marttunen (2017) investigated how MADM software can be tailored to particular purposes and they compared 23 software for supporting environmental planning processes (Table 3).

Many of the software tools are stand-alone applications, but in some cases (1000Minds, D-Sight, Expert Choice, Web-HIPRE) a web interface is implemented. In addition, most of the software listed support MCDM methods in general, instead PlanEval is specific for forest planning, PUrE2 for indoor air quality, and mDSS for environmental management. Furthermore, Analytica and Diviz are not just generic software, as they can be considered as visual programming languages. Diviz provides an XML-based encoding standard for MCDA data and it also allows linking MCDA with GIS (Geographic Information System) and R programming environments. In software supporting AHP analysis, there are two main approaches: (i) the pairwise comparison amongst the criteria is given in a matrix where each criterion is evaluated against each other (e.g. in Web-HIPRE); (ii) all the possible combinations of criteria pairs are evaluated by the decision-maker who defines the importance of the first criterion compared to the second one (e.g. in Expert Choice).

Table 3 Main features of MADM software.

Software	Supported methods	UA	SA	Generic/ specific	Notes	Licence	References
1000Minds	AHP/ Pairwise (PAPRIKA), MAVT	No	Yes	Generic	Web-based interface, decision surveys, and online vote	Commercial; free academic use	Martelli et al. (2016) and Howard, Scott, Ju, McQueen, and Scuffham (2019)
Analytica	MAVT, Swing	Yes	Yes	Generic	Matlab/Excel-like interface that can be implemented	Commercial	Weistroffer and Li (2016)
Criterium Decision Plus	AHP/ Pairwise, MAVT, Swing	No	Yes	Generic	Brainstorming for problem organisation	Commercial; free academic use	Weistroffer and Li (2016) and Haerer (2000)
Decerns	AHP/ Pairwise, MAVT, Swing, Outranking	Yes	Yes	Generic	Various group decision analysis and survey tools	Commercial; unregistered version has limitations	Yatsalo, Didenko, Gritsyuk, and Sullivan (2015) and Linkov, Moberg, Trump, Yatsalo, and Keisler (2020)
DecideIT	AHP/ Pairwise, MAVT	Yes	Yes	Generic	Modelling of uncertainties with intervals with DELTA method, probabilistic rankings	Commercial; free academic use	Pires, Martinho, Rodrigues, and Gomes (2019) and Danielson, Ekenberg, and Larsson (2020)
Diviz	MAVT, Outranking	Yes	Yes	Generic	Weighted aggregation, modular interface for executing the process implemented with XML-based XMCDA standard	Free	Kiciński and Solecka (2018) and Butchart-Kuhlmann, Kralisch, Fleischer, Meinhardt, and Brenning (2018)

Table 3 Main features of MADM software. *Continued*

Software	Supported methods	UA	SA	Generic/ specific	Notes	Licence	References
D-Sight	MAVT, Outranking	No	Yes	Generic	Support for PROMETHE method, web-based platform for group collaboration	Commercial	Zardari, Yusop, Shirazi, and Roslan (2015) and Sapkota, Arora, Malano, Sharma, and Moglia (2018)
Expert Choice	AHP/ Pairwise	No	Yes	Generic	Process diagram, web-based interface, comparison of stakeholders' weights	Commercial; 12-month term for academic use	Abudeif, Moneim, and Farrag (2015) and Khorsandi, Faramarzi, Aghapour, and Jafari (2019)
GMAA	MAVT	Yes	Yes	Generic	Certainty equivalent and probability equivalent methods	Free	Jiménez, Ríos-Insua, and Mateos (2006)
Hiview 3	MAVT, Swing	No	Yes	Generic	MACBETH and MAVT methods	Commercial; reduced cost for academic use	de Moraes, Garcia, Ensslin, da Conceição, and de Carvalho (2010) and Pozo-Martin (2015)
Logical Decisions	AHP/ Pairwise, MAVT, Swing	No	Yes	Generic	Automatic help pages	Free	Dowie, Kjer Kaltoft, Salkeld, and Cunich (2015)
mDSS	MAVT, Swing, Outranking	No	Yes	Specific on environmental management	Direct weighting, TOPSIS, ELECTRE and ordered weighting average, DPSIR framework for problem structuring		Petrillo, Carotenuto, Baffo, and De Felice (2018) and Figueroa-Perez, Leyva-Lopez, Pérez-Contreras, Sánchez, and Ramirez-Noriega (2020)

Continued

Table 3 Main features of MADM software. *Continued*

Software	Supported methods	UA	SA	Generic/ specific	Notes	Licence	References
M-MACBETH	AHP/ Pairwise, MAVT	No	Yes	Generic	MACBETH method	Commercial; reduced cost for academic use	Demesouka, Vavatsikos, and Anagnostopoulos (2016) and Ferreira, Spahr, and Sunderman (2016)
On Balance	MAVT, Swing	No	Yes	Generic	Process diagram, comparison of stakeholders' weights	Commercial	Eastlick and Haapala (2012)
PlanEval	AHP/ Pairwise, MAVT, Swing	No	No	Specific on forest planning	Tab-panel interface, comparison of stakeholders' weights	Free	Korosuo, Wikström, Öhman, and Eriksson (2011) and Nilsson, Nordström, and Öhman (2016)
PUrE2	AHP/ Pairwise, MAVT, Swing	Yes	Yes	Specific on indoor air quality	Tab-panel interface, comparison of stakeholders' weights	Free	Wu et al. (2017)
Smart Decision	AHP/ Pairwise, MAVT, Swing	No	Yes	Generic	System mapping graphs	Commercial	Satir, Alkan, Can, and Bak (2007)
Transparent Choice	AHP/ Pairwise, MAVT	No	Yes	Generic	Tab-panel interface	Commercial	Freire, Frantz, Roos-Frantz, and Sawicki (2019)
V.I.P. Analysis	MAVT	Yes	Yes	Generic	Aggregation model with constraints on the weights, optimality domain graphs		Dias and Clímaco (2000) and Angelo, Saraiva, Clímaco, Infante, and Valle (2017)

Table 3 Main features of MADM software. *Continued*

Software	Supported methods	UA	SA	Generic/ specific	Notes	Licence	References
V.I.S.A. Decisions	MAVT, Swing	No	Yes	Generic	Decision wizard	Commercial	Mondlane (2017)
Visual PROMETHEE	Outranking	Yes	Yes	Generic	Possibility to define scenarios	Commercial; free academic use	Talukder and Hipel (2018)
Web-HIPRE	AHP/ Pairwise, MAVT, Swing	No	Yes	Generic	Aggregated group model	Free	Geldermann et al. (2009) and Marttunen (2011)
WINPRE	AHP/ Pairwise, MAVT, Swing	Yes	Yes	Generic	Dominance relation graphs	Free	Pietersen (2006) and Linkov et al. (2007)

Adapted from Mustajoki, J., Marttunen, M. (2017). Comparison of multi-criteria decision analytical software for supporting environmental planning processes. Environmental Modelling & Software, 93, *78–91.*

MADM analysis can involve various kinds of uncertainties related to the input data but also to the methodology. For an explicit calculation of uncertainties, some software tools provide either interval (e.g. DecideIT, V.I.P. Analysis, WINPRE) or statistical uncertainty modelling methods (i.e. Decerns, DecideIT, PurE2, Smart Decisions), or support for scenarios (e.g. Visual PROMETHEE). Sensitivity analysis is an ex-post tool to analyse how robust the model is to possible uncertainties and the traditional one-way sensitivity analysis is available in most of the tools.

4 Data requirements

Data employed in multi-criteria analyses must be reliable and from recognised sources. Only in this way, final MCDM outcomes can be trustworthy and accepted by the involved stakeholders. Every measurement implies an error and uncertainty. A reliable measure for a parameter uncertainty can be obtained by calculating a standard error and confidence intervals along with the average value. Stochastic uncertainty can be calculated with the standard deviation of the value judgements, which sometimes may differ amongst decision-makers. In this case, the final value function can be based on central measures of these judgements (average, median, mode).

In general, every measurement is developed in compliance with reliable international standards and official laboratory methods. However, sometimes data need to be validated and discussed if information of recognised sources is not available. De Gisi et al. (2015) aimed at defining an integrated approach to monitor the efficiency and investments of activated sludge-based wastewater treatment plants (WWTPs) located in a large spatial area, useful to the decision-makers. The study was carried out by using data acquired by the Italian National Institute of Statistics (ISTAT) in 2008 by submitting questionnaires to WWTP operators. At this regard, after the data acquisition, a validation step was required since ISTAT surveys do not always provide clear information about the applied treatment scheme. In this case, the validation step was performed by means of an investigation based also on aerial photos. For other data, such as flow rate, inlet and outlet concentrations related to the principal parameters, the validation process consisted of two controls, one carried out by ISTAT to verify coherency with ISTAT time series, and one with a site-specific analysis to verify consistency with literature data. It is necessary to define a timeframe for collection and validation of data. However, the time span and the total cost of a high level of investigation may be unsuitable to the timescales and budgets agreed for an evaluation. Sometimes there may be lack of reliable data over a period of time not sufficient to organise and validate the methodologies (Guarini, Battisti, & Chiovitti, 2018).

5 Scale of analyses and target audience

The scale of the evaluation defines the details and boundaries of the system to be investigated. The performance indicators will consider the scale of the system, which can be for instance local or global. De Gisi, Petta, Farina, and De Feo (2014) compared the performance of several WWTPs at global scale by means of specific indexes. The available data input derives from official questionnaires as previously mentioned (Section 4), adopted as a standardised data acquisition tool, valid for each water utility operator. The data in the questionnaires referred to average values, e.g. the average plant inlet flow rate, or the average generic contaminant concentration. At the local plant level, more data is generally available. Average values can still be calculated, but taking into account more information. At the local scale, it is also possible to define evaluation criteria, which will be adopted in a MCDM, in addition to those designed for a global scale, as they can benefit from more information. Relevant international standards such as ISO 14031 (ISO 14031, 2021) include the categories of performance indicators to facilitate environmental and management decisions. Also for what concerns the target audience there is an important difference according to the scale of the problem to investigate. For instance, at plant level the audience interested in the MCDM results is represented by the facility manager of the WWTP in order to improve the WWTP daily performance. Instead, at larger scale, management bodies such as the Ministry of the Environment or regional organisations represent the target audience, which aims at evaluating and comparing the performances of various WWTPs in order to plan and verify financial resources designed to boost plants' performances.

6 Discussion

As stated in Section 2, MCDM can combine both quantitative and qualitative data, but also subjective judgements in absence of 'harder' data. MADM consists of generating and evaluating alternatives by establishing criteria (attributes) and assessing their weights, and the application of a ranking system. Each criterion is related to an objective in the given decision context, and the normalisation process is used for transforming different criteria into comparable measures. Therefore the normalisation of the matrix of alternatives is a crucial step in most MADM techniques and different normalisation tools can modify the ranking and final decision. Yazdani, Jahan, and Zavadskas (2017) investigated the influence of normalisation tools in several specific material evaluation studies. The results showed that ranking could vary when a different normalisation tool was considered depending on the number of criteria and number of alternatives material. Chakraborty and Yeh (2007) compared the effect of four commonly known normalisation procedures on SAW method by a simulation process in terms of their ranking consistency and overall preference value consistency. To achieve this, vector normalisation, linear scale transformation, and max method were more suitable in decision settings where attribute measurement units were diverse in range. If cost, benefit, and target criteria are necessary for a MADM problem, attention should be paid to the normalisation techniques to properly compare different values (Jahan, 2018).

MCDM should not be considered as one single objective tool to reach an unequivocal truth, but it should help individuals to learn more and discuss about the problem in a transparent way (Myllyviita, Holma, Antikainen, Lähtinen, & Leskinen, 2012).

For instance, the definition of a remediation project requires a deepening of the study area about hydro-geomorphological and chemical characterisation (US EPA, 2004), performance duration, and total costs. The project itself needs to fulfil several objectives, sometimes of conflicting nature (e.g. sustainability, economy, and public acceptance). In this context, Azapagic, Stamford, Youds, and Barteczko-Hibbert (2016) proposed a novel framework called DESIRES (DEcision Support IntegRating Economic Environmental and Social Sustainability) taking into account the economic, environmental, and social concerns of different stakeholders, by means of scenario analysis, life cycle costing, life cycle assessment (LCA), social sustainability assessment, system optimisation, and multi-attribute analysis. Social impact assessment (SIA) can be useful to anticipate the effects on local stakeholders of proposed projects (Vanclay, Esteves, Aucamp, & Franks, 2015). Aberilla, Gallego-Schmid, Stamford, and Azapagic (2020) developed a multi-disciplinary approach by integrating energy systems design, techno-economic analysis, MCDA, LCA, and SIA.

In addition, the management of errors is a crucial point because often data are inaccurate, imprecise, or unknown, due to measurement errors, data staleness, and measurement repetition (Tsang, Kao, Yip, Ho, & Lee, 2009). For instance, in the water, soil, and sanitation sector, data uncertainty can be intrinsic to each collection method (Ezbakhe & Perez-Foguet, 2018). By reporting not only the point estimation of the value but also its standard error or confidence interval, the parameter uncertainty is made visible. Therefore there are many sources of uncertainty that can influence the MCDM results in different ways, each of which deserves particular attention whilst interpreting the MCDM outcome. Policy-makers and decision-makers must consider the uncertainty involved in the framework in order to avoid making decisions based on misleading assumptions (Giné-Garriga, de Palencia, & Pérez-Foguet, 2013).

In addition, since the construction of the composite indicator is subjected to several aspects, such as standardisation, weighting, or aggregation, the methodological approach has to be clearly set in relation to the final purpose. Also the selection of reliable sub-indicators together with the dimensionality of the phenomena can influence the composite indicator (Langhans et al., 2014; OECD-JRC, 2008). Hence, the indicator reliability and robustness depend strictly on the alternative methodologies applied, which can imply a certain degree of uncertainty, which can propagate through the calculation of the composite indicator (Sabia et al., 2016). Therefore they should be interpreted very carefully, especially when important decisions are to be drawn on the basis of these outcomes (e.g. by policy-makers, media, the public, etc.) (Greco, Ishizaka, Tasiou, & Torrisi, 2019).

Finally, stakeholders are gaining increased access to the decision process where all information are available for consultation and transparency, but given the data uncertainty, they sometimes could have biased opinions (Marttunen, Mustajoki, Dufva, & Karjalainen, 2015). In particular, Borrero and Henao (2017) demonstrated through an experimental study that a completely rational decision-making process with a totally objective, analytical, and evidence-based management is a utopic endeavour. However, Rabiee, Aslani, and Rezaei (2021) presented a structured algorithm to detect the sources of biases systematically and to handle the partially biased decision-makers, demonstrating the reliability of the proposed tool through testing with several scenarios with different sizes.

7 Conclusion and future research

MCDM can provide a methodology that supports the decision-makers to understand and interpret better the stakeholders' preferences and be unequivocal about the critical aspects amongst different elements of value. Sustainable decision-making represents a powerful tool having a direct impact on the economy, the environment, and the society.

Researchers are continuously trying to apply MCDM techniques in their research domain, addressing mathematical, theoretical, and behavioural applications mainly in the field of supply chain, healthcare system, transport system, environmental management, and business intelligence. However, most of the real-life problems are multidisciplinary and cannot be limited to single criterion decision-making models. Future researches need to cover this gap by developing new frameworks, which aim at connecting multiple areas in the context of uncertain environments. The evolution of MCDM should include integrated sustainability assessments of the social, environmental, and economic areas in the final vision of the Sustainable Development Goals.

References

Aberilla, J. M., Gallego-Schmid, A., Stamford, L., & Azapagic, A. (2020). An integrated sustainability assessment of synergistic supply of energy and water in remote communities. *Sustainable Production and Consumption*, *22*, 1–21.

Abudeif, A. M., Moneim, A. A., & Farrag, A. F. (2015). Multicriteria decision analysis based on analytic hierarchy process in GIS environment for siting nuclear power plant in Egypt. *Annals of Nuclear Energy*, *75*, 682–692.

Afshari, A., Mojahed, M., & Yusuff, R. M. (2010). Simple additive weighting approach to personnel selection problem. *International Journal of Innovation and Technology Management*, *1*(5), 511.

Ahmadi, M., Behzadian, K., Ardeshir, A., & Kapelan, Z. (2017). Comprehensive risk management using fuzzy FMEA and MCDA techniques in highway construction projects. *Journal of Civil Engineering and Management*, *23*(2), 300–310.

Angelo, A. C. M., Saraiva, A. B., Clímaco, J. C. N., Infante, C. E., & Valle, R. (2017). Life cycle assessment and multi-criteria decision analysis: Selection of a strategy for domestic food waste management in Rio de Janeiro. *Journal of Cleaner Production*, *143*, 744–756.

Ashby, M. F., Brechet, Y. J. M., Cebon, D., & Salvo, L. (2004). Selection strategies for materials and processes. *Materials & Design*, *25*(1), 51–67.

Azapagic, A., Stamford, L., Youds, L., & Barteczko-Hibbert, C. (2016). Towards sustainable production and consumption: A novel decision-support framework integrating economic, environmental and social sustainability (DESIRES). *Computers & Chemical Engineering*, *91*, 93–103.

Borrero, S., & Henao, F. (2017). Can managers be really objective? Bias in multi-criteria decision analysis. *Academy of Strategic Management Journal*, *16*(1), 244.

Bottomley, P. A., Doyle, J. R., & Green, R. H. (2000). Testing the reliability of weight elicitation methods: Direct rating versus point allocation. *Journal of Marketing Research*, *37* (4), 508–513.

Burgass, M. J., Halpern, B. S., Nicholson, E., & Milner-Gulland, E. J. (2017). Navigating uncertainty in environmental composite indicators. *Ecological Indicators*, *75*, 268–278.

Butchart-Kuhlmann, D., Kralisch, S., Fleischer, M., Meinhardt, M., & Brenning, A. (2018). Multicriteria decision analysis framework for hydrological decision support using environmental flow components. *Ecological Indicators*, *93*, 470–480.

Çalı, S., & Balaman, Ş. Y. (2019). A novel outranking based multi criteria group decision making methodology integrating ELECTRE and VIKOR under intuitionistic fuzzy environment. *Expert Systems with Applications*, *119*, 36–50.

Cappuyns, V. (2016). Inclusion of social indicators in decision support tools for the selection of sustainable site remediation options. *Journal of Environmental Management*, *184*, 45–56.

Chakraborty, S., & Yeh, C. H. (2007). A simulation based comparative study of normalization procedures in multiattribute decision making. In *Vol. 6. Proceedings of the 6th conference on 6th WSEAS international conference on artificial intelligence, knowledge engineering and data bases* (pp. 102–109).

Danielson, M., Ekenberg, L., & Larsson, A. (2020). A second-order-based decision tool for evaluating decisions under conditions of severe uncertainty. *Knowledge-Based Systems*, *191*, 105219.

de Bekker-Grob, E. W., Ryan, M., & Gerard, K. (2012). Discrete choice experiments in health economics: A review of the literature. *Health Economics*, *21*(2), 145–172.

De Feo, G., & De Gisi, S. (2010a). Using an innovative criteria weighting tool for stakeholders involvement to rank MSW facility sites with the AHP. *Waste Management*, *30*(11), 2370–2382.

De Feo, G., & De Gisi, S. (2010b). Public opinion and awareness towards MSW and separate collection programmes: A sociological procedure for selecting areas and citizens with a low level of knowledge. *Waste Management*, *30*(6), 958–976.

De Feo, G., & De Gisi, S. (2014). Using MCDA and GIS for hazardous waste landfill siting considering land scarcity for waste disposal. *Waste Management*, *34*(11), 2225–2238.

De Feo, G., De Gisi, S., & Galasso, M. (2008). Definition of a practical multi-criteria procedure for selecting the best coagulant in a chemically assisted primary sedimentation process for the treatment of urban wastewater. *Desalination*, *230*(1–3), 229–238.

De Gisi, S., Petta, L., Farina, R., & De Feo, G. (2014). Development and application of a planning support tool in the municipal wastewater sector: The case study of Italy. *Land Use Policy*, *41*, 260–273.

De Gisi, S., Sabia, G., Casella, P., & Farina, R. (2015). An integrated approach for monitoring efficiency and investments of activated sludge-based wastewater treatment plants at large spatial scale. *Science of the Total Environment*, *523*, 201–218.

de Moraes, L., Garcia, R., Ensslin, L., da Conceição, M. J., & de Carvalho, S. M. (2010). The multicriteria analysis for construction of benchmarkers to support the clinical engineering in the healthcare technology management. *European Journal of Operational Research*, *200*(2), 607–615.

Demesouka, O. E., Vavatsikos, A. P., & Anagnostopoulos, K. P. (2016). Using MACBETH multicriteria technique for GIS-based landfill suitability analysis. *Journal of Environmental Engineering*, *142*(10), 04016042.

Dias, L. C., & Clímaco, J. N. (2000). Additive aggregation with variable interdependent parameters: The VIP analysis software. *Journal of the Operational Research Society*, *51*(9), 1070–1082.

Dowie, J., Kjer Kaltoft, M., Salkeld, G., & Cunich, M. (2015). Towards generic online multicriteria decision support in patient-centred health care. *Health Expectations*, *18*(5), 689–702.

Eastlick, D. D., & Haapala, K. R. (2012). Increasing the utility of sustainability assessment in product design. In *Vol. 45042. International design engineering technical conferences and computers and information in engineering conference* (pp. 713–722). American Society of Mechanical Engineers.

Edwards, W. (1977). How to use multi-attribute utility measurement for social decision making. *IEEE Transactions on Systems, Man, and Cybernetics*, *7*(5), 326–340.

Ervural, B. C., Evren, R., & Delen, D. (2018). A multi-objective decision-making approach for sustainable energy investment planning. *Renewable Energy*, *126*, 387–402.

EU Technical Working Groups. (2021). *Environmental technology verification (ETV)*. https:/ec.europa.eu/environment/ecoap/etv/technical-working-groups_en. (Accessed 7 June 2021).

Ezbakhe, F., & Perez-Foguet, A. (2018). Multi-criteria decision analysis under uncertainty: Two approaches to incorporating data uncertainty into water, sanitation and hygiene planning. *Water Resources Management*, *32*(15), 5169–5182.

Ferreira, F. A., Spahr, R. W., & Sunderman, M. A. (2016). Using multiple criteria decision analysis (MCDA) to assist in estimating residential housing values. *International Journal of Strategic Property Management*, *20*(4), 354–370.

Figueroa-Perez, J. F., Leyva-Lopez, J. C., Pérez-Contreras, E. O., Sánchez, P. J., & Ramirez-Noriega, A. D. (2020). An agent-based system for the design of new products using a fuzzy multicriteria approach. *International Journal of Fuzzy Systems*, *22*(8), 2691–2707.

Freeman, E. R. (1984). Strategic management: a stakeholder approach, ser. *Pitman series in business and public policy*. Pitman.

Freire, D. L., Frantz, Z. R., Roos-Frantz, F., & Sawicki, S. (2019). A methodology to rank enterprise application integration platforms from a performance perspective: An analytic hierarchy process-based approach. *Enterprise Information Systems*, *13*(9), 1292–1322.

Gbededo, M. A., Liyanage, K., & Garza-Reyes, J. A. (2018). Towards a life cycle sustainability analysis: A systematic review of approaches to sustainable manufacturing. *Journal of Cleaner Production*, *184*, 1002–1015.

Geldermann, J., Bertsch, V., Treitz, M., French, S., Papamichail, K. N., & Hämäläinen, R. P. (2009). Multi-criteria decision support and evaluation of strategies for nuclear remediation management. *Omega*, *37*(1), 238–251.

Giné-Garriga, R., de Palencia, A. J. F., & Pérez-Foguet, A. (2013). Water–sanitation–hygiene mapping: An improved approach for data collection at local level. *Science of the Total Environment*, *463*, 700–711.

Golge, M., Yenilmez, F., & Aksoy, A. (2013). Development of pollution indices for the middle section of the lower Seyhan Basin (Turkey). *Ecological Indicators*, *29*, 6–17.

Greco, S., Ishizaka, A., Tasiou, M., & Torrisi, G. (2019). On the methodological framework of composite indices: A review of the issues of weighting, aggregation, and robustness. *Social Indicators Research*, *141*(1), 61–94.

Green, P. E., Krieger, A. M., & Wind, Y. (2001). Thirty years of conjoint analysis: Reflections and prospects. *Interfaces*, *31*(3_supplement), S56–S73.

Guarini, M. R., Battisti, F., & Chiovitti, A. (2018). A methodology for the selection of multi-criteria decision analysis methods in real estate and land management processes. *Sustainability*, *10*(2), 507.

Haerer, W. (2000). Criterium decision plus 3.0. *ORMS Today*, *27*(1), 40–43.

Halisçelik, E., & Soytas, M. A. (2019). Sustainable development from millennium 2015 to sustainable development goals 2030. *Sustainable Development*, *27*(4), 545–572.

Hansen, P., & Ombler, F. (2008). A new method for scoring additive multi-attribute value models using pairwise rankings of alternatives. *Journal of Multi-Criteria Decision Analysis*, *15*(3–4), 87–107.

Howard, S., Scott, I. A., Ju, H., McQueen, L., & Scuffham, P. A. (2019). Multicriteria decision analysis (MCDA) for health technology assessment: The Queensland health experience. *Australian Health Review*, *43*(5), 591–599.

ISO 14031. (2021). *Environmental management. Environmental performance evaluation. Guidelines*.

Jahan, A. (2018). Developing WASPAS-RTB method for range target-based criteria: Toward selection for robust design. *Technological and Economic Development of Economy*, *24*(4), 1362–1387.

Jahan, A., Edwards, K. L., & Bahraminasab, M. (2016). *Multi-criteria decision analysis for supporting the selection of engineering materials in product design*. Butterworth-Heinemann.

Jiménez, A., Ríos-Insua, S., & Mateos, A. (2006). A generic multi-attribute analysis system. *Computers & Operations Research*, *33*(4), 1081–1101.

Kazimieras Zavadskas, E., Antucheviciene, J., & Kar, S. (2019). Multi-objective and multi-attribute optimization for sustainable development decision aiding. *Sustainability*, *11* (11), 3069.

Khorsandi, H., Faramarzi, A., Aghapour, A. A., & Jafari, S. J. (2019). Landfill site selection via integrating multi-criteria decision techniques with geographic information systems: A case study in Naqadeh, Iran. *Environmental Monitoring and Assessment*, *191*(12), 1–16.

Kiciński, M., & Solecka, K. (2018). Application of MCDA/MCDM methods for an integrated urban public transportation system—Case study, city of Cracow. *Archives of Transport*, *46*, 71–84.

Köne, A.Ç., & Büke, T. (2014). The evaluation of the air pollution index in Turkey. *Ecological Indicators*, *45*, 350–354.

Korosuo, A., Wikström, P., Öhman, K., & Eriksson, L. O. (2011). An integrated MCDA software application for forest planning: A case study in southwestern Sweden. *Mathematical and Computational Forestry and Natural-Resource Sciences*, *3*, 75–86.

Labianca, C., De Gisi, S., Todaro, F., & Notarnicola, M. (2020). Evaluation of remediation technologies for contaminated marine sediments through multi criteria decision analysis. *Environmental Engineering and Management Journal*, *19*(10), 1891–1903.

Langemeyer, J., Gómez-Baggethun, E., Haase, D., Scheuer, S., & Elmqvist, T. (2016). Bridging the gap between ecosystem service assessments and land-use planning through multi-criteria decision analysis (MCDA). *Environmental Science & Policy*, *62*, 45–56.

Langhans, S. D., Reichert, P., & Schuwirth, N. (2014). The method matters: A guide for indicator aggregation in ecological assessments. *Ecological Indicators*, *45*, 494–507.

Lindell, B. (2017). *Multi-criteria analysis in legal reasoning*. Edward Elgar Publishing.

Linkov, I., Moberg, E., Trump, B. D., Yatsalo, B., & Keisler, J. M. (2020). *Multi-criteria decision analysis: Case studies in engineering and the environment*. CRC Press.
Linkov, I., Satterstrom, F. K., Yatsalo, B., Tkachuk, A., Kiker, G. A., Kim, J., et al. (2007). Comparative assessment of several multi-criteria decision analysis tools for management of contaminated sediments. In *Environmental security in harbors and coastal areas* (pp. 195–215). Dordrecht: Springer.
Lolli, F., Ishizaka, A., Gamberini, R., Rimini, B., Ferrari, A. M., Marinelli, S., et al. (2016). Waste treatment: An environmental, economic and social analysis with a new group fuzzy PROMETHEE approach. *Clean Technologies and Environmental Policy*, *18*(5), 1317–1332.
Lozano-Oyola, M., Blancas, F. J., González, M., & Caballero, R. (2012). Sustainable tourism indicators as planning tools in cultural destinations. *Ecological Indicators*, *18*, 659–675.
Macharis, C., Milan, L., & Verlinde, S. (2014). A stakeholder-based multicriteria evaluation framework for city distribution. *Research in Transportation Business & Management*, *11*, 75–84.
Martelli, N., Hansen, P., Van Den Brink, H., Boudard, A., Cordonnier, A. L., Devaux, C., et al. (2016). Combining multi-criteria decision analysis and mini-health technology assessment: A funding decision-support tool for medical devices in a university hospital setting. *Journal of Biomedical Informatics*, *59*, 201–208.
Marttunen, M. (2011). *Interactive multi-criteria decision analysis in the collaborative management of watercourses*. Aalto University.
Marttunen, M., Mustajoki, J., Dufva, M., & Karjalainen, T. P. (2015). How to design and realize participation of stakeholders in MCDA processes? A framework for selecting an appropriate approach. *EURO Journal on Decision Processes*, *3*(1), 187–214.
Mauricio-Iglesias, M., Longo, S., & Hospido, A. (2020). Designing a robust index for WWTP energy efficiency: The ENERWATER water treatment energy index. *Science of the Total Environment*, *713*, 136642.
Molinos-Senante, M., Gómez, T., Garrido-Baserba, M., Caballero, R., & Sala-Garrido, R. (2014). Assessing the sustainability of small wastewater treatment systems: A composite indicator approach. *Science of the Total Environment*, *497*, 607–617.
Mondlane, A. I. (2017). Multicriteria decision analysis for flood risk management: The case of the Mapai dam at the Limpopo River Basin, Mozambique. In *Flood risk management* IntechOpen.
Mustajoki, J., & Marttunen, M. (2017). Comparison of multi-criteria decision analytical software for supporting environmental planning processes. *Environmental Modelling & Software*, *93*, 78–91.
Myllyviita, T., Holma, A., Antikainen, R., Lähtinen, K., & Leskinen, P. (2012). Assessing environmental impacts of biomass production chains–application of life cycle assessment (LCA) and multi-criteria decision analysis (MCDA). *Journal of Cleaner Production*, *29*, 238–245.
Nilsson, H., Nordström, E. M., & Öhman, K. (2016). Decision support for participatory forest planning using AHP and TOPSIS. *Forests*, *7*(5), 100.
OECD-JRC. (2008). *Handbook on constructing composite indicators methodology and user guide*. Paris: OECD Publications.
Özkan, B., Sarıçiçek, İ., & Özceylan, E. (2020). Evaluation of landfill sites using GIS-based MCDA with hesitant fuzzy linguistic term sets. *Environmental Science and Pollution Research*, *27*(34), 42908–42932.

Ozturk, D., & Batuk, F. (2011). Technique for order preference by similarity to ideal solution (TOPSIS) for spatial decision problems. *Proceedings ISPRS*, *1*(4).

Paolucci, M. (2000). *Metodi Decisionali multi Criterio. Appunti per il Corso di Metodi e Modelli per il Supporto alle Decisioni (in English "multi criteria decision methods. Notes for the course of methods and models for decision support")*. University of Genoa. 2001.

Paruolo, P., Saisana, M., & Saltelli, A. (2013). Ratings and rankings: Voodoo or science? *Journal of the Royal Statistical Society: Series A (Statistics in Society)*, *176*(3), 609–634.

Petrillo, A., Carotenuto, P., Baffo, I., & De Felice, F. (2018). A web-based multiple criteria decision support system for evaluation analysis of carpooling. *Environment, Development and Sustainability*, *20*(5), 2321–2341.

Pietersen, K. (2006). Multiple criteria decision analysis (MCDA): A tool to support sustainable management of groundwater resources in South Africa. *Water SA*, *32*(2), 119–128.

Pires, A., Martinho, G., Rodrigues, S., & Gomes, M. I. (2019). Multi-criteria decision-making in waste collection to reach sustainable waste management. In *Sustainable Solid Waste Collection and Management* (pp. 239–260). Cham: Springer.

Podvezko, V. (2011). The comparative analysis of MCDA methods SAW and COPRAS. *Engineering Economics*, *22*(2), 134–146.

Post, J. E., Preston, L. E., & Sauter-Sachs, S. (2002). *Redefining the corporation: Stakeholder management and organizational wealth*. Stanford University Press.

Pozo-Martin, F. (2015). *Multi-criteria decision analysis (MCDA) as the basis for the development, implementation and evaluation of interactive patient decision aids (Doctoral dissertation)*. London School of Hygiene & Tropical Medicine.

Rabiee, M., Aslani, B., & Rezaei, J. (2021). A decision support system for detecting and handling biased decision-makers in multi criteria group decision-making problems. *Expert Systems with Applications*, *171*, 114597.

Ryan, M., Gerard, K., & Currie, G. (2012). Using discrete choice experiments in health economics. In *The Elgar companion to health economics* (2nd ed.). Edward Elgar Publishing.

Saaty, T. L. (1977). A scaling method for priorities in hierarchical structures. *Journal of Mathematical Psychology*, *15*(3), 234–281.

Saaty, T. L. (1990). *Multicriteria decision making: The analytic hierarchy process: Planning, priority setting resource allocation*. Pittsburgh: RWS Publications.

Sabia, G., De Gisi, S., & Farina, R. (2016). Implementing a composite indicator approach for prioritizing activated sludge-based wastewater treatment plants at large spatial scale. *Ecological Indicators*, *71*, 1–18.

Saltelli, A., Aleksankina, K., Becker, W., Fennell, P., Ferretti, F., Holst, N., et al. (2019). Why so many published sensitivity analyses are false: A systematic review of sensitivity analysis practices. *Environmental Modelling & Software*, *114*, 29–39.

Sapkota, M., Arora, M., Malano, H., Sharma, A., & Moglia, M. (2018). Integrated evaluation of hybrid water supply systems using a PROMETHEE–GAIA approach. *Water*, *10* (5), 610.

Satir, T., Alkan, G. B., Can, S., & Bak, O. A. (2007). Port reception facilities: Using multi criteria decision making. In *Proceedings 2nd international conference on marine research and transportation, Naples, Italy*.

Talukder, B., & Hipel, K. W. (2018). The PROMETHEE framework for comparing the sustainability of agricultural systems. *Resources*, *7*(4), 74.

Tsang, S., Kao, B., Yip, K. Y., Ho, W. S., & Lee, S. D. (2009). Decision trees for uncertain data. *IEEE Transactions on Knowledge and Data Engineering*, *23*(1), 64–78.

US EPA. (2004). *Cleaning up the nation's waste sites: Markets and technology trends* (2004 ed.). Washington, DC: United States Environmental Protection Agency.

Vanclay, F., Esteves, A. M., Aucamp, I., & Franks, D. M. (2015). *Social impact assessment: Guidance for assessing and managing the social impacts of projects.*

Vinogradova, I. (2019). Multi-attribute decision-making methods as a part of mathematical optimization. *Mathematics*, *7*(10), 915.

Von Winterfeldt, D., & Edwards, W. (1993). *Decision analysis and behavioral research*. Cambridge University Press.

Wang, Y., Lam, K. C., Harder, M. K., Ma, W. C., & Yu, Q. (2013). Developing an indicator system to foster sustainability in strategic planning in China: A case study of Pudong new area, Shanghai. *Ecological Indicators*, *29*, 376–389.

Weistroffer, H. R., & Li, Y. (2016). Multiple criteria decision analysis software. In *Multiple criteria decision analysis* (pp. 1301–1341). New York, NY: Springer.

Wu, Y., Chen, K., Zeng, B., Yang, M., Li, L., & Zhang, H. (2017). A cloud decision framework in pure 2-tuple linguistic setting and its application for low-speed wind farm site selection. *Journal of Cleaner Production*, *142*, 2154–2165.

Yatsalo, B., Didenko, V., Gritsyuk, S., & Sullivan, T. (2015). Decerns: A framework for multi-criteria decision analysis. *International Journal of Computational Intelligence Systems*, *8* (3), 467–489.

Yazdani, M., Jahan, A., & Zavadskas, E. (2017). Analysis in material selection: Influence of normalization tools on COPRAS-G. *Economic Computation and Economic Cybernetics Studies and Research*, *51*(1), 59–74.

Zardari, N. H., Yusop, Z., Shirazi, S. M., & Roslan, N. A. B. (2015). Prioritization of farmlands in a multicriteria irrigation water allocation: PROMETHEE and GAIA applications. *Transactions of the ASABE*, *58*(1), 73–82.

SECTION

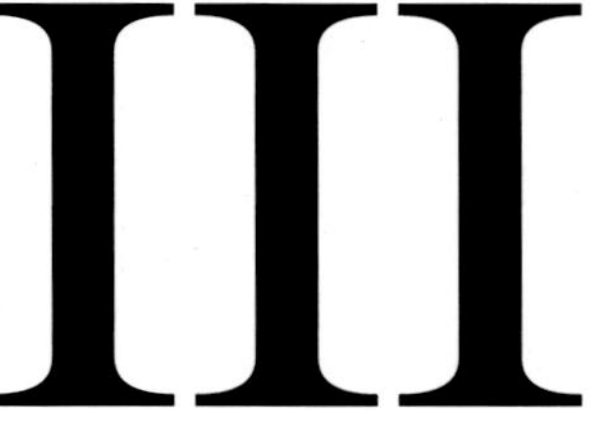

Case studies for sustainability assessments

CHAPTER

Life cycle assessment for eco-design in product development 12

George Barjoveanu[a], Carmen Teodosiu[a], Marcela Mihai[a,b], Irina Morosanu[a], Daniela Fighir[a], Ana-Maria Vasiliu[b], and Florin Bucatariu[a,b]

[a]*Department of Environmental Engineering and Management, "Gheorghe Asachi" Technical University of Iasi, Iasi, Romania* [b]*"Petru Poni" Institute of Macromolecular Chemistry, Iasi, Romania*

1 Aims

This chapter presents two case studies where LCA is used as an eco-design support tool to identify and quantify potential environmental impacts during the early stages of product development. These refer to laboratory synthesis and testing of new materials for development for environmental applications: (1) LCA evaluation of novel organic/inorganic composites for heavy metal ions removal from water and (2) LCA evaluation of novel sorbents obtained from waste biomass for the same purpose.

2 State of the art

Emerging technologies and materials used in water/wastewater treatment need to be thoroughly investigated from an environmental sustainability standpoint in order to gain a comprehensive image of their environmental impacts and to compare these to the technical benefits they bring in reducing pollution. Several knowledge and cooperation gaps have been identified in this field: definition of better environmental impact assessment methods and metrics for novel materials used in water and wastewater applications throughout their whole life cycle; better communication and cooperation between material designers, testers, and environmental engineers; better definition of sustainability targets regarding the new/waste materials used in water and wastewater applications.

The need for novel and more efficient water and wastewater applications stems from ever increasing number and types of emerging pollutants as water contaminants. Emerging pollutants include a plethora of chemical species with toxic, carcinogenic, and mutagenic characteristics and bio-accumulative behaviour in the

Assessing Progress towards Sustainability. https://doi.org/10.1016/B978-0-323-85851-9.00012-2

environment (Teodosiu, Gilca, Barjoveanu, & Fiore, 2018). They are found in water/wastewater at low concentrations (frequently in the ppb range) which means that their removal from water is quite difficult to achieve by means of conventional treatment technologies. Their behaviour together with some particular properties poses some important engineering challenges for their removal which in the end translate into higher economic costs and of course higher secondary environmental impacts.

Any development of these technologies or materials should consider a complex assessment of their technical efficiency (e.g. pollutant removal rates) and environmental performance.

These aspects are particularly important for the development and use of novel engineered materials for environmental applications which are usually used as catalysts, sorbents, or filtration materials (Debnath et al., 2019; Vajihinejad, Gumfekar, Bazoubandi, Rostami Najafabadi, & Soares, 2019; Wu et al., 2019). Herein we refer to granular engineered materials which are used in sorption processes for the removal of different classes of emerging pollutants from water/wastewater. Granular sorbents used for environmental applications include activated carbon, sand, biochar, ion exchange resins, various types of (waste) biomass, organic/inorganic composites, and other functionalised materials, and are usually employed as column fillers in water and wastewater treatment applications (Hamoda, Al-Ghusain, & Al-Mutairi, 2004; Hoai, Yoo, & Kim, 2010; Silva, Jaria, Otero, Esteves, & Calisto, 2019; Tong, McNamara, & Mayer, 2019).

The efficiency of these materials for pollutant removal is determined by the mechanism involved which, in turn, is determined by the material properties and the pollutant–material interactions. This rationale usually forms the main focus of researchers in material development which try to optimise sorption properties by increasing the number and accessibility of functional groups for pollutants (Bucatariu, Ghiorghita, Schwarz, Boita, & Mihai, 2019). Thus, from a technical perspective, an optimal material for sorption applications in water and wastewater treatment needs to have a good structural durability and to provide a large number of functional groups available for interaction with various pollutant species. Material scientists try to control the chemical functionality of surfaces at nanometric level and to create specific moieties onto solid composite surfaces (magnetic particles, membranes, carbon nanosheets) for subsequent molecular/ionic interactions. These properties are essential in chromatography, controlled drug release, water/wastewater treatment, etc. (Anandhakumar & Raichur, 2013; Boudou et al., 2012; Hamza et al., 2019; Zhang, Zhao, & Yuan, 2018). Almost the same approach is considered for the waste derived materials, where the treatment applied may influence the sorption and desorption properties and removal efficiencies of various pollutants (Morosanu, Teodosiu, Tofan, Fighir, & Paduraru, 2020). Moreover, the wastewater matrix and the sequence of pollutants removal induce other technical challenges that might affect the technical selection criteria (Morosanu, Teodosiu, Coroaba, & Paduraru, 2019; Morosanu, Teodosiu, Fighir, & Paduraru, 2019).

From an environmental sustainability standpoint, the main criteria that are relevant for materials used in environmental applications refer to ensuring high pollutant removal efficiencies together with low environmental impacts in the production, use, and post-use phases of these materials (Cucurachi & Blanco Rocha, 2019; Hischier & Walser, 2012). These simple requirements (should) translate into reliable eco-design criteria for these materials. However, until now, research efforts were targeted mostly towards measuring performance (i.e. by LCA) (Cinelli, Coles, Sadik, Karn, & Kirwan, 2016; Windsor, Cinelli, & Coles, 2018) and less towards to defining and applying eco-design criteria for novel materials development, use, and elimination.

Eco-design is defined as a systematic approach, which considers environmental aspects in design and development of products with the aim to reduce adverse environmental impacts throughout their life cycle (ISO, 2020). Thus the main goal of eco-design is to reduce the environmental impacts in the entire product life cycle without compromising fundamental parameters such as functionality, quality, and cost (Maccioni, Borgianni, & Pigosso, 2019). Eco-design principles (EDPs) have been developed to approach this goal and they act as guidelines to support designers from the earliest product development stages. These principles or strategies have been developed since the mid-1990s (Brezet et al., 1997) and cover all the life cycle stages of a product addressing strategies for design, production, use (consumption), and end of life, as presented in Table 1 together with some examples from the field of material development for environmental applications.

It has to be noted that the eco-design strategies presented in Table 1 have to be thought about in the materials design phase and addressed by the material developer as design objectives or (environmental) performance outcomes which will occur as benefits and improvements in the corresponding life cycle stages of the respective materials. This means that these strategies can be used as design criteria based on which performance indicators and metrics may be developed to measure progress towards materials sustainability.

Also, it is important to note that almost exclusively, all studies on eco-design of new engineered materials (including nanomaterials) approach only partially the eco-design principles, strategies or measures and miss a systematic approach to these practices. Thus the success of eco-design practices in product development faces some barriers: eco-design principles often conflict with other product characteristics or functions which leads to worse environmental performance or cost increase (Maccioni et al., 2019).

Among various instruments used to evaluate (potential) environmental impacts, LCA has been recognised and used as the main instrument for the evaluation of materials for environmental applications (Cinelli et al., 2016; Gottschalk & Nowack, 2011; Windsor et al., 2018). This is because LCA has few characteristics as an assessment instrument which can clearly support (eco)-design and product optimisation: (a) it evaluates a product throughout its life cycle, (b) it provides a holistic approach by generating complex environmental profiles, and (c) it can be used as a tool to explore various scenarios related to product development (including scale-up) and their associated potential environmental impacts (or improvements). This last type of LCA application is usually known as ex-ante or prospective LCA (Cucurachi, Van Der Giesen, & Guinée, 2018) and it is gaining increasing

Table 1 Application of eco-design principles and strategies in material development for environmental applications.

Life cycle stage where strategy works	Eco-design principle/strategies	Materials	Study objective	References
Design/ conceptualisation	**Optimise product function:** enhance product function, multifunctionality[a]	All studies try to improve technical performances	–	–
	Use low impact materials: renewables, recycled, avoid dangerous, materials produced by eco-friendly methods	Use renewable natural sources	Biosorbents	Current study
		Rapeseed waste biosorbents	Evaluate environmental sustainability	Current study
		Metal–organic frameworks		Grande, Blom, Spjelkavik, Moreau, and Payet (2017)
Production	**Production optimisation:** alternative production technologies, minimise no. of production steps, minimise energy consumption, reduce waste and pollution, cleaner consumables in production	(e.g. Green chemistry) Antimicrobial nanomaterials	Framework for sustainable material selection	Babbitt and Moore (2018); Falinski et al. (2018)
		Semiconductors, metal oxides, noble metals, and multi-composition alloys	Green chemistry	Duan, Wang, and Li (2015)
		TiO_2	LCA of synthesis routes	Caramazana-González et al. (2017)
		Graphene	LCA of synthesis routes	Cossutta, McKechnie, and Pickering (2017)
		Magnetic coated nanoparticles	Investigate sustainability during synthesis	Feijoo et al. (2017)
		Organic/inorganic composites	Investigate eco-design scenarios during synthesis	Barjoveanu, Teodosiu, Bucatariu, and Mihai (2020), Current study

Use/ consumption	**Optimisation of life time**: extend use life, improve reusability, reduce maintenance needs, minimise energy use, reduce consumables, use clean consumables	ZnO self-cleaning coating	Eliminate maintenance	Caramazana-González et al. (2017)
		Graphite nanoplatelets	Energy savings during the use	Pizza, Metz, Hassanzadeh, and Bantignies (2014)
End-of-life	**Optimisation of end-of-life**: minimise environment release impacts, safe-by-design, enhanced recyclability, improved treatability (separation, destruction, etc.)	Nano TiO_2 Carbon nanotubes	Characterisation factors development to evaluate release impacts	Eckelman, Mauter, Isaacs, ans Elimelech (2012); Ettrup et al. (2017)
		Magnetic nanoparticles	Investigate sustainability scenarios during synthesis	Feijoo et al. (2017)

[a]*Multifunctionality refers to the sorption availability for various emerging pollutants (inorganic, e.g. heavy metals or organic, e.g. pharmaceuticals, pesticides, etc.).*

use (Cucurachi & Blanco Rocha, 2019; Hischier & Walser, 2012; Salieri, Turner, Nowack, & Hischier, 2018) in the field of materials eco-design. This is contrary to the traditional LCA studies which have been used to assess full-scale and fully operational product systems considering a retrospective approach.

In the case of new materials development (e.g. nanomaterials or waste bio-based products) where product development and optimisation always occurs at laboratory scale, such ex-ante LCA studies have several methodological challenges that have to be overcome in order to benefit from the prospective features of LCA, these issues being briefly discussed as follows:

For all LCA studies, the definition of a functional unit and a relevant scope in close relation to the objectives of the study is of paramount importance. For new (or waste derived) materials this is more challenging because of the many new additional functionalities that have to be considered and the unclear or difficult scope definition of their studies (Hischier & Walser, 2012). With regard to the scope of the LCA on nanomaterials, most of these focus on the synthesis part of their life cycle and do not take into consideration a full life cycle perspective of their assessment (i.e. the use or disposal) or design (Livingston et al., 2020).

There is almost always a lack of available inventory data on these materials, because their manufacturing processes are often new and occur at a very small scale and thus, average industry data that forms the well-established LCA databases are of little help (Hischier, Salieri, & Pini, 2017). This generates rather significant issues related to data availability and to how uncertainty affects results. New materials development relies on new production methods which constantly change and this calls for attention when comparing their performances by LCA. Furthermore, scaling-up production processes of engineered materials is not similar to other materials because of their composition and specific production processes (Windsor et al., 2018).

Lastly, but maybe most importantly, current impact assessment methodologies do not incorporate environmental mechanisms and characterisation factors for calculating the health and environmental effects of nanomaterials, because there is a knowledge gap regarding this field, especially related to environmental releases (quantities that reach the environment), impact pathways (fate and exposure), and ecotoxicity effects (Salieri et al., 2018).

3 Novelty

Considering the aspects discussed before, this case study contributes with new insight related to the use of LCA as a support tool to investigate several eco-design scenarios for materials development. More concretely, the scenario-based LCA is targeted towards highly engineered nano-structured $CaCO_3$-cored micro-spheres coated with several layers of various polymers and to a biosorbent obtained from waste biomass. The LCA and eco-design scenarios are focused towards the very early stages of product development (lab-scale synthesis and testing) which bring important information related to the environmental profiles of the new products.

4 LCA evaluation of novel organic/inorganic composites

This study presents a comparing investigation of the technical and environmental performances of several organic/inorganic composites types. These particles were obtained by layer-by-layer (LbL) deposition of polyethylenimine (PEI) over inorganic $CaCO_3$ microparticles. The technical details related to their synthesis, characterisation, and testing have been previously published (Barjoveanu et al., 2020; Bucatariu et al., 2019; Zaharia et al., 2021), but their synthesis starts with an in situ generated inorganic particle onto which multiple polymeric layers are deposited. This synthesis design is a spin-off from a recent study (Barjoveanu et al., 2020), in which similar materials (PEI coated silica-cored microparticles) were compared for the removal of Cu(II) ions from aqueous solutions. In that case, the results pointed out that the highly engineered support material (porous silica microparticle) was the major contributor in most of the categories evaluated, so for this a different core material was investigated—the in situ generated $CaCO_3$ microparticles. The evaluation reported here considers 2 synthesis pathways for improving the environmental performance of the PEI-coated $CaCO_3$ microcrystals for the removal of Cu(II) ions from aqueous solutions.

4.1 Material development, synthesis, and testing

The synthesis goals were to obtain an organic/inorganic composite material that would combine the structural stability of an inorganic support with the high affinity for pollutant interaction given by a polymeric phase with high affinity for heavy metal ions and other pollutant species. For this purpose and considering our previous experience (Barjoveanu et al., 2020; Bucatariu et al., 2019) the inorganic support was changed from activated silica to $CaCO_3$ crystals which were obtained in situ by crystallisation from a supersaturated solution using different polymers (branched PEI (PEIB), linear PEI (PEIL), poly(acrylic acid) (PAA)) as crystal modulators. Afterwards, to increase the polymeric content of these particles, the LbL technique was employed for a total number of 7 PEIB or PEIL and PAA polyelectrolyte layers.

Starting from the synthesis goal of developing and improving inorganic/organic composite materials for the removal of heavy metals from synthetic wastewaters, two synthesis routes were investigated: one in which layer by layer of water-soluble PEIB and PAA were alternatively deposited onto $CaCO_3$ microparticles and the second one in which a PEIB-Cu polymer was used (replacing the PEIB from route 1) in the LbL synthesis followed by the extraction of the initial Cu ions. As proven in the previous study, replacing the PEI with a PEI-Cu has led to improving the composites' subsequent sorption capacity for Cu(II) ions. One should note that these syntheses

Table 2 Sorption capacity for Cu(II) (mg Cu(II)/g composite) in simulated solutions.

Sorbed metal ion	$CaCO_3$ microparticles	MS1a, (PEIB/PAA)$_3$PEIB on $CaCO_3$ microparticles	MS1b, (PEIB/PAA)$_3$PEIB on $CaCO_3$/PAA microparticles	MS1c, (PEIB/PAA)$_3$PEIB on $CaCO_3$/PEIB microparticles	MS1d, (PEIB/PAA)$_3$PEIB on $CaCO_3$/PEIL microparticles
Cu(II)	4.2	40.2	34.8	31.2	40.2

occur at ambient temperature and that no organic solvents are required, as PEI and PAA are among the water-soluble polymers and only electricity was used for the initial mixing. The two methods differ only in terms of the chemicals used, the B strategy uses a $CuSO_4{\cdot}5H_2O$ solution as a Cu(II) source and then a tertiary amine solution to eliminate the Cu(II) from the LbL polymeric composite matrix. It is important to note that the chemical quantities have accounted for the various excess mass of reactants.

These composite materials were tested as sorbents for the removal of Cu(II) and the results are presented in Table 2 for microparticles with the same LbL coatings, but with different support microparticles ($CaCO_3$ without (sample a) or with polymers (samples b-d)). These results show different affinities for Cu(II) ions in dependence to the chemical composition of the composites.

4.2 LCA planning: Objectives, system limits, functional units, LCIA methodology

This study aims to evaluate the environmental impacts caused by adsorption on $CaCO_3$ support microparticles obtained by two synthesis routes (obtained with PEI or PEI-Cu complex in LbLs, denoted as samples MS1a and MS1aCu) used for Cu(II) ion removal from aqueous solutions.

The functional unit in the LCA study was chosen as 1 mg of Cu(II) ions removed from a synthetic wastewater to evaluate the metal ion removal efficiency. This is according to the literature recommendations (Hischier & Walser, 2012) that LCA studies on nanotechnologies have to consider the nanomaterial function performance in the definition of the functional unit.

The scope of the LCA study includes as the foreground processes the laboratory scale operations for the synthesis of inorganic/organic composites, and the testing (Cu(II) ions sorption processes). The background system includes processes for general chemicals and electricity production. Because the analysis is focused on lab-scale processes, it did not include transport and disposal or infrastructure processes to minimise related uncertainty.

The inventory data used to model the synthesis of the inorganic/organic composite materials was measured during their synthesis (1 batch), as well as for their testing (regarding their performance in removing Cu(II) ions, 10 repeated experiments), as exemplified in Table 3 for MS1a and MS1aCu (samples obtained with the same polymers but applying the two synthesis strategies). Generic chemicals were sourced from the Ecoinvent 3.3 database, and no allocation was performed because no co-products were obtained from the syntheses.

ReCiPe 2016 method considering all 18 midpoint impact categories (Huijbregts et al., 2017) was applied and SimaPro 9.02 software was used to compile the inventory and to perform the life cycle impact assessment. There are several attempts to define the characterisation factors to calculate impacts due to the environmental release of nanostructured materials, but these refer mainly to nano-TiO_2 which have a completely different behaviour in the environmental compartments. As pointed out previously, lack of end-of-life fate and toxicity of these materials form a major knowledge gap on this topic, but defining some are outside the scope of this study and this remains a limitation of our analysis.

4.3 LCA results and discussion

The environmental profiles obtained from the characterisation stage are presented in Table 4 and Fig. 2. These are referred herein as the default scenario against which eco-design measures are compared.

In Table 4, one may notice that higher impact values were recorded in all categories for MS1aCu/$CaCO_3$ materials as compared to the MS1a/$CaCO_3$ material. This is due to the lower sorption capacity for Cu(II) ions in the case of the first materials (15.8 mg Cu(II)/g composite compared to 40.2 mg Cu(II) for the latter one), which furthermore can be linked to the much smaller amount of deposited polymeric material onto the $CaCO_3$ microparticles. This is contrary to what it was expected, because in the previous study (Barjoveanu et al., 2020), the use of Cu-PEI composite has led to improved sorption capacity. In this case, the extraction of Cu(II) ions has probably conducted to the $CaCO_3$ partial dissolution followed by the superficial recrystallisation of some of the $CaCO_3$, thus blocking some of the amine sites which otherwise would have been available for Cu(II) sorption (Zaharia et al., 2021). Besides functionality, the environmental performance was influenced by the synthesis method as well, which in the case of MS1aCu/CaCO3 generates higher environmental impacts, especially in the toxicity categories due to the extraction of Cu(II) ions.

The impact profile structures for the synthesised materials are presented in Fig. 1. It may be noticed that the important contributors in most impact categories are the inorganic chemicals used for the $CaCO_3$ microparticle synthesis: for $CaCl_2$ ranging from 5.1% in Freshwater toxicity to 58.8% in Terrestrial acidification and 60.0% in land use and for Na_2CO_2 ranging from 6.6 in Freshwater toxicity % to 66.2% in Marine eutrophication. These values stand out as environmental hotspots and replacing these substances may form an effective eco-design strategy. The organic phase has only minor contributions to the overall impact: between 0.2% in marine eutrophication and 8% in freshwater toxicity for PAA and from 1.7% in freshwater eutrophication to 29.9% in Freshwater toxicity for PEI. Despite the fact that in the inventory, the electricity consumption is very small (only 1.86E-4 kWh/g composite

Table 3 Inventory data for $(PEIB/PAA)_3PEIB/CaCO_3$ microparticles obtained by both synthesis strategies (sample MS1a and sample MS1aCu).

Inventory entry	Unit	$MS1a/CaCO_3$		$MS1aCu/CaCO_3$		Data source (Ecoinvent process)	Comments
		/g of material	/mg sorbed Cu(II)	/g of material	/mg sorbed Cu(II)		
$MS1a/CaCO_3$	g	1	0.0248				Considering 40.2 mg Cu(II)/g of composite material
$MS1aCu/CaCO_3$	g	–		1	0.0632		
Inputs							
Ultrapure Water	g	76.6	1.90563	76.6	4.8420	Water, ultrapure, at plant/GLO U	For synthesis and rinsing (excess)
Polyethylenimine (linear or branched)	g	0.0249	0.00062	0.0249	0.0016	Ethylenediamine {GLO}| market for | Alloc Def, U	Modelled as Ethylenediamine
Poly(acrylic acid) (PAA)	g	0.031	0.00076	0.031	0.0020		Modelled according to (Barjoveanu et al., 2020)
Glutaraldehyde	g	0.025	0.00062	0.025	0.0016	Acetic anhydride from acetaldehyde, at plant/RER U	Modelled as acetic anhydride
$CaCl_2$ anhydrous	g	1.11	0.02761	1.11	0.0702	Calcium chloride, $CaCl_2$, at plant/RER U	
Sodium Carbonate	g	1.06	0.02636	1.06	0.067	Sodium carbonate, at plant/GLO U	

Copper sulphate	g	–	–	0.0625	0.0040	Copper sulphate, CuSO4, at plant	
Electricity	kWh	0.00018	4.629E-06	0.00018	1.175E-05	Electricity, low voltage {RO} 2020	For stirring
Outputs (emissions to water)							
Amine, tertiary	g	0.01719	0.00042	0.01719	0.0011	Estimated	Considers a 10% washout of the PEI
Glutaraldehyde	g	0.025	0.00062	0.25	0.0016	Estimated	Accounts for the excess glutaraldehyde
Chemically polluted water	g	76.60	1.90	76.60	1.90	Measured	
Copper ions	G	–	–	0.016	0.0010	Estimated	

Table 4 Environmental impacts of composite $CaCO_3$/PEI materials (per functional unit).

Impact category	Unit	IMP	MS1a/ $CaCO_3$	MS1aCu/ $CaCO_3$	(MS1a/ $CaCO_3$)/ (MS1aCu/ $CaCO_3$)
Global warming	kg CO_2 eq	CC	6.48E-05	0.000185918	34.88%
Stratospheric ozone depletion	kg CFC11 eq	OD	1.94E-11	8.2557 E-11	23.53%
Ionising radiation	kBq Co-60 eq	IR	1.29E-05	3.7415 E-05	34.39%
Ozone formation, Human health	kg NOx eq	OF-HH	1.17E-07	3.97336E-07	29.39%
Fine particulate matter formation	kg PM2.5 eq	PM	1.01E-07	4.64569E-07	21.80%
Ozone formation, Terrestrial ecosystems	kg NOx eq	OF-ECO	1.19E-07	4.05121E-07	29.35%
Terrestrial acidification	kg SO_2 eq	TA	3.37E-07	1.44687E-06	23.27%
Freshwater eutrophication	kg P eq	FE	9.31E-08	3.77997E-07	24.62%
Marine eutrophication	kg N eq	ME	3.92E-08	1.10137E-07	35.60%
Terrestrial ecotoxicity	kg 1,4-DCB	TTOX	0.000236	0.004111562	5.74%
Freshwater ecotoxicity	kg 1,4-DCB	FTOX	3.68E-07	2.75623E-05	1.33%
Marine ecotoxicity	kg 1,4-DCB	MTOX	6.09E-07	3.97897E-05	1.53%
Human carcinogenic toxicity	kg 1,4-DCB	HC-TOX	2.86E-06	1.36998E-05	20.91%
Human non-carcinogenic toxicity	kg 1,4-DCB	HNonC-TOX	1.41E-05	0.000966142	1.46%
Land use	m2a crop eq	LAND	1.23E-06	4.64651E-06	26.47%
Mineral resource scarcity	kg Cu eq	MIN	7.86E-08	2.22532E-06	3.53%
Fossil resource scarcity	kg oil eq	FOS	1.87E-05	5.38288E-05	34.69%
Water consumption	m3	WAT	0.000176	0.000455628	38.54%

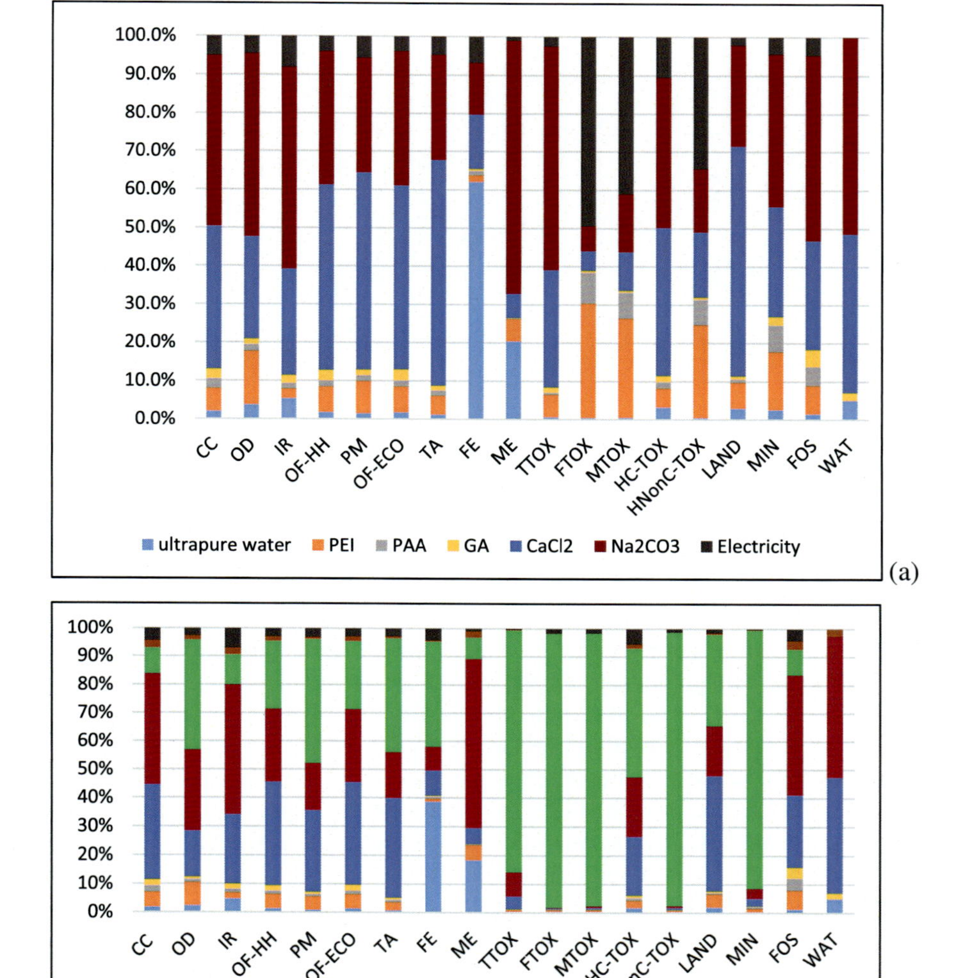

FIG. 1

MS1a/$CaCO_3$ (A) and MS1aCu/$CaCO_3$ (B) impact profile structure.

for a 1 min stirring operation) it generates considerable impacts in the toxicity-related categories (49.4% in FTOX, 41.0% in MTOX, 34.4% in the human non-carcinogenic toxicity categories).

In case of the other composite material (Fig. 1B), which was obtained by using a PEICu composite instead of just PEI, the impact profile is dominated by the $CuSO_4$ production, which contributes mostly in the toxicity-related categories and

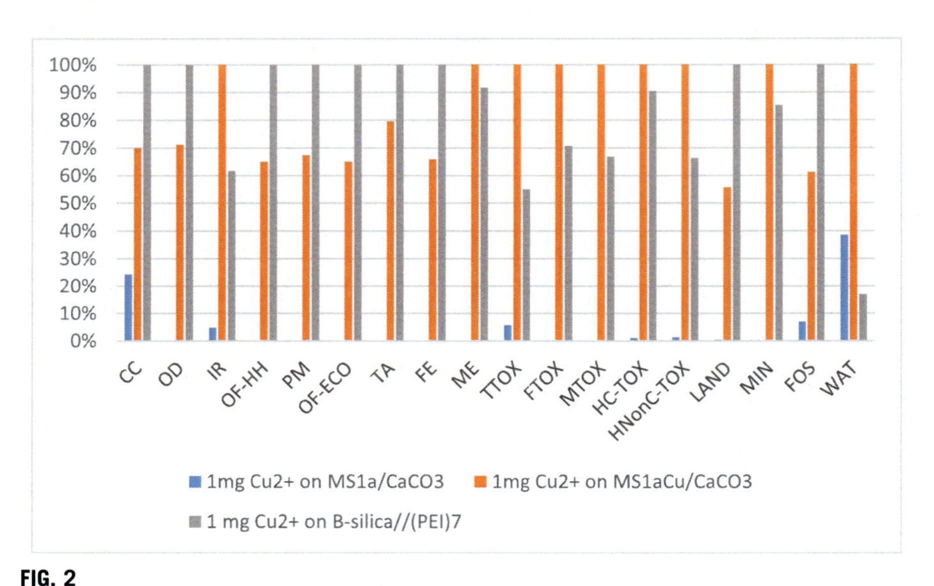

FIG. 2

Organic/inorganic microparticles comparison (impacts per mg Cu(II) ions removed).

the mineral resource scarcity (90.4%). Another noteworthy contributor for both these materials is the ultrapure water, whose impacts are predominant in Eutrophication categories due to the removal of nutrient ions from tap water.

These results may be used to evaluate the initial environmental performance of these materials, in the process of removing Cu(II) ions from wastewaters. In this context, in Fig. 2 a comparison among the environmental impacts of these composite materials is presented. From a materials synthesis perspective, the silica-cored particles (Barjoveanu et al., 2020) generate higher impacts than the $CaCO_3$ microparticles, except some categories, where the MS1aCu/$CaCO_3$ has higher impacts, due to the use of $CuSO_4$ in the synthesis phase. This comparison shows that in this particular case, following the simplest synthesis route leads to the best environmental performance (MS1a/$CaCO_3$).

In Table 5, the laboratory scale activities related to the design of PEI-coated $CaCO_3$ microparticles are evaluated **from a qualitative standpoint** by considering the following:

- the eco-design principles which are applicable for the conceptualisation and production phases of novel materials.
- The life cycle impact assessment results that were obtained for these novel materials in the lab-scale synthesis phase.

This evaluation considers the previously presented LCA results and it is focused on the technical and environmental performance brought by MS1a, $(PEIB/PAA)_3$PEIB on $CaCO_3$ microparticles which had the best environmental performance among the

Table 5 Matrix for evaluation of eco-design efforts.

Eco-design Principle	Eco-design Strategies/ design goal	Performed actions	Effects on technical performance	Effects on environmental performance
Improve product function	Enhance product function/ increase affinity for M(II) ions	Increase polymeric content by increasing LbL steps	Improved sorption capacity (up to 40.2 mg Cu(II)/g)	Improved environmental performance (decrease of impacts values)
	Dematerialisation/reduce no of LbL steps	Not achieved		
	Multifunctionality	Applied for multiple ions	M(II) ions sorbed based on competition	Not evaluated
Use low impact materials	Use renewable natural sources	N/A	N/A	N/A
	Maximise recycled materials	N/A	N/A	N/A
	Replace dangerous substances	Replaced core silica with $CaCO_3$	Improved performance	Improved performance/ significant impacts decrease
	Materials produced by eco-friendly methods	In situ $CaCO_3$ micro-spheres generation	improved performance (compared to silica)	Improved performance/ significant impacts decrease
	Safe-by-design for end of life	Not achieved	Not tested	Not tested (lack of EoL fate data)
Materials reduction	Reduce specific consumption of materials/ reduce quantity/functional unit	Improved specific consumption/not achieved for ultrapure water consumption	Increased performance considering improved functionality	Increased performance considering improved functionality
Production optimisation	Alternative production technologies	N/A	N/A	N/A

Continued

Table 5 Matrix for evaluation of eco-design efforts. *Continued*

Eco-design Principle	Eco-design Strategies/ design goal	Performed actions	Effects on technical performance	Effects on environmental performance
	Minimise no. of production steps/decrease no of LbL steps	Not achieved	Not achieved	Not achieved
	Minimise energy consumption	Synthesis at ambient temperature/low electricity consumption	N/A	Low impacts, but very sensitive to electricity consumption
	Reduce waste and pollution	Not achieved	N/A	Not achieved
Distribution optimisation	N/A	N/A	N/A	N/A
Optimisation of life time	Extend use life/(e.g. improve reusability)	Not achieved/not tested	Not achieved	Not achieved/not enough info
	Safe-by-design/avoid EoL toxic releases	Design stable material	Stable, Depending on pH,	
	Enhanced recyclability	Not achieved/not tested	N/A	N/A

Notes: N/A not available/not applicable.

$CaCO_3$-cored particles and the silica//$(PEI)_7$ particles (developed and studied in the previous study and used here as reference). To the authors' best knowledge this is one of the first exercises proposed to evaluate the new materials' synthesis options against eco-design criteria. One should notice that most of the environmental improvement comes from the improved functionality of the product, and this should form the focus of material designers. Of course, this matrix can be parameterised to transform it into a quantitative evaluation instrument which can be used as a component in a MCDA support tool. Furthermore, it has to be noted that because the moment in the technological development of these materials (and processes) is so early, the eco-design strategies and measures cannot be fully identified and assessed, especially for the post-use phases in the materials' life cycle. Additional data related to process design, options, parameters, as well as data on environmental behaviour of the materials is needed to develop a complete eco-design performance check.

5 LCA evaluation of novel sorbents obtained from rapeseed waste biomass

This study investigates another eco-design strategy to improve the environmental performance of water and wastewater treatment. Here, instead of targeting the material functionality (better performance per unit of product), the strategy is to avoid (or minimise as much as possible) the environmental burdens of materials used in treatment processes. This is achieved by developing and using recycled materials or secondary materials for the removal of emerging pollutants from wastewaters. Concretely, here we investigate by means of LCA the environmental impacts associated to the conditioning of rapeseed (RS) waste and its use for the Pb(II) ions removal from aqueous solutions.

Relatively recently some important research focus has been awarded to the use of biomass (especially waste biomass) to be used as adsorbents for sequential or simultaneous emerging pollutants from wastewater (Morosanu, Teodosiu, Coroaba, & Paduraru, 2019). Advantages like availability and cheap prices of waste biomass as biosorbent have to be proven against issues like adsorption and regeneration capacity, stability, and removal efficiencies.

5.1 Material development, synthesis, and testing

The biosorbent precursor was provided by a company that produces rapeseed oil. The rapeseed meal (RSM) was obtained by pressing the plant seeds, followed by solvent (n-hexane) extraction of oil. The biosorbent precursor (rapeseed meal) was then crushed with an electric grinder and sieved manually to particle sizes below 0.6 mm. The granules were then washed several times with deionised water, dried at 40°C for 6 h, and left overnight to fully dry. The chemical treatment of RSM was carried out in two ways: by using a solution of 1 M HNO_3 and a solution of 1 M NaOH, respectively. The procedure consisted of: (i) 6.5 g of the cleaned RSM was mixed with a volume of 650 mL solution 1 M of HNO_3 or NaOH on a hotplate for 8 h at 50°C, then left overnight at room temperature; (ii) the supernatant was then discarded and the RSM was washed several times with deionised water; (iii) finally, the RSM was dried in an oven at 40°C for 3 h and then placed in a desiccator until further use.

The efficiency of the RSM–pristine, treated with acid (RSM-A) and with base (RSM-B), respectively, was determined by using a solution containing Pb(II) ions at different initial concentrations (30–146 mg/L). The tests have proven almost a double sorption capacity for Pb^{2+} ions in the case of the base-treated RSM (38.358 mg/g), as compared to the acid-treated one (19.414 mg/g).

5.2 LCA planning: Objectives, system limits, functional unit, LCIA methodology

The life cycle assessment study is based on laboratory scale sorption experiments during which parameters like contact time and initial pollutant concentrations, and biosorbent selectivity were investigated based on the experimental data and modelling approach. The life cycle inventory has included the biosorbent conditioning phase during which rapeseed waste was treated with various concentrations of acids and bases, as presented in Table 6. The actual biosorption phase has provided information related to the functional unit of the study which was chosen as 1 mg

Table 6 Life cycle inventory for treated rapeseed meal.

Products	RSM HNO_3 treated	RSM NaOH treated		Ecoinvent process	Data source
Biosorbent	1	1	g		Measured
Sorption capacity for Pb(II) ions	19.414	38.358	mg/g		Measured
Inputs					
Rapeseed waste biomass	1.46	4.47	g	Rapeseed meal, from crushing (solvent), at plant/PL Economic	Measured
NaOH	–	160.1363	g	Sodium hydroxide, 50% in H_2O, membrane cell, at plant/RER U	Measured

Table 6 Life cycle inventory for treated rapeseed meal. *Continued*

Products	RSM HNO_3 treated	RSM NaOH treated		Ecoinvent process	Data source
HNO_3	37.5			Nitric acid, 50% in H2O, at plant/RER U	Measured
Water	1.24	4.48	kg	Water, ultrapure, at plant/GLO U	Measured
Electricity	1.0592	3.289		Electricity, low voltage {RO}	Measured for grinding, heating, and drying
Outputs					
Biomass waste	0.46	3.47		Waste, organic	

Pb(II) ions removed from synthetic wastewater. The life cycle impact assessment was performed with ReCiPe 2016 at midpoint using all the 18 environmental impact categories.

5.3 LCA results and discussion

The results of the life cycle impact assessment are presented in Fig. 3 and show that the environmental profiles of both biosorbents (acid- and base-treated rapeseed waste) are dominated by the electricity consumption (done in laboratory for operations like grinding, mixing, heating, and drying in lab ovens) and by the production of chemicals.

It is important to note that although it was considered in the life cycle inventory (as a by-product of rapeseed oil production and accounting for 23.72% of the total impacts considering an economic perspective for this allocation), the contribution of the rapeseed waste is negligible as compared to that of other contributors, with the exception of the land use category where it contributes up to 50% for the acid-treated biosorbent and 55% for the base-treated biosorbent. These highlight the environmental-friendly character of these biosorbents.

When comparing the impact values obtained for the two biosorbents (Table 7), it is clear that the hydroxide-treated rapeseed biomass had much higher impacts in all categories, except the stratospheric ozone depletion, where the production of nitric acid leads to N_2O emissions and higher impacts in this category. This is despite the fact that the base-treated biomass has almost double the sorption capacity for lead ions and the reason is the much lower productivity in the case of the base-treated

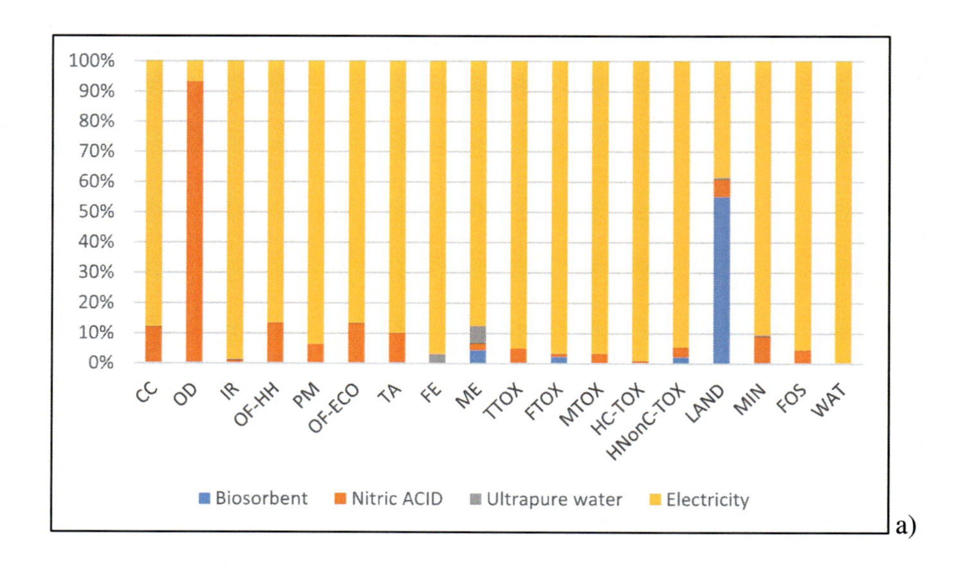

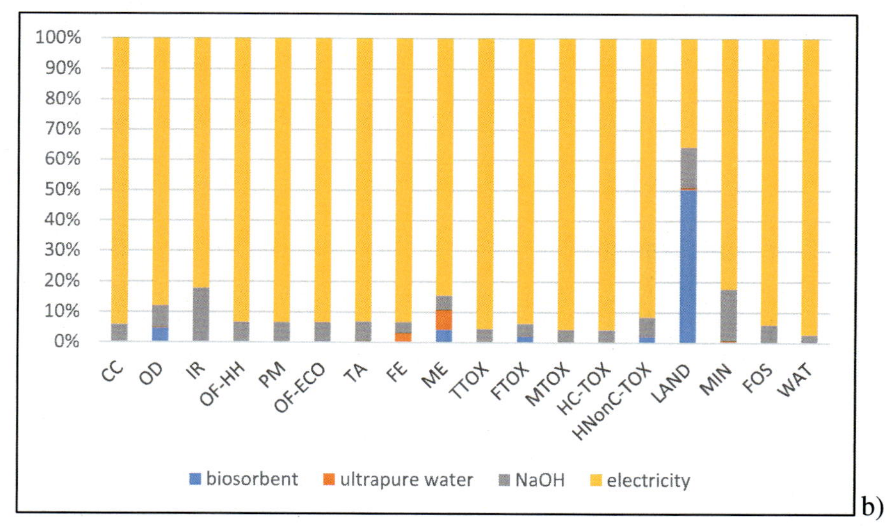

FIG. 3

Environmental profiles for conditioning of 1 g RSM (A—acid treated and B—base treated).

waste biomass (4.47 g biomass processed for 1 g of biosorbent), compared to the acid-treated one (1.46 g/g biosorbent). This situation proves that the way in which the production processes are conducted may have considerable impacts on the overall efficiency and environmental sustainability of the products, and may ultimately reduce the advantages brought by using wastes as resources.

Table 7 Life cycle impact assessment results obtained for per mg Pb(II) ions sorbed.

Impact category	Unit	Abbr.	Absolute values		Relative values, % of max	
			HNO_3 treated	NaOH treated	HNO_3 treated	NaOH treated
Global warming	kg CO_2 eq	CC	0.051017	0.074762	68.2%	100.0%
Stratospheric ozone depletion	kg CFC11 eq	OD	1.92E-07	2.30E-08	100.0%	12.0%
Ionising radiation	kBq Co-60 eq	IR	0.008126	0.015324	53.0%	100.0%
Ozone formation, Human health	kg NOx eq	OF-HH	8.14E-05	0.000118	68.7%	100.0%
Fine particulate matter formation	kg PM2.5 eq	PM	5.76E-05	9.08E-05	63.5%	100.0%
Ozone formation, Terrestrial ecosystems	kg NOx eq	OF-ECO	8.23E-05	0.00012	68.6%	100.0%
Terrestrial acidification	kg SO_2 eq	TA	0.000168	0.000254	65.9%	100.0%
Freshwater eutrophication	kg P eq	FE	6.90E-05	0.000113	61.3%	100.0%
Marine eutrophication	kg N eq	ME	4.63E-06	7.53E-06	61.5%	100.0%
Terrestrial ecotoxicity	kg 1,4-DCB	TTOX	0.159204	0.248779	64.0%	100.0%
Freshwater ecotoxicity	kg 1,4-DCB	FTOX	7.25E-05	0.000117	61.7%	100.0%
Marine ecotoxicity	kg 1,4-DCB	MTOX	0.00019	0.000302	63.0%	100.0%
Human carcinogenic toxicity	kg 1,4-DCB	HC-TOX	0.003934	0.00639	61.6%	100.0%
Human non-carcinogenic toxicity	kg 1,4-DCB	HNonC-TOX	0.003442	0.005583	61.7%	100.0%
Land use	m^2a crop eq	LAND	0.00023	0.00039	59.0%	100.0%
Mineral resource scarcity	kg Cu eq	MIN	1.49E-05	2.57E-05	57.9%	100.0%
Fossil resource scarcity	kg oil eq	FOS	0.011885	0.018943	62.7%	100.0%
Water consumption	m^3	WAT	0.720262	1.159375	62.1%	100.0%

6 Conclusions

This chapter presented two case studies of using LCA as an eco-design support tool to evaluate the environmental sustainability of new materials used in water/wastewater treatment processes.

In the first case study, based on the LCA results, a systematic, qualitative evaluation of how the material conceptualisation, synthesis, and testing according to eco-design principles and practices has been performed. This evaluation has pointed out that the functionality (which is similar to technical performances) represents a key aspect regarding the environmental performance of the new product. This was confirmed also for the second case study, where aspects related to the production efficiency have proven important for the overall efficiency.

Both case studies have shown that it is important to consider as earliest as possible in the product development process eco-design criteria to evaluate the prospective environmental sustainability of novel materials. At the same time, the results have pointed out that aspects like functional unit definition, inventory data gaps and uncertainties, and interpretation of LCIA results have to be carefully considered when evaluating eco-design options. This early environmental evaluation has to carefully used because some eco-design strategies and measures cannot be fully identified and assessed at this point, for example the post-use phases in the life cycle of materials. Additional data related to process design, options, parameters, as well as data on environmental behaviour of the materials are still needed and all these aspects require further research.

Acknowledgements

This work was supported by:

1. a grant of the Romanian Ministry of Education and Research, CNCS-UEFISCDI, project number PN-III-P4-ID-PCE-2020-1199, contract PCE 56/2021, "Innovative and sustainable solutions for priority and emerging pollutants removal through advanced wastewater treatment processes" (SUSTINWATER) within PNCDI III and
2. a grant of the Romanian Ministry of Research and Innovation, CCCDI-UEFISCDI, project number PN-III-P1-1.2-PCCDI-2017-0245, contract 26PCCDI/01.03.2018, "Integrated and sustainable processes for environmental clean-up, wastewater reuse and waste valorization" (SUSTENVPRO), within PNCDI III.

References

Anandhakumar, S., & Raichur, A. M. (2013). Polyelectrolyte/silver nanocomposite multilayer films as multifunctional thin film platforms for remote activated protein and drug delivery. *Acta Biomaterialia*, *9*, 8864–8874. https://doi.org/10.1016/j.actbio.2013.06.012.

Babbitt, C. W., & Moore, E. A. (2018). Sustainable nanomaterials by design A new material selection tool offers a proactive perspective on nanotechnology research and development. *Nature Nanotechnology*, *13*, 621–623.

Barjoveanu, G., Teodosiu, C., Bucatariu, F., & Mihai, M. (2020). Prospective life cycle assessment for sustainable synthesis design of organic/inorganic composites for water treatment. *Journal of Cleaner Production*, *272*, 122672. https://doi.org/10.1016/j.jclepro.2020.122672.

Boudou, T., Kharkar, P., Jing, J., Guillot, R., Pignot-Paintrand, I., Auzely-Velty, R., et al. (2012). Polyelectrolyte multilayer nanoshells with hydrophobic nanodomains for delivery of paclitaxel. In *Vol. 159. ASME 2012 summer bioeng. conf. SBC 2012* (pp. 407–408). https://doi.org/10.1115/SBC2012-80206.

Brezet, H., van Hemel, C., & UNEP IE Cleaner Production Network.; Rathenau Instituut.; Technische Universiteit Delft. (1997). *Ecodesign: A promising approach to sustainable production and consumption*. United Nations Publication. United Nations Environment Programme, Industry and Environment, Cleaner Production.

Bucatariu, F., Ghiorghita, C.-A., Schwarz, D., Boita, T., & Mihai, M. (2019). Layer-by-layer polyelectrolyte architectures with ultra-fast and high loading/release properties for copper ions. *Colloids and Surfaces A: Physicochemical and Engineering Aspects*, *579*, 123704. https://doi.org/10.1016/J.COLSURFA.2019.123704.

Caramazana-González, P., Dunne, P. W., Gimeno-Fabra, M., Zilka, M., Ticha, M., Stieberova, B., et al. (2017). Assessing the life cycle environmental impacts of titania nanoparticle production by continuous flow solvo/hydrothermal syntheses. *Green Chemistry*, *19*, 1536–1547. https://doi.org/10.1039/c6gc03357a.

Cinelli, M., Coles, S. R., Sadik, O., Karn, B., & Kirwan, K. (2016). A framework of criteria for the sustainability assessment of nanoproducts. *Journal of Cleaner Production*, *126*, 277–287. https://doi.org/10.1016/j.jclepro.2016.02.118.

Cossutta, M., McKechnie, J., & Pickering, S. J. (2017). A comparative LCA of different graphene production routes. *Green Chemistry*, *19*, 5874–5884. https://doi.org/10.1039/c7gc02444d.

Cucurachi, S., & Blanco Rocha, C. F. (2019). Life-cycle assessment of engineered nanomaterials. In *Nanotechnology in eco-efficient construction* Elsevier Ltd. https://doi.org/10.1016/b978-0-08-102641-0.00031-1.

Cucurachi, S., Van Der Giesen, C., & Guinée, J. (2018). Ex-ante LCA of emerging technologies. *Procedia CIRP*, *69*, 463–468. https://doi.org/10.1016/j.procir.2017.11.005.

Debnath, M. K., Rahman, M. A., Minami, H., Rahman, M. M., Alam, M. A., Sharafat, M. K., et al. (2019). Single step modification of micrometer-sized polystyrene particles by electromagnetic polyaniline and sorption of chromium(VI) metal ions from water. *Journal of Applied Polymer Science*, *136*, 23–25. https://doi.org/10.1002/app.47524.

Duan, H., Wang, D., & Li, Y. (2015). Green chemistry for nanoparticle synthesis. *Chemical Society Reviews*, *44*, 5778–5792. https://doi.org/10.1039/c4cs00363b.

Eckelman, M. J., Mauter, M. S., Isaacs, J. A., & Elimelech, M. (2012). New perspectives on nanomaterial aquatic ecotoxicity: Production impacts exceed direct exposure impacts for carbon nanotoubes. *Environmental Science & Technology*, *46*, 2902–2910. https://doi.org/10.1021/es203409a.

Ettrup, K., Kounina, A., Hansen, S. F., Meesters, J. A. J., Vea, E. B., & Laurent, A. (2017). Development of comparative toxicity potentials of TiO_2 nanoparticles for use in life cycle assessment. *Environmental Science & Technology*, *51*, 4027–4037. https://doi.org/10.1021/acs.est.6b05049.

Falinski, M. M., Plata, D. L., Chopra, S. S., Theis, T. L., Gilbertson, L. M., & Zimmerman, J. B. (2018). A framework for sustainable nanomaterial selection and design based on performance, hazard, and economic considerations. *Nature Nanotechnology*, *13*, 708–714. https://doi.org/10.1038/s41565-018-0120-4.

Feijoo, S., González-García, S., Moldes-Diz, Y., Vazquez-Vazquez, C., Feijoo, G., & Moreira, M. T. (2017). Comparative life cycle assessment of different synthesis routes of magnetic nanoparticles. *Journal of Cleaner Production*, *143*, 528–538. https://doi.org/10.1016/J.JCLEPRO.2016.12.079.

Gottschalk, F., & Nowack, B. (2011). The release of engineered nanomaterials to the environment. *Journal of Environmental Monitoring*, *13*, 1145–1155. https://doi.org/10.1039/c0em00547a.

Grande, C. A., Blom, R., Spjelkavik, A., Moreau, V., & Payet, J. (2017). Life-cycle assessment as a tool for eco-design of metal-organic frameworks (MOFs). *Sustainable Materials and Technologies*, *14*, 11–18.

Hamoda, M., Al-Ghusain, I., & Al-Mutairi, N. (2004). Sand filtration of wastewater for tertiary treatment and water reuse. *Desalination*, *164*, 203–211. https://doi.org/10.1016/S0011-9164(04)00189-4.

Hamza, M. F., Gamal, A., Hussein, G., Nagar, M. S., Abdel-Rahman, A. A., Wei, Y., et al. (2019). Uranium(VI) and zirconium(IV) sorption on magnetic chitosan derivatives—Effect of different functional groups on separation properties. *Journal of Chemical Technology and Biotechnology*. https://doi.org/10.1002/jctb.6185.

Hischier, R., Salieri, B., & Pini, M. (2017). Most important factors of variability and uncertainty in an LCA study of nanomaterials—Findings from a case study with nano titanium dioxide. *NanoImpact*, *7*, 17–26. https://doi.org/10.1016/J.IMPACT.2017.05.001.

Hischier, R., & Walser, T. (2012). Life cycle assessment of engineered nanomaterials: State of the art and strategies to overcome existing gaps. *Science of the Total Environment*, *425*, 271–282. https://doi.org/10.1016/J.SCITOTENV.2012.03.001.

Hoai, N. T., Yoo, D. K., & Kim, D. (2010). Batch and column separation characteristics of copper-imprinted porous polymer micro-beads synthesized by a direct imprinting method. *Journal of Hazardous Materials*, *173*, 462–467. https://doi.org/10.1016/j.jhazmat.2009.08.107.

Huijbregts, M. A. J., Steinmann, Z. J. N., Elshout, P. M. F., Stam, G., Verones, F., Vieira, M., et al. (2017). ReCiPe2016: A harmonised life cycle impact assessment method at midpoint and endpoint level. *International Journal of Life Cycle Assessment*, *22*, 138–147. https://doi.org/10.1007/s11367-016-1246-y.

ISO. (2020). *ISO 14006:2020(en) Environmental management systems—Guidelines for incorporating ecodesign.*

Livingston, A., Trout, B. L., Horvath, I. T., Johnson, M. D., Vaccaro, L., Coronas, J., et al. (2020). Challenges and directions for green chemical engineering-role of nanoscale materials. In *Sustainable Nanoscale Engineering: From Materials Design to Chemical Processing* (pp. 1–18). https://doi.org/10.1016/B978-0-12-814681-1.00001-1.

Maccioni, L., Borgianni, Y., & Pigosso, D. C. A. (2019). Can the choice of eco-design principles affect products' success? *Design Science*, *5*, 1–31. https://doi.org/10.1017/dsj.2019.24.

Morosanu, I., Teodosiu, C., Coroaba, A., & Paduraru, C. (2019). Sequencing batch biosorption of micropollutants from aqueous effluents by rapeseed waste: Experimental assessment and statistical modelling. *Journal of Environmental Management*, *230*, 110–118. https://doi.org/10.1016/j.jenvman.2018.09.075.

Morosanu, I., Teodosiu, C., Fighir, D., & Paduraru, C. (2019). Simultaneous biosorption of micropollutants from aqueous effluents by rapeseed waste. *Process Safety and Environment Protection*, *132*, 231–239. https://doi.org/10.1016/j.psep.2019.09.029.

Morosanu, I., Teodosiu, C., Tofan, L., Fighir, D., & Paduraru, C. (2020). *Valorization of rapeseed waste biomass in sorption processes for wastewater treatment*. https://doi.org/10.5772/intechopen.94942.

Pizza, A., Metz, R., Hassanzadeh, M., & Bantignies, J. L. (2014). Life cycle assessment of nanocomposites made of thermally conductive graphite nanoplatelets. *International Journal of Life Cycle Assessment*, *19*, 1226–1237. https://doi.org/10.1007/s11367-014-0733-2.

Salieri, B., Turner, D. A., Nowack, B., & Hischier, R. (2018). Life cycle assessment of manufactured nanomaterials: Where are we? *NanoImpact*, *10*, 108–120. https://doi.org/10.1016/J.IMPACT.2017.12.003.

Silva, C. P., Jaria, G., Otero, M., Esteves, V. I., & Calisto, V. (2019). Adsorption of pharmaceuticals from biologically treated municipal wastewater using paper mill sludge-based activated carbon. *Environmental Science and Pollution Research*, *26*, 13173–13184. https://doi.org/10.1007/s11356-019-04823-w.

Teodosiu, C., Gilca, A.-F., Barjoveanu, G., & Fiore, S. (2018). Emerging pollutants removal through advanced drinking water treatment: A review on processes and environmental performances assessment. *Journal of Cleaner Production*, *197*. https://doi.org/10.1016/j.jclepro.2018.06.247.

Tong, Y., McNamara, P. J., & Mayer, B. K. (2019). Adsorption of organic micropollutants onto biochar: A review of relevant kinetics, mechanisms and equilibrium. *Environmental Science: Water Research & Technology*, *5*, 821–838. https://doi.org/10.1039/C8EW00938D.

Vajihinejad, V., Gumfekar, S. P., Bazoubandi, B., Rostami Najafabadi, Z., & Soares, J. B. P. (2019). Water soluble polymer flocculants: Synthesis, characterization, and performance assessment. *Macromolecular Materials and Engineering*, *304*, 1–43. https://doi.org/10.1002/mame.201800526.

Windsor, R., Cinelli, M., & Coles, S. R. (2018). Comparison of tools for the sustainability assessment of nanomaterials. *Current Opinion in Green and Sustainable Chemistry*, *12*, 69–75. https://doi.org/10.1016/j.cogsc.2018.06.010.

Wu, J., Xu, F., Li, S., Ma, P., Zhang, X., Liu, Q., et al. (2019). Porous polymers as multifunctional material platforms toward task-specific applications. *Advanced Materials*, *31*, 1802922. https://doi.org/10.1002/adma.201802922.

Zaharia, M. M., Bucatariu, F., Doroftei, F., Loghin, D. F., Vasiliu, A. L., & Mihai, M. (2021). Multifunctional $CaCO_3$/polyelectrolyte sorbents for heavy metal ions decontamination of synthetic waters. *Colloids and Surfaces A: Physicochemical and Engineering Aspects*, *613*. https://doi.org/10.1016/j.colsurfa.2020.126084.

Zhang, W., Zhao, Q., & Yuan, J. (2018). Porous polyelectrolytes: The interplay of charge and pores for new functionalities. *Angewandte Chemie International Edition*, *57*, 6754–6773. https://doi.org/10.1002/anie.201710272.

CHAPTER

Life Cycle Assessment for the design of a pilot recovery plant

13

Olatz Pombo, Andrew Ferdinando, Ana Belén de Isla, and Jose Miguel Martínez
LKS Krean, KREAN Group, Mondragón, Spain

1 Aims

The aim of this chapter is to develop a low-impact environmental recovery plant to value brewery by-products for aquaculture feed ingredients. To do so, LCA methodology was used from the early stage design. A variety of energy saving measures as well as material alternatives were assessed. The preferred solutions were implemented getting a proposed design. LCA at this stage allowed detection of the materials with the most impact, proposing alternative solutions and reaching a higher environmental reduction. The focus was placed on the architectural design of the building and passive strategies, since they are more efficient in terms of energy savings than active strategies.

2 State of the art

The construction sector in the EU accounts for 40% of the total energy consumption (Aïssani, Chateauneuf, Fontaine, & Audebert, 2014; Terés-Zubiaga, Campos-Celador, González-Pino, & Escudero-Revilla, 2015). The sector also generates about one-third of all waste (European Commission, 2018) and is associated with environmental pressures that arise at different stages of a building's life cycle, including the manufacturing of construction products, building construction, use, renovation, and the management of building waste (European Commission, 2014).

In the last decades, the awareness of environmental concerns has led to more stringent energy policies aimed at reducing the need for operational energy and hence the need for oil-based heating and electricity (Hauschild, Rosenbaum, & Olsen, 2018). As buildings become more energy efficient, there has been a shift to a more holistic life cycle analysis of buildings (Nwodo & Anumba, 2019). Recent studies reflect the rising recognition that, without adopting a holistic life cycle perspective and without addressing major restraints in this area, the chance to achieve buildings with low environmental impacts is limited (Azari, 2019). The use of such

Assessing Progress towards Sustainability. https://doi.org/10.1016/B978-0-323-85851-9.00006-7

an approach at the beginning of a design process has been identified as a decisive tool in the pursuit of sustainable construction (Pombo, Rivela, & Neila, 2016). The earlier the assessment, the higher is the potential to effectively influence the life cycle performance of the building (Kohler & Moffatt, 2003).

Life Cycle Assessment (LCA) has developed briskly over the last three decades (Chang, Lee, & Chen, 2014; Laurin, 2017). In the building sector this methodology has become more and more important during the last decades, in order to take into account the whole energy used starting from the construction up to the demolition (Schiavoni, D'Alessandro, Bianchi, & Asdrubali, 2016).

However, most case studies focus on residential/commercial buildings (Bahramian & Yetilmezsoy, 2020; Li et al., 2020), while very few can be found focused on industrial buildings (Rai, Sodagar, Fieldson, & Hu, 2011; Tulevech et al., 2018).

In the industrial area, LCA has been used for the optimisation on product level (Luz, De Francisco, & Piekarski, 2015) or on manufacturing process (Shin, Suh, Stroud, & Yoon, 2017). Nevertheless, LCA has rarely been used for the optimisation of industrial facilities and limited work has been published (Heravi, Fathi, & Faeghi, 2015). The highest priority aims in the design and construction of industrial facilities have primarily been minimal investments costs, flexibility and expandability of the building structure. Meanwhile, the resources and energy optimisation have been regarded as secondary issues. However, with the sharpening of building codes, the upcoming number of European regulations on the climate protection and energy efficiency, as well as the raising awareness of a corporate social responsibility, life cycle optimisation is starting to gain importance among industrial investors (Kovacic, Waltenbereger, & Gourlis, 2016).

3 Novelty

The literature review stated that energy efficiency has not been a priority aim in the design and construction of industrial buildings until the sharpening of building codes. Although LCA has been broadly applied in commercial and residential buildings, very few studies have been found on industrial buildings. The novelty of the present chapter is the application of LCA methodology for the design of a low environmental impact recovery plant for the valuation of brewery by-products. The study allows to see the relevance of buildings' material selection and energy saving measures in the reduction of the environmental impacts during design process.

4 Case study description

The current study is part of a European funded LIFE project, Life BREWERY, which has the objective of demonstrating an innovative and highly replicable integrated solution to recover brewery by-products for aquaculture feed ingredients. The construction of a recovery plant is necessary to value brewer by-products. The present chapter focuses on the eco-design of a low environmental recovery plant, replicable across Europe.

LCA methodology was used from the preliminary design steps for decision making considering different energy saving measures as well as material alternatives to reduce the impacts of the building. The LCA in the present work was conducted according to European EN 15978 standard (EN 15978:2011, 2012).

GIS methodology has been used to identify the specific and optimised locations where the recovery plant could be implemented in the future across Europe. The objective was to reduce the transport impact from the brewery producer plant to the new recovery plant. In this chapter, one of the case studies is analysed in depth. The location selected was Lleida (NE Spain), with the most important breweries: Heineken, Mahou-San Miguel, and La Zaragozana.

4.1 Goal and scope

The aim of this study was to design a low environmental impact Pilot Recovery Plant of beer by-products for aquaculture feed ingredients, located in Lerida, Spain.

Functional unit

The functional unit was defined as 1 m^2 of gross internal floor area of a warehouse located in Lleida and built in 2020. A lifespan of 60 years was considered. To reach this goal, a comparative analysis of different energy saving measures and materials was made, considering the whole life cycle.

System boundaries

The analysis included the production of construction materials, transport of the materials to the construction site, construction, use stage (maintenance, replacements, heating and cooling energy consumption), and end of life (EoL).

Impact categories

According to the scope of the assessment, the following impact categories have been selected: GWP: global warming potential, in kg CO_2e; AP: acidification of land and water sources, in kg SO_2e; EP: eutrophication, in kg $(PO4)^{3-}e$; ODP: depletion of the stratospheric ozone layer, in kg CFC-11e; POCP: formation of tropospheric ozone, in kg C_2H_4; and ADP: abiotic depletion potential, in MJ.

4.2 Life cycle inventory

The LCA covered the complete building envelope and structural elements, including the material components of footings and foundations, structural wall assembly (from cladding to interior finishes), structural floors and ceilings (including finishes), roof assemblies and additional building elements, such as interior non-structural walls or finishes. Electrical and mechanical equipment and controls, plumbing fixtures, fire detection and alarm system fixtures, elevators, and conveying systems, and renewable energy systems were excluded from the analysis.

The calculations were performed with One Click LCA calculation tool. The software is fully compliant with EN 15978 standard (EN 15978:2011, 2012). The tool supports CML (2002–November 2012 or newer) methodology and all assessed impact categories. All the datasets in the tool follow EN 15804 standard (EN 15804:2012+A1:2013, 2014). Tool provides also different scenarios for the construction site emissions based on the location and gross floor area.

For the use phase, material maintenance and replacements of the materials with a life span lower than 60 years were considered. For the energy calculation, electricity used for heating and cooling was considered. Consumption was obtained from energy simulations. Spanish electricity mix was used to determine the impacts. With respect to the end-of-Life phase this information is taken by the tool from the EPDs.

4.3 Life cycle impact assessment and interpretation

Preliminary building design

The building is composed of two main blocks. The largest, unconditioned, which integrates the industrial processes and a smaller auxiliary office building, which was oriented to the South to allow for greater control of solar gains. The building has an internal gross floor area of 3200 m^2. Although industrial buildings are excluded from the energy efficiency requirements established by the Spanish Building Regulation (Ministerio de Fomento, 2019), the building envelope was defined according to the requirements of the regulation for new buildings.

At early design stage, the focus was placed on the reduction of heating and cooling consumption and the selection of low environmental construction systems. According to literature, energy consumption of the building during use phase has a great influence in the life cycle energy and carbon impacts of a building (Russell-Smith, Lepech, Fruchter, & Meyer, 2015).

Once energy consumption was reduced, and therefore, the impact of the use phase, different materials have been analysed for each main building element, to reduce the environmental impacts of the construction process.

Energy saving measures

Regarding energy issues, energy consumption was assessed. Building energy simulation software DesignBuilder was used to calculate heating and cooling energy consumption together with the weather data of Lleida. EnergyPlus calculation engine is used by this software.

Table 1 summarises passive measures assessed and annual energy consumption. As stated before, the warehouse area is unconditioned, while the office area is conditioned by a heat pump. Therefore electricity is used for heating and cooling.

Results revealed that the combination of measures M2, M7, and M8 allowed for a higher reduction of energy consumption. Moreover, sun-shading elements (M9) implied a high reduction in cooling while a higher consumption in heating. However, it should be noted that they helped to reduce glare in workstation areas providing

Table 1 Passive measures assessed to be implemented in the building design.

	Passive measure	Implemented in: Warehouse	Implemented in: Offices	Annual energy consumption (kWh/year): Heating	Annual energy consumption (kWh/year): Cooling
	Preliminary design			12,177.01	4309.41
M1	Reduction of U-value 25% in external walls			11,438.70	4301.23
M2	Reduction of U-value 50% in external walls			8613.58	3954.07
M3	Reduction of U-value of rooflight polycarbonates adding aerogel			12,177.01	4309.41
M4	Replacement of standard skylights by northlights			12,177.01	4309.41
M5	Increase of roof reflectance to RSI 85			12,439.89	4142.58
M6	Reduction of U-value 25% in roof			11,824.89	4292.01
M7	Reduction of U-value 50% in roof			11,611.58	4219.12
M8	Triple glazing in office and auxiliary windows			12,060.57	3890.49
M9	Solar protection in south façade			14,341.00	2714.72

higher comfort to the occupants. Measures M3 and M4 did not imply any energy reduction as the warehouse was not conditioned. However, internal temperature analyses showed that M3 did not seem to vary the internal temperatures, while M4 could help to avoid overheating problems in summer. Finally, M5 did not represent any improvement in the energy consumption of the building. It could be explained because the warehouse was not conditioned, and it represented the 89% of the total roof. However, this measure was selected as it contributed to Heat Island reduction.

The implementation of energy saving measures allowed for reduced heating and cooling consumptions to 9457.54 kWh/year and 2507.59 kWh/year, respectively. Therefore a reduction of 30% was achieved, which corresponded 22% to heating and 42% to cooling.

Material alternatives

At this early stage, different alternatives were analysed for each building element, to select the most suitable solutions for this building. The equivalence in energy performance and exposure conditions was ensured by the following aspects: (1) structural elements were modelled with required load bearing capacity; (2) all alternative envelope (roof, wall, and slab) materials were modelled with similar R-value /U-value/F-Factor to ensure similar energy loss through the structure; (3) the size and location of the windows and skylights were kept similar in both models to ensure similar energy loss through the structure.

Based on previous projects, external studies, and the fact that the building needs to be replicated in other countries, prefabricated, industrialised, and dry solutions have been favoured to minimise onsite impacts and consumptions, reduce material consumption and waste, and cut construction duration onsite. These solutions also facilitate the possibility for future dismantling and recycling.

Structure

In the case of the structure, steel and precast concrete were analysed (Fig. 1). Timber was discarded on food process sanitary issues. Both offer possibilities for clean future dismantling although steel has a greater possibility to be recycled. However, the overall life cycle was considered to see the overall impact and make design choices early on. Some elements such as the skylights and auxiliary structures must be steel. Therefore a steel solution was compared to a hybrid steel and concrete solution (concrete for all major structural elements). The results showed that the hybrid solution had less impact on all impact categories.

From these results, design decisions were made in favour of the hybrid solution. Spain has a strong precast concrete market, which backed up this decision. The location of prefabrication plants could be critical in other locations. Therefore it was included as a parameter in the GIS development for the project. The design of the structural grid and solutions also allowed the use of steel in the case that no precast solutions were available.

Based on these results, the following structural elements were proposed (Table 2).

External walls

The façade of the warehouse area is made from precast concrete elements up to 4.5 m due to functional requirements. Insulated precast concrete panels and aluminium sandwich panels were compared to reduce environmental impacts of the façade.

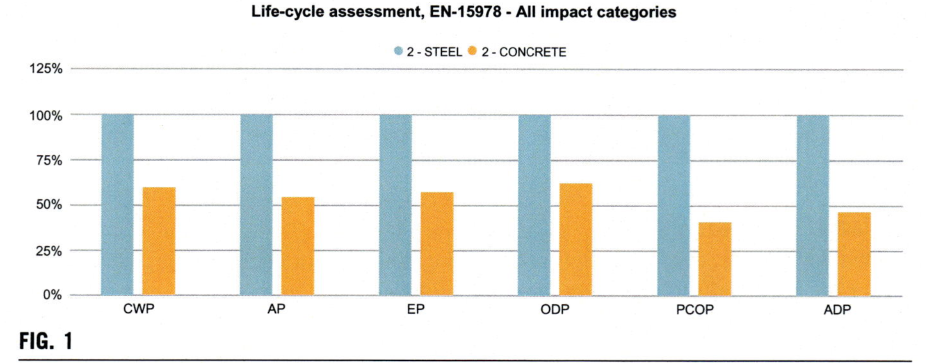

FIG. 1

LCA of steel structure (steel) vs hybrid steel and concrete structure (concrete).

Table 2 Structure definition based on LCA results.

Area	Element	Structure material	Dimension (mm)
Warehouse	Columns	Precast concrete	500 × 500
	Beams	Precast concrete	500 × 1000
	Rafters	Precast concrete	200 × 400
	Skylights	Steel tube	200 × 200
	Auxiliary façade structure	Steel tube	100 × 150
Offices	Columns	Steel tube	200 × 200
	Beams	Steel profile	HEB 240
	Floor	Concrete/Steel	

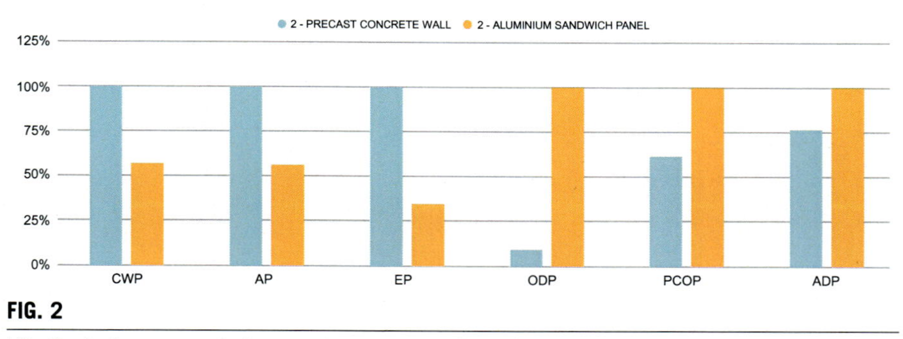

FIG. 2

Life Cycle Assessment of precast concrete wall vs aluminium sandwich panel.

Results (Fig. 2) revealed that aluminium sandwich panels had less impact for GWP, AP, and EP categories, while increased the impact of OPD, POCP, and primary energy categories. Due to the high reduction on GWP category, part of the façade was replaced by sandwich panels.

In the case of office area, there was more flexibility for the design and material selection. Four types of façades were assessed: (1) timber cladding, (2) compact rockwool panels, (3) ceramic elements, and (4) phenolic panels (Fig. 3). To be comparable these solutions had same load bearing capacity, no impact on energy performance, and service life was estimated to be as long as building for every solution.

Timber façade had the lowest impact for all categories except for Ozone Depletion Potential (ODP) category, followed by rockwool panel façade (Fig. 3). In this category, rockwool panel façade and phenolic façades had the lowest impact, while ceramic façade had the highest. Timber and rockwool panel façade solutions were selected for the building design.

Windows

Timber, aluminium, and PVC windows were analysed. Timber windows presented a great reduction compared with the other solutions (Fig. 4). Therefore timber windows were specified for the office block.

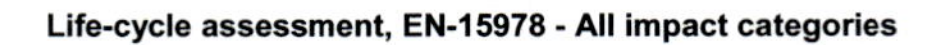

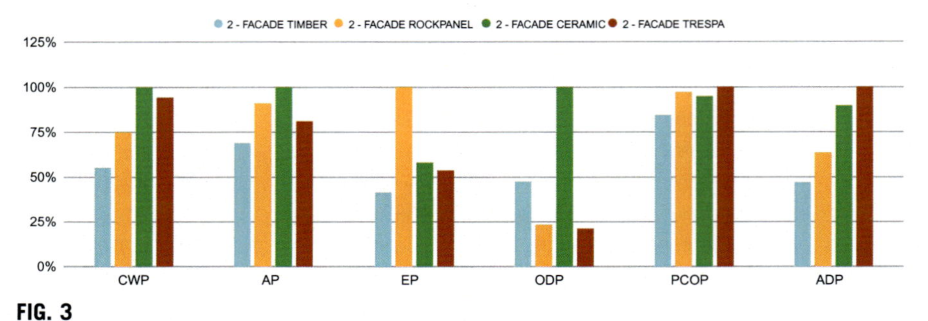

FIG. 3

Life Cycle Assessment of different façade alternatives for office area.

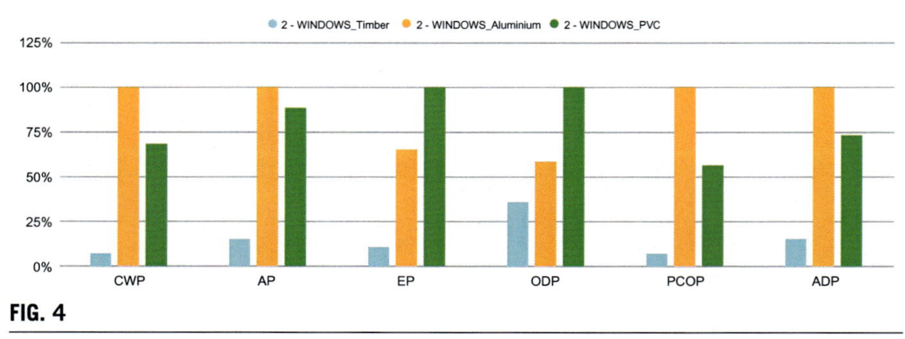

FIG. 4

Life Cycle Assessment of timber, aluminium, and PCV windows.

Thermal insulation

Thermal insulation was an important part of the design in order to reduce energy demands and maintain comfort levels. There are a wide range of insulation solutions which offer different thermal performances and an analysis can be made relating these thermal properties with the embedded energy that each has in its fabrication. It has been demonstrated that as buildings go closer to achieving nearly zero-energy building (NZEB), where energy use consumptions are greatly reduced, the impact of the embedded energy of a building becomes more important.

In terms of insulations, the embedded energy should therefore be related to its energy performance and therefore according to requirements different solutions can be established.

The results of this specific study can be observed in Table 3. In line with the initial energy simulation results, an improved U-value of $0.17\,W/m^2k$ for the façade has been contemplated, which requires a thermal resistance of insulation material of $5.88\ m^2k/W$.

Table 3 Embodied energy of 1 m^2 of insulation material with a thermal resistance of 5.88 m^2k/W.

	Insulation type	Thermal conductivity (W/mk)	Embodied energy (MJ/kg)	Density (kg/m^3)	Thickness needed for a R of 5.88 m^2k/W (mm)	Volume (m^3)	Embodied energy of solution (MJ/m^2)
Mineral	Rockwool rigid DD	0.039	16.8	180	229	0.23	693.74
	Rigid	0.034	16.8	120	200	0.2	403.20
	Semirigid	0.034	16.8	70	200	0.2	235.20
	Glasswool	0.032	49.6	80	188	0.19	746.92
	Cellular glass	0.038	26	110	224	0.22	639.29
	Aerogel	0.018	53	1.1	106	0.11	6.17
Oil based	EPS	0.029	108	20	171	0.17	368.47
	XPS	0.029	95	32	171	0.17	518.59
	PUR	0.02	101	45	118	0.12	534.71
	PIR	0.018	101	45	106	0.11	481.24
Naturals	Cork	0.04	1.38	180	235	0.24	58.45
	Cellulose	0.04	16.64	60	235	0.24	234.92
	Sheep's swool	0.036	20.9	31	212	0.21	137.20
	Wood fibre	0.038	10.8	120	224	0.22	289.69
	Hemp	0.039	20	30	229	0.23	137.65
Others	VIP	0.005	100	100	29	0.03	294.12

Final design

According to preliminary simulations and material alternatives analysed, a design was proposed (Fig. 5). The thermal envelope was optimised according to the results of the simulations (Table 4).

As the warehouse was unconditioned, U-value of the façade of the preliminary design was maintained, improving only the U-value of the office façade. This allowed reducing the amount of materials, which had a direct influence on the environmental impact of the building and the total cost.

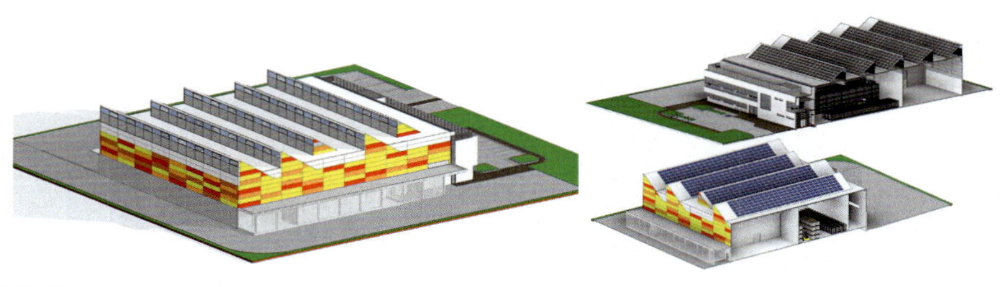

FIG. 5

3D images of the proposed building.

Table 4 Building envelope description of the proposed design.

Building element	Type	Description	*U-values*
Roof	Deck roof (whole building)	TPO + 120 mm PIR + Single profile steel sheet	0.186
	Skylights	Polycarbonate skylights	4.46
Façade	Warehouse 1	Steel sandwich panel with mineral wool (60 mm)	0.35
	Warehouse 2	Precast concrete wall element (160 mm) + PIRM refrigeration panels (40 mm)	0.35
	Office	CLT timber sandwich panel + semi dense mineral wool (100 mm) + Rockpanel ventilated façade	0.12
Windows	Warehouse	Double glazing windows with aluminium frame	1.8
	Office	Triple glazing windows with wooden frame	1
Floor	Whole building	Ready-mix concrete slab with XPS insulation (50 mm)	0.65

LCA results showed that the proposed building reached a reduction between 6% and 11% depending on the impact category analysed, compared with the preliminary design (Fig. 6).

The proposed building was analysed in detail to improve the design of the building, reducing the environmental impact. As illustrated in Fig. 7, material production (A1–A3) had the highest influence on the results, which represented between 68% and 70% for AP, EP, and POCP; 63% for GWP; and 54% for ADP. The contribution of construction, transport, and deconstruction was lower than 10% for all categories. HVAC energy consumption represented between 20% and 25% of the impact for OPD and ADP; between 10% and 20% for GWP, AP, and EP; and 9% for POCP. As energy consumption was optimised by the assessment of diverse passive measures, focus was placed on the impact of building materials.

Concerning building elements (Fig. 8), floor slabs, ceilings, beams, and roofs had the highest impact, followed by external walls and façade, internal walls, other structures, foundations, and vertical structures. As far as material type is concerned (Fig. 9), metals, precast concrete, and coatings had the highest impact. Conversely, wood and bricks had the lowest impact.

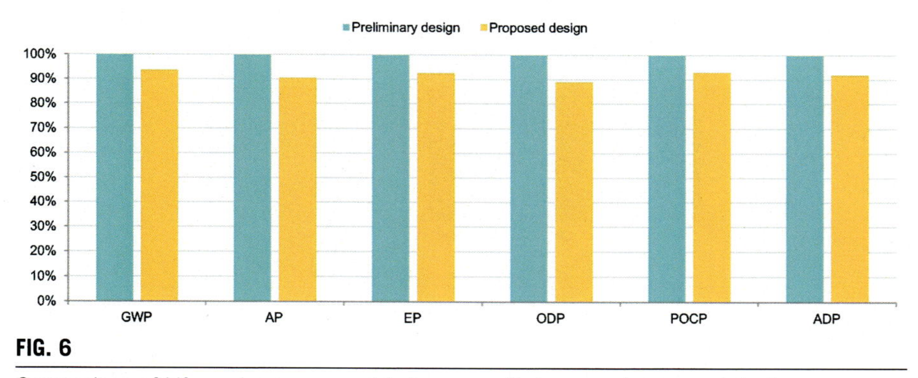

FIG. 6

Comparison of Life cycle assessment between preliminary and proposed design.

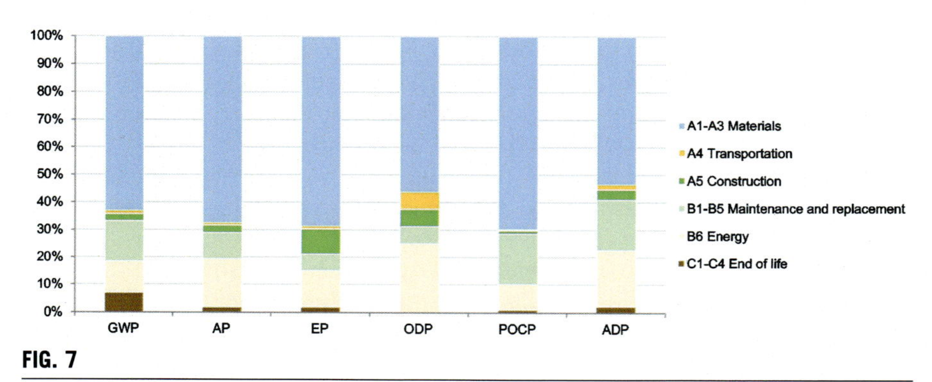

FIG. 7

Contribution of each life cycle phase to the total category impact.

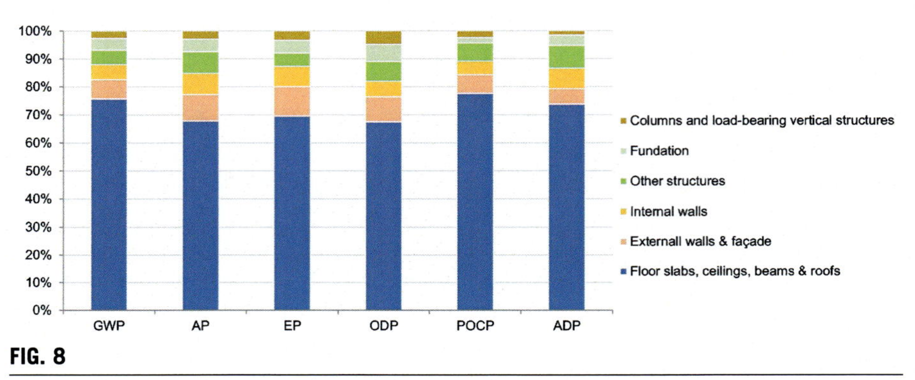

FIG. 8

Contribution of each building element to the total category impact.

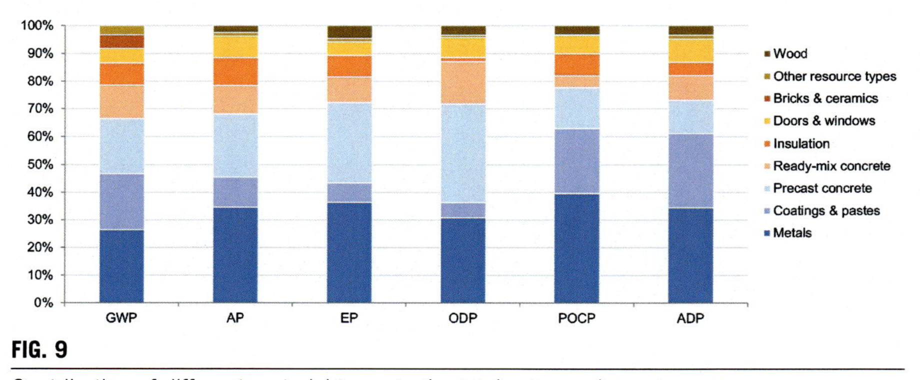

FIG. 9

Contribution of different material types to the total category impact.

Finally, Table 5 presents the most contributing materials to the GWP category. These eight materials together constitute more than 80% of impacts. The rest of materials had a contribution lower than 4%. Steel structure in façade and rooflights, precast concrete structure, and ready-mix concrete were the most contributing materials with an impact of 18.9%, 18.1%, and 14.4%, respectively.

According to these results, the following improvements were implemented: (1) replacement of business-as-usual steel structure used in Spain by 90% recycled steel structure; (2) low impact ready-mix concrete.

Results showed that these measures allowed a reduction in all impact categories except in EP (Fig. 10). However, the comparison of preliminary design and final design showed a reduction of life cycle environmental impact in all categories, achieving almost 30% of reduction in ODP and POCP, 25% in AP and ADP, nearly 20% in GWP and 6% in EP.

Table 5 Most contributing materials (GWP).

Resource	Cradle to gate impacts (A1–A3)
1. Steel structure in façade and rooflights	18.9%
2. Precast concrete structure (beams + columns)	18.1%
3. Ready-mix concrete	14.4%
4. PIR insulation panels	8.3%
5. Sandwich panel with PIR core and double steel siding	6.7%
6. Epoxy-based reactive resin	5.5%
7. Single profiled steel sheet cover/cladding	5%
8. Precast concrete wall elements	4%

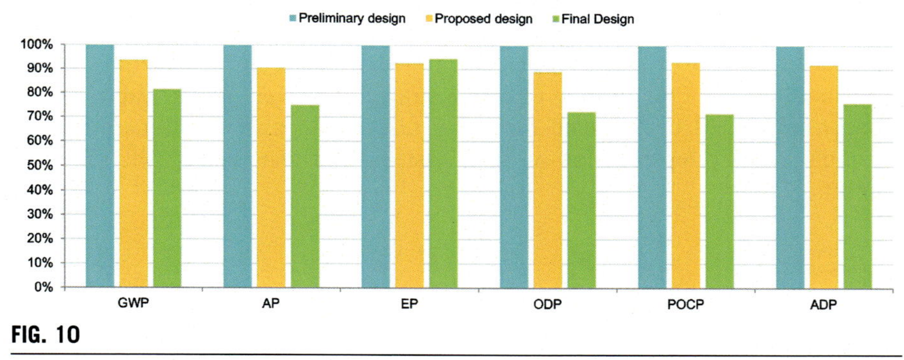

FIG. 10

Comparison of Life cycle assessment between preliminary design, proposed design, and final design.

5 Conclusions

In this chapter, the methodology for the eco-design of a recovery plant of brewery by-products for aquaculture feed ingredients was presented. LCA was used for the decision making from the early stage design of the building, focusing mainly on two strategies: on the one hand, the reduction of heating and cooling energy consumption during use phase, and on the other hand, the selection of construction materials with lower environmental impacts, applicable in an industrial building located in the North-East of Spain. For the building analysed, located in Lleida, energy consumption during use phase was reduced almost 30% from the preliminary design. Moreover, LCA results of the proposed design revealed that heating and cooling energy consumption during use phase represented from 9% to 25% depending on the environmental category, while the impact of construction materials ranged from 54% to 70%. These results highlighted the need to consider the whole life cycle of the building in the design of low environmental buildings. When final design was compared to preliminary design, a reduction of life cycle environmental impact was achieved in all categories with reductions higher than 20% in GWP, AP, ADP, OPC, and POCP and higher than 5% in EP.

Spanish regulation is not yet asking to reduce the environmental impact through the life cycle approach. Only total and non-renewable energy consumption is limited as well as the minimum amount of renewable energy to be produced on site. In the case of industrial buildings, they are excluded from energy efficiency requirements. However, LEED certification scheme has a credit called Building Life Cycle Impact reduction, where LCA of the project's structure and enclosure must demonstrate a minimum of 10% reduction, compared with a baseline building, in at least three of the six impact categories analysed in this chapter, one of which must be global warming potential. The reductions achieved in this research could allow to meet LEED requirements in case of certification would be pursued.

Regarding limitations of the study, the analysis and the design of the building have been limited to passive strategies allowing reducing and cooling energy consumption. The optimisation of active strategies as well as renewable generation onsite should be included in further research to assess the energy saving potential and the repercussion of these systems and materials on the life cycle impacts.

Acknowledgements

Life BREWERY project (LIFE16ENV/ES/000160) aims to demonstrate the viability of reusing brewer by-products as new feed ingredients for aquaculture. It is funded by LIFE European Environment Program, which is the EU's financial instrument supporting environmental, nature conservation, and climate action projects throughout the EU.

References

Aïssani, A., Chateauneuf, A., Fontaine, J.-P., & Audebert, P. (2014). Cost model for optimum thicknesses of insulated walls considering indirect impacts and uncertainties. *Energy and Buildings*, *84*, 21–32. https://doi.org/10.1016/j.enbuild.2014.07.090.

Azari, R. (2019). Life cycle energy consumption of buildings; embodied + operational. In *Sustainable construction technologies* (pp. 123–144). Elsevier. https://doi.org/10.1016/b978-0-12-811749-1.00004-3.

Bahramian, M., & Yetilmezsoy, K. (2020). Life cycle assessment of the building industry: An overview of two decades of research (1995–2018). *Energy and Buildings*, *219*, 109917. https://doi.org/10.1016/j.enbuild.2020.109917.

Chang, D., Lee, C. K. M., & Chen, C. H. (2014). Review of life cycle assessment towards sustainable product development. *Journal of Cleaner Production*, *83*, 48–60. https://doi.org/10.1016/j.jclepro.2014.07.050.

EN 15804:2012+A1:2013. (2014). *Sustainability of construction works—Environmental product declarations—Core rules for the product category of construction products.*

EN 15978:2011. (2012). *Sustainability of construction works—Assessment of environmental performance of buildings—Calculation method.*

European Commission. (2014). *On resource efficiency opportunities in the building sector.* COM(2014) 445 Final (pp. 1–10). Retrieved from http://eur-lex.europa.eu/legal-content/EN/TXT/PDF/?uri=CELEX:52014DC0445&from=EN.

European Commission. (2018). *EU construction & demolition waste management protocol.* Official Journal of the European Union. September.

Hauschild, M. Z., Rosenbaum, R. K., & Olsen, S. I. (2018). *Life cycle assessment: Theory and practice*. Springer International Publishing. https://doi.org/10.1007/978-3-319-56475-3.

Heravi, G., Fathi, M., & Faeghi, S. (2015). Evaluation of sustainability indicators of industrial buildings focused on petrochemical projects. *Journal of Cleaner Production*, *109*, 92–107. https://doi.org/10.1016/j.jclepro.2015.06.133.

Kohler, N., & Moffatt, S. (2003). Life-cycle analysis of the built environment. In *UNEP industry and environment* (pp. 17–21).

Kovacic, I., Waltenbereger, L., & Gourlis, G. (2016). Tool for life cycle analysis of facade-systems for industrial buildings. *Journal of Cleaner Production*, *130*, 260–272. https://doi.org/10.1016/j.jclepro.2015.10.063.

Laurin, L. (2017). Overview of LCA—History, concept, and methodology. In *Encyclopedia of sustainable technologies* (pp. 217–222). https://doi.org/10.1016/B978-0-12-409548-9.10058-2.

Li, C. Z., Lai, X., Xiao, B., Tam, V. W. Y., Guo, S., & Zhao, Y. (2020). A holistic review on life cycle energy of buildings: An analysis from 2009 to 2019. *Renewable and Sustainable Energy Reviews*, *134*(April), 110372. https://doi.org/10.1016/j.rser.2020.110372.

Luz, L. M., De Francisco, A. C., & Piekarski, C. M. (2015). Proposed model for assessing the contribution of the indicators obtained from the analysis of life-cycle inventory to the generation of industry innovation. *Journal of Cleaner Production*, *96*, 339–348. https://doi.org/10.1016/j.jclepro.2014.03.004.

Ministerio de Fomento. (2019). *Código Técnico de la Edificicación, Documento Básico HE Ahorro de Energía, 2019 §*. España.

Nwodo, M. N., & Anumba, C. J. (2019). A review of life cycle assessment of buildings using a systematic approach. In *Building and environment* Elsevier Ltd. https://doi.org/10.1016/j.buildenv.2019.106290.

Pombo, O., Rivela, B., & Neila, J. (2016). The challenge of sustainable building renovation: Assessment of current criteria and future outlook. *Journal of Cleaner Production*, *123*, 88–100. https://doi.org/10.1016/j.jclepro.2015.06.137.

Rai, D., Sodagar, B., Fieldson, R., & Hu, X. (2011). Assessment of CO_2 emissions reduction in a distribution warehouse. *Energy*, *36*(4), 2271–2277. https://doi.org/10.1016/j.energy.2010.05.006.

Russell-Smith, S. V., Lepech, M. D., Fruchter, R., & Meyer, Y. B. (2015). Sustainable target value design: Integrating life cycle assessment and target value design to improve building energy and environmental performance. *Journal of Cleaner Production*, *88*, 43–51. https://doi.org/10.1016/j.jclepro.2014.03.025.

Schiavoni, S., D'Alessandro, F., Bianchi, F., & Asdrubali, F. (2016). Insulation materials for the building sector: A review and comparative analysis. *Renewable and Sustainable Energy Reviews*, *62*, 988–1011. https://doi.org/10.1016/j.rser.2016.05.045.

Shin, S. J., Suh, S. H., Stroud, I., & Yoon, S. C. (2017). Process-oriented life cycle assessment framework for environmentally conscious manufacturing. *Journal of Intelligent Manufacturing*, *28*(6), 1481–1499. https://doi.org/10.1007/s10845-015-1062-4.

Terés-Zubiaga, J., Campos-Celador, A., González-Pino, I., & Escudero-Revilla, C. (2015). Energy and economic assessment of the envelope retrofitting in residential buildings in Northern Spain. *Energy and Buildings*, *86*, 194–202. https://doi.org/10.1016/j.enbuild.2014.10.018.

Tulevech, S. M., Hage, D. J., Jorgensen, S. K., Guensler, C. L., Himmler, R., & Gheewala, S. H. (2018). Life cycle assessment: A multi-scenario case study of a low-energy industrial building in Thailand. *Energy and Buildings*, *168*, 191–200. https://doi.org/10.1016/j.enbuild.2018.03.011.

CHAPTER

LCA and food and personal care products sustainability: Case studies of Thai riceberry rice products

14

Rattanawan Tam Mungkung[a,b] **and Shabbir H. Gheewala**[c,d]

[a]*Centre of Excellence on enVironmental strategy for GREEN business (VGREEN), Faculty of Environment, Kasetsart University, Bangkok, Thailand* [b]*Department of Environmental Technology and Management, Faculty of Environment, Kasetsart University, Bangkok, Thailand* [c]*The Joint Graduate School of Energy and Environment (JGSEE), King Mongkut's University of Technology Thonburi (KMUTT), Bangkok, Thailand* [d]*Centre of Excellence on Energy Technology and Environment (CEE), PERDO, Ministry of Higher Education, Science, Research and Innovation, Bangkok, Thailand*

1 Aims

LCA was applied as an analytical tool to assess the potential environmental impacts of various riceberry rice products. This study demonstrated the application of LCA for identifying improvement options through hotspot analysis and quantifying the potential reduction in life cycle environmental impacts. Moreover, the integration of environmental and economic performances expressed as eco-efficiency was used to assess the sustainability level for recommendations to industry as well as policy-makers towards a value-added economy.

2 State of the art

LCA has been widely applied for assessing the environmental performance of agri-food products. LCA results have been particularly useful in deriving strategies for minimising the potential life cycle environmental impacts of food production and food packaging. Possible approaches to improve the environmental profiles of food products could be selecting local and low-impact materials, considering energy-efficient production processes, re-designing eco-packaging and green logistics, changing cooking methods, reducing food wastes, and optimising the end-of-life

Assessing Progress towards Sustainability. https://doi.org/10.1016/B978-0-323-85851-9.00002-X

treatment (Castellani, Sala, & Benini, 2017; Del Borghi, Strazza, Magrassi, Taramasso, & Gallo, 2018; Kulatunga, Karunatilake, Weerasinghe, & Ihalawatta, 2015; Molina-Besch, Wikström, & Williams, 2019; Zufia & Arana, 2008). With the focus on rice products, LCA was adopted to assess the environmental impacts of milled rice produced in Thailand. It was reported that the global warming potential of rice production per kg was 2.9269E+03 gCO_2-e, along with 3.1869 gSO_2-e of acidification and 12.896 gNO_3-e of eutrophication and 95% of the global warming inputs to the system are associated with the cultivation process, 2% with the harvesting process and 2% with the seeding and milling processes (Kasmaprapruet, Paengjuntuek, Saikhwan, & Phungrassami, 2009). LCA has also been applied to calculate the carbon footprint value of rice products in Italy and Thailand (Blengini & Busto, 2009; Mungkung et al., 2010, 2012; Office of Agricultural Economics, Thailand Ministry of Agriculture and Cooperatives, 2014). To move towards sustainable agriculture and food production systems, it has been suggested to combine LCA with economic analysis for policy recommendations. Eco-efficiency, the integration of environmental indicators from LCA results with the economic analysis, has often been used to identify the most sustainable management system in agricultural products. The sustainability of kiwi fruit production in organic and integrated orchards in New Zealand was assessed by applying eco-efficiency. The eco-efficiency value of organic orchards was equal to 25.5 ($\pm$8.9) NZD net profit per $kgCO_2e$, which was significantly higher than that of integrated orchards with 7.2 ($\pm$2.7) NZD net profit per $kgCO_2e$. The results indicated that the rate of fertiliser application was linked to higher GHG emissions of integrated orchards as compared to the organic orchard systems (Müller, Holmes, Deurer, & Clothier, 2015). Ho, Hoang, Wilson, and Nguyen (2018) analysed the eco-efficiency of coffee production in Vietnam, comparing between conventional and sustainability-certified coffee; the result showed that sustainability-certified coffee farms are more eco-efficient than conventional farms due to 50% reduction of environmental impacts in each year whilst maintaining the constant value added of products. Eco-efficiency was also used to evaluate the environmental and economic performances of paddy rice production in Thailand; it was reported that the eco-efficiency value of rain-fed farming system was equal to 4.04 baht/$kgCO_2e$, whereas the irrigated rice farming system in the dry season and the irrigated rice farming system in the wet season equal 2.16 and 2.46 baht/$kgCO_2e$, respectively (Thanawong, Perret, & Basset-Mens, 2014). Considering the utility of LCA and eco-efficiency assessment in rice production and agricultural products, it is of interest to apply these tools for studying riceberry rice processed products to support the policy decision on whether or not the cultivation areas should be expanded and which value-added products should be produced to support the national policy on value-added economy system.

3 Novelty

This study demonstrated the application of LCA as an analytical tool to assess the environmental performance of various riceberry rice products. Various environmental improvement options were proposed by using the life cycle management concept in order to minimise the environmental impacts. Eco-efficiency was also applied to evaluate the combined environmental and economic performance of riceberry rice products. The integration of LCA and eco-efficiency results has not yet been applied for policy recommendations to support the national policy to move towards a value-added economy.

4 Case study description

Rice is a staple food for more than half of the people in the world. There are more than 120,000 varieties of rice at the global scale. Rice is often characterised by length and shape (long grain, medium grain, or short grain rice), texture (sticky or parboiled rice), colour (polished, brown, forbidden, and wild rice), and aroma (basmati or jasmine rice). Riceberry rice is a newly registered rice variety since 2002 which was developed by the National Research Committee of Thailand and the Rice Science Center, Kasetsart University through a natural crossbreeding process between Jasmine rice called Khao Dawk Mali 105 or *Oryza sativa* L. cv. KDML 105 (white colour) and Hom Nil rice or *Oryza sativa* (reddish purple colour) (Fig. 1). Hom Nil rice was developed by cross-breeding of Black glutinous rice and KDML 105. It is a highly nutritious rice with approximately 10%–12.5% protein and 12%–13% amylose, and contains polyphenolic compounds which have antioxidant properties. As it is not sensitive to light, Hom Nil rice can be planted all year long. More importantly, Hom Nil rice has a pleasant smell when cooked. KDML 105 rice is the most famous Jasmine rice ('Hom Mali' rice, the local name). The specialty of KDML 105 is that

FIG. 1

Some rice varieties in Thailand.

Source: Asian Inspirations. (2021). Riceberry, Thailand' super grain. *https:/asianinspirations.com.au/food-knowledge/riceberry-thailands-super-grain/. (Accessed 18 May 2021); Ministry of Science and Technology. (2017).* Hom Nil rice product. *https:/most.go.th/main/th/knowledge/portfolio/technology-integration/165-practical-rad/7483-aromatic-black-rice-hom-nin-rice. (Accessed 18 May 2021); Farm Channel Co., Ltd. (2020).* Hom Mali 105 rice. *https:/farmchannelthailand.com/main/rice/hom-mali-rice/hom-mali-105/. (Accessed 18 May 2021).*

the cooked rice is soft textured with a natural fragrant smell. It has the advantages of being mildly tolerant to drought, salinity, and acid sulphate soil whereas the rice is vulnerable to dry leaf edge disease, orange leaf disease, brown spot leaf disease, burn disease, and leaf curl disease. Breeding for a high nutrition rice, the beneficial characteristics of KDML 105 rice and Hom Nil rice were selected and their disadvantages were minimised using biotechnology.

Riceberry rice is a long grain purple rice with a tolerance to diseases, clean husk, and high yield. It offers a lower level of amylose (15.3% amylose) as compared to KDML 105 rice (21.3% amylose), whilst containing higher total polyphenolics and anthocyanins. It can provide a lower swelling power and higher gelatinisation temperatures of starch; the higher gelatinisation temperatures are positively correlated with the swelling power which is responsible for the unique characteristics of finished products (RGD and RSC, 2015). In addition, riceberry rice contains a high level of beneficial resistant starch and lower levels of slowly digestible starch (SDS) with medium predicted glycaemic index (pGI) that highlights the health-promoting characteristics (RGD and RSC, 2015). As a consequence, it has especially been encouraged for developing innovative value-added products with an economic system capable of adjusting to climate change and low carbon society.

4.1 Goal and scope

LCA and eco-efficiency were applied as the analytical tools to evaluate the environmental and economic performance of various riceberry rice products, covering both food and non-food products (Fig. 2). The selected riceberry rice products included riceberry cracker, riceberry porridge, riceberry snack bar, riceberry soap, and riceberry shampoo (Fig. 2). The unit of analysis for the LCA studies was based on the sold unit of each riceberry product, viz., 50 g of riceberry snack in linear low density polyethylene (LLDPE) retort pouch, 35 g of riceberry porridge in laminated polypropylene and polyethylene terephthalate plastics with aluminium bag, 50 g of riceberry snack bar in oriented polypropylene plastic bag, 90 g of riceberry soap in kraft paper, and 250 mL of riceberry shampoo in PET bottle.

The scope of analysis was a full-scale LCA study from cradle to grave (Fig. 3). The details of each step are explained as follows; the raw materials used for each product (i.e. agricultural ingredients, packaging) were sourced from local suppliers. The production process of each product is described in Fig. 3. All riceberry products were distributed to the distribution centre or retail store where the consumer can purchase and order via online shipping. In the use and consumption stage, the assumption for food products consumption is to follow the product instructions, whilst the use phase of other products was assumed to follow the method specified in the Product Category Rules for Household Cleaning Products (PCR) (TGO, 2017). For the final waste disposal, it was assumed that packaging waste after use and consumption was transported to landfill.

(a) Riceberry snack bar

(b) Riceberry porridge

(c) Riceberry snack

(d) Riceberry soap

(e) Riceberry shampoo

FIG. 2

Studied riceberry rice products.

4.2 Methodology

The scope of LCA study was from cradle to grave, covering raw material production, transport of raw materials to the factory, processing, packaging, distribution, consumption, and post-consumption waste management. Along the whole life cycle of riceberry rice products, the inputs and outputs associated with each life cycle stage were identified. The inventory data were obtained from both primary and secondary sources. Primary inventory data of inputs (land, seed, fertiliser, petrol, energy, and water use) and outputs (paddy rice and waste) for riceberry farming stage were gathered through farmer interviews. The direct GHG emissions from riceberry field were obtained via in-field measurements. The secondary inventory data (i.e. primary energy production, fertiliser production, petrol production, etc.) were taken from the ecoinvent v.3.0 database (Ecoinvent center, 2018; Mungkung, Dangsiri, & Gheewala, 2021). The inventory data of riceberry rice production processing, i.e. the foreground data, were collected based on the annual production in 2019 (Chancharoonpong, Mungkung, & Gheewala, 2021; Mungkung et al., 2021). The

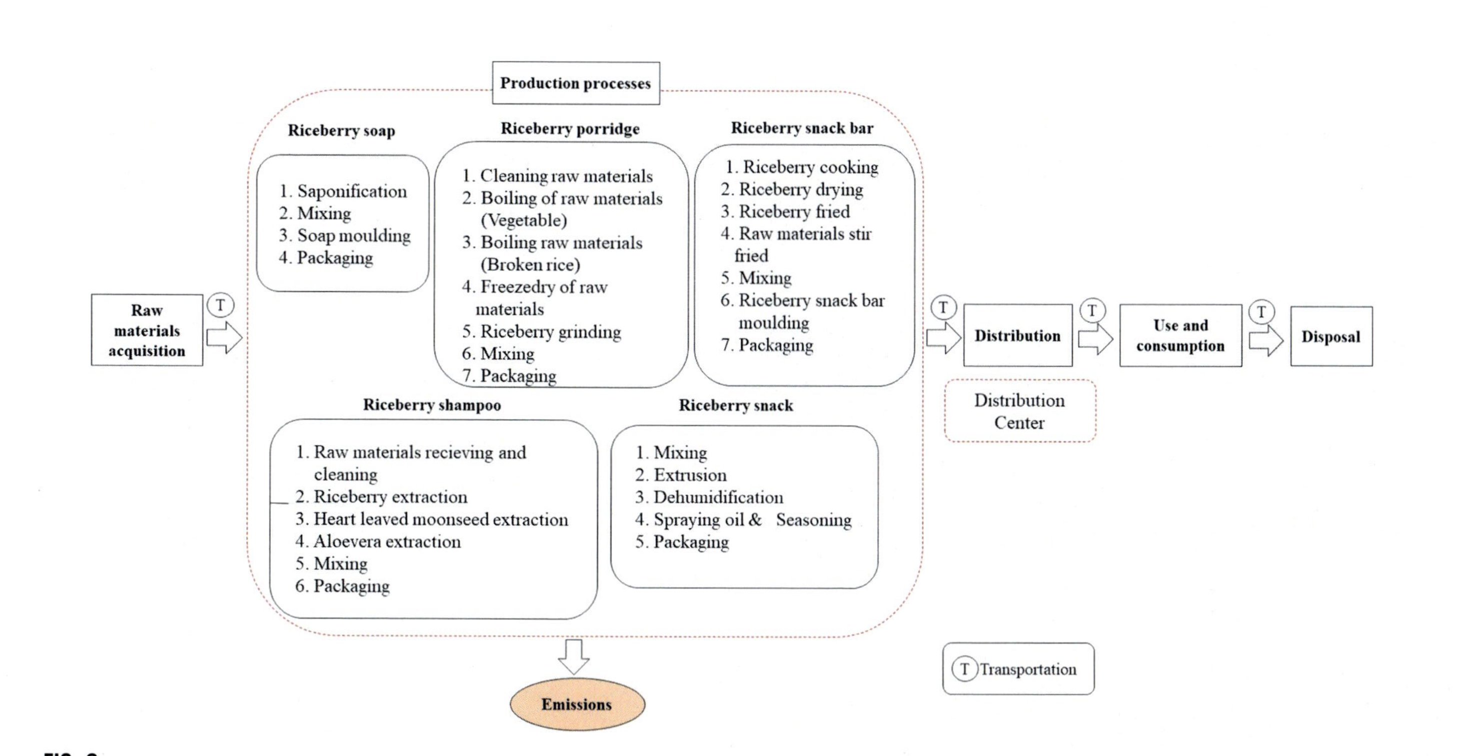

FIG. 3

System boundary of LCA studies for riceberry rice products.

background data, such as the production of fertiliser, the production of diesel, the production of fertiliser, were primarily sourced from the Thai national life cycle inventory databases and from relevant articles published in peer-reviewed journals (Thailand Greenhouse Gas Management Organization (Public Organization), 2020, EC, The European Commission, 2016a, 2016b; Ecoinvent center, 2018). This was supplemented by international databases where necessary, for example, all emissions related to transportation and distribution were computed with road transportation by passenger car, small size with a full load, and empty load for a round trip (Ecoinvent center, 2018). The waste remaining after post-consumption was packaging waste which was assumed to be landfilled without energy recovery. Global warming (GW), Human health cancer effect (HHC), Acidification (AC), Fresh water aquatic eco-toxicity (FWAE), Water scarcity (WS), and Energy use (EU) were selected as the impact categories of interest and would facilitate the comparison with previous LCA studies due to their widespread use in LCA of agri-food products (Weidema, Thrane, Christensen, Schmidt, & Løkke, 2008). Many past studies have also considered these impacts in rice production, especially global warming (Kamalakkannan, Kulatunga, & Kassel, 2020; Mungkung, Gheewala, Silalertruksa, & Dangsiri, 2019; Yodkhum, Gheewala, & Sampattagul, 2017). The life cycle impact assessment method used to calculate the potential life cycle environmental impacts and carbon footprint was the Product Environmental Footprint (PEF) (EC, The European Commission, 2016a, 2016b). Mass allocation was applied at the rice milling stage.

To evaluate the environmental and economic performances of riceberry rice products, the eco-efficiency calculation was then performed according to the method described step by step in ISO 14045 (ISO, 2012). The product system value was based on the selling price in the market. However, it should be noted that the prices of products varied depending on the market positioning of different products. From the perspective of the producer, nevertheless, the selling price can still serve as proxy of the product value. The quality of products, the features, and the aesthetics of packaging materials are highlighted as the key factors affecting the willingness to pay from consumers. In particular, the function of packaging plays a very important factor for consumers willing to pay a high price (Atagan & Yükçü, 2013). The eco-efficiency indicator was calculated by dividing product system value (selling price) with the selected (six) environmental impact indicators from LCA results, as shown in Eq. (1).

$$\text{Eco-efficiency} = \text{Selling price}/\text{Environmental impact indicator} \quad (1)$$

This equation provides a ratio of the selling price per unit of environmental impact. The higher the numerator value means higher economic performance leading to higher eco-efficiency. Similarly, a lower denominator value means lower environmental impacts also leading to higher eco-efficiency value. The eco-efficiency values were adopted as the indicator to identify the sustainability level of riceberry rice products.

4.3 Results and discussion

4.3.1 LCA results

Fig. 4 shows the LCA results indicating the environmental performance of studied riceberry rice products. In general, the raw materials acquisition stage was the key hotspot causing significant contribution in global warming as well as other impacts (Fig. 5). For riceberry snack bar, the highest contribution to the impacts on freshwater aquatic eco-toxicity (7.48 CTUe, 99%) was from the riceberry rice cultivation stage and the highest contribution to the impact on human health cancer effect (3.11E−09 CTUh, 92%) was linked to the production of peanut butter. This corresponded with the results from a previous study by Wongkaew and Jogloy (2011) who found that the carcinogens in peanuts were caused by Aflatoxin B1. For riceberry porridge, the major contribution of climate change was associated with the production of riceberry broken rice from riceberry farming and the production of packaging bag (1.37E−01 $kgCO_2e$, 41%), followed by the electricity consumption in freeze dry machine during production processing activities (1.28E−01 $kgCO_2e$, 38%). A similar trend was found in the impact on acidification, the major cause of impact was linked to the raw materials acquisition stage, especially from broken riceberry rice (1.21E−03 mol H^+e, 62%) followed by the production process from electricity used in freeze dry machine during production processing activities (4.85E−04 mol H^+e, 25%). The high energy used was, in addition, attached to the packaging production from raw materials acquisition stage (1.51E+00 MJ, 40%) and from electricity used in freeze dry machine during production processing activities (1.38E+00 MJ, 37%).

In a similar manner, the key hotspots of riceberry snack were the raw material acquisition stage, varying between 83% and 99% for all impact categories (Fig. 5). The riceberry rice cultivation and packaging production significantly contributed to the impact on climate change (1.84E−01 $kgCO_2e$, 71%) followed by the electricity consumption for production processing activities (4.02E−02 $kgCO_2e$, 15%). For acidification, similar results were observed; the key hotspots were attached to riceberry rice cultivation (6.70E−04 mol H^+e, 50%), the production of packaging (1.87E−04 mol H^+e, 13%), and the production of corrugated box (1.70E−04 mol H^+e, 12%). The highest contributor to the impact on freshwater aquatic eco-toxicity was the raw material acquisition stage (3.84E−01 CTUe, 96%), especially from the production of broken riceberry rice (1.31E−01 CTUe, 33%) followed by the production of low-fat soy flour (6.26E−02 CTUe, 16%), and the production of full-fat soy flour (5.09E−02 CTUe, 13%). In terms of water scarcity, the key issues were associated with the production of broken riceberry rice (1.05 E+01 m^3 depriv., 79%) and the production of corrugated box (2.21E+00 m^3 depriv., 17%). The contribution of energy use in raw material acquisition stage was also the highest (3.09E−01 MJ, 94%), followed by the energy use for transportation (1.82E−02 MJ, 6%).

For non-food products, it was revealed that the production of rice bran oil (especially the rice farming stage) resulted in the highest contribution to the impacts on

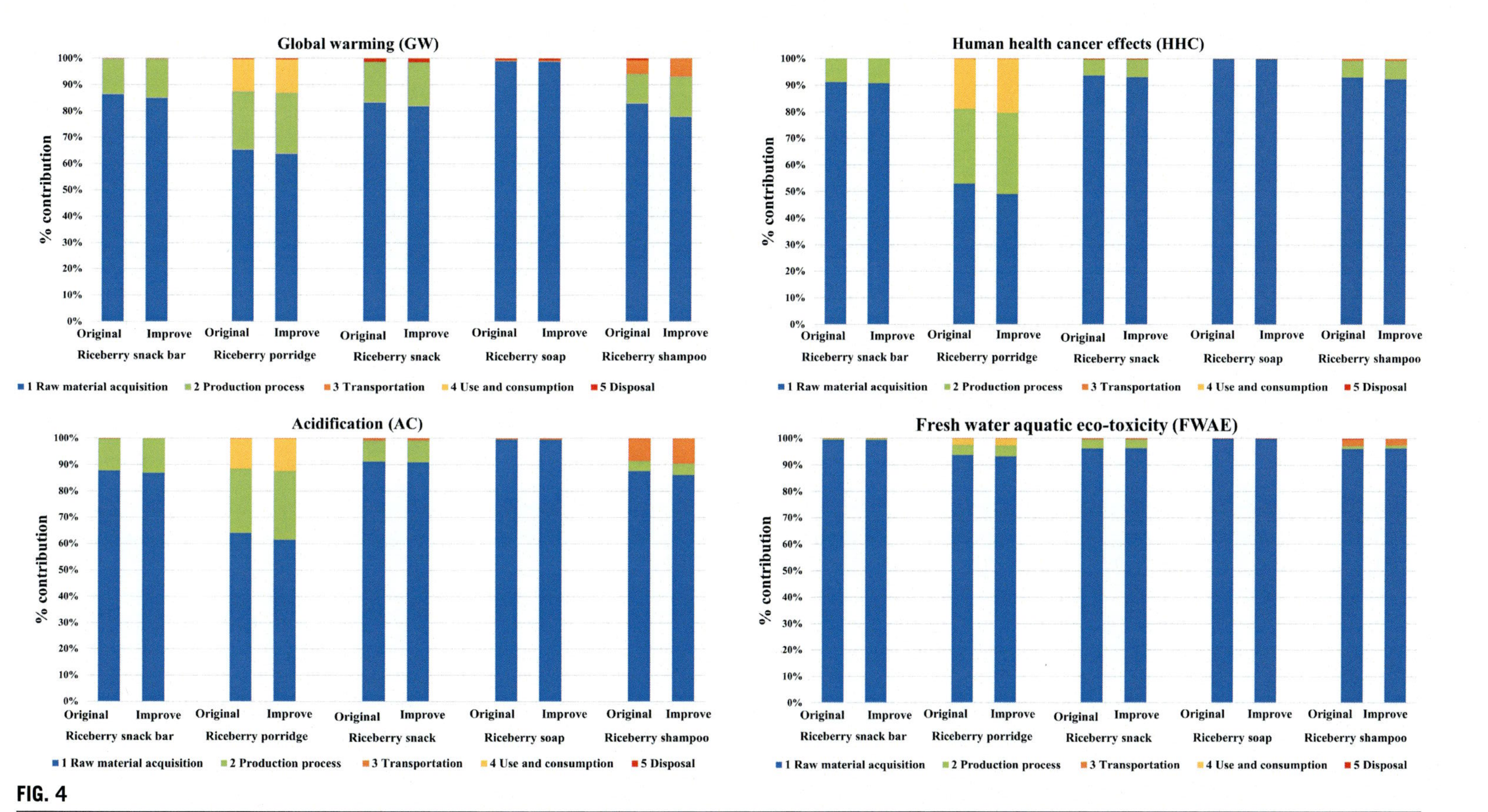

FIG. 4

LCA results of riceberry products (before and after improvement): snack bar, porridge, snack, soap, and shampoo.

continued

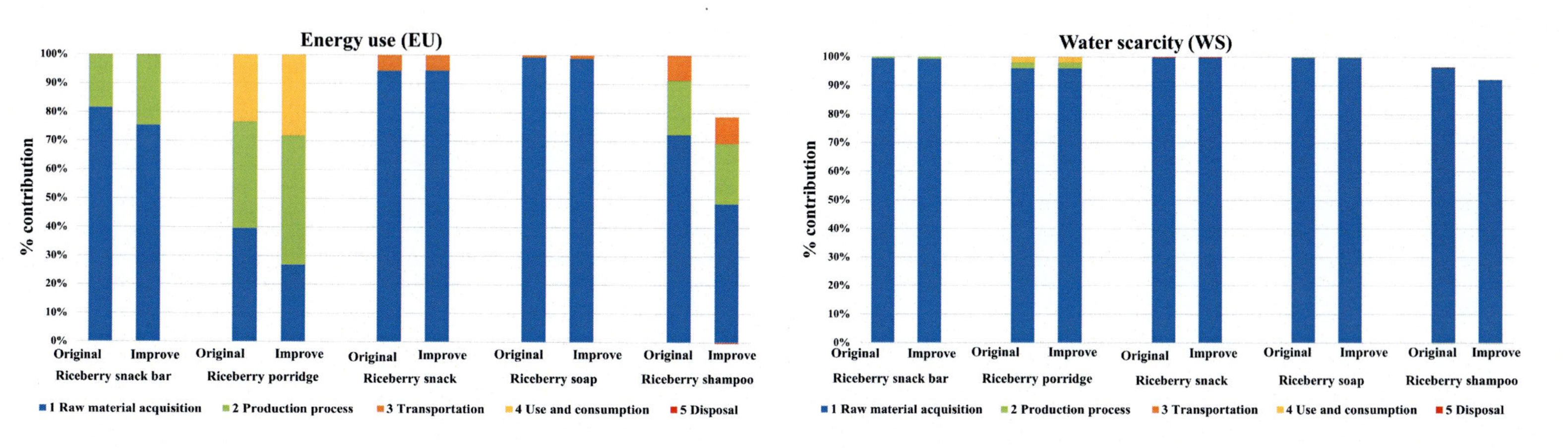
Energy use (EU)
% contribution
100%
90%
80%
70%
60%
50%
40%
30%
20%
10%
0%
Original
Improve
Riceberry snack bar
Riceberry porridge
Riceberry snack
Riceberry soap
Riceberry shampoo
1 Raw material acquisition
2 Production process
3 Transportation
4 Use and consumption
5 Disposal
Water scarcity (WS)
% contribution
100%
90%
80%
70%
60%
50%
40%
30%
20%
10%
0%
Original
Improve
Riceberry snack bar
Riceberry porridge
Riceberry snack
Riceberry soap
Riceberry shampoo
1 Raw material acquisition
2 Production process
3 Transportation
4 Use and consumption
5 Disposal

FIG. 4—cont'd

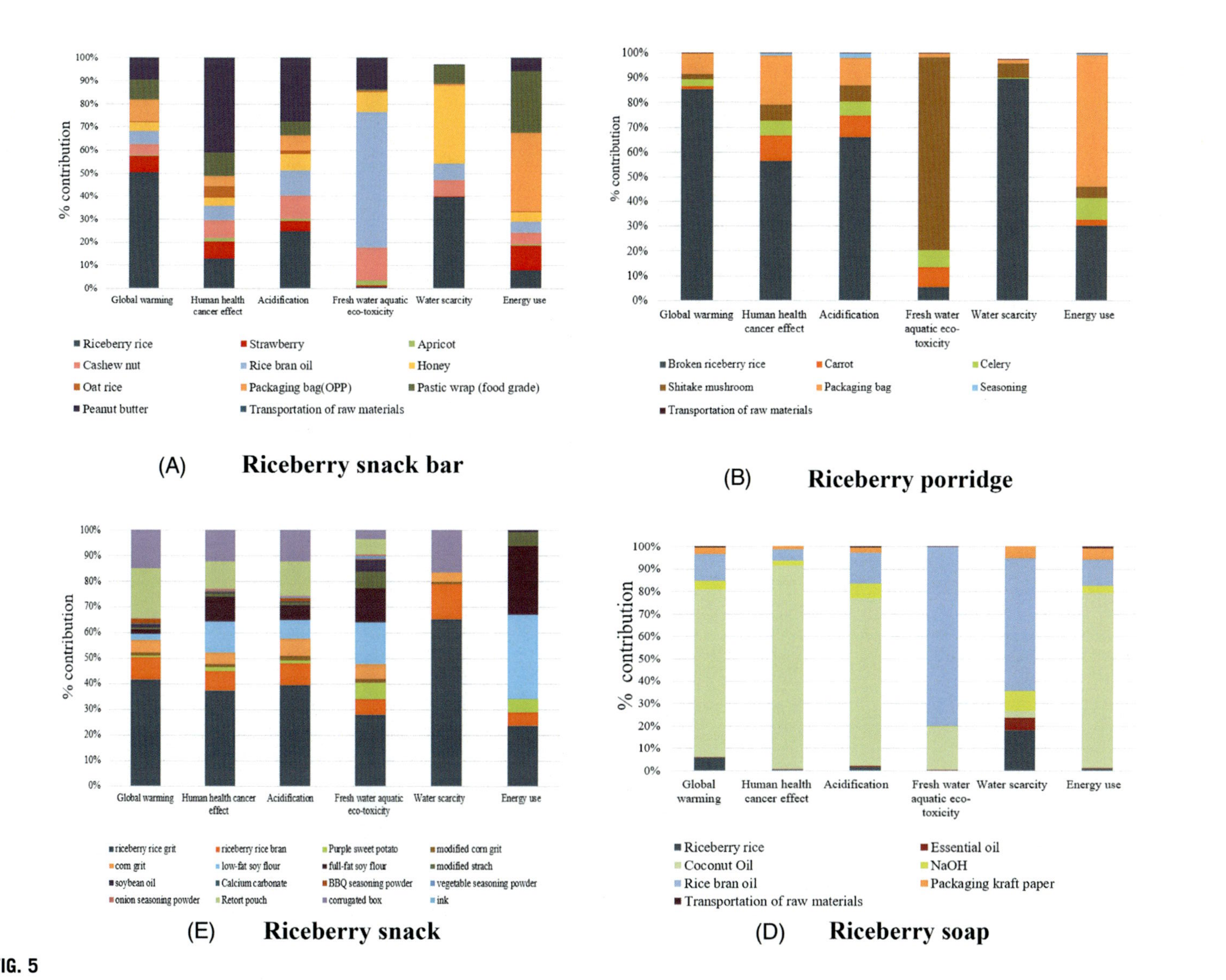

FIG. 5

Contribution analysis of life cycle environmental impacts associated with raw materials for (A) riceberry snack bar, (B) riceberry porridge, (C) riceberry snack, (D) riceberry soap, and

continued

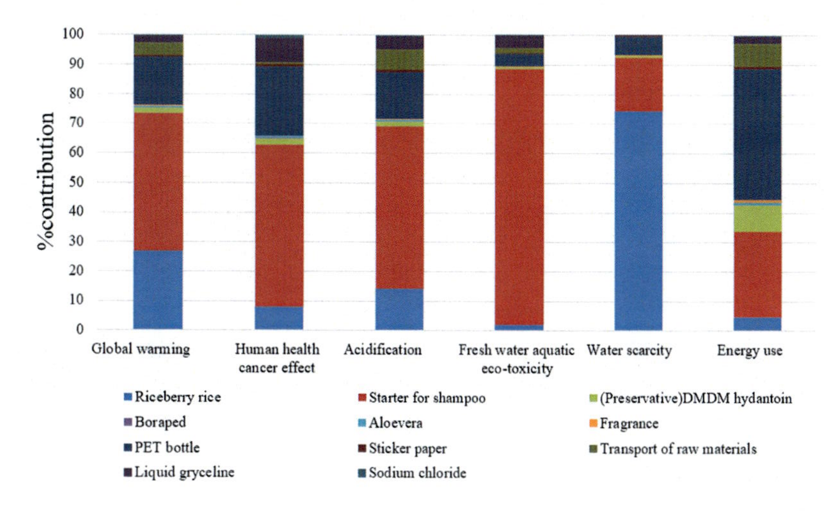

FIG. 5—cont'd (E) riceberry shampoo.

freshwater aquatic eco-toxicity (8.13 CTUe, 99%) and water scarcity (3.90E−01 m^3 depriv., 99%) whilst the production of coconut oil (especially the crude coconut oil production) resulted in the highest contribution to the impacts on global warming (4.22E−01 $kgCO_2e$, 97%), human health cancer effect (1.35E−08 CTUh, 99%), acidification (3.70E−03 mol H^+e, 99%), and energy use (3.34E+00 MJ, 96%) for riceberry soap. For riceberry shampoo, the production of sodium lauryl sulphate (SLS) which is a surfactant used to remove oily stains and residues has the highest contribution in human health cancer effect (3.82E−09 CTUh, 55%), freshwater aquatic eco-toxicity (1.35E+00 CTUe, 87%), acidification (1.69E−03 mol H^+e, 55%), and global warming impact (3.03E−01, 47%) whilst rice cultivation resulted in the highest contribution to the impact on water scarcity (5.17E−01, 74%) (Fig. 5). The production of polyethylene terephthalate (PET) bottle had the highest contribution in energy use (2.23E+00 MJ, 44%). In summary, the results of impacts contribution from riceberry rice particular to food products are in line with the case study of breakfast cereals and snacks by Jeswani, Burkinshaw, and Azapagic (2015). The findings indicate that the major hotspots of each product are at raw materials acquisition which should be targeted for the environmental improvements.

The life cycle impacts of the riceberry rice products before and after improvement of the environmental performance are presented in Table 1. In general, the possible options for improving the environmental performances had a potential to reduce the environmental impacts on GW for 4%–28%, HHC for 4%–43%, Ac for 1%–28%, FWAE for 3%–24%, WS for 1%–84%, and EU for 17%–48%. As commonly observed for food or other agricultural products, the raw material stage was a significant contributor to the impacts for all the products. Rice cultivation contributed the most to the life cycle impacts of riceberry snack; hence, this stage was targeted for considering improvement options. One of the first options considered was to shift from conventional to organic rice farming. Mechanisation was another option which matched well with the operational issue of labour shortage which is quite common; mechanised seeding and the use of drones for spraying pesticides/herbicides were the two strategies considered. Other improvement strategies included re-design of the packaging from 9 to 4 colours whilst maintaining attractiveness, changing the material of the corrugated box to reduce thickness whilst maintaining functional strength, and recycling the corrugated boxes to enhance resource efficiency. A low-fat and low-sodium formula was also used, but this was more for improving the health aspects. Incorporating all the improvement options mentioned before resulted in a significant improvement in the environmental profile of the riceberry snack, the most outstanding being 84% for water scarcity.

Replacement of ingredients contributing significantly to the water scarcity impact was considered for the riceberry snack bar. Two ingredients, peanuts and apricot, could potentially be replaced by cashew nuts and pineapple, respectively, as the alternatives provided similar nutrients, texture, and flavour as the original ones. Pineapple is a locally cultivated fruit, very prominent in Thailand.

Table 1 The eco-efficiency indicators before and after the environmental performance improvement of riceberry rice products.

Eco-efficiency	Riceberry snack bar			Riceberry porridge			Riceberry snack			Riceberry soap			Riceberry shampoo		
	original	Improved		Original	Improved		Original	Improved		Original	Improved		Original	Improved	
Selling price (USD?)	**1.1**	**1.1**	**% Increase**	**0.6**	**0.6**	**% Increase**	**1.05**	**1.05**	**% Increase**	**2.7**	**2.7**	**% Increase**	**4.5**	**4.5**	**% Increase**
GW (USD/kg CO_2 eq)	4.02E+00	4.42E+00	10	1.02E+00	1.06E+00	5	4.04E+00	4.51E+00	12	6.22E+00	7.46E+00	20	5.80E+00	8.04E+00	39
HHC (USD/CTUh)	3.46E+08	3.60E+08	4	1.73E+08	1.88E+08	8	2.64E+08	3.77E+08	43	2.04E+08	3.55E+08	74	6.57E+08	8.17E+08	24
AC (USD/mol H^+ eq)	7.95E+02	8.50E+02	7	3.07E+02	3.29E+02	7	1.53E−03	1.52E−03	1	7.38E+02	1.02E+03	39	1.34E+03	1.60E+03	20
FWAE (USD/CTUe)	4.26E−01	5.08E−01	19	5.36E−01	5.78E−01	8	2.64E+00	3.46E+00	31	3.79E−01	4.14E−01	9	2.89E+00	2.97E+00	3
WS (USD/m^3 depriv.)	1.31E+00	1.60E+00	22	5.62E−01	5.66E−01	1	7.91E−02	5.09E−01	543	6.80E+00	6.84E+00	1	6.72E+00	7.49E+00	11
EU (USD/MJ)	4.32E−01	5.77E−01	34	1.63E−01	1.97E−01	21	3.22E+00	4.22E+00	31	8.41E−01	1.31E+00	55	6.61E−01	1.27E+00	92

Replacement could potentially reduce the water scarcity impact by 35%. The additional advantage was the reduction in cost as the replacement alternatives were cheaper. Thus the selling price of the product could be reconsidered after the implementation of the modified recipe and re-designing of the packaging.

Similar to the riceberry snack bar, replacing some of the ingredients of riceberry porridge also yields environmental benefits. For example, celery could be replaced by adding more shiitake mushroom; this served an additional advantage of providing a better taste option as it has been observed that many people do not like the smell of celery. In addition to that, replacing the existing packaging comprising normal laminated polypropylene and polyethylene terephthalate plastics in aluminium bag with a paper packaging cup could potentially reduce the EU impact by 18%. The other impacts also reduced in a range of 1%–8%.

A new recipe could be considered for riceberry soap where the currently used coconut oil was replaced by palm kernel oil. Both these oils have quite similar properties in terms of high lauric acid content which makes them suitable for making soap (Bhattacharya, 2019). In fact, the palm kernel oil also has other similar properties such as bubbliness, hardness, cleansing condition, and a lower saponification value which is useful for soaps, reducing the requirement of NaOH, a harmful chemical. The replacement of coconut oil by palm kernel oil could potentially reduce human health cancer effect by 41% and also most of the other impacts in the range of 7%–33%. Only the water scarcity impact would increase by 10%.

In case of the riceberry shampoo, rice cultivation was targeted as the improvement option. Replacing the conventional farming system (continual flooding) with the alternate wet and dry (AWD) system could help reducing the water consumption (Mungkung et al., 2019). Another option was to consider the use of recycled PET for the packaging instead of the virgin material. The combined effect of these two improvement options could potentially reduce energy use by 48% and global warming by 28%. Other impact categories had quite significant benefits too, including human health cancer by 20%, acidification by 17%, water scarcity by 10%, and freshwater aquatic eco-toxicity by 3%.

In conclusion, as can be anticipated, the LCA results clearly showed that the raw material acquisition stage was the significant hotspot contributing more than 50% to all the impact categories considered for all the products studied. Consequently, the improvement options were focused on this stage including alterations to the product recipe and re-design of the packaging. It must be noted that when considering alternative materials for the product recipes, care had to be to ensure that the replacement ingredients could maintain the nutrients, texture, flavour, etc. similar to the original so that the taste and quality would not be compromised at the expense of improving the environmental profile.

4.4 Eco-efficiency results

Table 1 presents the eco-efficiency indicators before and after the environmental performance improvement of riceberry snack bar, riceberry porridge, riceberry snack, riceberry soap, and riceberry shampoo. The comparative eco-efficiency values of riceberry snack bar, riceberry porridge, riceberry snack, riceberry soap, and riceberry shampoo indicated that the non-food products have the higher values in all impact indicators. The riceberry soap and riceberry shampoo products are sold both in Thailand and overseas countries and the consumers are more willing to pay for riceberry soap and riceberry shampoo for health and wellness. As a consequence, the producers have set a higher selling price because they have foreseen that health-conscious consumers are willing to buy good quality products at relatively higher prices (Kongdechakul, 2018). Moreover, it was discovered that the profit margin was almost 90% by considering the production cost based on raw materials cost and labour. Amongst food products, riceberry snack bar was found to have the highest eco-efficiency value, followed by riceberry snack, whilst the eco-efficiency value of riceberry porridge was lowest. Riceberry snack bar was sold at the local coffee shops and its selling price was set lower as compared to other kinds of snacks to draw the attention of potential buyers who are travellers and teenagers. Interestingly, the calculation of production cost and selling price showed a gross profit margin of almost 86%. For riceberry snack, the use of riceberry rice grit and bran instead of whole grain rice could reduce production cost. For riceberry porridge, the selling price was set rather low because the product was sold to local retailers. It was revealed that the gross profit margin of the rice porridge was only about 30%.

After the environmental performance improvements, non-food products showed the higher eco-efficiency improvement especially in EU, GW, HHC, and AC whereas food products showed the higher eco-efficiency improvement particularly to WS, EU, HHC, and GW indicators. Riceberry soap showed that the reduction of life cycle environmental impacts resulted in the higher eco-efficiency improvement ranging from 3% to 92% whilst the riceberry soap showed increasing eco-efficiency values from 1% to 74%. The improved eco-efficiency values for riceberry snack bar was 4%–34%, riceberry porridge was 1%–21%, and riceberry snack was 1%–543%. The highest improvement of eco-efficiency value in GW was found in riceberry shampoo and riceberry soap. This was mainly associated with the reduction of methane emission in riceberry field during farming stage by shifting from traditional farming practice to AWD farming practice. The highest improvement of eco-efficiency value in HHC and AC attached to riceberry soap and riceberry snack resulted from lower impact on environment from the modification of recipe by using palm kernel oil instead of coconut oil in riceberry soap and using low-fat and low-sodium formula seasoning in riceberry snack. The highest improvement of eco-efficiency value in FWAE and WS attached to riceberry snack resulted from reduction in environmental impacts due to converting from non-organic to organic rice farming systems, using a seeding machine instead of manual seeding, and applying a drone for spraying pesticides/herbicides to replace the ground-based spraying. EU riceberry shampoo had the highest improvement in eco-efficiency value because of the reduction of energy consumption for recycled PET preform production as compared to the virgin PET preform production.

5 Conclusions

This study has demonstrated the application of LCA and eco-efficiency to various kinds of Thai riceberry rice products, aiming to support the decision on the development of healthy, innovative, low-carbon and high value-added product to move towards the value-based economy driven by innovation for sustainable development. Selected riceberry food and non-food products covered in this study were riceberry snack, riceberry porridge, riceberry snack bar, riceberry soap, and riceberry shampoo. The environmental performance of the studied riceberry rice products was evaluated by using LCA, and the integration of environmental and economic performance in terms of eco-efficiency. Based on the LCA results of the original products, the hotspots generating significant environmental impacts were identified and improvement options derived for better environmental performance. It was revealed that the raw material acquisition stage was the major contributor to climate change as well as other environmental impacts. The improvement options were then proposed by converting from conventional to organic riceberry rice farming along with a new design of recipe and packaging. The eco-efficiency values of climate change were generally increased by that improvement. The increased eco-efficiency values of non-food products (e.g. riceberry soap and riceberry shampoo) far exceeded that of food products (e.g. riceberry snack bar, riceberry porridge, and riceberry snack) due mainly to the high selling price of riceberry soap and shampoo compared to the other products. In this connection, it was recommended that non-food riceberry rice products should be more encouraged because of their higher value added and eco-efficiency. In addition, the manufacturers could design for a new business model after the improvement of environmental performances. However, this study is limited to the existing riceberry products. In order to achieve the national policy goals for value-based economy, inventing new products by applying high innovation technology and creative life cycle design is vital for the Thai riceberry industry. Proposed strategies for reforming the Thai riceberry industry towards Thailand 4.0 could be enhanced via the following actions: (1) uplifting the riceberry farming practices to be more sustainable according to the international framework such as SRP (Sustainable Rice Platform) by considering environmental, social, and economic aspects; (2) improving the sustainability performance of the whole riceberry supply chain according to the international framework such as SAFA (Sustainability Assessment of Food and Agriculture) by considering environmental, social, economic, and governance dimensions; (3) developing innovative and high value-added riceberry products with advanced technology; (4) promoting the learning of life cycle thinking both in environmental and economic aspects and transferring the life cycle design concept to the research and development teams of riceberry processors; and (5) communicating the life cycle environmental and economic performances of riceberry rice and riceberry processed products to the national policy makers leading to the development of supportive policy and strategies. These will support and contribute to the Thailand 4.0 policy aiming at security, prosperity, and sustainability especially in the field of sustainable agriculture.

Acknowledgements

The case studies were supported by the (National Science and Technology Development Agency) Research Chair Grant programme under the project "Network for Research and Innovation for Trade and Production of Sustainable Food and Bioenergy" (Grant No. FDA-CO-2559-3268-TH) and the research project entitled "Sustainability assessment of riceberry farming & processing for trade and niche market access" (Grant No. FDA-CO-2561-7910-TH). R. M. is the principal researcher involved in the research development, the data collection, the validation of collected data, the data analysis, and conclusions of the results, including leading the writing of manuscript. S.H.G. helped with the data analysis, data validation, and reviewing the manuscript. All authors have read and agreed to the published version of the manuscript.

References

Atagan, G., & Yükçü, S. (2013). Effect of packing cost on the sales price and contribution margin/Ambalaj Maliyetinin Satis Fiyati ve Katki Payina Etkisi. *Ege Akademik Bakis*, *13*(1), 1.

Bhattacharya, A. (2019). *Effect of high temperature on crop productivity and metabolism of macro molecules*. Academic Press.

Blengini, G. A., & Busto, M. (2009). The life cycle of rice: LCA of alternative agri-food chain management systems in Vercelli (Italy). *Journal of Environmental Management*, *90*(3), 1512–1522.

Castellani, V., Sala, S., & Benini, L. (2017). Hotspots analysis and critical interpretation of food life cycle assessment studies for selecting eco-innovation options and for policy support. *Journal of Cleaner Production*, *140*, 556–568.

Chancharoonpong, P., Mungkung, R., & Gheewala, S. H. (2021). Life cycle assessment and eco-efficiency of high value-added riceberry rice products to support Thailand 4.0 policy decisions. *Journal of Cleaner Production*, *292*, 126061.

Del Borghi, A., Strazza, C., Magrassi, F., Taramasso, A. C., & Gallo, M. (2018). Life cycle assessment for eco-design of product–package systems in the food industry—The case of legumes. *Sustainable Production and Consumption*, *13*, 24–36.

EC, The European Commission. (2016a). *Product environmental footprint pilot guidance. Guidance for the implementation of the EU product environmental footprint (PEF) during the environmental footprint (EF) pilot phase*. http:/ec.europa.eu/environment/eussd/smgp/pdf/Guidance_products.pdf. Accessed 10 May 2019.

EC, The European Commission. (2016b). *Product environmental footprint pilot guidance. Guidance for the implementation of the EU product environmental footprint (PEF) during the environmental footprint (EF) pilot phase*. http:/ec.europa.eu/environment/eussd/smgp/pdf/Guidance_products.pdf. Accessed 10 May 2020.

Ecoinvent center. (2018). *Ecoinvent V3.5 database*. Dübendorf, Switzerland: Swiss Centre for Life Cycle Inventories.

Ho, T. Q., Hoang, V. N., Wilson, C., & Nguyen, T. T. (2018). Eco-efficiency analysis of sustainability-certified coffee production in Vietnam. *Journal of Cleaner Production*, *183*, 251–260.

ISO. (2012). *Environmental managemente—Eco-efficiency assessment of product systemse—Principles, requirements and guidelines (ISO 14045:2012)* (2nd ed.). Brussels: International Organization for Standardization.

Jeswani, H. K., Burkinshaw, R., & Azapagic, A. (2015). Environmental sustainability issues in the food–energy–water nexus: Breakfast cereals and snacks. *Sustainable Production and Consumption*, *2*, 17–28.

Kamalakkannan, S., Kulatunga, A. K., & Kassel, N. C. (2020). Environmental and social sustainability of the tea industry in the wake of global market challenges: A case study in Sri Lanka. *International Journal of Sustainable Manufacturing*, *4*(2–4), 379–395.

Kasmaprapruet, S., Paengjuntuek, W., Saikhwan, P., & Phungrassami, H. (2009). Life cycle assessment of milled rice production: Case study in Thailand. *European Journal of Scientific Research*, *30*(2), 195–203.

Kongdechakul, M. C. (2018). *Factors that influence Thai health-conscious consumers in their purchasing decision towards healthy snack foods*. Doctoral dissertation Thammasat University.

Kulatunga, A. K., Karunatilake, N., Weerasinghe, N., & Ihalawatta, R. K. (2015). Sustainable manufacturing based decision support model for product design and development process. *Procedia CIRP*, *26*, 87–92.

Molina-Besch, K., Wikström, F., & Williams, H. (2019). The environmental impact of packaging in food supply chains—Does life cycle assessment of food provide the full picture? *The International Journal of Life Cycle Assessment*, *24*(1), 37–50.

Müller, K., Holmes, A., Deurer, M., & Clothier, B. E. (2015). Eco-efficiency as a sustainability measure for kiwifruit production in New Zealand. *Journal of Cleaner Production*, *106*, 333–342.

Mungkung, R., Dangsiri, S., & Gheewala, S. H. (2021). Development of a low-carbon, healthy and innovative value-added riceberry rice product through life cycle design. *Clean Technologies and Environmental Policy*, *23*, 1–11.

Mungkung, R., Gheewala, S. H., Kanyarushoki, C., Hospido, A., van der Werf, H., Poovarodom, N., et al. (2012). Product carbon footprinting in Thailand: A step towards sustainable consumption and production? *Environmental Development*, *3*, 100–108.

Mungkung, R., Gheewala, S. H., Silalertruksa, T., & Dangsiri, S. (2019). Water footprint inventory database of Thai rice farming for water policy decisions and water scarcity footprint label. *The International Journal of Life Cycle Assessment*, *24*(12), 2128–2139.

Mungkung, R., Gheewala, S. H., Towprayoon, S., Poovarodom, N., Tanaparisutti, P., & Chupairote, T. (2010, November). Carbon footprinting of Thai rice products: Practical issues for a harmonised methodology. In *9th international conference on ecobalance. Tokyo, Japan* (pp. 9–12).

Office of Agricultural Economics, & Thailand Ministry of Agriculture and Cooperatives. (2014). *The study of carbon footprint of rice products*. Thailand.

RGD & RSC, Rice Science Center, & Rice Gene Discovery. (2015). *Riceberry*. https:/bit.ly/2YJBkkH (Accessed 10 May 2019) (in Thai).

Thailand Greenhouse gas management Organization (Public organization). (2017). *Product category rules household cleaning products (PCR) in Thai*. Approval http://thaicarbonlabel.tgo.or.th/tools/files.php?mod=Y25Wc1pYTT0&type=WDBaSlRFVlQ&files=TWpFeQ. Accessed 7 December 2020.

Thailand Greenhouse Gas Management Organization (Public Organization). (2020). *Company and product approval*. http://thaicarbonlabel.tgo.or.th/products_approval/products_approval.pnc. Accessed 7 December 2020.

Thanawong, K., Perret, S. R., & Basset-Mens, C. (2014). Eco-efficiency of paddy rice production in Northeastern Thailand: A comparison of rain-fed and irrigated cropping systems. *Journal of Cleaner Production*, *73*, 204–217.

Weidema, B. P., Thrane, M., Christensen, P., Schmidt, J., & Løkke, S. (2008). Carbon footprint: A catalyst for life cycle assessment. *Journal of Industrial Ecology*, *12*(1), 3–6.

Wongkaew, S., & Jogloy, S. (2011). Aflatoxins in peanuts: A proposed solution. *Khon Kaen Agriculture Journal*, *39*(3), 1–11.

Yodkhum, S., Gheewala, S. H., & Sampattagul, S. (2017). Life cycle GHG evaluation of organic rice production in northern Thailand. *Journal of Environmental Management*, *196*, 217–223. https://doi.org/10.1016/j.jenvman.2017.03.004.

Zufia, J., & Arana, L. (2008). Life cycle assessment to eco-design food products: Industrial cooked dish case study. *Journal of Cleaner Production*, *16*(17), 1915–1921.

CHAPTER

Environmental and economic sustainability of cocoa production in west sub-Saharan Africa

15

Marta Tuninetti[a], Francesco Laio[a], and Tiziano Distefano[b]

[a]*Department of Environment, Land and Infrastructure Engineering (DIATI), Politecnico di Torino, Torino, Italy* [b]*DEM (Department of Economics and Management), University of Pisa, Pisa, Italy*

1 Aims

We aim at assessing the water use related to the cultivation of cocoa plants at a high spatial resolution (5 arc minute), from 2010 to 2017, in Western sub-Saharan Africa. We have analysed the water footprint at both unitary (crop's water use efficiency) and aggregated (annual production's water) scale.

Further, we have assessed the economic aspects related to the production and distribution of cocoa beans and derivatives. Then, we have analysed the international market composition, in terms of top importer and exporter countries, provided insights regarding the price dynamic, and showed the utilisation of cocoa beans along the supply chain.

2 State of the art

In 2015 the United Nations defined 17 Sustainable Development Goals (SDGs) based upon three pillars: eradicating poverty, protecting the planet, and promoting shared prosperity (A New Era in Global Health, 2017). Each SDG is relevant when considered separately from the others, but it is emerging very clearly the need to consider their reciprocal interdependence, in terms of synergies and trade-offs, to ensure a joint attainment of the 2030 Agenda (Philippidis et al., 2020; Ronzon & Sanjuán, 2020). The need of applying multi-dimensional and interdisciplinary approaches was already underlined far before the definition of the SDGs, as exemplified by the Water–Food–Energy Nexus introduced during the Bonn Nexus Conference in

Assessing Progress towards Sustainability. https://doi.org/10.1016/B978-0-323-85851-9.00005-5

2011. In this framework, a great attention has been devoted to the agricultural sector where different sustainability dimensions coexist, including for example water resources, land management, poverty eradication, and starvation. Indeed, the agricultural sector is responsible for 70% of the global freshwater withdrawal (FAO, 2011), it requires five billion hectares (38%) of the global land surface (see www.fao.org/sustainability/news/detail/en/c/1274219), it is responsible of biodiversity loss (OECD & Analysis, 2020), and it is linked with rural poverty and income inequality mostly in the Global South (Barbier, 2010; Distefano, 2020).

These reasons, together with the need to define a path towards sustainability, have led an increasing number of scholars to provide a quantification of the physical volume of natural resources (mostly, land and water) required along the supply chain of agricultural products, also by including the critical role of international trade (Duarte, Pinilla, & Serrano, 2014; Kastner, Erb, & Nonhebel, 2011). One of the most successful applications is represented by the introduction of the concepts of virtual water trade and water footprint (Mekonnen & Hoekstra, 2011). The findings suggest a progressive diversion of water resources ('embedded' in the agricultural products) from the developing (e.g. China, India, and Brazil) to the developed countries (Distefano, Riccaboni, & Marin, 2018). For instance, Dutch consumption often relies on external water resources throughout the world, with significant impacts in water-scarce regions in many cases (van Oel, Mekonnen, & Hoekstra, 2009).

Most literature estimated the amount of water associated with the international trade of agricultural products, with a particular attention to staple crops (D'Odorico et al., 2019; Konar et al., 2011). Although the market for *luxury* foods (such as tea, coffee, and cocoa) is expanding, and even though water scarcity already affects most of the areas where these plants are cultivated, there is a paucity of estimations of their water footprint (Carr & Lockwood, 2011). Among the few attempts, (Tamea et al., 2013) showed that almost half of the water embedded in exported goods from Italy was due to luxury foods in 2010; (Carr & Lockwood, 2011) estimated that about a third of global virtual water trade is generated by luxury foods; and Naranjo-Merino, Ortíz-Rodriguez, and Villamizar (2018) showed that the green and blue water associated to the cocoa production in Colombia was about 13,189 m^3/ton and 5687 m^3/ton, respectively, thus much higher than the water behind grains and staple crops production. However, it remains poorly understood what are the spatio-temporal patterns of water use for cocoa production in relatively poorer countries (i.e. the core of global cocoa production) and how it is distributed through the international markets.

2.1 Novelty

This case study advances the state-of-the-art literature by providing the most up-to-date Water Footprint assessment of cocoa production in Western African countries at a very high spatial resolution. The evaluation also integrates analysis of the evolution of the patterns of cocoa plantations to monitor intensification and extensification production practices. As a second effort, we analyse the international market composition of cocoa beans and derivatives in top importer and exporter countries, thus providing insights into the price dynamics and supply chain. This double-facet framework of environmental and economic sustainability sets grounds for new investigations on the social sustainability of cocoa farmers' job.

3 Case study description

Since the Spanish conquerors drank the cocoa and chocolate drinks (whose etymology derives from 'xocoatl' and 'cacahuatl') after conquering the Aztec empire in Mexico in the XVI century, the diffusion and production of this plant became global. The increasing demand of this food is also due to the nutritional and health properties, other than its peculiar taste (Afoakwa, Paterson, Fowler, & Vieira, 2008). Nowadays, the tropical regions in Africa, Asia, and Latin America are the primary suppliers, with around 50 million people that depend on cocoa farming for their income, contributing annual cocoa production worldwide at some 5 million tons (Kroeger, Koenig, Thomson, & Streck, 2017). However, this sector is affected by inequality in the distribution of the value added: growers and workers in West Africa are receiving less than 6% of the final value of a chocolate bar (www.fairtrade.org.uk/farmers-and-workers/cocoa/) and farmers are paid less than a third of the final price of cocoa beans (Cappelle, 2009). This unbalance is a heritage of the colonisation process that also promoted the replacement of forest areas with the intensive cultivation of highly profitable plants (i.e. rent crops) to satisfy the consumption in richer countries. West African post-colonial land development policies compromised the sustainable management of forest ecosystems with an estimated 80% of the new agricultural lands (including plantations) taken from the forestland (Foley et al., 2011). This entails that little forest remains in the traditional growing areas and also the possibility of replanting former trees is hampered by the high costs of conversion, the loss in profitability (Ongolo, Kouamé Kouassi, Chérif, & Giessen, 2018), and by the lack of opportunity to create alternative jobs.

3.1 Economic assessment: On production and distribution

The globalisation process has led to the expansion of international markets, and the trends of trade of luxury foods followed the same fate. In this section, we focus on cocoa beans (i.e. a raw food) and the first derivatives in terms of industrial processing, viz. cocoa paste, by analysing the productional and distributional dynamics, from 2000 to 2017, in the food international market, by looking at both a global perspective and on the main exporter (Ivory Coast) and importer (the Netherlands). We collect data on the production and international trade (import and export) in terms of physical quantities (tons) and monetary values (USD) at the country scale. We also provide some insights on the chain of processing and distribution by taking advantage of the new data, from 2014 to 2017, on the so-called *supply utilisation accounts* (SUAs) that allow to unravel the specific utilisation of cocoa beans, namely: domestic consumption, industrial processing, and exports. The aim is to collect information on how a specific product is used along the supply chain, to wit whether it is sold to other firms (transformational purpose) or to the final users (consumption purpose). Although the present study only draws a picture of the first passages along the international cocoa supply chain, from raw to processed items, it poses the first breaks to future studies on cascade effects, shock propagation, etc. Then, the SUAs could be linked with the economic input–output analysis—see Miller and Blair (2009) for an exhaustive introduction—that assesses the monetary flows—as reported in the national accounts—across all sectors (e.g. agriculture, manufacturing, energy, etc.) that composes the economic system.

All the data are collected from the publicly available online database provided by the Food and Agricultural Organisation of the United Nations (FAOSTAT).

3.1.1 Cocoa's production and export

Fig. 1 reports the dynamic of the production and export of cocoa beans and paste, with a focus on some key countries. Top-left panel shows that the export of cocoa paste (orange lines, right y-axis) has more than doubled, passing from 0.3 Mton in 2000 to almost 0.7 Mton in 2017. Instead, the production and export of cocoa beans (blue lines, left y-axis) has increased by about 57.8% and 54.8%, respectively. Note that, on average, the exports of cocoa beans represent almost 70% of the total production, entailing that most of the consumption happens outside the place of origins of this luxury product. This is not by chance: the top producers come from West sub-Saharan Africa (Côte d'Ivoire, Ghana, and Nigeria), Asia (Indonesia), and Latin America (Brazil), all of which have experienced a long period of colonisation that determined their agricultural reforms towards the conversion of lands (and forest areas) to monocultures of crops that are appealing in richer (mostly Europeans) countries. This is also reflected by the unbalance in the distribution of the value added along the supply chain, which significantly questions the economic sustainability of cocoa production.

Fig. 1 (top-right panel) shows the annual shares of cocoa beans production among the main countries.

The specific geographical and climatic conditions required to produce cocoa limit the possibility to extend the production outside the tropical zones. Indeed, almost 80% of the global production is concentrated in four countries, with the leadership of Ivory Coast that covers almost 40%. Interestingly, in the case of exports (bottom-left panel), Indonesia disappeared from the list of top countries in the last years, passing from a share of about 30% in 2000 to less than 0.02% in 2017. This is due to the contraction of the production from 2010 to 2017 (−30%, from 0.85 Mton to less than 0.6 Mton) and, probably, to an increase of domestic consumption and/or industrial processing. The composition of the key exporters drastically changes when looking at cocoa paste. Although Ivory Coast still represents the main player (39% in 2017), the rest of top countries are Europeans (where cocoa beans are wholly imported) with a predominant presence of the Netherlands (23.2% in 2017, with a pick of more than 30% in 2010). This reflects the discrepancy between the level

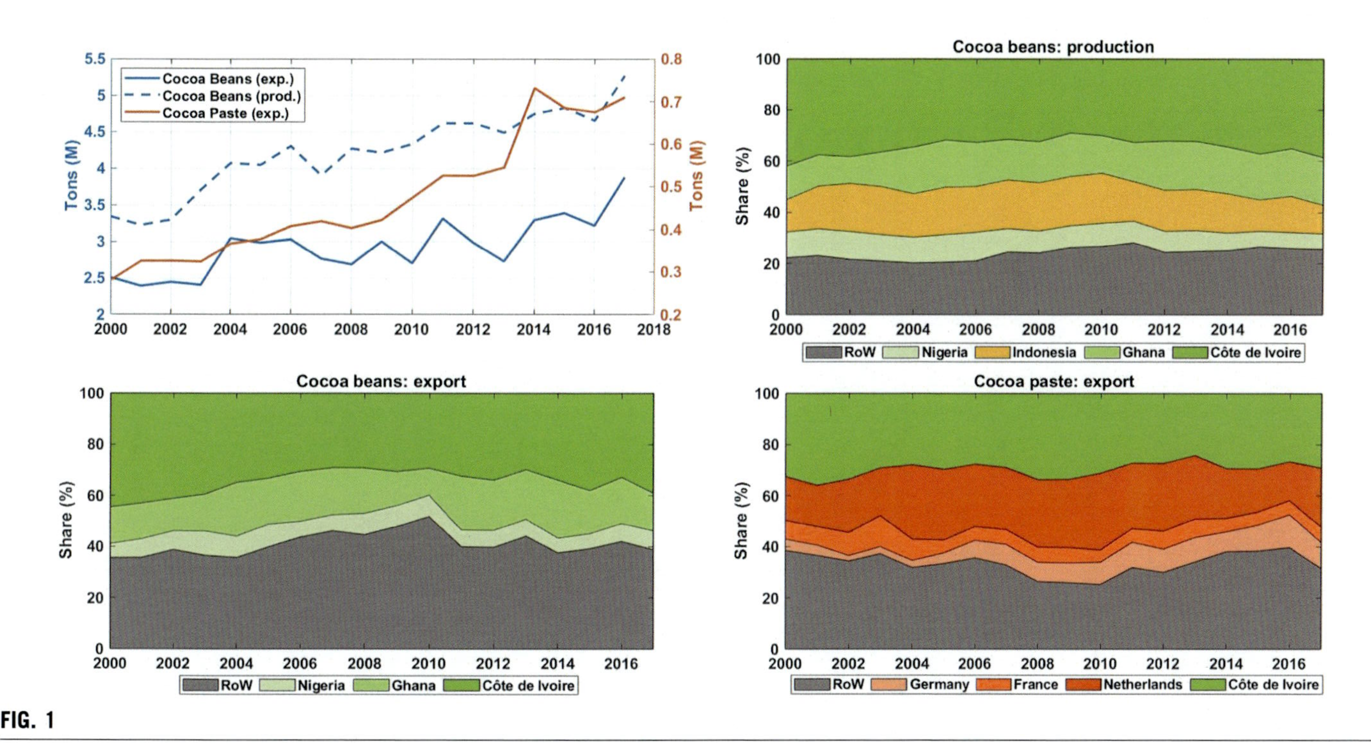

FIG. 1

Production and export of cocoa beans and paste (2000–2017). *Top-left panel* shows the dynamic of global production of beans (*left* y-axis) and paste (*right* y-axis) compared with global production of cocoa beans *(dashed blue line)*. The remaining three panels plot the change in the percentages of production of cocoa beans *(top-right)*, and export of cocoa beans *(bottom-left)* and paste *(bottom-right)* among the top countries. The acronym 'RoW' stands for the 'Rest of the World'.

of industrial development of African and European countries: the former has economies highly dependent on rural production and they lack the infrastructures needed to implement the industrial process (and possibly escape from poverty). Hence, the combination of colonial legacy, low domestic consumption of luxury foods, and the lack of industries oblige these countries to export their precious cocoa beans towards richer countries that have the possibility to implement the industrial processing.

3.1.2 Insights from a supply chain analysis

Fig. 2 shows the volumes and the ranking of the main importers of cocoa beans (top-left) and paste (bottom-left) from Ivory Coast. In both cases, the Netherlands represents the key partner by importing, in 2017, around 23% and 37% of cocoa beans and paste, respectively. However, we observe contrasting trends: in the case of cocoa beans, the Netherlands reduced its import compared to year 2000 (−6%, from 0.36 to 0.34 Mton), while it almost tripled the imports of cocoa paste in absolute terms (from 0.027 to 0.076 Mton). The other main importers of cocoa beans are the US (20%) and other European countries (Germany and Belgium), while France is ranking second (18.2%) in case of cocoa paste. The right side of Fig. 2 shows the average exporting deflated price of cocoa from Ivory Coast (USD per ton). In both cases, we observe a cyclical dynamic of the average weighted exporting prices with ascending trends. Over the period, the average price of cocoa beans and paste increased by about 51% and 41%, respectively. Note that, the distribution of prices applied to the main importers is very close to the average price (dashed black lines): this might be an insight of the presence of asymmetric power in the international trade markets (Distefano, Chiarotti, Laio, & Ridolfi, 2019). Indeed, after the liberalisation of the African cocoa markets, the power of multinational exporter/processors increased substantially, determining the emergence of an oligopsony—where few big buyers influence the price—structure (Wilcox, 2004).

Fig. 3 reports the supply utilisation accounts of Ivory Coast (top) and the Netherlands (bottom) for cocoa beans (left) and paste (right). In each year, the total available crops (i.e. production + import + stocks) can be used for different purposes. Most of the production of cocoa beans from Ivory Coast is exported, with less than 20% being processed domestically. In case of cocoa's derivatives (i.e. cocoa paste) the distribution splits almost in half between exports and further processing. On the contrary, the Netherlands processes most of the available (i.e. imported) cocoa beans and paste, while only a third is exported. This fact represents a further confirmation of the different economic and industrial structures of these countries, and in general of the poorer South (where resources and raw materials are abundant) and richer North (where industrialisation is more advanced). This issue must be considered when dealing with water footprint analysis. On the one hand, crops behave differently during the production and commercialisation process, with the price of staple crops maintaining a significant imprint of water used in their production (Falsetti, Vallino, Ridolfi, & Laio, 2020). On the other hand, it is necessary to consider the

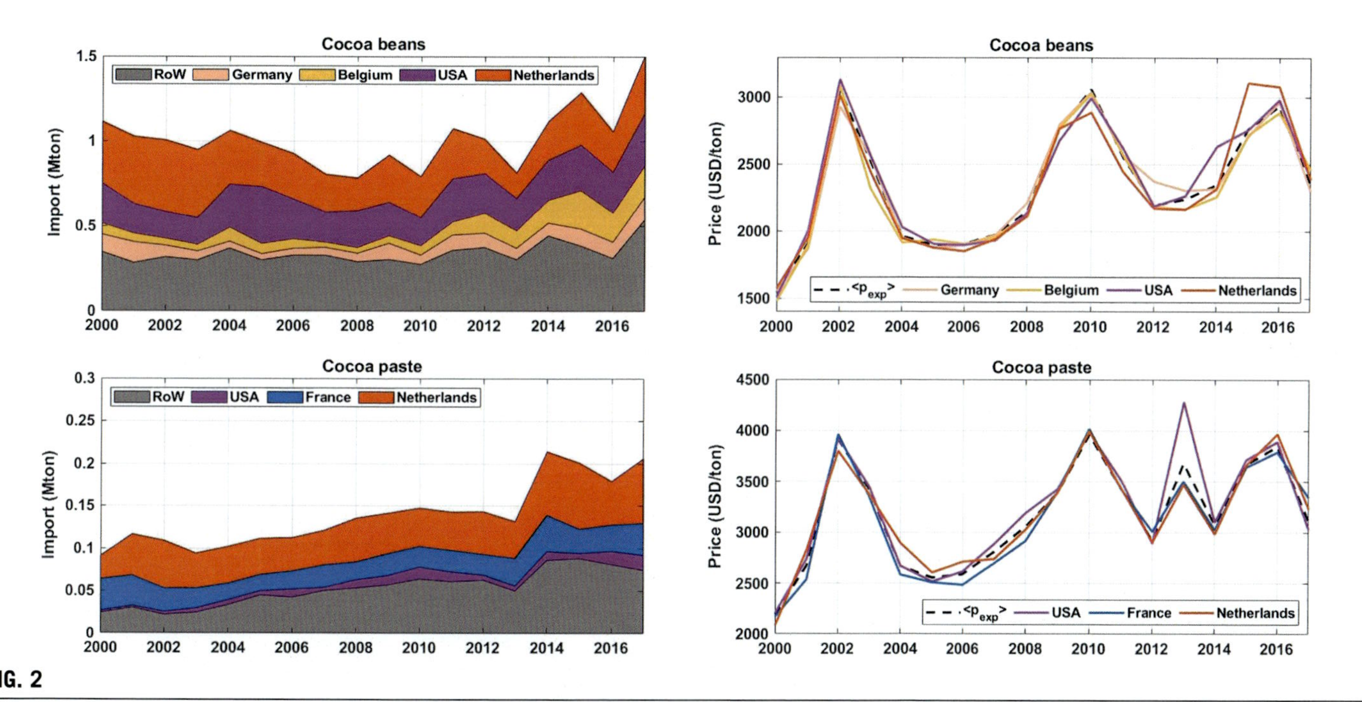

FIG. 2

Top importers from Côte d'Ivoire (2000–2017). *Left panels* show the dynamic and the composition of the main importers of cocoa beans *(top)* and paste *(bottom)* from Ivory Coast. *Right panels* show the average exporting deflated price—of cocoa beans *(top)* and paste *(bottom)*—from Ivory Coast towards the main importers, compared with the average exporting deflated price of Ivory Coast ($<p_{exp}>$, *dashed black line*).

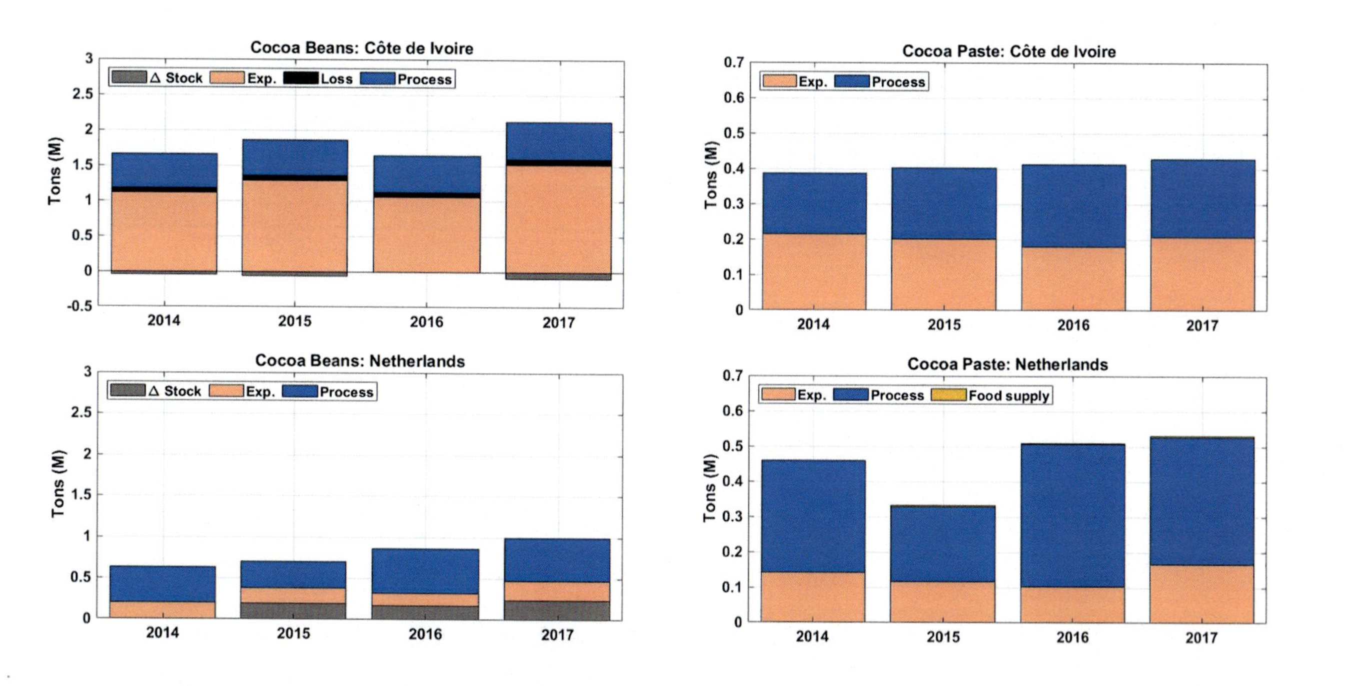

FIG. 3

Supply utilisation accounts of Côte d'Ivoire and the Netherlands (2014–2017). Decomposition of source of destination of available (production + import + stocks) cocoa beans *(left panels)* and paste *(right panels)*, for Ivory Coast *(top)* and the Netherlands *(bottom)*.

whole supply chain to evaluate the actual burden on natural resources due to agricultural and industrial activities, at both domestic and international scales.

3.2 Environmental sustainability assessment of cocoa production

A description of what is going to be presented here should be included (as done at the economic part).

3.2.1 The land footprint of cocoa

Top Western African producers of cocoa are Ivory Coast, Ghana, Nigeria, and Cameroon accounting for 65% of the global production in 2017 (FAOSTAT). Ivory Coast and Ghana show the highest land productivity (or crop yield, 0.5 ton/ha in 2010), also compared to the world average. Between 2010 and 2017, cocoa production has increased by 40% and 30% in these countries, but the growth was boosted by different factors, with different implications for the sustainability of cocoa production (Tuninetti, Ridolfi, & Laio, 2020). Ivory Coast has consistently increased its harvested areas (+57%) through an extensification process with important implications for forest loss (see also Yao Sadaiou Sabas, Gislain Danmo, Akoua Tamia Madeleine, & Jan, 2020), while losing land productivity in many locations (Fig. 4, D or −11% on a country average). Instead, Ghana mostly boosted the land productivity (+22%) though an intensification process and increased the cultivated area only by 8%. Previous studies have shown that the main factors that have contributed to the increase in Ghana's cocoa production were the support measures of the government-owned cocoa marketing board COCOBOD, such as the introduction of free pest and disease control programmes, the introduction of packages of hybrid seeds, fertilisers, insecticides, and fungicides (Wessel & Quist-Wessel, 2015).

During the second half of the twentieth century, the cocoa frontier moved from the drier east to the wetter southwest of the country, fuelled by massive immigration of prospective cocoa farmers from the Savanna (Fig. 4A). It has been argued that the climate gradient was a major driver of these east–west migrations and that cocoa farmers, by replacing forest with farmland over vast areas, contributed to the further drying of the climate in a positive feedback cycle (Ruf, Schroth, & Doffangui, 2015).

Over the period 2010–2017 we still observe clues of cocoa plant migration towards the Eastern Guinean Forests (Fig. 4C), as in the cases of Ghana and Nigeria. In particular, in Ghana this cross-country migration determines a 10% reduction of the cocoa's transpiration demand and an increase of the yield around 20%. Reasons behind this yield improvement may be explained by some trees replanting that has been boosted by the National Cocoa Rehabilitation Programme that has provided 20 million cocoa seedlings to farmers for free in 2012 (Wessel & Quist-Wessel, 2015).

In the case of Ivory Coast, most migration happened in the direction of the North-West Savanna region and this caused a reduction of land productivity (or yield)

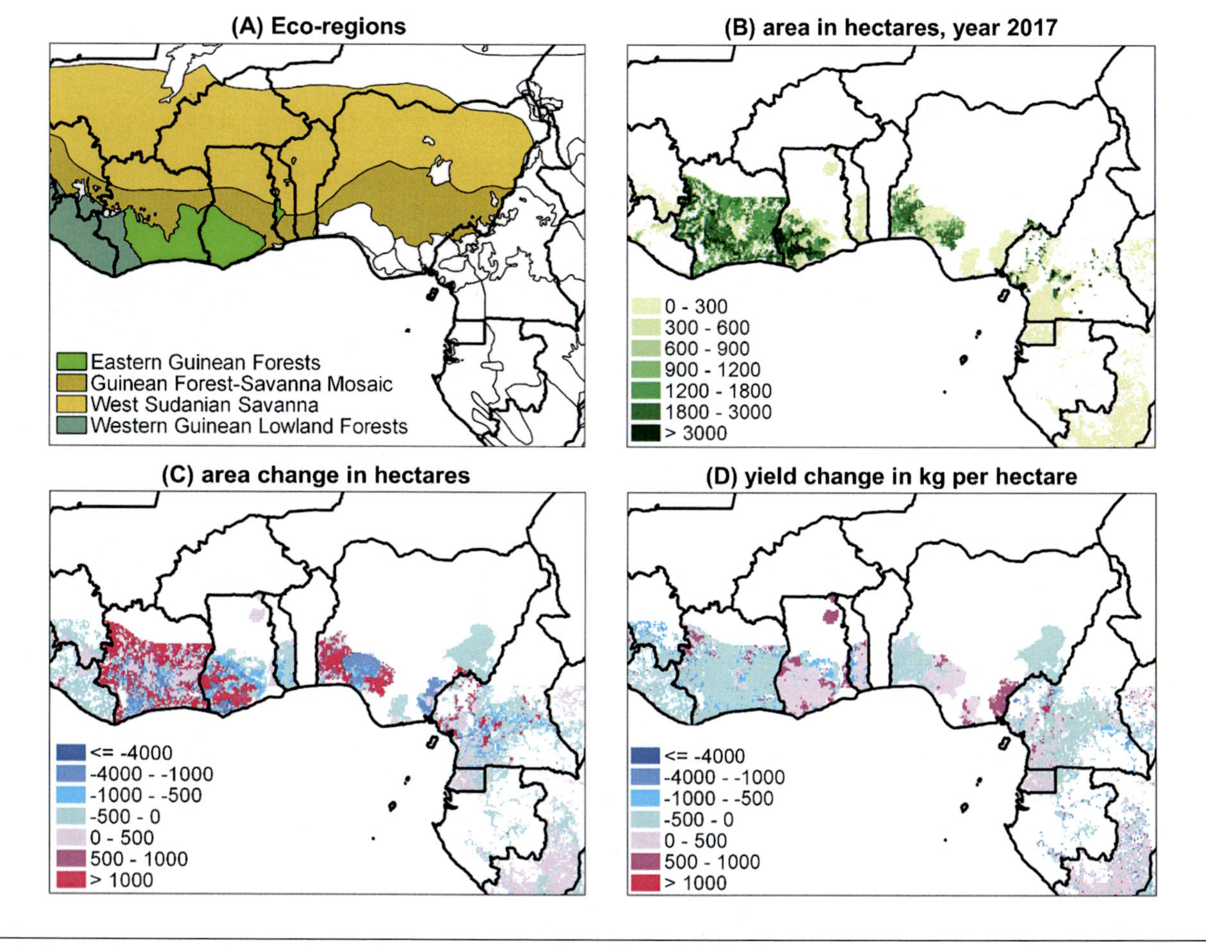

FIG. 4

Cocoa production in Western African countries. Key eco-regions delimited by the WWF (A), harvested area in year 2017 (B), change in harvested area between 2017 and 2010 (C), change in coco yield between 2017 and 2010 (D). Data in panels (B), (C), (D) are shown at a spatial resolution of 5 arc minute (10 km × 10 km at the equator) and are derived from the SPAM dataset.

From (N.d.).

around 10% on a country average, with pixels losing up to 500 kg of cocoa per hectare of land (Fig. 4D). Nevertheless, cocoa production increased by 40% over 2010–2017 thanks to the extensification process. Also Cameroon shows an increase of production (+20%), thanks to the coupled effect of intensification and extensification, mostly across the Western part of the country.

3.2.2 The water footprint of cocoa

Methodological approach

Agricultural water footprint quantifies the amount of water required to produce a certain quantity of agricultural product. It is calculated as the ratio between the crop water requirement and the actual crop yield (D'Odorico et al., 2019; Mekonnen & Hoekstra, 2011).

The crop water requirement (CWR) quantifies the amount of water required by crops along the cropping season, namely the amount of water lost through the plants' stomata (Allen, 1998). It is a function of climatic and phenological properties, and agricultural practices (e.g. irrigated versus rainfed agriculture, length of the cropping season). The daily CWR is calculated following the approach proposed by (Allen, 1998). Accordingly, CWR is equal to the product between a daily crop coefficient (K_c); the daily reference evapotranspiration (ET_0) from a hypothetical well-watered grass surface with fixed crop height, albedo, and canopy resistance; and a daily water stress coefficient (K_s). The crop coefficient incorporates the crop characteristics (e.g. canopy cover, albedo, height) and the effect of evaporation from the soil. In the case of cocoa, the value of K_c is nearly equal to 1 along the whole cropping period, thus the transpiration demand of cocoa is very close to ET_0. The water stress coefficient diminishes the reference evapotranspiration demand of the plant by a factor that accounts for the amount of water available in the root zone. The value of K_s ranges between 0 and 1. It is equal to zero when the plant's transpiration is reduced to zero due to lack of water. Under this condition (i.e. wilting point) the plant closes its stomata and limits the water dispersion. When K_s is close to 1, enough water is available in the soil for proper transpiration and, thus, for carbon uptake from the atmosphere.

In this case study, the CWR of cocoa is annually evaluated along the period 1990–2014 using the data on reference evapotranspiration and rainfall provided by the CRU dataset. Then a long-term average is used to evaluate the unit water footprint (uWF) of cocoa, namely $uWF = 10 \cdot CWR \cdot Y^{-1}$, where Y is the cocoa yield in years 2010 and 2017 (SPAM database), expressed in ton·ha^{-1}. The cell-wise WF assessment is performed at a high spatial resolution of $5' \times 5'$, corresponding to a cell of $10\,\text{km} \times 10\,\text{km}$ at the equator.

We notice that cocoa is grown under rainfed conditions in West sub-Saharan Africa. Thus, depending on the soil water content available in the root zone the plant can evapotranspire at sub-optimal to optimal rates. In particular, when daily rainfall is not sufficient the potential evapotranspiration is reduced by the water stress factor, of a fraction between 0 and 1.

Maps of cocoa's water footprint

Climate and agronomic practices are key in determining the water footprint of any crop. Fig. 5 shows the unitary water footprint (or water efficiency, panels A, B) and the total water footprint (panels C, D) of cocoa production in years 2010 and 2017. Cocoa production is rainfed and irrigation equipment is missing over these lands (according to the SPAM database). Thus rainfall is the key factor in determining cocoa productivity, which can be impacted by the occurrence of water stress along the cropping period. Water stress implies a reduction of the potential evapotranspiration. On average, we find that cocoa plants in these plantations evapotranspire at 70% to 80% of their potential rate.

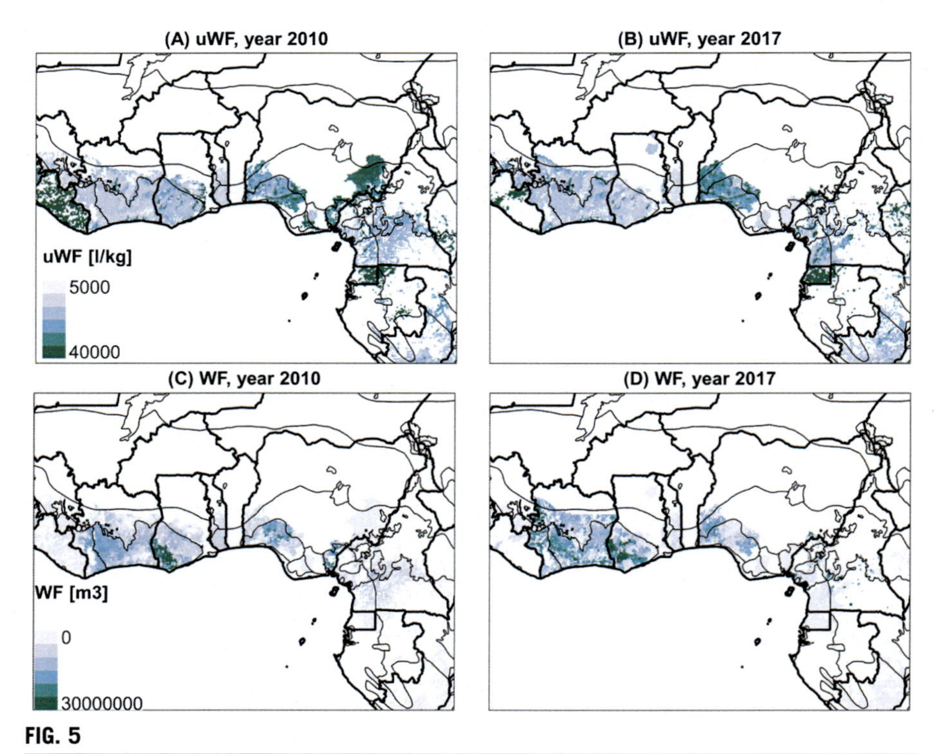

FIG. 5

Water use for cocoa production in Western African countries. Unitary water footprint (*uWF*) of cocoa [l/kg or m^3/ton] in years 2010 (A) and 2017 (B); total Water Footprint (WF) of cocoa [m^3] in years 2010 (C) and 2017 (D). Water footprint values refer to green WF (i.e. rainfall infiltrated water) as cocoa production is grown under rainfed conditions in these areas.

From (N.d.).

In 2017 between 14,500 (Ghana) and 28,500 (Nigeria) l/kg of water were required to produce cocoa. Significant heterogeneities are visible at the sub-national scale, especially in Nigeria with values lower than 15,000 L/kg close to the Niger Delta and larger than 25,000 L/kg across the Nigerian lowland forest (Fig. 5B). Importantly, we observe a decrease of the unit water footprint (*uWF*) larger than 70% over the production sites in the Niger Delta between 2010 and 2017 (Fig. 6), which nearly followed the productivity boost of this region (Fig. 4D). We notice significant decreases in *uWF* also in Ghana (Fig. 6A) thanks to yield increase (Fig. 4D). However, differently from Nigeria, where a gain in land productivity generally produces important water savings (Fig. 6B), in Ghana we observe an important increase of the total water volume despite the increased water use efficiency. This is mostly due to changes in the cocoa's harvested area, whose patterns well compare with those of water footprint.

In a short time period, as the one analysed in this study, we conclude that water use efficiency is mostly influenced by land productivity and the total water footprint nearly overlaps the areas' patterns. However, as we will show in the next section, the relation between water footprint and cocoa yield follows a power law distribution (by definition), which has some critical implications for farmers to improve the efficiency of their plantations. Moreover, over a long-term period this relation can be impacted by the effects of climate change due to temperatures' trends and rainfall variations, which can impact both the cocoa transpiration demand (potential and actual) and the land productivity.

3.2.3 Implications of the power law relation between yield and water footprint

We analyse the relation between cocoa yield and the unitary water footprint across Western African countries (Fig. 7). Data are fitted by a power law distribution (coefficient of determination: 0.97). The analysis of this curve has important implications for diminishing the unitary water footprint, hence increasing the water use efficiency. The results shown in Fig. 7B originate from different cropping systems under varying management conditions and different hydro-climatic zones, ranging from the Western Guinean lowland forest to the Western Sudanian Savanna (Fig. 4A). For yields larger than 0.5 ton/ha, the *uWF* values remain nearly constant, while great gains in water productivity can be easily obtained for lower yields. This is the case of the plantations in Western Guinean lowland forest (Fig. 7C), where most pixels show a unitary water footprint (or water productivity) below the threshold required before generating any yield increase (see Falkenmark, Folke, & Rockström, 2003 for details). Conversely, the forest-Savanna mosaic (Fig. 7E) shows a bipartite pattern where a number of cocoa's areas being still below the threshold but others having overcome this threshold.

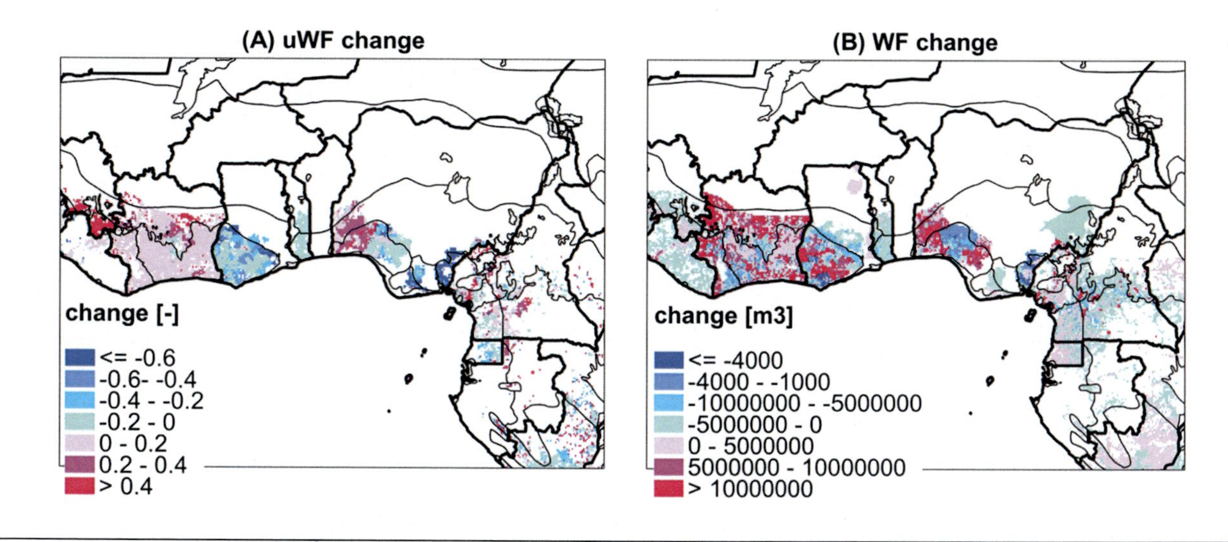

FIG. 6

Change on water use for cocoa production in Western African countries. Change of water use between 2010 and 2017 evaluated for the unitary Water Footprint (uWF, A) and total Water Footprint (WF, B).

From (N.d.).

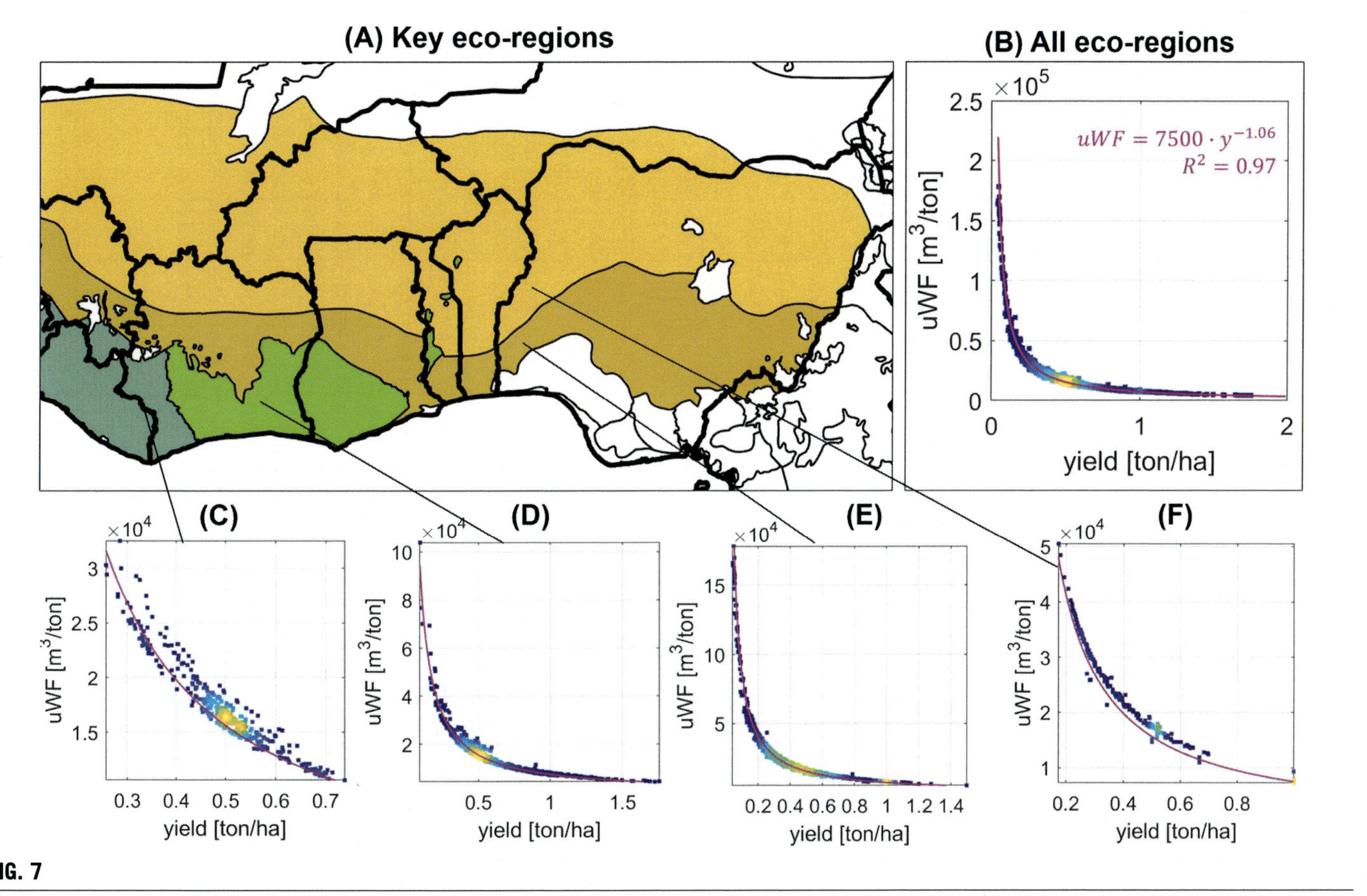

FIG. 7

Water Footprint analysis across key eco-regions in Western Africa. Relations between unitary water footprint (or water use efficiency) and cocoa yield across four key eco-regions: Western Guinean lowland forest (C), Eastern Guinean forest (D), Guinean forest-savanna mosaic (E), Western Sudanian Savanna (F).

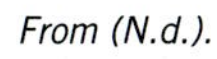

From (N.d.).

4 Conclusions

Cocoa represents a *luxury* food of consumption in richer countries, but it is also a fundamental agricultural product in the poorer tropical countries where it is cultivated. Although West sub-Saharan Africa represents the core region (nearly 70%) of global production, the lack of industries and the low local demand make this area the main exporter worldwide. In this case study, we have analysed the economic and environmental sustainability of cocoa production, by integrating cocoa's prices and trade analyses with land and water footprint assessment. Over the study period (2000–2017), the production and export of cocoa beans has increased by over 50%, with cocoa beans being almost processed outside the Western African countries. The lack of proper infrastructures for crop processing significantly questions the economic sustainability of cocoa production. As we have shown, the cocoa beans export price is 30% higher than the cocoa paste price, and prices are mostly determined by few multinational exporters.

The economic interest behind cocoa imports in the European market and to the US unavoidably challenges the environmental sustainability of cocoa production in the study areas. Most of the production increase over the past decades was driven by land extensification at the expense of forests. Cote d'Ivoire, the top African producer, increased its cultivated areas by nearly 60% between 2010 and 2017, especially at the expense of the Guinean Forest-Savanna Mosaic. Another pattern emerged for Ghana, the top-second producer, as it boosted land productivity by 22%, and reduced the average cocoa water demand by 10% thanks to a migration of the plantations towards the Eastern Guinean Forest. Cocoa plantations in these areas are entirely fed by rainfall. Hence, dry periods can significantly threaten cocoa's productivity and, thus, the economic revenues coming from the sale of cocoa beans. Our results show that, on an annual average, cocoa transpiration rates at 70% to 80% of its optimum value (i.e. the potential evapotranspiration). This suggests that some investments for irrigation infrastructure might be required in order to optimise the potentially achievable yield.

On average, Ghana is the most water-efficient country (14,500 L/kg) while Nigeria is the least efficient one (28,500 L/kg). The largest annual water use happens in Cote d'Ivoire (30 km^3) and Ghana (12 km^3) due to their large annual production of cocoa. Finally, by analysing in a unique framework the power law relation between yield and water footprint, we have shown that boosting water use efficiency might be significantly critical in all the pixels having yields larger than 500 kg per hectare. The largest improvement of water use efficiency seems to be achievable in the Western Guinean lowland forest.

By focusing on cocoa, we provide a first attempt towards an assessment of agricultural sustainability where the dimensions of environmental and economic impacts are jointly considered.

References

A New Era in Global Health. (2017). *Nursing and the United Nations 2030 agenda for sustainable development* (p. 593). Springer Publishing Company.

Afoakwa, E. O., Paterson, A., Fowler, M., & Vieira, J. (2008). Particle size distribution and compositional effects on textural properties and appearance of dark chocolates. *Journal of Food Engineering*, *87*(2), 181–190. https://doi.org/10.1016/j.jfoodeng.2007.11.025.

Allen, R. G. (1998). *Food and agriculture Organization of the United Nations, curatori. Crop evapotranspiration: Guidelines for computing crop water (requirements).*

Barbier, E. B. (2010). Poverty, development, and environment. *Environment and Development Economics*, *15*(6), 635–660. https://doi.org/10.1017/S1355770X1000032X.

Cappelle, J. (2009). *Towards a sustainable cocoa chain: Power and possibilities within the cocoa and chocolate sector*. Oxfam International.

Carr, M. K. V., & Lockwood, G. (2011). The water relations and irrigation requirements of cocoa (*Theobroma cacao* L.): A review. *Experimental Agriculture*, *47*(4), 653–676. https://doi.org/10.1017/S0014479711000421.

D'Odorico, P., Carr, J., Dalin, C., Dell'Angelo, J., Konar, M., Laio, F., … Tuninetti, M. (2019). Global virtual water trade and the hydrological cycle: Patterns, drivers, and socio-environmental impacts. *Environmental Research Letters*, *14*(5). https://doi.org/10.1088/1748-9326/ab05f4, 053001.

Distefano, T. (2020). *Water resources and economic processes* (1st ed.). Oxon and New York: Routledge.

Distefano, T., Riccaboni, M., & Marin, G. (2018). Systemic risk in the global water input-output network. *Water Resources and Economics*, *23*, 28–52. https://doi.org/10.1016/j.wre.2018.01.004.

Distefano, T., Chiarotti, G., Laio, F., & Ridolfi, L. (2019). Spatial distribution of the international food prices: Unexpected heterogeneity and randomness. *Ecological Economics*, *159*, 122–132. https://doi.org/10.1016/j.ecolecon.2019.01.010.

Duarte, R., Pinilla, V., & Serrano, A. (2014). The effect of globalisation on water consumption: A case study of the Spanish virtual water trade, 1849–1935. *Ecological Economics*, *100*, 96–105. https://doi.org/10.1016/j.ecolecon.2014.01.020.

Falkenmark, M., Folke, C., & Rockström, J. (2003). Water for food and nature in drought–prone tropics: Vapour shift in rain–fed agriculture. *Philosophical Transactions of the Royal Society of London. Series B: Biological Sciences*, *358*(1440), 1997–2009. https://doi.org/10.1098/rstb.2003.1400.

Falsetti, B., Vallino, E., Ridolfi, L., & Laio, F. (2020). Is water consumption embedded in crop prices? A global data-driven analysis. *Environmental Research Letters*, *15*(10). https://doi.org/10.1088/1748-9326/aba782, 104016.

FAO. (2011). *The state of the world's land and water resources for food and agriculture (SOLAW) – Managing systems at risk*. Food and Agriculture Organization of the United Nations.

Foley, J. A., Ramankutty, N., Brauman, K. A., Cassidy, E. S., Gerber, J. S., Johnston, M., Mueller, N. D., O'Connell, C., Ray, D. K., West, P. C., Balzer, C., Bennett, E. M., Carpenter, S. R., Hill, J., Monfreda, C., Polasky, S., Rockström, J., Sheehan, J., Siebert, S., … Zaks, D. P. M. (2011). Solutions for a cultivated planet. *Nature*, *478*(7369), 337–342. https://doi.org/10.1038/nature10452.

Kastner, T., Erb, K.-H., & Nonhebel, S. (2011). International wood trade and forest change: A global analysis. *Global Environmental Change*, *21*(3), 947–956. https://doi.org/10.1016/j.gloenvcha.2011.05.003.

Konar, M., Dalin, C., Suweis, S., Hanasaki, N., Rinaldo, A., & Rodriguez-Iturbe, I. (2011). Water for food: The global virtual water trade network: Network analysis of global virtual water. *Water Resources Research*, *47*(5). https://doi.org/10.1029/2010WR010307.

Kroeger, A., Koenig, S., Thomson, A., & Streck, C. (2017). *Forest- and climate-smart cocoa in Côte d'Ivoire and Ghana*. Washington, DC: World Bank. http:/hdl.handle.net/10986/29014.

Mekonnen, M. M., & Hoekstra, A. Y. (2011). The green, blue and grey water footprint of crops and derived crop products. *Hydrology and Earth System Sciences*, *15*(5), 1577–1600. https://doi.org/10.5194/hess-15-1577-2011.

Miller, R. E., & Blair, P. D. (2009). *Input-output analysis: Foundations and extensions* (2nd ed., pp. 1–750). Cambridge University Press. https://doi.org/10.1017/CBO9780511626982.

Naranjo-Merino, C., Ortíz-Rodriguez, O., & Villamizar, G. R. (2018). Assessing green and blue water footprints in the supply chain of cocoa production: A case study in the northeast of Colombia. *Sustainability*, *10*(1), 38. https://doi.org/10.3390/su10010038.

OECD, & Analysis, I. I. for A. S. (2020). *New approaches to economic challenges systemic thinking for policy making the potential of systems Analysis for addressing global policy challenges in the 21st century*. OECD Publishing. http:/public.eblib.com/choice/PublicFullRecord.aspx?p=6408934.

Ongolo, S., Kouamé Kouassi, S., Chérif, S., & Giessen, L. (2018). The tragedy of forestland sustainability in postcolonial Africa: Land development, cocoa, and politics in Côte d'Ivoire. *Sustainability*, *10*(12), 4611. https://doi.org/10.3390/su10124611.

Philippidis, G., Shutes, L., M'Barek, R., Ronzon, T., Tabeau, A., & van Meijl, H. (2020). Snakes and ladders: World development pathways' synergies and trade-offs through the lens of the sustainable development goals. *Journal of Cleaner Production*, *267*. https://doi.org/10.1016/j.jclepro.2020.122147, 122147.

Ronzon, T., & Sanjuán, A. I. (2020). Friends or foes? A compatibility assessment of bioeconomy-related sustainable development goals for European policy coherence. *Journal of Cleaner Production*, *254*. https://doi.org/10.1016/j.jclepro.2019.119832, 119832.

Ruf, F., Schroth, G., & Doffangui, K. (2015). Climate change, cocoa migrations and deforestation in West Africa: What does the past tell us about the future? *Sustainability Science*, *10*(1), 101–111. https://doi.org/10.1007/s11625-014-0282-4.

Tamea, S., Allamano, P., Carr, J. A., Claps, P., Laio, F., & Ridolfi, L. (2013). Local and global perspectives on the virtual water trade. *Hydrology and Earth System Sciences*, *17*(3), 1205–1215. https://doi.org/10.5194/hess-17-1205-2013.

Tuninetti, M., Ridolfi, L., & Laio, F. (2020). Ever-increasing agricultural land and water productivity: A global multi-crop analysis. *Environmental Research Letters*, *15*(9), 0940a2. https://doi.org/10.1088/1748-9326/abacf8.

van Oel, P. R., Mekonnen, M. M., & Hoekstra, A. Y. (2009). The external water footprint of the Netherlands: Geographically-explicit quantification and impact assessment. *Ecological Economics*, *69*(1), 82–92. https://doi.org/10.1016/j.ecolecon.2009.07.014.

Wessel, M., & Quist-Wessel, P. M. F. (2015). Cocoa production in West Africa, a review and analysis of recent developments. *NJAS-Wageningen Journal of Life Sciences*, *74–75*, 1–7. https://doi.org/10.1016/j.njas.2015.09.001.

Wilcox, M. D. (2004). *Market power and structural adjustment: The case of West African cocoa market liberalization*. DOI.Org (Datacite). https://doi.org/10.22004/AG.ECON.20084.

Yao Sadaiou Sabas, B., Gislain Danmo, K., Akoua Tamia Madeleine, K., & Jan, B. (2020). Cocoa production and Forest dynamics in Ivory Coast from 1985 to 2019. *Land*, *9*(12), 524. https://doi.org/10.3390/land9120524.

CHAPTER

Environmental assessment of urban water systems: LCA case studies 16

Diana M. Byrne[a], Philippe Roux[b,c], and Lluís Corominas[d,e]

[a]*Department of Civil Engineering, University of Kentucky, Lexington, KY, United States* [b]*ITAP, Univ Montpellier, INRAE, Institut Agro, Montpellier, France* [c]*Elsa, Research Group for Environmental Lifecycle and Sustainability Assessment, Montpellier, France* [d]*Catalan Institute for Water Research (ICRA), Girona, Spain* [e]*Universitat de Girona, Girona, Spain*

1 Aims

1. Describe UWS environmental impacts and the applicability of LCA for assessing UWS sustainability.
2. Review the state of the art for LCA applied to UWS.
3. Provide three case studies that apply LCA to UWS in order to highlight their methodological choices as well as the study's findings and implications.
4. Discuss challenges and opportunities for future applications of LCA to UWS.

2 State of the art

Urban water systems (UWS) include natural and human-engineered parts of the water cycle in urban areas. UWS comprises both the existing freshwater ecosystems and the water infrastructure that humans have built to (i) collect, store, treat, and supply water for potable and non-potable uses; (ii) collect, transport, and treat the generated wastewater; and (iii) to discharge treated or untreated water to freshwater ecosystems. The key infrastructure elements are thus drinking water and wastewater treatment plants, reclaimed water production plants, distribution and sewer systems, combined sewer overflow structures, stormwater tanks, etc. All these elements of the UWS are intrinsically linked by the movement of water, as well as any associated matter and energy fluxes, leading to a unique and vital socio-environmental system (as summarised in Fig. 1). Because of the obviously connected nature of the different parts of the system, management actions on any component influence the others in a cascading effect, with potentially negative or unpredictable outcomes. Therefore it is very important to push forward the integrated management of UWS to make more qualified decisions for both the environment, the society, and the economy (Garcia et al., 2016).

Assessing Progress towards Sustainability. https://doi.org/10.1016/B978-0-323-85851-9.00017-1

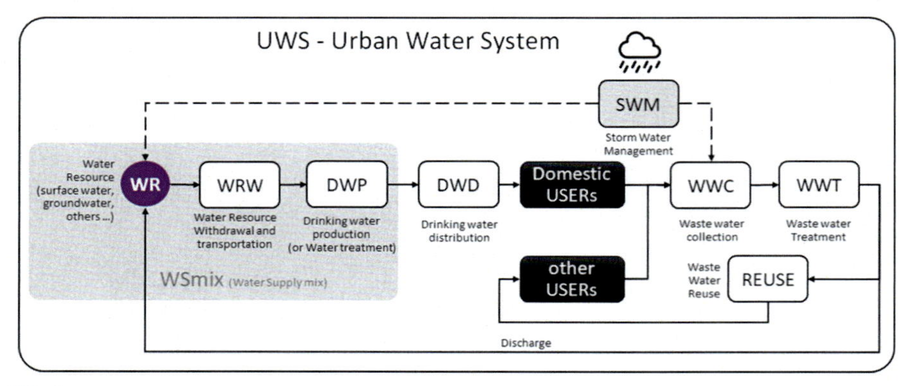

FIG. 1

Typical urban water systems.

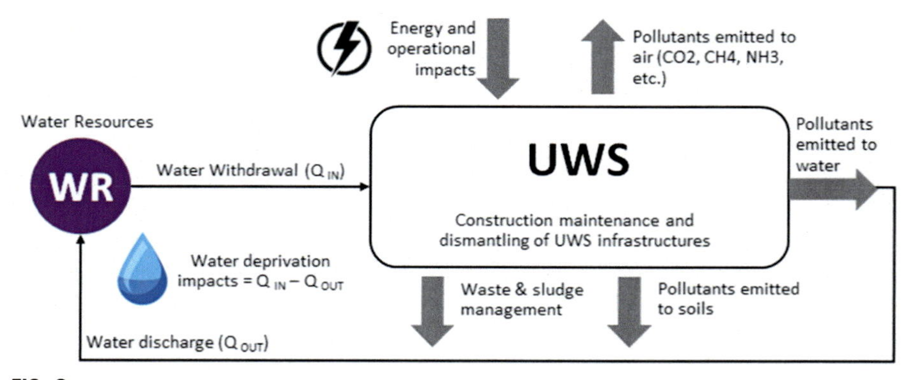

FIG. 2

Main environmental impacts to be considered when assessing an UWS.

UWS consume energy and resources in their construction and operation, and generate emissions which are discharged to water (soluble and particulate chemical compounds discharged to freshwater ecosystems), soil (soluble and chemical compounds discharged to soils), and gas phases (e.g., greenhouse gas emissions generated during wastewater transport and treatment or after discharge in freshwater ecosystems). Transporting, heating, and treating water involves energy consumption, which generates indirect impacts from its production and transport. Analysing all these emissions, resources, and energy consumption (as summarised in Fig. 2) in an integrated manner is of paramount importance to make wise decisions at the urban scale reaching optimal design, construction, and operation.

Life cycle assessment (LCA) has characteristics that make it very suitable for assessing such systems: (i) multi-criteria approach and (ii) life cycle perspective (i.e., impacts considered from cradle to grave, from raw material extraction to waste end-of-life management). In this context, LCA has naturally emerged over the years as an environmental assessment method capable of taking into account in a holistic way all the impacts of UWS (Loubet, Roux, Loiseau, & Bellon-Maurel, 2014). In addition, LCA is a standardised method (International Organization for Standardization, 2006a, 2006b) and is recommended by many environmental agencies around the world. LCA takes a global perspective and includes both direct and indirect impacts of a UWS that contribute to a defined set of environmental indicators (Teodosiu, Barjoveanu, Sluser, Popa, & Trofin, 2016). LCA can provide decision-makers with environmental performance information of UWS using a standardised, transparent approach that is based on the best available science. LCA can provide municipalities with a better understanding of their entire UWS, including upstream (supply chain) and downstream (waste disposal or beneficial reuse) components; help identify performance weaknesses and hotspots for targeted improvement/optimisation; inform decision-makers about potential and otherwise unforeseen trade-offs/burden-shifting within their UWS; provide evidence to help water authorities meet and verify corporate sustainability objectives (e.g., carbon neutrality, corporate social responsibility); and help water authorities to negotiate with relevant regulatory authorities regarding environmental regulations (to ensure appropriate balancing of global versus local impacts and avoid burden-shifting). Some of the current hot topics are decentralisation of water treatment and circular economy; the benefits and potential environmental impacts of the transition of UWS cannot be fully understood without tools like LCA.

In recent years, LCA has been increasingly applied to integrated UWS (i.e., UWS LCAs which address at least two of the three elements of wastewater, drinking water, or stormwater). To identify the most recent papers (limited to research articles including in press manuscripts since 2018, after the review published in Byrne, Lohman, Cook, Peters, & Guest, 2017) applying LCA to integrated UWS, articles were gathered through the Scopus database using title, abstract, and keyword search terms for combinations of wastewater, drinking water, and stormwater with LCA. A total of 31 papers were identified (summarised in Table 1) and focused on topics such as greywater reuse (e.g., Kobayashi et al., 2020; Opher et al., 2019), separate and combined systems (e.g., Risch et al., 2018; Yang et al., 2020), integrating management of water and wastewater systems (e.g., Xue et al., 2019), and rainwater/stormwater harvesting (e.g., Arden et al., 2021; Faragò et al., 2019; Ghimire et al., 2019). All identified studies included operation and almost all studies included at least one aspect of the related infrastructure.

The most common life cycle impact assessment (LCIA) method applied was ReCiPe (13 papers) while TRACI was also frequently applied (8 papers) typically for studies in the United States and Canada. Applying these LCIA methods, these studies quantified a variety of impact categories, many of which focused on climate impacts (94% of papers, e.g., climate change, global warming), resource depletion (77% of papers, e.g., fossil fuel depletion, metal depletion, abiotic depletion potential), acidification (68% of papers), and ozone depletion (65% of papers). Other common impact categories assessed were related to human health (61% of papers, e.g., carcinogenic toxicity, non-carcinogenic toxicity, respiratory effects), eutrophication (58% of papers, e.g., freshwater eutrophication, marine eutrophication), smog (58% of papers, e.g., smog formation potential, photochemical oxidant formation), and ecotoxicity (58% of papers, e.g., freshwater ecotoxicity, marine ecotoxicity).

Table 1 Outputs of the literature review performed.

Paper	Country	Inf	Op	Functional unit	LCIA Method
Andersson et al. (2020) Water Science and Technology	Sweden	N	Y	'(1) m^3 of treated water and (2) kg N removed'	Not specified
Antunes, Ghisi, and Severis (2020) Science of the Total Environment	Brazil	Y	Y	'water supply (in m^3) and the drainage of the parking area (in m^2) for the building over the time horizon of 20 years'	ReCiPe
Antunes et al. (2020) Water	United Kingdom	Y	Y	'water supply in m^3'	ReCiPe
Arden et al. (2021) Water Research	United States	Y	Y	'1 gal of NPR (non-potable reuse) water provided to the building'	IPCC, ReCiPe, other
Arden et al. (2020) Sustainability	United States	Y	Y	'delivery of NPR (non-potable reuse) water for the whole building'	CED, IPCC, TRACI, other
Barbosa de Jesus, Costa Kiperstok, and Borges Cohim (2020) Water and Environment Journal	Brazil	Y	Y	'1 m^3 of water produced by the system'	CED
Bonoli et al. (2019) Water	Italy	Y	Y	'1 m^3 of reclaimed water'	ReCiPe
Cashman et al. (2018) Bioresource Technology	United States	Y	Y	'one cubic meter of wastewater treated'	CED, TRACI
Faragò, Brudler, Godskesen, and Rygaard (2019) Journal of Cleaner Production	Denmark	Y	Y	'provision of 31,000 m^3/year of non-potable water, to 2000 inhabitants in Nye, and the prevention of flooding according to defined municipal safety standards'	ILCD
Friedrich, Poganietz, and Lehn (2020) Resources, Conservation and Recycling	Germany	Y	Y	'the sum of energy, water- and waste-related and material requirements for the performance of the primary functions for the analysed urban neighbourhood for one year'	ReCiPe
García-Sánchez and Güereca (2019) Science of the Total Environment	Mexico	N	Y	'one cubic meter (1 m^3) of water for consumption'	ReCiPe

Ghimire et al. (2019) Resources, Conservation and Recycling	United States	Y	Y	'1 m^3 of RWH (rainwater harvesting) and ACH (air-conditioning condensate harvesting) delivery for flushing toilets and urinals'	TRACI
Goga, Friedrich, and Buckley (2019) Water SA	South Africa	Y	Y	'1 kL of water at the specified standard for potable water produced over the lifespan of each process unit'	ReCiPe
Hsien, Low, Fuchen, and Han (2019) Resources, Conservation and Recycling	Singapore	N	Y	'1 m^3 of water (NEWater or tap water) delivered to the consumer'	ReCiPe
Jabri, Nolde, Ciroth, and Bousselmi (2020) International Journal of Environmental Science and Technology	Germany	Y	Y	'production of 1 m^3 of treated greywater per day for 365 working days of plant achieving a water effluent with COD concentration of 25 mg/L'	CML
Jeong, Broesicke, Drew, and Crittenden (2018) Journal of Cleaner Production	United States	Y	Y	'1 m^3 water used for outdoor irrigation and/or toilet flushing'	TRACI
Kobayashi, Ashbolt, Davies, and Liu (2020) Environment International	Canada	Y	Y	'annual treatment of greywater generated per person'	TRACI
Leong, Balan, Chong, and Poh (2019) Journal of Cleaner Production	Malaysia	Y	Y	'collection, storage, and distribution of 1 m^3 of non-potable water for both toilet flushing and irrigation over a project lifespan of 50 years'	CML, TRACI, other
Marinoski and Ghisi (2019) Resources, Conservation and Recycling	Brazil	Y	Y	'total volume of water consumed in the house during the defined time horizon'	ReCiPe
Opher, Friedler, and Shapira (2019) International Journal of Life Cycle Assessment	Israel	Y	Y	'annual supply, reclamation, and reuse of water consumed by Model City'	ILCD
Pinelli et al. (2020) Integrated Environmental Assessment and Management	Egypt	Y	Y	'1 m^3 of drainage canal water'	ILCD
Remy, Seis, Miehe, Orsoni, and Bortoli (2019) Water Science and Technology: Water Supply	France	Y	Y	'the provision of 1 m^3 of drinking water ready to be fed into the local drinking water network'	CED, IPCC, other

Continued

Table 1 Outputs of the literature review performed. *Continued*

Paper	Country	Inf	Op	Functional unit	LCIA Method
Risch et al. (2018) Water Research	France	N	Y	Not specified	ReCiPe, other
Ryberg, Bjerre, Nielsen, and Hauschild (2020) Journal of Industrial Ecology	Denmark	Y	Y	'annual supply of 8.36 million m^3 water and management and treatment of 29.3 million m^3 storm- and wastewater'	Other
Santana, Cornejo, Rodríguez-Roda, Buttiglieri, and Corominas (2019) Journal of Cleaner Production	Spain	Y	Y	'one year of operation of the entire water system of Lloret de Mar'	ReCiPe, other
Singh et al. (2018) Environmental Science and Pollution Research	India	Y	Y	'1 m^3 of treated wastewater'	CML
Tarpani, Lapolli, Recio, and Gallego-Schmid (2021) Journal of Cleaner Production	Brazil	Y	Y	'to increase the availability of drinking water in the local distribution network by 1000 m^3'	ReCiPe
Tavakol-Davani et al. (2018) Journal of Irrigation and Drainage Engineering	United States	Y	Y	'1 m^3 reduction of CSO volume over the life cycle of facilities'	TRACI
Xue et al. (2019) Water Research X	United States	Y	Y	drinking water: '1 m^3 of treated and distributed water meeting or exceeding National Primary Drinking Water Regulations'; wastewater: '1 m^3 of treated and discharged wastewater'; combined drinking water and wastewater: '1 m^3 of delivered drinking water, which is subsequently treated after use'	CED, TRACI
Yang, Jiang, and Jian (2020) Journal of Coastal Research	China	Y	Y	'urban drainage system required for 1 ha of service area within 1 year'	CML
Zanni et al. (2019) Sustainable Production and Consumption	Italy	Y	Y	'1 m^3 of reclaimed water'	ReCiPe

Inf, *infrastructure;* Op, *operation;* N, *no (not included);* Y, *yes (included).*

Finally, many studies (58% of papers) also included an impact category specific to water quantity (e.g., freshwater depletion, water scarcity, water stress index). Consistent with the multifunctionality of urban water systems (Section 3), functional units varied but typically used volume of water (with or without a treatment objective) as the basis.

3 Case studies description

This section includes a schematic description of 3 case studies which highlight the importance of LCA in UWS, and which are good examples of adequate balance in the inventory effort of the construction and operation phases. The 1st study involves water distribution networks, sewer systems, and wastewater treatment and analyses of different levels of decentralisation of reclaimed water production and distribution. The 2nd study shows the potential of the WaLa LCA modelling tool to model the impacts from withdrawal to discharge in a city. The 3rd study is a good example of the trade-offs between sewer systems construction and wastewater treatment operation to compare centralised vs decentralised systems. For each case study, the background, aim of study and novelty, goal and scope, functional unit, system boundaries, scenarios, life cycle inventory, impact assessment, and results are described.

Case-study 1. Assessing water REUSE of a touristic city (Spain) (Santana et al., 2019).

Background: Current trends point to tourism as a growing economic sector for many communities worldwide with possible environmental implications including water use. One solution for these communities is water reuse, which not only has been shown to minimise potable water consumption but also other environmental impacts.

Aims of Case Study and Novelty: This study is a life cycle assessment of the environmental impacts (carbon footprint, aquatic eutrophication, metals depletion, and water footprint) of four distinct water management scenarios in the Spanish tourism-dependent city of Lloret de Mar (northeastern Mediterranean coast of Spain). The city has a year-round population of 37,042 and a seasonal population equivalent of 16,337. The novelty lies in the study of water reuse integrated in the entire water management system, with detailed construction and operation inventories of all elements.

Goal and Scope Definition:

Goal of LCA: The objective of this study is to determine the environmental impacts of hotel water reuse in a tourism-dependent community.

Functional unit. 1 year of operation of the entire water management system of Lloret de Mar.

System boundaries: This LCA includes the following water management components: potable water treatment, potable water supply, onsite membrane bioreactor (MBR) collection, onsite MBR treatment, onsite MBR supply (for reuse), sewer system, wastewater treatment plant (WWTP), tertiary treatment plant, and the reused water distribution system.

Scenarios: One scenario is a business-as usual scenario where hotels are completely dependent on potable water, while the other three scenarios

(Decentralised, Hybrid, and Centralised) integrate water reuse (through varying degrees of centralisation) in all hotels located in the city.

Life Cycle Inventory: Generally, for each component, the construction and operation stage impacts were included within the boundaries to carry out the study. However, for certain components, for which the infrastructure is assumed to be the same for all scenarios, the construction stage impacts were excluded.

Life Cycle Impact Assessment: ReCiPe and AWARE. The study includes carbon footprint, marine eutrophication, metals depletion, and water footprint as potential impact categories.

Interpretation of Results: An LCA of water reuse integrated in the entire water management system gives a clear picture of the impacts of water reuse outside of the impacts specifically related to water reuse treatment and distribution technologies. Results (Fig. 3) show that the implementation of water reuse incurs net carbon footprint, metals depletion, and water footprint impacts, while there is a net benefit associated with marine eutrophication impacts. This is mainly due to the treatment technologies used for potable water and greywater/wastewater treatment for reuse. Comparing water reuse scenarios, an increasing degree of centralisation is associated with lower carbon and water footprints.

Case-study 2. Assessing UWS scenarios of a megalopolis (Paris, France) (Loubet, Roux, & Bellon-Maurel, 2016; Loubet, Roux, Guérin-Schneider, & Bellon-Maurel, 2016).

Background: A framework and an associated modelling tool to perform life cycle assessment of urban water systems, namely the WaLA model, were previously developed. The WaLA model is applied to the urban water system of the Paris suburban area, in France (4.3 M inhabitants). This study aims to verify the capacity of the model to provide environmental insights into stakeholder issues related to future trends influencing the system (e.g., evolution of water demand, increasing water scarcity) or policy responses (e.g., choices of water resources and technologies). This is achieved by evaluating a baseline scenario for 2012 and several forecasting scenarios for 2022 and 2050.

Aims of Case Study and Novelty: In 2016 when this work was published, the novelty of this case study was that it was the first time the environmental performance of an urban water system was assessed at the scale of the entire system. All previous studies were limited to the evaluation of the elementary components of a water system (drinking water production, distribution network, sewerage networks, wastewater treatment, etc.) whereas this work proposes a coherent modelling framework to carry out a holistic evaluation.

Goal and Scope Definition:

Goal of LCA: The objective of this study is to establish and investigate different scenarios and test the capacity of the model to address management questions such as comparing alternative choices of water resources or assessing various forecasting scenarios under global changes (climate, water availability and stress, population growth, etc.).

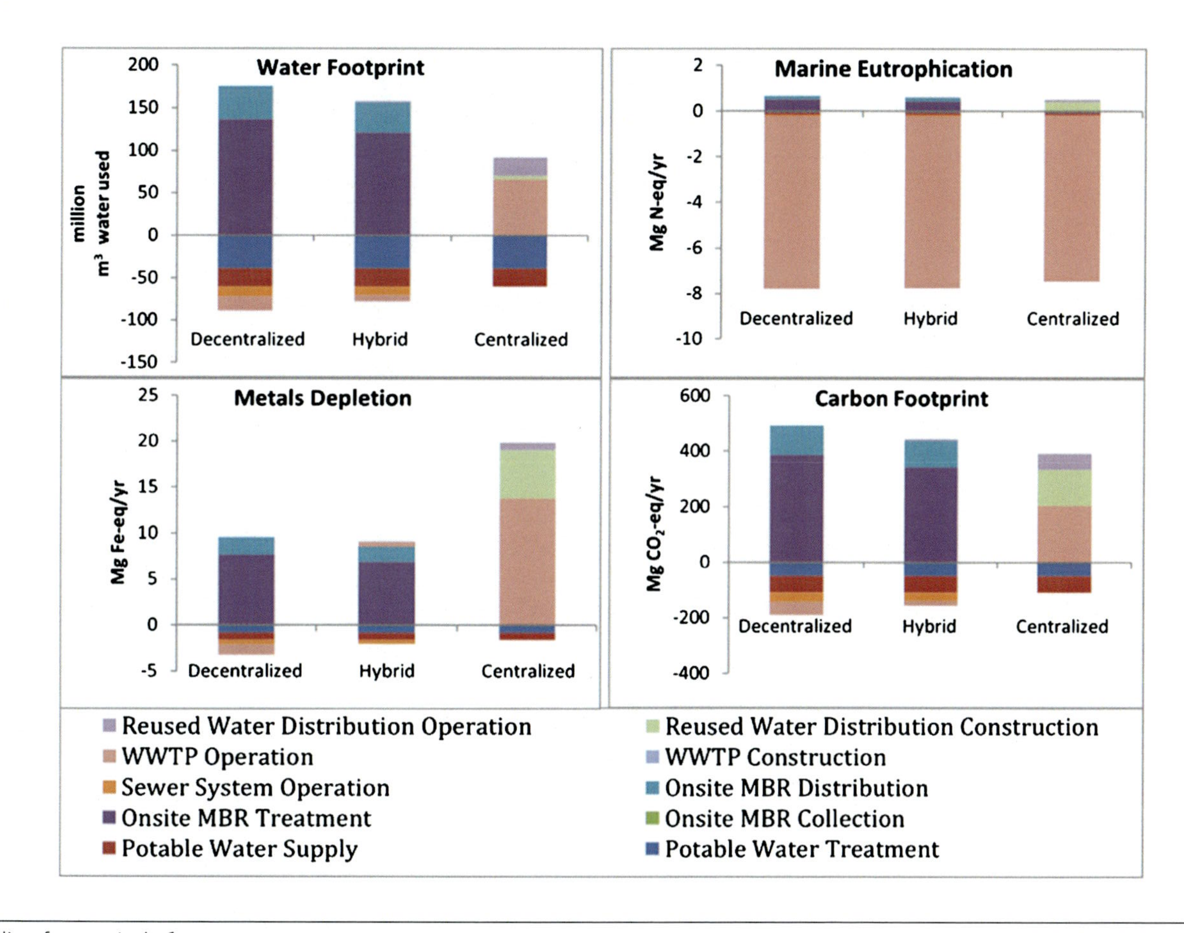

FIG. 3

LCA results of case study 1.

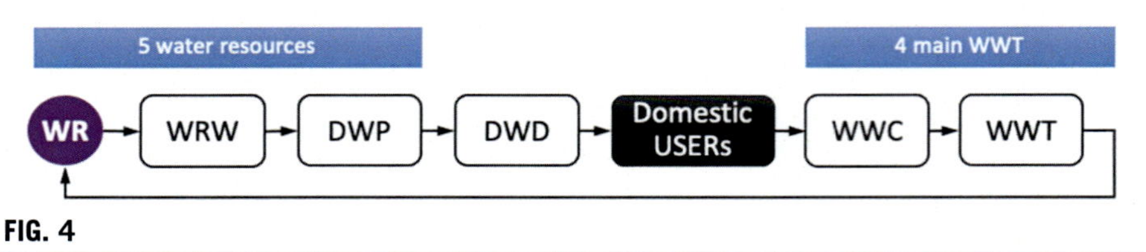

FIG. 4

System boundaries of case-study 2.

Functional Unit (FU): Three functional units were investigated: FU1—whole UWS/year, FU2—1 equivalent inhabitant/year, FU3—1 m^3 of drinking water at user's gate. System boundaries are shown in Fig. 4.
Scenarios: Several scenarios were tested including (i) short-term changes (optimisation of water withdrawal among the 5 available resources) and (ii) long-term scenarios (resource availability, users' consumption changes, population growth, technology changes).

Life Cycle Inventory: All components of an UWS including plants and networks as well as the end of life of sludge. Water consumption is also accounted for to provide an accurate water footprint.

Life Cycle Impact Assessment: LCIA method ILCD was used and adapted for accounting for water deprivation effects. Other impact categories (climate change, ozone depletion, etc.) are evaluated at the midpoint level with ILCD and at the end-point level with Impact 2002+.

Interpretation of Results: The results presented in Fig. 5 clearly show the environmental impacts are dominated by wastewater collection and treatment, except for the impact 'WD: water deprivation' and 'RD: resource depletion' for which the contributions are more balanced. Beyond this analysis of the contribution of the current system in Paris, this study demonstrates the feasibility of modelling different scenarios of prospective evolution of water management in a megalopolis and of comparing them in terms of public decision support.

Case-study 3. Comparison of the environmental performance of decentralised versus centralised wastewater systems (France) (Risch, Boutin, & Roux, 2021).

Background: Decentralised wastewater management (DWM) systems are deployed in areas where the topography does not allow for gravity flow to a centralised system or requires a complex and expensive pumping station network. Also, DWM systems are often the only option in rural areas where there are no sewage transport networks. LCA is used to address the question of the degree to which DWM systems can be considered as viable alternatives from an environmental point of view compared to centralised ones.

Aims of Case Study and Novelty: The question of the environmental performance of more or less centralised or decentralised wastewater management systems is becoming increasingly relevant. The originality of this publication compared to the

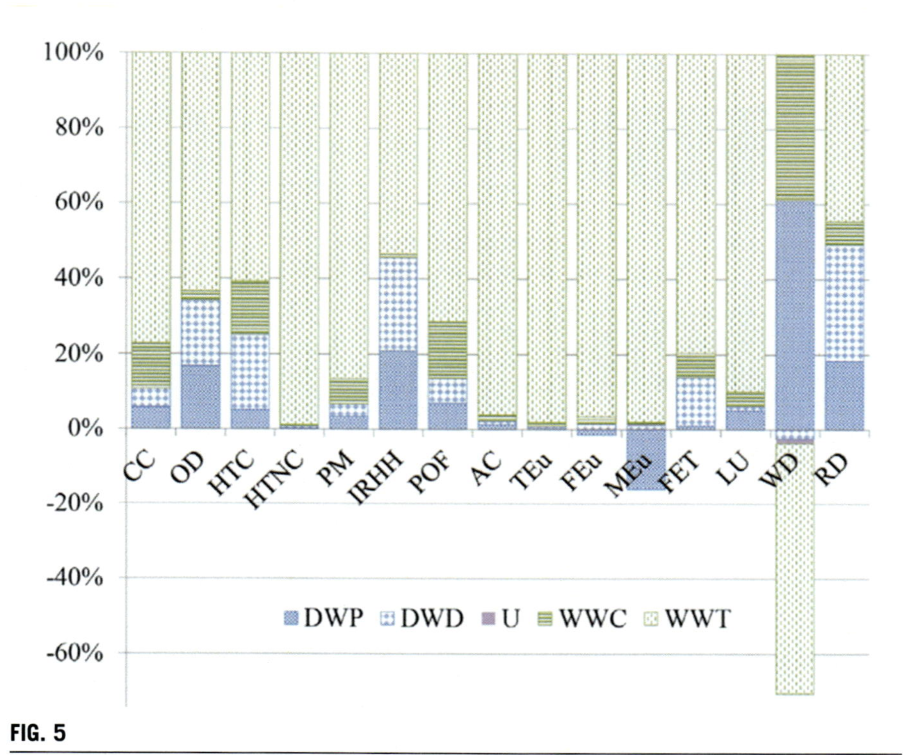

FIG. 5

LCA results of case study 2, relative contributions of UWS components for baseline scenario. LCIA method: ILCD.

state of the art is that it does not pose the question of comparing centralised versus decentralised systems in LCA in a Manichean way. This contribution does not claim to assert whether option A (centralised) is better than B (decentralised), but rather to answer the question 'under which conditions is A better than B?'. For this purpose, the main parameters that affect the results of the comparison are identified and charts are used to simply see the favourable and unfavourable situations for decentralised systems.

Goal and Scope Definition:

Goal of LCA: Assess under which conditions a decentralised scenario is better than a centralised scenario (including a longer or shorter connection to a conventional activated sludge plant). For this purpose, a two-step LCA approach is proposed:

1. a contribution analysis to the impacts to identify the two parameters that drive the results: length of the connection to the WWTP and number of households connected.
2. calculation of the thresholds for switching between the two solutions of centralised vs. decentralised.

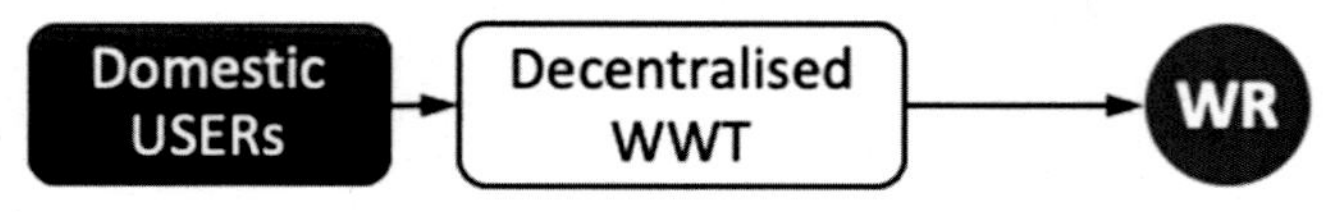

FIG. 6

System boundaries of case study 3.

Functional Unit (FU): Treatment of 46 g of BOD_5·(inhabitant·d)$^{-1}$. System boundaries are shown in Fig. 6.
Scenarios: Baseline scenario—connection of several houses to a centralised activated sludge WWTP by a pipe of length L (km) compared to 2 decentralised systems—vertical/horizontal flow constructed wetland (red bed filter) and septic tank + filter.

Life Cycle Inventory: All infrastructure, operation, and emissions of each system are included (cradle to grave). This includes in particular a very detailed mass balance of inputs (within wastewater) and outputs (discharges, air emissions, sludge content) for C—Carbon, N—Nitrogen, P—Phosphorus, 12 metal and 3 organic compounds.

Life Cycle Impact Assessment: ReCiPe 2016, v1.03 at midpoint or endpoint level.

Interpretation of Results: Fig. 7 presents the comparison of results of the midpoint LCA impacts for the two decentralised systems studied—WWT4 (septic tank and sand filter) and WWT3 (vertical/horizontal constructed wetland). Fig. 8 shows the comparison of all the solutions considered (centralised and decentralised) at the endpoint level (impacts on human health, ecosystems, and resources), including quantified uncertainty. It should be noted, however, that this comparison is based on a specific situation with a given number of isolated settlements and a given distance and elevation to the centralised station (reflecting pumping needs). The case study takes these two parameters, which are the drivers of the comparison, and varies them in order to obtain charts of situations favourable or unfavourable for decentralised solutions (see the publication for these charts).

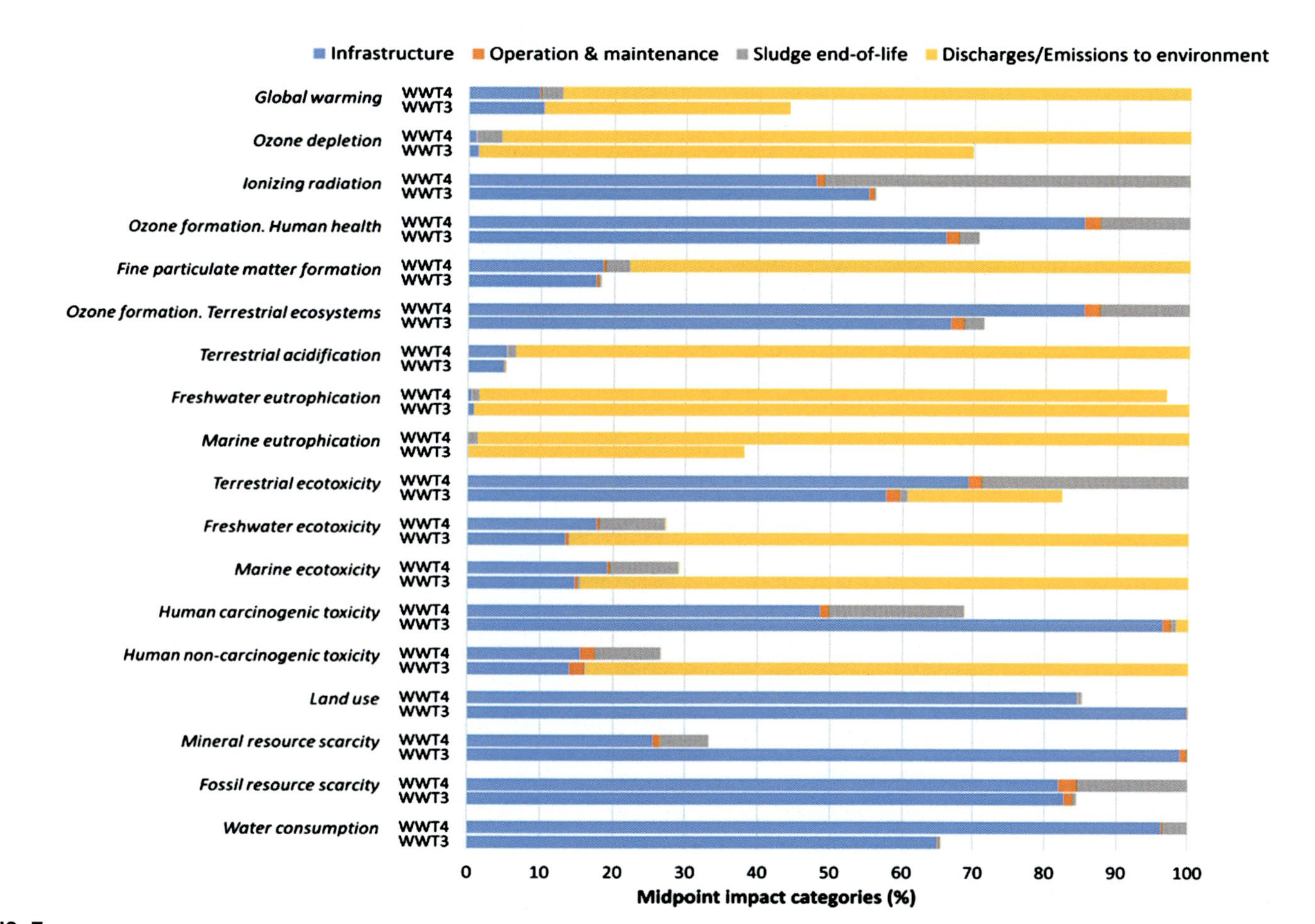

FIG. 7

LCA results of case study 3 (part I).

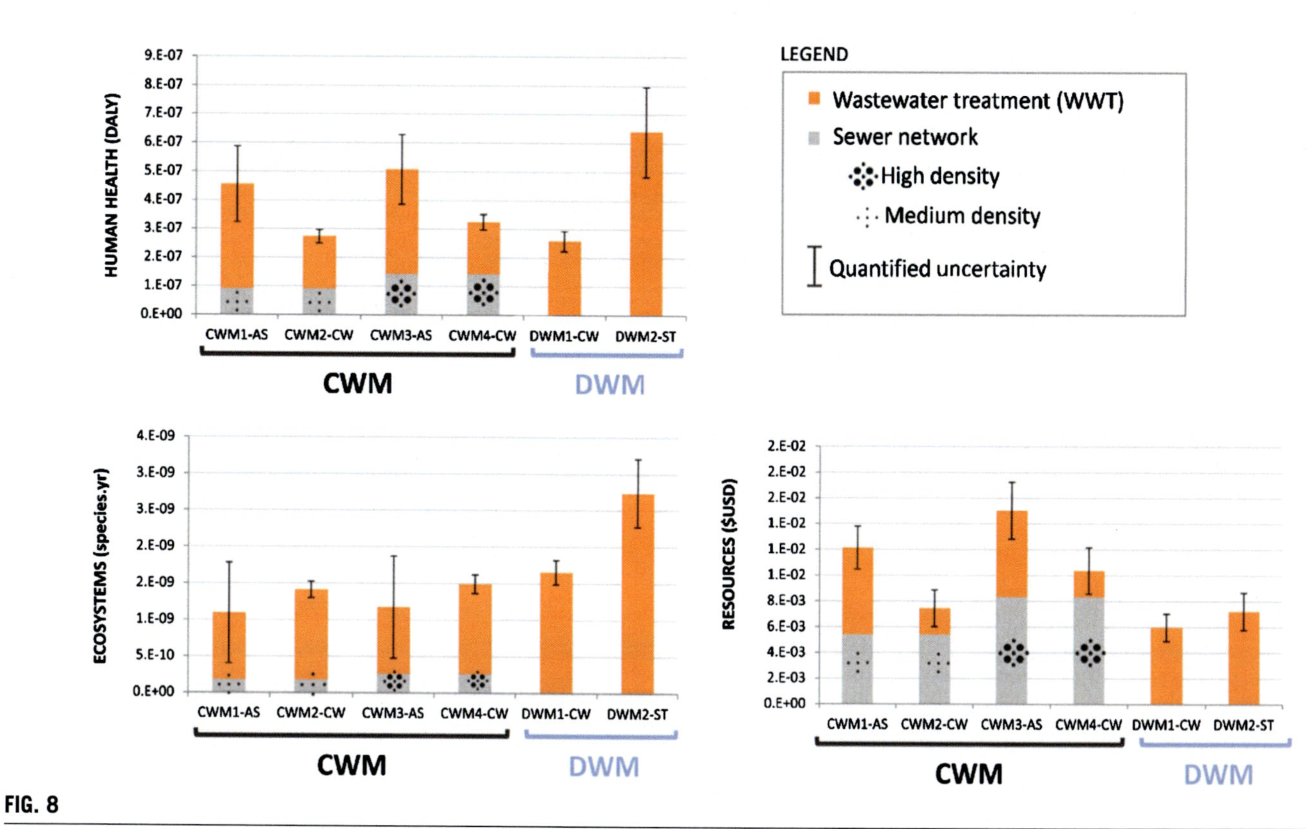

FIG. 8

LCA results of case study 3 (part II).

4 Conclusions

The literature review revealed that UWS LCA studies typically include operational impacts and apply a functional unit based on volume of water (with or without a treatment objective), though specific functional units varied, highlighting the multifunctionality of UWS. Through the examples provided we highlight the fact that the assessment of complex technological systems such as UWS requires reliable and detailed inventory data. Specifically, the construction of UWS infrastructure has been neglected in past LCA studies, probably because of the difficulties in gathering the construction information. The application of LCA has shown that construction of UWS elements should no longer be neglected (Morera, Remy, Comas, & Corominas, 2016; Risch, Gutierrez, Roux, Boutin, & Corominas, 2015), also as demonstrated in the 3 case studies presented. The construction phase plays a key role when using LCA to find the optimum between centralised and decentralised solutions at the scale of small (rural communities) or large (megalopolis) urban systems. This necessarily requires advanced water/wastewater network models that need precise inventory data. In this respect, the coupling of LCA with geographic information systems (providing information on networks and users) and with optimisation algorithms is a promising way to search for these trade-offs.

In the 3 case studies the interdependencies between UWS elements are taken into account by modelling the emissions and discharges from the different elements. Conventional systems (such as an activated sludge WWTP) are now well documented regarding their emissions and discharges (sludge, leaching, air emissions, etc.) and mechanistic models are available for their estimation when measured data is not available. However, limited inventories are available for non-conventional WWTPs and for other elements of the UWS, especially with regards to emerging solutions such as nature-based solutions. Modelling tools can be used to simulate potential UWS scenarios. Some platforms are available to model the performance of more than one UWS element (e.g., SIMBA# from inCTRL Solutions or WEST from DHI) which can improve understanding of interdependencies in terms of energy, resources consumption, and emissions.

The most important challenge for assessing and eco-designing future UWS systems with LCA will probably be managing multifunctionality. Indeed, at present, UWS components are mainly mono-functional and it is therefore possible to calculate the environmental impacts for each one using a functional unit (FU) representative of the provided service (e.g., FU = the production of one m^3 of drinking water or FU = the treatment of the discharges of one inhabitant equivalent during one day). The emergence of solutions linked to circular economy is already raising the issue of multifunctionality in LCA: energy recovery from sludge, reuse of wastewater and recovered nutrients in agriculture, etc. The emergence of nature-based solutions will further amplify this phenomenon, as these systems are almost all multifunctional (e.g., rainwater storage and regulation, wastewater treatment, nutrient recovery, water reuse, recreational areas, landscaping, biodiversity islands, continuity of ecological corridors). To be fair, a conventional LCA must compare the environmental performance of systems that provide the same service (quantified by a single functional unit). When we start comparing conventional UWS with emerging solutions including nature-based solutions and many circular economy loops, this will not be possible. The solution to these questions could be to move to analyses at meso scales (city, territory) to assess the complete system through the concept of territorial LCAs developed by Loiseau, Roux, Junqua, Maurel, and Bellon-Maurel (2013) and Loiseau et al. (2018). This would mean no longer calculating impacts per unit of provided service (e.g., impact in kg eqCO_2 per m^3 of treated water) but rather an eco-efficiency ratio (e.g., for 1 kg of CO_2 emitted, what is the set of provided services by each scenario in terms of functionality, landscape, energy provision, value or jobs created, etc.). Furthermore, progress is expected to be made in LCA impact assessment relevant to UWS, such as the development of impact indicators which take into account the effects of microplastics emissions to oceans.

Finally, LCA is well established for wastewater treatment and guidance is now available for practitioners (Corominas et al., 2020). Thus, on the basis of this guidance, the use of LCA for wastewater treatment entered a phase of routine use. This is not yet the case for other UWS elements (drinking water production, wastewater reuse, etc.) nor for the integration of UWS elements. Such guidance would harmonise the application of LCA to UWS, providing recommendations in the different steps: goal and scope (including functional unit), inventory, impact assessment, and interpretation.

Acknowledgements

Lluís Corominas acknowledges the Ministry of Economy and competitiveness for the Ramon and Cajal grant (RYC-2013-14595) and its corresponding I3 consolidation. The authors acknowledge the CLEaN-TOUR (CTM2017-85385-C2-1-R) and INVEST (RTI2018-097471-B-C21) projects from the Spanish Ministry of Economy and Competitiveness and thank Generalitat de Catalunya through Consolidated Research Group 2017 SGR 1318. ICRA researcher thanks funding from the CERCA program.

References

Andersson, S., Rahmberg, M., Nilsson, Å., Grundestam, C., Saagi, R., & Lindblom, E. (2020). Evaluation of environmental impacts for future influent scenarios using a model-based approach. *Water Science and Technology*, *81*(8), 1615–1622.

Antunes, L. N., Ghisi, E., & Severis, R. M. (2020). Environmental assessment of a permeable pavement system used to harvest stormwater for non-potable water uses in a building. *Science of the Total Environment*, *746*, 141087.

Antunes, L. N., Sydney, C., Ghisi, E., Phoenix, V. R., Thives, L. P., White, C., & Garcia, E. S. H. (2020). Reduction of environmental impacts due to using permeable pavements to harvest stormwater. *Water*, *12*(10), 2840.

Arden, S., Morelli, B., Cashman, S., Ma, X. C., Jahne, M., & Garland, J. (2021). Onsite non-potable reuse for large buildings: Environmental and economic suitability as a function of building characteristics and location. *Water Research*, *191*, 116635.

Arden, S., Morelli, B., Schoen, M., Cashman, S., Jahne, M., Ma, X. C., & Garland, J. (2020). Human health, economic and environmental assessment of onsite non-potable water reuse systems for a large, mixed-use urban building. *Sustainability*, *12*(13), 5459.

Barbosa de Jesus, T., Costa Kiperstok, A., & Borges Cohim, E. (2020). Life cycle assessment of rainwater harvesting systems for Brazilian semi-arid households. *Water Environment Journal*, *34*(3), 322–330.

Bonoli, A., Di Fusco, E., Zanni, S., Lauriola, I., Ciriello, V., & Di Federico, V. (2019). Green smart technology for water (GST4Water): Life cycle analysis of urban water consumption. *Water*, *11*(2), 389.

Byrne, D. M., Lohman, H. A., Cook, S. M., Peters, G. M., & Guest, J. S. (2017). Life cycle assessment (LCA) of urban water infrastructure: Emerging approaches to balance objectives and inform comprehensive decision-making. *Environmental Science: Water Research & Technology*, *3*(6), 1002–1014.

Cashman, S., Ma, X., Mosley, J., Garland, J., Crone, B., & Xue, X. (2018). Energy and greenhouse gas life cycle assessment and cost analysis of aerobic and anaerobic membrane bioreactor systems: Influence of scale, population density, climate, and methane recovery. *Bioresource Technology*, *254*, 56–66.

Corominas, L., Byrne, D., Guest, J. S., Hospido, A., Roux, P., Shaw, A., & Short, M. D. (2020). The application of life cycle assessment (LCA) to wastewater treatment: A best practice guide and critical review. *Water Research*, *184*, 116058.

Faragò, M., Brudler, S., Godskesen, B., & Rygaard, M. (2019). An eco-efficiency evaluation of community-scale rainwater and stormwater harvesting in Aarhus, Denmark. *Journal of Cleaner Production*, *219*, 601–612.

Friedrich, J., Poganietz, W. R., & Lehn, H. (2020). Life-cycle assessment of system alternatives for the water-energy-waste Nexus in the urban building stock. *Resources, Conservation and Recycling*, *158*, 104808.

Garcia, X., Barceló, D., Comas, J., Corominas, L., Hadjimichael, A., Page, T. J., & Acuña, V. (2016). Placing ecosystem services at the heart of urban water systems management. *Science of the Total Environment*, *563–564*, 1078–1085.

García-Sánchez, M., & Güereca, L. P. (2019). Environmental and social life cycle assessment of urban water systems: The case of Mexico City. *Science of the Total Environment*, *693*, 133464.

Ghimire, S. R., Johnston, J. M., Garland, J., Edelen, A., Ma, X. C., & Jahne, M. (2019). Life cycle assessment of a rainwater harvesting system compared with an AC condensate harvesting system. *Resources, Conservation and Recycling*, *146*, 536–548.

Goga, T., Friedrich, E., & Buckley, C. A. (2019). Environmental life cycle assessment for potable water production—A case study of seawater desalination and mine-water reclamation in South Africa. *Water SA*, *45*(4), 700–709.

Hsien, C., Low, J. S. C., Fuchen, S. C., & Han, T. W. (2019). Life cycle assessment of water supply in Singapore—A water-scarce urban city with multiple water sources. *Resources, Conservation and Recycling*, *151*, 104476.

International Organization for Standardization. (2006a). *ISO 14040: Environmental management—Life cycle assessment—Principles and framework (ISO 14040:2006)*.

International Organization for Standardization. (2006b). *ISO 14044: Environmental management—Life cycle assessment—Requirements and guidelines (ISO 14044:2006)*.

Jabri, K. M., Nolde, E., Ciroth, A., & Bousselmi, L. (2020). Life cycle assessment of a decentralized greywater treatment alternative for non-potable reuse application. *International Journal of Environmental Science and Technology*, *17*(1), 433–444.

Jeong, H., Broesicke, O. A., Drew, B., & Crittenden, J. C. (2018). Life cycle assessment of small-scale greywater reclamation systems combined with conventional centralized water systems for the City of Atlanta, Georgia. *Journal of Cleaner Production*, *174*, 333–342.

Kobayashi, Y., Ashbolt, N. J., Davies, E. G. R., & Liu, Y. (2020). Life cycle assessment of decentralized greywater treatment systems with reuse at different scales in cold regions. *Environment International*, *134*, 105215.

Leong, J. Y. C., Balan, P., Chong, M. N., & Poh, P. E. (2019). Life-cycle assessment and life-cycle cost analysis of decentralised rainwater harvesting, greywater recycling and hybrid rainwater-greywater systems. *Journal of Cleaner Production*, *229*, 1211–1224.

Loiseau, E., Aissani, L., Le Féon, S., Laurent, F., Cerceau, J., Sala, S., & Roux, P. (2018). Territorial life cycle assessment (LCA): What exactly is it about? A proposal towards using a common terminology and a research agenda. *Journal of Cleaner Production*, *176*, 474–485.

Loiseau, E., Roux, P., Junqua, G., Maurel, P., & Bellon-Maurel, V. (2013). Adapting the LCA framework to environmental assessment in land planning. *International Journal of Life Cycle Assessment*, *18*, 1533–1548.

Loubet, P., Roux, P., & Bellon-Maurel, V. (2016). WaLA, a versatile model for the life cycle assessment of urban water systems: Formalism and framework for a modular approach. *Water Research*, *88*, 69–82.

Loubet, P., Roux, P., Guérin-Schneider, L., & Bellon-Maurel, V. (2016). Life cycle assessment of forecasting scenarios for urban water management: A first implementation of the WaLA model on Paris suburban area. *Water Research*, *90*, 128–140.

Loubet, P., Roux, P., Loiseau, E., & Bellon-Maurel, V. (2014). Life cycle assessments of urban water systems: A comparative analysis of selected peer-reviewed literature. *Water Research*, *67*, 187–202.

Marinoski, A. K., & Ghisi, E. (2019). Environmental performance of hybrid rainwater-greywater systems in residential buildings. *Resources, Conservation and Recycling*, *144*, 100–114.

Morera, S., Remy, C., Comas, J., & Corominas, L. (2016). Life cycle assessment of construction and renovation of sewer systems using a detailed inventory tool. *International Journal of Life Cycle Assessment*, *21*, 1–13.

Opher, T., Friedler, E., & Shapira, A. (2019). Comparative life cycle sustainability assessment of urban water reuse at various centralization scales. *The International Journal of Life Cycle Assessment*, *24*(7), 1319–1332.

Pinelli, D., Zanaroli, G., Rashed, A. A., Oertlé, E., Wardenaar, T., Mancini, M., … Frascari, D. (2020). Comparative preliminary evaluation of 2 in-stream water treatment technologies for the agricultural reuse of drainage water in the Nile Delta. *Integrated Environmental Assessment and Management*, *16*(6), 920–933.

Remy, C., Seis, W., Miehe, U., Orsoni, J., & Bortoli, J. (2019). Risk management and environmental benefits of a prospective system for indirect potable reuse of municipal wastewater in France. *Water Supply*, *19*(5), 1533–1540.

Risch, E., Boutin, C., & Roux, P. (2021). Applying life cycle assessment to assess the environmental performance of decentralised versus centralised wastewater systems. *Water Research*, *196*, 116991.

Risch, E., Gasperi, J., Gromaire, M. C., Chebbo, G., Azimi, S., Rocher, V., ... Sinfort, C. (2018). Impacts from urban water systems on receiving waters—How to account for severe wet-weather events in LCA? *Water Research*, *128*, 412–423.

Risch, E., Gutierrez, O., Roux, P., Boutin, C., & Corominas, L. (2015). Life cycle assessment of urban wastewater systems: Quantifying the relative contribution of sewer systems. *Water Research*, *77*, 35–48.

Ryberg, M. W., Bjerre, T. K., Nielsen, P. H., & Hauschild, M. (2020). Absolute environmental sustainability assessment of a Danish utility company relative to the planetary boundaries. *Journal of Industrial Ecology*, *25*(3), 765–777.

Santana, M. V., Cornejo, P. K., Rodríguez-Roda, I., Buttiglieri, G., & Corominas, L. (2019). Holistic life cycle assessment of water reuse in a tourist-based community. *Journal of Cleaner Production*, *233*, 743–752.

Singh, A., Kamble, S. J., Sawant, M., Chakravarthy, Y., Kazmi, A., Aymerich, E., ... Philip, L. (2018). Technical, hygiene, economic, and life cycle assessment of full-scale moving bed biofilm reactors for wastewater treatment in India. *Environmental Science and Pollution Research*, *25*(3), 2552–2569.

Tarpani, R. R. Z., Lapolli, F. R., Recio, M.Á. L., & Gallego-Schmid, A. (2021). Comparative life cycle assessment of three alternative techniques for increasing potable water supply in cities in the global south. *Journal of Cleaner Production*, *290*, 125871.

Tavakol-Davani, H., Burian, S. J., Butler, D., Sample, D., Devkota, J., & Apul, D. (2018). Combining hydrologic analysis and life cycle assessment approaches to evaluate sustainability of water infrastructure. *Journal of Irrigation and Drainage Engineering*, *144*(11), 05018006.

Teodosiu, C., Barjoveanu, G., Sluser, B. R., Popa, S. A. E., & Trofin, O. (2016). Environmental assessment of municipal wastewater discharges: A comparative study of evaluation methods. *The International Journal of Life Cycle Assessment*, *21*(3), 395–411.

Xue, X., Cashman, S., Gaglione, A., Mosley, J., Weiss, L., Ma, X. C., ... Garland, J. (2019). Holistic analysis of urban water systems in the greater Cincinnati region: (1) life cycle assessment and cost implications. *Water Research X*, *2*, 100015.

Yang, C., Jiang, W., & Jian, N. (2020). Life cycle environmental impact assessment of three urban drainage systems for non-point source pollution control. *Journal of Coastal Research*, *115*(SI), 367–372.

Zanni, S., Cipolla, S. S., di Fusco, E., Lenci, A., Altobelli, M., Currado, A., ... Bonoli, A. (2019). Modeling for sustainability: Life cycle assessment application to evaluate environmental performance of water recycling solutions at the dwelling level. *Sustainable Production and Consumption*, *17*, 47–61.

CHAPTER

Environmental sustainability in energy production systems 17

Jacopo Bacenetti[a] and Sara González-García[b]

[a]Department of Environmental Science and Policy, Università degli Studi di Milano, Milan, Italy

[b]CRETUS, Department of Chemical Engineering, Universidade de Santiago de Compostela, Santiago de Compostela, Spain

1 Aims

This chapter focuses on agricultural renewable energy sources, and it aims at providing a general overview of the environmental impacts and benefits related to the renewable energy production in agriculture. To this purpose, some applications of Life Cycle Assessment (LCA) to different production pathways based on different feedstock (energy crops, by-products, and agricultural waste) are presented. Based on the results of the different case studies, the strengths and limits of the LCA approach application are discussed.

2 State of the art

The supply of sustainable energy is one of the main challenges that mankind will face over the coming decades, particularly because of the need to address global warming. Bioenergy refers to all types of energy derived from the conversion of biological organic sources referred to as feedstocks and biomass. Feedstocks include woody biomass (forestry and industrial residues), agricultural biomass (crops and residues), and biowastes such as solid municipal organic wastes, sewage, and livestock manure. Nevertheless, of all biomass materials, wood has been the most used in Europe. Biomass can make a substantial contribution to satisfy future energy demand in a sustainable way. In this regard, it is currently the largest global contributor of renewable energy and has significant potential to expand in the production of heat, electricity, as well as fuels for transport. One of the most promising sectors for growth in bioenergy production is in the form of residues from agriculture. Currently, this sector contributes less than 3% to the total bioenergy production. Data shows that utilising the residues from all major crops for energy can generate approx. 4.3 billion tonnes (low estimate) to 9.4 billion tonnes (high estimate) annually around the world. Utilising

Assessing Progress towards Sustainability. http://doi.org/10.1016/B978-0-323-85851-9.00013-4

standard energy conversion factors, the theoretical energy potential from residues can be in the range of 400–2000 Mtoe. The major contribution would be from cereals—mainly maize, rice, and wheat. Energy generation from agricultural residues could meet about 3%–14% of the total energy supply globally. The forestry sector is the largest contributor to the bioenergy mix globally. Forestry products including charcoal, fuelwood, pellets, and wood chips account for more than 85% of all the biomass used for energy purposes. One of the primary products from forests that are used for bioenergy production is wood fuel. Most of the wooden fuel is used for traditional cooking and heating in developing countries in Asia and Africa. Globally, 1.9 billion m^3 of wood fuel was used for energy purposes (IRENA, 2019).

In the last years, increasing attention was paid to the valorisation of agricultural waste and by-products for energy purpose. Agricultural wastes and by-products are produced during agro-food processes. More in detail, this biomass can be generated: (i) during the crop's biological cycle (e.g. pruning residues) and (ii) at the harvesting operations of the main product (e.g. cereals straw, corn stumps, etc.). The valorisation of by-products presents some issues related to their characteristics. In fact, despite the availability of large amount of biomass, the main critical aspects are the strong seasonality, the harvesting period tightness, and the high dispersion on territory (Fiala & Nonini, 2018). The opportunity to recover an agricultural by-product for its valorisation (raw material for industrial purposes, energy conversion, etc.) must always be carefully evaluated considering that nutrients and organic matter are removed from the soil and will not be available for the subsequent crops (Statuto, Frederiksen, & Picuno, 2019). With the repetition of the same production cycle (monocropping), the by-products' use should be generally avoided in all cases where their removal obliges the farmer to a subsequent soil resources reintegration (N–P–K fertilisation and/or livestock waste—manure and/or slurry distribution) (Ricciardi, Cillari, Carnevale Miino, & Collivignarelli, 2020).

In this context, this chapter presents the results of two different case studies focusing the attention on two promising residual agricultural biomass. In detail, the first case study analyses the environmental performances of electricity production in a medium size combined heat and power (CHP) plant while the second is focused on the valorisation of pruning residues from winemaking in small plants.

3 Novelty

Over the years, the environmental sustainability of different renewable energy production pathways was evaluated using the LCA approach. Despite this, most of these LCA studies focuses on the specific features of the evaluated process. The achieved results are useful to understand the impact of the different solutions and to identify mitigation strategies able to improve their environmental sustainability. However, despite this, from these studies it is difficult to have general and comprehensive information regarding the environmental performances of the different solutions for renewable energy production. Filling this gap represents the main novelty of this Chapter.

4 Case studies description

4.1 Wooden biomass-based systems analyzed from a life cycle perspective

Energy industries have contributed to ~32% of global CO_2 emission over the last 20 years (González-García & Bacenetti, 2019). Nevertheless, by 2018, total European greenhouse gases (GHG) emission decreased nearly 20% since 2005. Energy efficiency, renewables, and fuel shifting were crucial on that reduction specifically in the power sector (IEA, 2020). The use of renewable energy has grown 13% in 2010 to 18% in 2018 and solid biomass is one of the main driving forces of that increment (Sikkema, Proskurina, Banja, & Vakkilainen, 2021). Woody biomass, known as a carbon-neutral fuel, is progressively used for heating and power production, supported by national specific schemes (Sikkema et al., 2021). According to the World Energy Council (WEC, 1997), by 2050, energy consumption in the world will increase by more than 2 times. Consequently, the share of biomass in renewable energy will be about 32% (Kuznetsov, Syrodoy, Nigay, Maksimov, & Gutareva, 2021). Wooden biomass is the largest source of renewable energy globally (Springer, Kaliyan, Bobick, & Hill, 2017) and among woody sources, chips, sawdust, and forestry residues are quite common types of energy fuel. The last two examples are produced at the wooden factories and during harvesting activities in large volumes. Thus there is a clear potential for its use as renewable feedstock for energy purposed in the European Union although they could also be converted into different high value-added products including chemicals (Mateos, 2018). Since the waste-to-energy concept is being highly promoted as a part of the efforts into sustainable development in energy sector (Ferreira, Monteiro, Brito, & Vilarinho, 2017), this chapter aims to environmentally analyse by using Life Cycle Assessment (LCA) the production of electricity and heat from forestry residues considering as case study an example of district heating system in Italy. The interest behind this study is promoting the use of this type of wooden biomass in small combustion as substitute for conventional fuels (González-García & Bacenetti, 2019). In this study, forestry residues from dedicated forests focused on the production of industrial roundwood have been considered for analysis.

4.1.1 Goal and scope definition

The main goal of this case study is to provide a global overview of the environmental consequences associated with the production of bioenergy (electricity and heat in a Combined Heat and Power plant—CHP) considering forestry residues as feedstock. An attributional cradle-to-bioenergy factory gate approach is considered and energy allocation is selected to allocate the environmental impacts derived from the energy-based conversion system between both co-products.

4.1.2 Functional unit, system boundaries, and assumptions

The functional unit considered to report the environmental impacts is 1 kWh of electricity produced in the CHP. The choice of this energy-based unit allows the comparison with those profiles identified in the literature for bioenergy production system where not only forestry residues but also agricultural were considered as feedstock (González-García et al., 2014; González-García & Bacenetti, 2019; Muench & Guenther, 2013).

As far as the system boundaries are concerned, all the stages from raw materials extraction, distribution, chipping of forestry residues, and final combustion in the CHP have been included within the system boundaries. The feedstock considered in this case study is based on forestry residues derived from industrial forest systems destined to the production of industrial roundwood for further uses. In addition, other wooden residues derived from related industries (mostly those from boards factories) have been also considered (but in a lesser extent) to make available a greater volume of biomass.

Once the biomass is collected, it is directly chipped in the forest site by means of a self-propelled chipper. This step should not be required for industrial wooden residues such as chips and sawdust. Next, wooden biomass is distributed up to the bioenergy plant by a diesel lorry for 350 km.

As difference to other studies where wooden residues were used as feedstock and no environmental burdens were allocated to their production (González-García & Bacenetti, 2019), this approach has not been considered here. Mass allocation approach has applied to forest activities for roundwood and residues, since both co-products have market interest.

Once in the plant, biomass is sent to a boiler with a thermal power of 6 MW to produce thermal energy which is them transformed into mechanical energy and finally into electric energy in a turbine with 1 MW of electric power. Fig. 1 displays the system boundaries of the system under study.

Here two valuable co-products are obtained: thermal energy, which can be used in the district heating, and electricity, which is sent to the national grid. Although there is demand of electricity in the CHP which is directly taken from the grid, it has been assumed no recycling the one produced because of technical reasons (Fiala, 2012). As

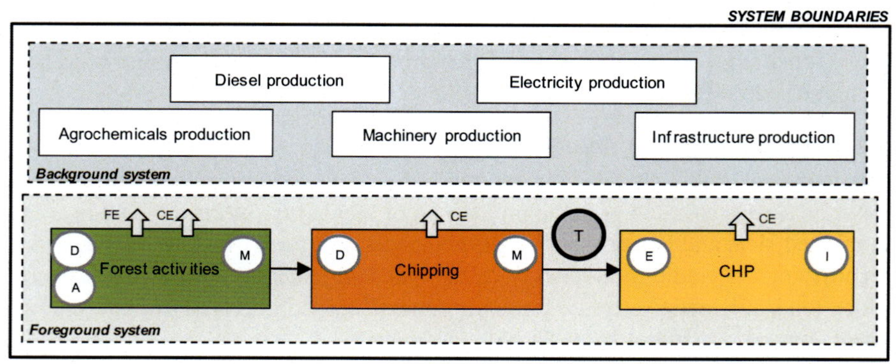

FIG. 1

System boundaries of the bioenergy system under study.

detailed before, energy allocation has been considered to distribute the impacts from the system between both co-products. Allocation factors of 20% and 80% have been assigned to electricity and heat, respectively.

In this study, it has been assumed that the biogenic CO_2 emissions derived from the combustion of wooden biomass in the boiler are compensated by the CO_2 absorbed during the growth of forest biomass. In addition, distribution of heat and electricity to final users has not been included under analysis.

4.1.3 Life cycle inventory

Table 1 summarises the main inventory data corresponding to the foreground system. These data correspond to mass and energy flows modelled according to information detailed in González-García and Bacenetti (2019) but adapted to the specific feedstock. A moisture content of 35% and a density of 280 kg/m^3 have been assumed for the forestry residues (González-García et al., 2014). Background processes involved in the system such as those related with the production of the different supplies have been taken from Ecoinvent ® database v3.5 (Wernet et al., 2016).

4.1.4 Environmental assessment methods

The ReCiPe 2016 hierarchist Midpoint method V1.03 World (2010) (Huijbregts et al., 2017) has been used for the selection of characterisation factors required to estimate the environmental impacts, the profile being reported in terms of nine midpoint categories considered relevant in terms of communication purposes: global warming (GW), stratospheric ozone depletion (SOD), terrestrial acidification (TA), freshwater eutrophication (FE), marine eutrophication (ME), terrestrial ecotoxicity (TET), freshwater ecotoxicity (FET), marine ecotoxicity (MET), and fossil resource scarcity (FRS). The SimaPro software v9.0 (PRé Consultants, 2020) has been considered for the computational implementation of the life cycle inventory and the life cycle impact assessment calculations.

Table 1 Data summary associated with the system.

Inputs			**Outputs**		
Materials			*Energy*		
Forestry residues	dm^3	4.50	Electricity	kWh	1.00
Diesel	dm^3	$2.51 \cdot 10^{-3}$	Heat	MJ	14.11
Energy			*Emissions to air*		
Electricity	kWh	0.236	Sulphur dioxide	mg	10.50
Transport			***Particulates***	***mg***	***189***
Lorry	kg·km	679	Nitrogen oxides	mg	371
			Waste to treatment		
			Ash to sanitary landfill	g	77.5

4.1.5 Results and discussion

Assessment of global profile

The results of the life cycle impact assessment using ReCiPe 2016 hierarchist Midpoint method v1.03 are detailed in Table 2.

Fig. 2 depicts the distribution of environmental impacts between activities involved in the life cycle: forest activities, chipping, transport, and CHP. Bearing

Table 2 Characterisation results per functional unit (1 kWh).

Impact category	Unit	Value
Global Warming (GW)	g CO_2 eq	45.6
Stratospheric Ozone Depletion (SOD)	mg CFC11 eq	0.029
Terrestrial Acidification (TA)	g SO_2 eq	0.201
Freshwater Eutrophication (FE)	mg P eq	1.19
Marine Eutrophication (ME)	mg N eq	1.09
Terrestrial Ecotoxicity (TET)	g 1,4-DCB	325
Freshwater Ecotoxicity (FET)	g 1,4-DCB	5.87
Marine Ecotoxicity (MET)	g 1,4-DCB	8.31
Fossil Resource Scarcity (FRS)	g oil eq	14.2

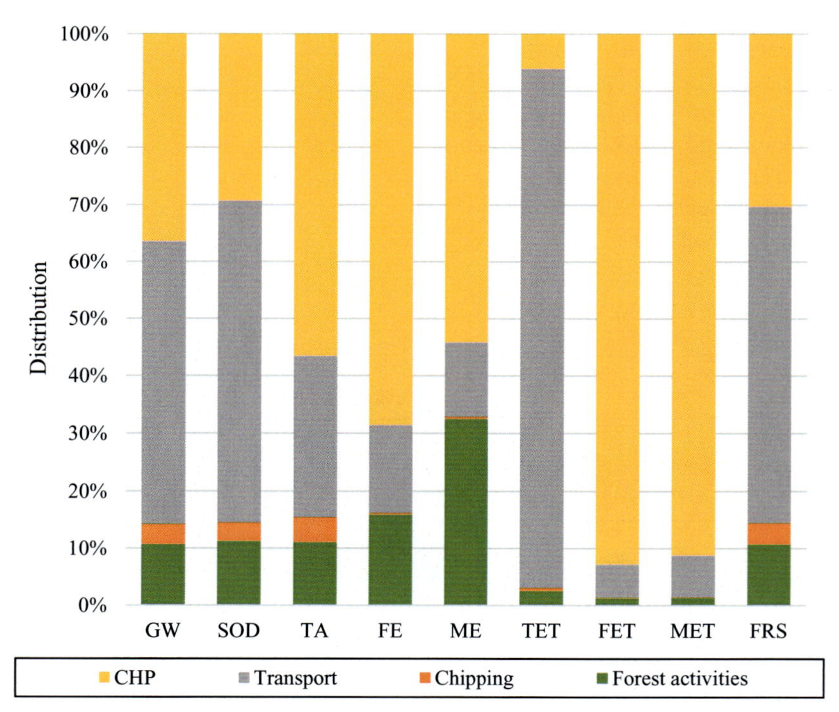

FIG. 2

Distribution of environmental burdens between processes involved.

in mind the profile, two stages can be considered as environmental hotspots. Thus the distribution of forest residues from forest site to the bioenergy plant involves contribution ratios ranging from 6% to 91% depending on the category. The effect is considerably remarkable in these categories directly affected by fossil fuels use such as GW, SOD, TET, and FRS. The activities performed in the plant focused on the biomass transformation into electricity and heat are responsible for outstanding contributing ratios in categories such as FET, MET, TA, FE, and ME.

Transport plays an important role in numerous impact categories. It includes the distribution by lorry from the forest site to the power plant for 350 km. In this study, we have considered an average transport distance although this assumption is quite sensitive. González-García and Bacenetti (2019) also identified this step as an environmental hotspot but considering 800 km (the average distance for an Italian case study). In this regard, a sensitivity analysis was performed by the authors regarding the transport effect over the global profile, concluding that transport distance could be decisive in decision-making strategies related with forest biomass-based systems. In this sense and with the aim of promoting wood-based energy systems, attention should be paid into the use of cleaner lorries such as Euro 5 and Euro 6, which involve considerable reductions in terms of, e.g. nitrogen oxides emission. In our case study, changing from Euro 5 to Euro 6 lorry should involve a reduction of 4% in SOD and 7% in TA, while in other categories it should be no higher than 1%. Of course, shifting from diesel to green vehicles should be an alternative. Nevertheless, more research is required on this issue with the aim of developing competitive electric or hybrid lorries.

The effect over the profile from the power plant is also remarkable. This stage includes the CHP-related activities as well as management of waste produced. Activities related to the power plant and mainly, the production of the high demand for electric energy is the main responsible for the contributions of this section to GW, TA, FE, and ME.Nevertheless, the disposal and treatment of the residual ashes in a sanitary landfill is behind the contributions to FET and MET. Forest activities involve contributions of 11% in GW, SOD, TA, and FRS; of 16% in FE; and of 33% in ME. This stage includes all processes performed in the forest such as site establishment, fertilisation, thinning, cleaning, harvesting, etc. All these activities require the use of forest machines and the corresponding consumption of fossil fuels such as diesel and lubricating oil, as well as mineral fertilisers. Consequently, fertilisation activities are behind the contributions in eutrophication-related categories and the consumption of fossil fuels in the remaining impact categories. Finally, chipping of forestry residues does not have a remarkable effect over the global profile. This stage includes the requirement of diesel and corresponding combustion emissions. Accordingly, the contributions from this stage are light in categories such as GW, SOD, TA, and FRS directly linked with the use of fossil fuels.

Comparison with national electricity profile

Since the final goal of bioenergy production is reducing the use of fossil-based energy and thus, promoting the use of renewable sources, the profile corresponding to the electricity produced in the case study has been compared with those associated

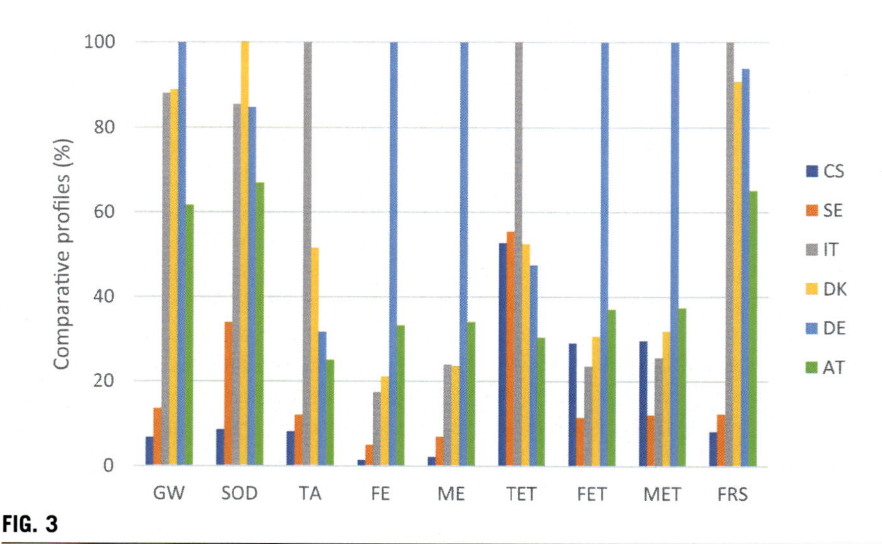

FIG. 3

Comparative profiles per kWh of electricity. *CS*, case study; *SE*, Sweden; *IT*, Italy; *DK*, Denmark; *DE*, Germany; *AT*, Austria.

with the electricity supplied by the national grid in EU countries that have reported a growth in the rate of district heating consumption in recent years (Sayegh et al., 2018) such as Austria, Sweden, Denmark, and Germany as well as Italy, where the plant is located. Fig. 3 depicts the comparative profile between the different scenarios per functional unit. In general terms, it can be concluded that the scenario under assessment involves environmental improvements regardless of the impact category considered for comparison. This scenario should be quite competitive from an environmental approach with the Swedish electricity profile, reporting improvements in all the categories except in FET and MET. Outstanding improvements are achieved in comparison with the electricity profiles associated to countries such as Italy, Denmark, and Germany. Thus it can be stated that the use of wooden residues in energy sector can contribute to solving environmental problems directly associated with fossil energy.

4.2 Agro-waste based systems analyzed from a life cycle perspective

Concerning renewable energy production, the European Unit sets ambitious targets aiming at increasing the European share of renewable energy (Cucchiella, D'Adamo, & Gastaldi, 2018). Thanks to this legislative framework, of the global total installed capacity of renewable energy in 2012 (1440 GW), about 22% was in the European Union (Cucchiella et al., 2018).

In this context, producing energy from the valorisation of agricultural and agro-food by-products is an effective solution because, while helping to reach the EU targets can optimise the management of these feedstock. In the last years, strong attention has been paid on the renewable energy production from woody biomass, but with focus on the exploitation of biomass from forestry (Kanematsu, Oosawa, Okubo, & Kikuchi, 2017; Schmidt, Leduc, Dotzauer, Kindermann, & Schmid, 2010) or from dedicated plantations (Dias et al., 2017; González-García & Bacenetti, 2019). However, agricultural woody residues are available in massive quantities and provide a considerable potential for energy production. Their exploitation for renewable energy production can also be considered as an extra income source. However, each agricultural sector produces biomass wastes and by-products characterised by specific chemical and physical features but also by different availability in terms of amounts and timing. In grape production, the management of the pruning residues is often problematic due to economic and phytosanitary reasons. Although considerable amounts of pruning residues are available, the biomass is spread on a large area, its harvesting has a high cost due to orographic characteristics of the vineyards and has a low economic value and poor market demand (Boschiero, Cherubini, Nati, & Zerbe, 2016). Consequently, pruning residues are usually collected outside the vineyards to be burnt or chopped and left in the inter-rows to be chopped. However, this second solution is not optimal, because it involves phytosanitary concerns related to the possible overwintering of pests and pathogens from pruning residues left on field. Therefore the valorisation of pruning residues of grapevine can also have an additional benefit. However, it is feasible only if the produced energy is fully exploited and locally utilised (if possible, directly from the wine producers) (Picchi, Silvestri, & Cristoforetti, 2013; Zanetti et al., 2017).

In the last years, several studies focused on the energetic valorisation of pruning residues for energy purposes. To select the best strategies for bioenergy production, the analytic evaluation of their environmental performances is needed. Up to now, few studies were carried out on the possibility to valorise the pruning residues for the contextual generation of heat and cold for the supply of the energy requirements of the agro-food industry. The aim of this study was to evaluate the environmental performance related to the valorisation of pruning residues to produce heat and cold by means of a biomass boiler and an absorption chiller. To do so, LCA was used and three different management scenarios of pruning residues were evaluated.

4.2.1 Goal and scope definition

The goal of this study is to compare three different management systems for vineyard pruning residues in the context of Northern Italy wine production system. Three different management systems of pruning residues were considered. In the first one, Baseline Scenario—BS, the residues are managed traditionally and, consequently, chopped using a multi-purpose mulcher with interchangeable knives and left in the inter-rows to be incorporate into the soil using a disc plough. In BS, no energetic valorisation of pruning residues occurs and, consequently, the heat and cold requirements of the winery are fully supplied using fossil fuels. More in details, a boiler fed with natural gas and an electric chiller are used to satisfy the heat and cold demand, respectively. In the Alternative Scenario 1 (AS1), the field operations involve the chipping and the collection of the pruning residues using a picker up-shredder as well as the transport of the produced wood chip to the winery. The transport is performed (distance 3 km) using a farm trailer coupled with a tractor. The wood chip is burnt in a biomass boiler to produce heat that is partially used to supply the thermal energy requirement of the winery and, partially, to feed an adsorption chiller. Both heat and cold are used at the winery during the winemaking process. Among the different solutions for energetic valorisation of the pruning residues, the coupling of a biomass boiler with an adsorption chiller is particularly suitable for winemaking process (Dutilh & Kramer, 2000). The thermal energy produced in BS by natural gas is replaced by that produced in the biomass boiler (nominal thermal power of 150 kW) while the 'cold demand' is satisfied with the adsorption chiller instead of with an electric refrigeration unit.

In the Alternative Scenario 2 (AS2), the pruning residues are collected as in AS1 but, about energetic valorisation, only the production of heat by the combustion of wood chips in a biomass boiler is considered. As in AS1, heat produced replaces the thermal energy that in BS is generated using natural gas but, unlike AS1, the 'demand for cooling' is satisfied using a traditional refrigeration unit that consumes the electricity extracted from the national grid.

4.2.2 Functional unit, system boundaries, and assumptions

The scenarios described in the previous section provide different functions; however, the common function is the management of the pruning residues. For this reason, the selected functional unit is the management of the pruning residues produced on 1 ha (1 ha) of vineyard. While, about the system boundary, a 'from cradle to gate' approach was considered. All the activities from pruning residues management into the vineyards to the winery's utilities system were included. Grapevine cultivation, winemaking (except the energy supply system), wine distribution, use, and end of life were excluded. Fig. 4 displays the system boundaries of the system under study.

In this study, it has been assumed that the biogenic CO_2 emissions derived from the combustion of wooden biomass in the boiler are compensated by the CO_2

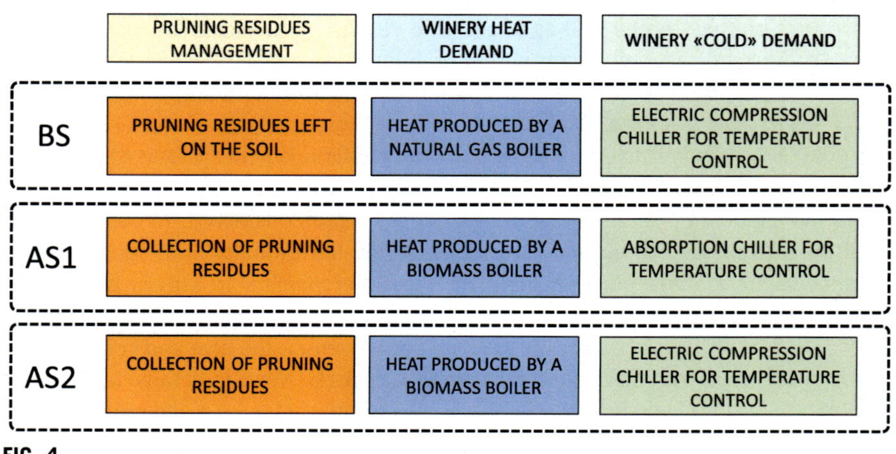

FIG. 4

Schematisation of the three scenarios evaluated.

absorbed during the growth of grape. In addition, no changes in the soil organic carbon content were considered.

4.2.3 Life cycle inventory

Primary inventory data concerning the amount of pruning residues available and their management as well as the energy consumption (heat and electricity) during winemaking were collected by means of surveys and interviews in a social winery. Secondary data were taken from the literature: (i) the energy efficiency of the boiler and for the coefficient of performance (cop) of the adsorption chiller (Fiala, 2012); (ii) the physico-chemical characteristics of pruning residues (Duca et al., 2016); (iii) the emissions from pruning residues' combustion (González-García et al., 2014; Prando et al., 2016); (iv) for pruning residues, a harvest loss of 15% (Spinelli, Magagnotti, & Nati, 2010; Spinelli, Nati, Pari, Mescalchin, & Magagnotti, 2012). Background data to produce diesel fuel, electricity and heat, tractors and agricultural machines, boiler and chiller were obtained from the Ecoinvent database® v.3.5 (Moreno Ruiz et al., 2018; Weidema et al., 2013; Wernet et al., 2016).

4.2.4 Environmental assessment methods

Using the characterisation factors reported by the midpoint ILCD method (EC-JRC-European Commission, Joint Research Centre, Institute for Environment and Sustainability, 2012), the following impact categories were considered: Climate change (CC), Ozone depletion (OD), Human toxicity, cancer effects (HTc), Human toxicity, non-cancer effects (HTnoc), Particulate matter (PM), Photochemical oxidant formation (POF), Terrestrial acidification (TA), Freshwater eutrophication (FE), Terrestrial eutrophication (TE), Marine eutrophication (ME), Freshwater ecotoxicity (FEx), Mineral fossil, and renewable resource depletion (MFRD).

4.2.5 Results and discussion

Assessment of global profile

The results of the life cycle impact assessment are reported in Table 3 while Fig. 5 shows the relative comparison among the three scenarios.

The identification of the best scenario depends on the considered impact category. As regard to the energetic valorisation of wood chips produced by pruning residues in AS1, the production of thermal and electric energy involves environmental benefits for some impact categories, such as climate change—CC, ozone depletion—OD, human toxicity-cancer effects—HT-c, terrestrial acidification—TA, and freshwater eutrophication—FE. These benefits are related to the avoided production of heat from natural gas and to the avoided consumption of electricity for cooling at the winery. For both the AS, wood chips combustion worsens the results for all the impact categories (e.g. HT-noc, PM, POF, TE, ME, FEx, and MFRD) affected by the pollutants emitted (e.g. particulate matters, NMVOC) in the exhaust gases of biomass boiler. This worsening is remarkable, and it is due to the combustion of wood in a small size biomass boiler not equipped with specific devices for exhaust gas treatment and cleaning. When the pruning residues are collected to produce only heat (AS2), the

Table 3 Characterisation results per functional unit (1 ha).

Impact category	Unit	BS	AS1	AS2
CC	kg CO_2 eq	89.45	−807.39	−258.21
OD	mg CFC-11 eq	5.46	−42.893	0.234
HT-noc	CTUh	$3.71 \cdot 10^{-5}$	$-4.29 \cdot 10^{-5}$	$-2.55 \cdot 10^{-5}$
HT-c	CTUh	$9.57 \cdot 10^{-7}$	$-1.08 \cdot 10^{-6}$	$-3.66 \cdot 10^{-6}$
PM	kg PM2.5 eq	0.063	2.437	0.359
POF	kg NMVOC eq	1.112	1.157	0.888
TA	molc H+ eq	0.842	−1.931	−0.722
TE	molc N eq	4.280	6.705	4.255
FE	g P eq	3.667	−105.27	−0.078
ME	kg N eq	0.391	0.503	0.345
FEx	CTUe	136.90	1611.56	−338.32
MFRD	g Sb eq	2.645	25.017	3.052

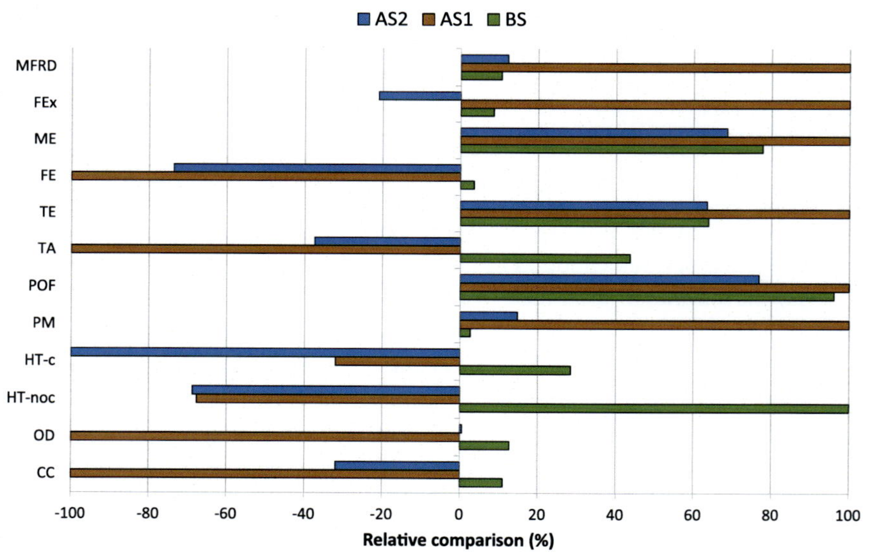

FIG. 5

Relative comparison among the three evaluated scenarios.

impact related to the manufacturing of adsorption chiller is not accounted and, consequently, the impact for the toxicity-related impact categories (HT-C, HT-noc, and FEx) is reduced.

For BS, the environmental impact could be reduced, for the CC, thanks to the increased soil carbon content related to the soil incorporation of chopped residues. However, this aspect is hardly quantifiable (Cowie, Smith, & Johnson, 2006; Repullo,

Carbonell, Hidalgo, Rodríguez-Lizana, & Ordóñez, 2012) and, consequently, it was not considered in the analysis. Concerning the contribution analysis, for BS the main contributors are (i) the consumption of diesel (from 3% in HT-noc to 97% of OD); (ii) the emissions related to diesel combustion in the tractor (from 0.3% in HT-c to 96% of TE); (iii) tractor consumption for HT-c, FE, and FEx (37%, 42%, and 50%) and MFRD (89%). Agricultural machinery consumption shows a low impact: <20% except than for HT-c (23%), Fe (15%), and FEx (11%). For the two alternative scenarios, the results for contribution analysis are reported in Fig. 6 (AS1 on the top, AS2 on the bottom).

4.3 Other sustainability issues

Beyond environmental sustainability, the analysis of impacts is used throughout different sectors on the goal of determining the potential damages and benefits to the society that an activity or production system could cause (Vis, Dörnbrack, & Haye, 2014). Providing a global world population with sustainable bioenergy sources is a challenge. In this regard, EU decision-makers embraced an unprecedented focus on the climate emergency and sustainable growth. In addition, on the horizon it is the achievement of carbon neutrality in the EU by 2050 but this lengthy journey requires satisfying not only environmental issues but also social and economic challenges. Bioenergy sits at the nexus of two of the main environmental crises of the 21st century that are biodiversity and climate emergencies, but it should not be at any economic and social cost. Therefore the possibility of using bioenergy as a climate change mitigation measure has triggered a discussion of whether or how bioenergy systems contribute to sustainable development (Robledo-Abad et al., 2017). The integration of environmental and economic analysis of bioenergy systems has been concentrated in some studies, mostly focused on agricultural biomass (Robledo-Abad et al., 2017). Nevertheless, social dimension of sustainability is commonly forgotten although many efforts have been performed to standardise and provide the procedures to include it under the multidimensionality of life cycle assessment (Alidoosti et al., 2021). According to Rafiaani et al. (2018), the social sustainability of bioindustries requires an analysis of how the bio-product can be accepted, how big is the domain of acceptance by the society and what are the advantages of bioindustry for various societies. The understanding of sustainability analysis is complex since it requires the assessment or ongoing monitoring of impacts on human well-being such as food security, conflicts or social tension, health impacts, deforestation of forest degradation, displacement of activities or land use, poverty, market opportunities, fair salary, and employment among others. Therefore, and based on Alidoosti et al. (2021), there is rarely a comprehensive social sustainability definition with balancing other aspects of sustainability, including economic and environmental aspects.

Further research should be conducted with the aim of acquiring a full overview of the sustainability index of bioenergy systems paying special attention on the

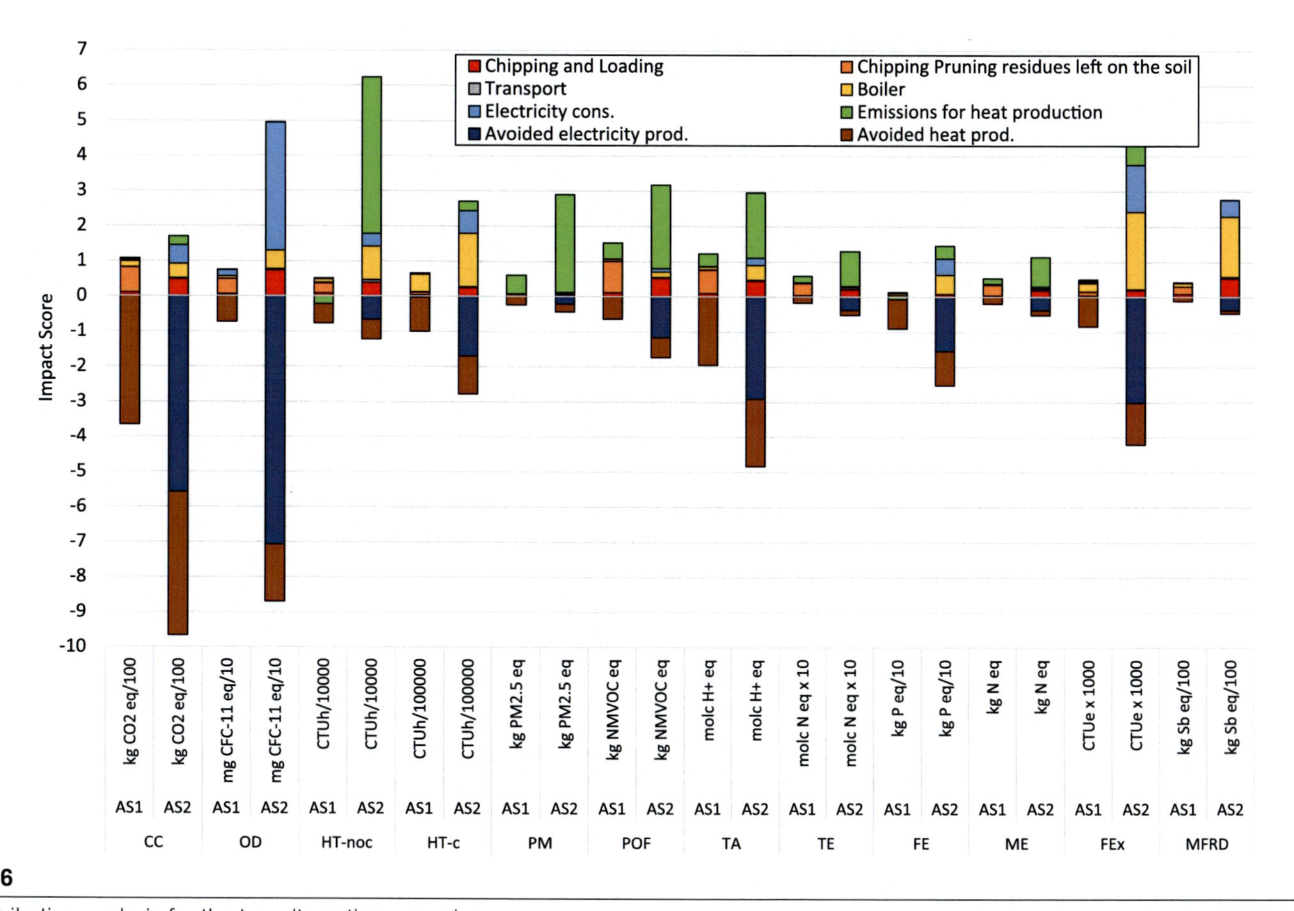

FIG. 6

Contribution analysis for the two alternative scenarios.

selection of representative indicators or impacts to analyse these specific systems since the preferences of the stakeholders involved in the analysis considerably affect the results (Alidoosti et al., 2021). In addition, there is no consensus regarding the selection of impacts required to perform a sustainability analysis in bioenergy systems. In this regard, report context conditions and criteria for attributing development impacts transparently are required to provide a solid scientific basis for policymaking and governance agreements in the field of sustainable bioenergy.

5 Conclusions

Sustainability mainly relies on three pillars: economic, environmental, and social aspects. This is true also for renewable energy sources and, among them, also for woody biomass-to-energy pathways. Economic aspect was widely studied and, for the renewable electricity production, is usually supported by the presence of subsidy framework, the social one was less investigated also because the set-up of a widely accepted assessment methodology is still under development. Conversely, the environmental sustainability was deeply evaluated mainly using the LCA approach. Regarding the renewable energy production from wooden biomass, a huge potential for energy production is still unexploited because the economic aspect is still unsatisfactory. When not supported by subsidies, the generation of renewable energy from wooden biomass cannot be economically viable and, therefore, is still underdeveloped. This is true for heat production from secondary biomass such as agricultural by-products. In addition, it can contribute to the creation of new jobs in regions with developed forest production, vineyards, and wood processing. Therefore additional research should be required not only in the environmental issue of the sustainability but also in the social and economic pillars to obtain a complete overview. Moreover, future research activities should consider an expansion of the system boundary to consider the possible variation of the soil carbon content as well as the effect due to the higher nutrients removal related to pruning residues collection, when applicable.

More in detail from the two case studies presented in this chapter it could be concluded that when electricity production considers wood chips in medium-large size power plant (case study 1), the production and/or collection of the biomass is the main driver of the environmental results. On the contrary, when heat derives from agricultural by-products in small size devices, pollutant emissions produced during the biomass combustion play a key role. Unfortunately, the link between plant size and emissions is quite strict and affects deeply the economic and environmental performances. More in detail, for medium-large size power plants, the installation of devices for the pollutant abatement is economically sustainable and, for some impact categories, can offset the environmental load related to the consumption of dedicated biomass. Concerning small-medium size plants, the installation of the pollutant abatement devices is not affordable from an economic point of view and, consequently, for the impact categories affected by these pollutants the impact is higher, sometimes higher than the one related to the energy production from fossil fuel.

Acknowledgements

This work has been partially supported by the Forest related Life Cycle Assessment Network which is financed by Samnordisk skogsforskning (SNS).

Authors' contribution

JB and SGG developed the investigation, conceptualised, writing–review, and editing the chapter.

References

Alidoosti, Z., Sadegheih, A., Govindan, K., Pishvaee, M. S., Mostafaeipour, A., & Hossain, A. K. (2021). Social sustainability of treatment technologies for bioenergy generation from the municipal solid waste using best worst method. *Journal of Cleaner Production*, *288*, 125592.

Boschiero, M., Cherubini, F., Nati, C., & Zerbe, S. (2016). Life cycle assessment of bioenergy production from orchards woody residues in northern Italy. *Journal of Cleaner Production*, *112*, 2569–2580.

Cowie, A. L., Smith, P., & Johnson, D. (2006). Does soil carbon loss in biomass production systems negate the greenhouse benefits of bioenergy? *Mitigation and Adaptation Strategies for Global Change*, *11*(5–6), 979–1002.

Cucchiella, F., D'Adamo, I., & Gastaldi, M. (2018). Future trajectories of renewable energy consumption in the European Union. *Resources*, *7*, 10.

Dias, G. M., Ayer, N. W., Kariyapperuma, K., Thevathasan, N., Gordon, A., Sidders, D., et al. (2017). Life cycle assessment of thermal energy production from short-rotation willow biomass in Southern Ontario, Canada. *Applied Energy*, *204*, 343–352.

Duca, D., Toscano, G., Pizzi, A., Rossini, G., Fabrizi, S., Lucesoli, G., et al. (2016). Evaluation of the characteristics of vineyard pruning residues for energy applications: Effect of different copper-based treatments. *Journal of Agricultural Engineering*, *47*(1), 22–27.

Dutilh, C. E., & Kramer, K. J. (2000). Energy consumption in the food chain: Comparing alternative options in food production and consumption. *Ambio: A Journal of the Human Environment*, *29*(2), 98–101.

EC-JRC-European Commission, Joint Research Centre, Institute for Environment and Sustainability. (2012). *Characterisation factors of the ILCD recommended life cycle impact assessment methods. Database and supporting information* (1st ed.). Luxembourg: Publications Office of the European Union. February 2012. EUR 25167.

Ferreira, S., Monteiro, E., Brito, P., & Vilarinho, C. (2017). Biomass resources in Portugal: Current status and prospects. *Renewable and Sustainable Energy Reviews*, *78*, 1221–1235.

Fiala, M. (2012). *Energia da biomasse, Maggioli Editore* (pp. 1–437).

Fiala, M., & Nonini, L. (2018). Biomass and biofuels. In *Vol. 189. EPJ web of conferences* (p. 00006). EDP Sciences.

González-García, S., & Bacenetti, J. (2019). Exploring the production of bio-energy from wood biomass. Italian case study. *Science of the Total Environment*, *647*, 158–168.

González-García, S., Dias, A. C., Clermidy, S., Benoist, A., Maurel, V. B., Gasol, C. M., et al. (2014). Comparative environmental and energy profiles of potential bioenergy production chains in southern Europe. *Journal of Cleaner Production*, *76*, 42–54.

Huijbregts, M. A. J., Steinmann, Z. J. N., Elshout, P. M. F., Stam, G., Verones, F., Vieira, M. D. M., et al. (2017). *ReCiPe 2016 v1.1*.

IEA. (2020). https:/www.iea.org/reports/european-union-2020.

IRENA. (2019). *Renewable energy and jobs – annual review 2019*. Abu Dhabi: IRENA.

Kanematsu, Y., Oosawa, K., Okubo, T., & Kikuchi, Y. (2017). Designing the scale of a woody biomass CHP considering local forestry reformation: A case study of Tanegashima, Japan. *Applied Energy*, *198*, 160–172.

Kuznetsov, G. V., Syrodoy, S. V., Nigay, N. A., Maksimov, V. I., & Gutareva, N. Y. (2021). Features of the processes of heat and mass transfer when drying a large thickness layer of wood biomass. *Renewable Energy*, *169*, 498–511.

Mateos, E. (2018). Study on the potential of Forest biomass residues for bio-energy. *Proceedings*, *2*, 1420. https:/doi.org/10.3390/proceedings2231420.

Moreno Ruiz, E., Valsasina, L., Brunner, F., Symeonidis, A., FitzGerald, D., Treyer, K., et al. (2018). *Documentation of changes implemented in ecoinvent database v3.5*. Zürich, Switzerland: Ecoinvent.

Muench, S., & Guenther, E. (2013). A systematic review of bioenergy life cycle assessments. *Applied Energy*, *112*, 257–273.

Picchi, G., Silvestri, S., & Cristoforetti, A. (2013). Vineyard residues as a fuel for domestic boilers in Trento Province (Italy): Comparison to wood chips and means of polluting emissions control. *Fuel*, *113*, 43–49.

Prando, D., Boschiero, M., Campana, D., Gallo, R., Vakalis, S., Baratieri, M., et al. (2016). Assessment of different feedstocks in South Tyrol (Northern Italy): Energy potential and suitability for domestic pellet boilers. *Biomass and Bioenergy*, *90*, 155–162.

PRé Consultants. (2020). *SimaPro database manual*. The Netherlands: Methods Library.

Rafiaani, P., Kuppens, T., Van Dael, M., Azadia, H., Lebailly, P., & Van Passel, S. (2018). Social sustainability assessments in the biobased economy: Towards a systemic approach. *Renewable and Sustainable Energy Reviews*, *82*, 1839–1853.

Repullo, M. A., Carbonell, R., Hidalgo, J., Rodríguez-Lizana, A., & Ordóñez, R. (2012). Using olive pruning residues to cover soil and improve fertility. *Soil and Tillage Research*, *124*, 36–46.

Ricciardi, P., Cillari, G., Carnevale Miino, M., & Collivignarelli, M. C. (2020). Valorization of agro-industry residues in the building and environmental sector: A review. *Waste Management & Research*, *38*(5), 487–513.

Robledo-Abad, C., Althaus, H.-J., Berndes, G., Bolwig, S., Corbera, E., et al. (2017). Bioenergy production and sustainable development: Science base for policymaking remains limited. *GCB Bioenergy*, *9*, 541–556.

Sayegh, M. A., Jadwiszczak, P., Axcell, B. P., Niemierka, E., Brys, K., & Jouhara, H. (2018). Heat pump placement, connection and operational modes in European district heating. *Energy and Buildings*, *166*, 122–144.

Schmidt, J., Leduc, S., Dotzauer, E., Kindermann, G., & Schmid, E. (2010). Cost-effective CO_2 emission reduction through heat, power and biofuel production from woody biomass: A spatially explicit comparison of conversion technologies. *Applied Energy*, *87*(7), 2128–2141.

Sikkema, R., Proskurina, S., Banja, M., & Vakkilainen, E. (2021). How can solid biomass contribute to the EU's renewable energy targets in 2020, 2030 and what are the GHG drivers and safeguards in energy- and forestry sectors? *Renewable Energy*, *165*, 758–772.

Spinelli, R., Magagnotti, N., & Nati, C. (2010). Harvesting vineyard pruning residues for energy use. *Biosystems Engineering*, *105*(3), 316–322.

Spinelli, R., Nati, C., Pari, L., Mescalchin, E., & Magagnotti, N. (2012). Production and quality of biomass fuels from mechanized collection and processing of vineyard pruning residues. *Applied Energy*, *89*(1), 374–379.

Springer, N., Kaliyan, N., Bobick, B., & Hill, J. (2017). Seeing the forest for the trees: How much woody biomass can the Midwest United States sustainably produce? *Biomass and Bioenergy*, *105*, 266–277.

Statuto, D., Frederiksen, P., & Picuno, P. (2019). Valorization of agricultural by-products within the "Energyscapes": Renewable energy as driving force in modeling rural landscape. *Natural Resources Research*, *28*(1), 111–124.

Vis, M., Dörnbrack, A.-S., & Haye, S. (2014). Socio-economic impact assessment tools. In *Socio-economic impacts of bioenergy production* (pp. 1–16). Switzerland: Springer International Publishing. https:/doi.org/10.1007/978-3-319-03829-2_1.

WEC. (1997). *Energy for the future: Renewable sources of energy, white paper for a community strategy and action plan* (p. 53). Bruxelles: WEC.

Weidema, B. P., Bauer, C., Hischier, R., Mutel, C., Nemecek, T., Reinhard, J., et al. (2013). Overview and methodology. In *Vol. 3. Data quality guideline for the ecoinvent database version 3. Ecoinvent Report 1*. St. Gallen: The Ecoinvent Centre.

Wernet, G., Bauer, C., Steubing, B., Reinhard, J., Moreno-Ruiz, E., & Weidema, B. (2016). The ecoinvent database version 3 (part I): Overview and methodology. *International Journal of Life Cycle Assessment*, *21*, 1218–1230.

Zanetti, M., Brandelet, B., Marini, D., Sgarbossa, A., Giorio, C., Badocco, D., et al. (2017). Vineyard pruning residues pellets for use in domestic appliances: A quality assessment according to the EN ISO 17225. *Journal of Agricultural Engineering*, *48*(2), 99–108.

CHAPTER

Sustainability assessment of biotechnological processes: LCA and LCC of second-generation biobutanol production

18

Antonio Marzocchella[a], Roberto Chirone[a,b], Andrea Paulillo[b,c], Paola Lettieri[c], and Piero Salatino[a]

[a]*Department of Chemical, Materials and Production Engineering, University of Naples Federico II, Naples, Italy* [b]*eLoop S.r.l., Naples, Italy* [c]*Department of Chemical Engineering, University College London, London, United Kingdom*

1 Aims

This chapter focuses on a novel process to produce second-generation biobutanol from agro-food waste, as a sustainable alternative to fossil fuels. The case study has been selected so as to accomplish the following objectives:

- Demonstrate the combined application of Life Cycle Assessment (LCA) and Life Cycle Costing (LCC) to evaluate the environmental and economic performance of second-generation biobutanol from potato peel.
- Identify hotspots along the life cycle and propose improvements.
- Compare biobutanol with conventional fossil fuel alternatives, identifying and discussing relevant trade-offs.

2 State of the art

Biofuels have three main advantages over the conventional fossil counterpart: (i) they are produced from renewable resources; (ii) they minimise environmental impacts due to low SO_x, NO_x emissions, and other toxic emissions release during the use phase (Briens, Piskorz, & Berruti, 2008; Littlejohns, Rehmann, Murdy, Oo, & Neill, 2018); and (iii) they typically yield 60%–140% reduction in carbon footprint, depending on raw materials, technology, and methodological assumptions (Colling Klein, Bonomi, & Maciel Filho, 2018; Hanaki & Portugal-Pereira, 2018; Humbird et al., 2011b; Macedo, Seabra, & Silva, 2008; Xue et al., 2016).

Assessing Progress towards Sustainability. http://doi.org/10.1016/B978-0-323-85851-9.00015-8

Biofuels are classified based on biomass origin. Biofuels of first generation are produced from food crops that can also be used, directly and indirectly, for human/animal consumption. Second-generation biofuels are produced from residual biomass and non-food crops. A vast proportion of second-generation biofuel source is of lignocellulosic nature. The third and the fourth generation biofuels are produced from microalgae, either wild or genetically modified (European Union, 2018).

The main limitation of first-generation biofuels is the food-vs-fuel competition of agricultural commodities. Second-generation biofuels may overcome this limitation, but they still face a number of technical limitations (van Dyk, Su, McMillan, & Saddler, 2019; Walter, Seabra, Machado, de Barros Correia, & de Oliveira, 2018). The production of lignocellulosic biomass requires fertilisers and pesticides, a source of substantial environmental impacts. The harvesting and transportation of biomass are other critical issues for environmental and social aspects. Furthermore, biorefinery processes to produce second-generation biofuels generally yield smaller fuel production and larger water footprint compared to both conventional counterparts and first-generation biofuels (Walter et al., 2018); of note, water consumption for biofuel production reported in the literature spans several orders of magnitude (Berndes et al., 2015).

Biobutanol is one of the most promising biofuels due to its superior fuel features: its energy content exceeds by nearly 30% than that of ethanol and its combustion performance is similar to that of gasoline. Biobutanol can be used as direct transportation fuel in blends of gasoline or diesel, with greater amounts and with minor or no engine modifications (Ezeji & Blaschek, 2010) compared to bioethanol and biodiesel (Niemistö et al., 2013). A biotechnological route to produce butanol is the Acetone–Butanol–Ethanol (ABE) fermentation (Jones & Woods, 1986). Although the ABE fermentation process is well known, many challenges still need to be addressed for its industrial application, such as high feedstock costs, low alcohol yield, small reactor productivity, high operational and equipment costs, and water consumption. Research efforts are made along a twofold pathway: technology development; cheaper and more sustainable feedstock (Amiri & Karimi, 2018; García, Päkkilä, Ojamo, Muurinen, & Keiski, 2011; Nanda et al., 2017; Procentese et al., 2017; Raganati et al., 2018).

Despite the extensive scientific literature of LCA studies on biofuel production (Muench & Guenther, 2013; Ubando, Rivera, Chen, & Culaba, 2019), very few studies address biobutanol production. Pereira, Chagas, Dias, Cavalett, and Bonomi (2014) investigated the environmental performance of biobutanol production from the hemicellulosic fraction of bagasse and straw. The comparison with the oil-based production demonstrated the environmental superiority of the biotechnological route to biobutanol in several environmental categories: remarkable reductions of 88%, 22%, and 18% were reported in the categories abiotic depletion, global warming, and ozone layer depletion, respectively. Brito and Martins (2017) applied LCA to compare the environmental impacts of three biobutanol production methods, including oxo-synthesis and ABE fermentation from wheat straw and corn. The authors used the impact assessment 'IMPACT 2002+' which implements a midpoint analysis in a combined approach to obtain a final damage score considering four categories of damage: human health, ecosystem quality, climate change, and resources. The analysis showed that biobutanol from wheat straw has the lowest environmental impacts, with a reduction of 20% compared to butanol production from oxo-synthesis. More recently, Meng et al. (2021) estimated the carbon footprint of biobutanol production from municipal solid waste at 12.57 g CO_2eq./MJ, with a net saving of greenhouse gases of 115% compared to gasoline. Desta, Lee, and Wu (2021) calculated the well-to-wheel greenhouse gas emissions of biobutanol from corn stover. The authors found a reduction of 64%, 81%, and 78% in carbon emissions compared to first-generation butanol, gasoline, and diesel, respectively.

The scientific literature on combined economic and environmental analysis on biobutanol production is even scanter. Tao et al. (2014) investigated the biobutanol production from cellulosic biomass corn stover. They found that biobutanol cost is slightly higher (around 12%) than that of cellulosic ethanol, but the two yield similar—4.3–4.5 kgCO_2eq./GGE (gasoline gallon equivalent)—field-to-wheel carbon emissions. Levasseur, Bahn, Beloin-saint-pierre, Marinova, and Vaillancourt (2017) studied the biobutanol produced from pre-hydrolysate in a Canadian Kraft dissolving pulp mill. They found that carbon emissions range from −5% to +20% compared to gasoline according to the scenario considered, and that the cost of feedstock is the main hotspot from an economic standpoint. Pereira, Dias, Mariano, Filho, and Bonomi (2015) suggest that field-to-wheel greenhouse gas emissions per km travelled by a gasoline engine car propelled by biobutanol from sugarcane are significantly smaller than the average baseline of GHG emissions from gasoline (0.04–0.07 kgCO_2eq./km compared to 0.32 kgCO_2eq./km), whilst second-generation production path is more economically attractive compared to first-generation path.

3 Novelty

This case study presents an economic and environmental evaluation of a novel process to produce second-generation biobutanol from agro-food wastes, using LCA and LCC in line with the ISO 14040/44 standards (ISO, 2006a, 2006b). The novel process integrates the steps for the conversion of agro-food wastes: pre-treatment, hydrolysis, fermentation, in situ solvent recovery, and distillation. Several agro-food waste residues were investigated as part of the H2020 project Waste2Fuels, including apple pomace, potato peel, brewer's spent grain, and coffee silver skin (Giacobbe, Pezzella, Lettera, Sannia, & Piscitelli, 2018; Hijosa-Valsero, Garita-Cambronero, Paniagua-García, & Díez-Antolínez, 2018; Hijosa-Valsero, Paniagua-García, & Díez-Antolínez, 2017, 2018; Lettieri & Chirone, 2019; Niglio, Procentese, Russo, Sannia, & Marzocchella, 2019). The present analysis focuses on one of the most promising cases: production of biobutanol from potato peel. To the best of the authors' knowledge, no studies were focused on second-generation butanol production from industrial agro-food waste.

4 Case study description

4.1 Process description

The production of second-generation biobutanol from agro-food waste—including potato peel—involves four core steps (Tao et al., 2014) (Fig. 1): biomass pre-treatment and enzymatic hydrolysis, ABE fermentation with (in situ) butanol recovery, and distillation. Ancillary units—in situ enzyme production, wastewater treatment, and electricity/steam production from lignin-rich residues of biomass and from biogas obtained in the wastewater treatment—are also included.

Pre-treatment

The pre-treatment consists of auto-hydrolysis (for potato peel, or chemical hydrolysis for other biomasses like apple residues) followed by enzymatic hydrolysis. The objective of the chemical/auto-hydrolysis is to render the cellulosic structure of biomass available whilst reducing the degree of polymerisation of carbohydrates. This step enhances sugar production in the subsequent enzymatic hydrolysis, limits carbohydrate degradation and formation of undesired products that can inhibit the subsequent enzymatic hydrolysis and fermentation. Enzymatic hydrolysis completes the conversion of the carbohydrate fraction of biomass into fermentable sugars (García et al., 2011).

ABE fermentation

The sugars from the pre-treatment are fermented by anaerobic bacteria to produce acetone, butanol, and ethanol, typically in a 3:6:1 M ratio. This bioconversion, known as ABE fermentation, involves two phases: acidogenesis and solventogenesis (Jones & Woods, 1986). The acidogenesis phase produces acids (acetic and butyric

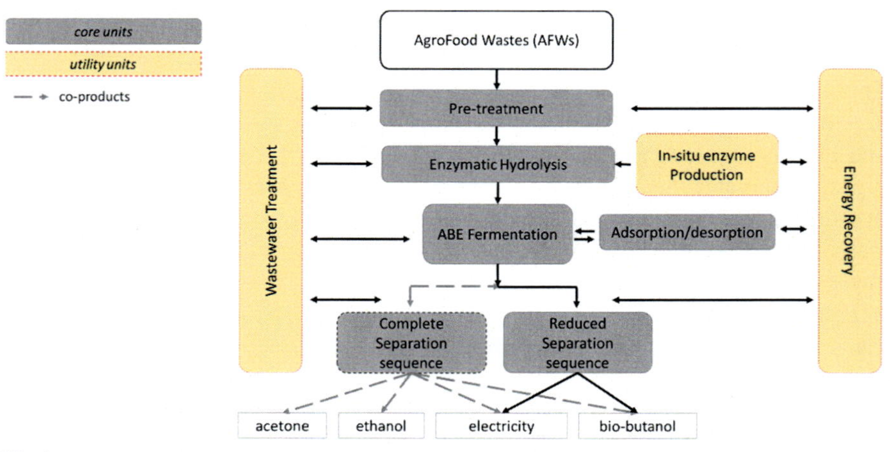

FIG. 1

Synoptic diagram of the biobutanol production process.

acid), H_2, and CO_2 and is associated to bacteria growth. Increasing the acid concentration the pH decreases and the cell metabolism shifts to solvent production: the solventogenesis phase. Under this phase, the rate of production of acids diminishes and acids are converted by microorganisms into solvents: acetone, butanol, and ethanol. The product of the ABE fermentation reactor is a broth, composed of water, unconverted substrates and metabolites—e.g. acetone, butanol, ethanol, acetic/butyric acids—at low concentrations. A significant limitation of the ABE fermentation is the product inhibition, extensive as butanol concentration approaches 13 g of butanol per litre (Papoutsakis, 2008). For this reason, recovery of butanol coupled with fermentation is beneficial and critical to increase butanol production, hence the overall efficiency of the process (Setlhaku, Heitmann, Górak, & Wichmann, 2013). Multiple process options may be considered, including gas stripping, pervaporation, adsorption, or liquid–liquid extraction. In this study, an hybrid approach combining physical adsorption and distillation is used (Wukovits, Kirchbacher, Miltner, & Friedl, 2018).

Distillation

Distillation is the final step for recovery and concentration at a specific target purity of the solvents from the fermentation broth. Two different scenarios are investigated: a simple configuration of two distillation columns that enables recovery of butanol only, and a more complex configuration of five columns for recovery of acetone and ethanol, in addition to butanol.

Utilities

The aqueous waste streams produced during the pre-treatment and purification steps are treated in a wastewater treatment plant consisting of aerobic and anaerobic digestion steps. The former generates a stream of clean water that is reused in the process, whilst the latter produces biogas (Humbird et al., 2011a). The biogas, together with lignin (from pre-treatment) and other combustible streams (including unconverted carbohydrate fraction of biomass), is valorised via generation of steam and electricity, which are used to satisfy the requirements of the entire plant. It is worth to note that electricity is generated in excess of what required and is assumed to be sold to the grid.

4.2 Methods

Life cycle assessment

Goal and scope definition

The LCA analysis has three objectives: (i) compare two process layouts, consisting of a reduced and a complete distillation sequences (see Fig. 1 and Section 4.1); (ii) identify key hotspots along the supply chain of biobutanol production from potato peel; and (iii) compare the environmental performance of producing biobutanol with that of conventional, fossil-based alternatives including gasoline and E85—a commercial blend of 85% ethanol and 15% gasoline. The study is framed in the European context. The analysis adopts an attributional approach. This entails that the potential effects of implementing the process and introducing biobutanol on the market are not investigated.

The system boundaries (Fig. 2) are cradle to gate: they include all activities from the extraction of raw materials up to the production of biobutanol. In line with other studies, the functional unit corresponds to 1 MJ of biobutanol produced (Levasseur et al., 2017; Meng et al., 2021; Pereira et al., 2015).

The biobutanol process delivers more than one function (Fig. 2). Besides producing biobutanol, the process provides a means to manage potato peel and generates a surplus electricity which is assumed to be sent to the electric grid. Furthermore, when a complete distillation sequence is considered, the process also co-produces ethanol and acetone. We applied the crediting approach to allocate the environmental impacts to biobutanol, with the following assumptions:

- agro-food waste would be treated via anaerobic digestion or industrial composting according to a 50–50 split (European Environment Agency, 2020; European Union, 2008);
- acetone and ethanol displace equivalent amounts of the same products produced from fossil resources;
- surplus electricity displaces that generated according to the EU-28 grid mix in 2019 (GaBi ts database).

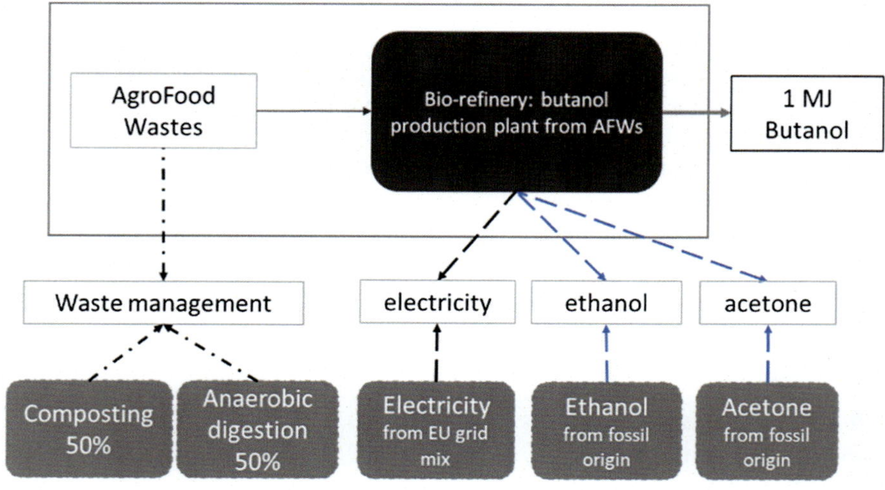

FIG. 2

System boundaries of LCA study. *Grey line* represents the functional unit. *Black dashed lines* identify co-products in common to complete and reduced distillation sequences. *Blue dashed lines* identify additional co-products from the complete sequence.

Life cycle inventory analysis

For inventory collection purposes, we divide the system boundaries in foreground and background (Clift, Doig, & Finnveden, 2000). Life cycle inventory data for the former is based on mass and energy balances from Aspen Plus simulations of the process (Fig. 1) and literature data; thermal energy and electricity demand, chemicals consumption, and waste stream amounts of the utility units were obtained from the literature for a similar process, ethanol production, at a similar scale (Humbird et al., 2011a; Tao et al., 2014). Construction and decommissioning of the plant are also modelled using data for an ethanol plant available in the EcoInvent database. Key inventory data is reported in Table 1, whilst the complete inventory is provided by Lettieri and Chirone (2019). The background system is modelled using the EcoInvent database version 3.6.

Life cycle impact assessment

The Environmental Footprint (EF) 2.0 method developed by the Joint Research Centre (JRC) of the European Commission (Fazio et al., 2018; JRC, 2018) is used for quantifying the environmental impacts. This study considers 14 environmental impact categories, including climate change, water scarcity, and resource use, amongst others. A description of these environmental impact categories is provided in (Hauschild, Rosenbaum, & Olsen, 2017).

For the comparative analysis between biobutanol, gasoline, and E85, the resulting environmental impacts have been normalised and weighed to obtain a single environmental score. Normalisation has been performed using global factors developed by Sala, Crenna, Secchi, and Pant (2017). Weighting has been based on the

Table 1 Key inventory data referred for the production of 1 MJ of biobutanol.

	Potato peel	
	Complete	**Reduced**
Inputs		
Potato peel, kg	4.17E−01	6.22E−01
Enzyme, kg	1.53E−02	2.30E−02
Citric acid, kg	4.00E−02	5.97E−02
Sodium hydroxide, kg	2.08E−02	3.11E−02
Water, kg	3.98E+00	5.97E+00
Nitrogen, kg	2.11E−01	2.87E−01
Amberlite XAD-7, kg	6.11E−03	8.89E−03
Ammonia, kg	2.78E−04	5.56E−04
Outputs, co-product		
Ethanol, kg	4.44E−03	–
Acetone, kg	5.56E−03	–
Electricity surplus, kWh	3.94E−01	6.71E−01

approach developed by Sala, Cerutti, and Pant (2018). This developed weighting factors taking into consideration public and LCA expert scorings on various aspects and criteria for each environmental impact category, and the robustness of the categories' underlying models and that of normalisation factors based on global emissions.

Life cycle costing

The goal and scope of the LCC study coincide with that of the LCA (Fig. 2). The LCC is based on the estimation of the total cost of production (TPC): the sum of total capital investment (TCI) and operating costs including fixed (FOC) and variable (VOC) ones (Eq. 1).

$$TPC = TCI + VOC + FOC \tag{1}$$

The estimation was carried out with reference to a plant lifetime of 25 years and capacity of 1.7 kton/h of potato peel.

Total capital investment (TIC) was calculated as the sum of direct costs (DC), indirect costs (IC), and land and working capital costs (LWCC), as reported in Eq. (2), and estimated by processing the data for similar plant for second-generation ethanol production (Humbird et al., 2011a).

$$TCI = DC + IC + LWCC \tag{2}$$

Direct costs (DC) included equipment costs, warehouse, site development, and additional piping. Notably, warehouse, site development, and additional piping cost were calculated as percentage of the equipment costs following the inside-battery-limits (ISBL) approach (Humbird et al., 2011a; Turton, Bailie, Whiting, & Shaeiwitz, 2008). Indirect costs (IC), including field expenses, home office and construction activities, project contingency and other costs were calculated as percentage of direct costs (DC). Land and Working Capital (LWCC) were estimated as percentage of the total capital investment (TCI).

Fixed operating costs (FOC) covered the cost of labour and other overheads, like maintenance and property insurance. The number of employees was estimated according to the degree of automation of each step of the process and considering, in addition, a reasonable number of management and support employees. The cost of labour was the actual cost of an employ for a company: it includes the salary, payroll taxes, pension costs, health insurance, dental insurance, and any other benefits that a company provides to an employee. Maintenance and property insurance were estimated as 3% of ISBL and 0.7% of Fixed Capital Investment, respectively.

Variable operating costs (VOC) incorporated the cost of all material and energy flows, including the costs of waste handling and the revenues associated with by-products. The amounts of raw material and wastes were obtained from the life cycle inventory, whilst the costs (€/kg) were based from literature sources (Humbird et al., 2011a, 2011b; Turton et al., 2008) or obtained from prices and markets reports developed by ICIS (Independent Commodity Intelligence Services; www.icis.com). In computing revenues, the disposal cost for potato peel was assumed to be 0.1 €/kg (Kaza, Yao, Bhada-Tata, & Van Woerden, 2018), whilst surplus electricity, acetone, and ethanol were assumed to be sold on the market for a price of 0.06 €/kWh

(Renewable & Agency, 2019), 0.95 €/kg and 0.75 €/kg, respectively (Google search, September 2020).

4.3 Results and discussion

Life cycle assessment

To fulfil the first goal of the study, Fig. 3 compares the environmental performance of reduced and complete distillation sequences, expressed in terms of percentage difference. Positive values imply that the complete scenario outperforms the reduced one. The reduced distillation sequence yields lower butanol productivity (−33%) and it is characterised by high requirements of adsorbent, used in the in situ butanol recovery. These factors determine the low environmental performance of this process option, the complete distillation sequence being advantageous in most categories, apart from ionising radiations and climate change. However, the reduced distillation sequence is also characterised by lower electricity consumption when compared to the complete option, and therefore a higher amount of surplus electricity is sent to the grid. The credits associated with surplus electricity are particularly significant in the categories of ionising radiation (due to nuclear and coal power plants) and climate change (due to fossil-based power plants), which explains why this process option is preferable in these categories. The complete distillation sequence also generates other co-products (acetone and ethanol) but, interestingly, the credits associated with these are small when compared to the other credits and environmental impacts (see Fig. 4).

Regarding the second goal of this study, we focus on the operational phase, which is the major contributor to all environmental categories. The hotspot analysis (Fig. 4) shows that the pre-treatment step generates the largest environmental impacts in most categories, with contributions ranging from 11% in the category cancer human health and up to 80% in the category ozone depletion. The environmental impacts of the pre-treatment primarily originate from the production of citric acid, which is used for breaking down the cellulose and hemicellulose polymers in lignocellulosic biomass to form individual fermentable sugars. The production of enzymes, which are used in the fermentation step, is the largest contributor to the categories land use and cancer human health effects, whilst direct NO_x emissions from energy recovery dominate marine and terrestrial eutrophication impacts.

As a multi-functional system, the biobutanol production process receives credits for managing agri-food waste, and for generating surplus electricity and, in the case of a complete distillation sequence, also co-products including ethanol and acetone ("Life cycle assessment" section). These credits are particularly significant in the categories ionising radiations, climate change, photochemical ozone formation, resource use—energy carriers and non-cancer human health effects. The credits associated with surplus electricity generation dominate ionising radiations and resource—energy carriers. The remaining categories receive similar contributions from electricity surplus and avoided processes for agri-food waste management, i.e. landfill and incineration. As noted before, the credits for co-products generation are negligible in all environmental categories.

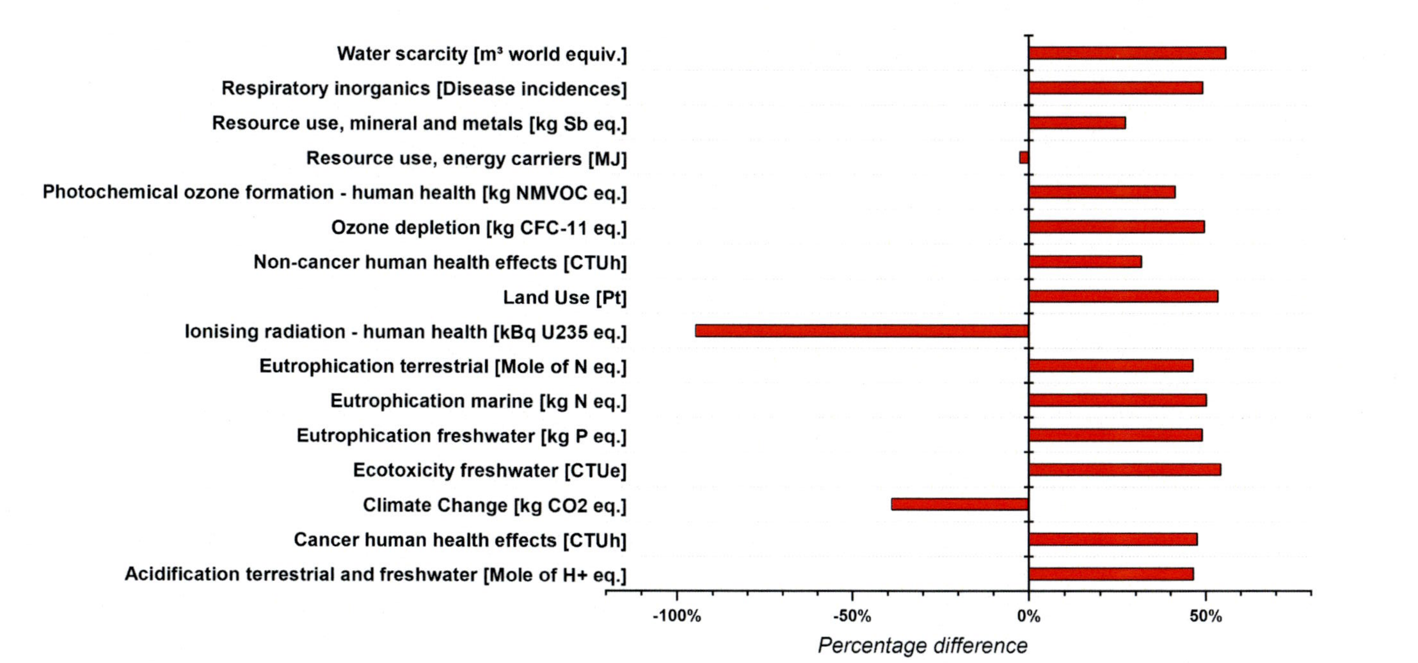

FIG. 3

Percentage difference between environmental impacts of reduced and complete distillation sequences.

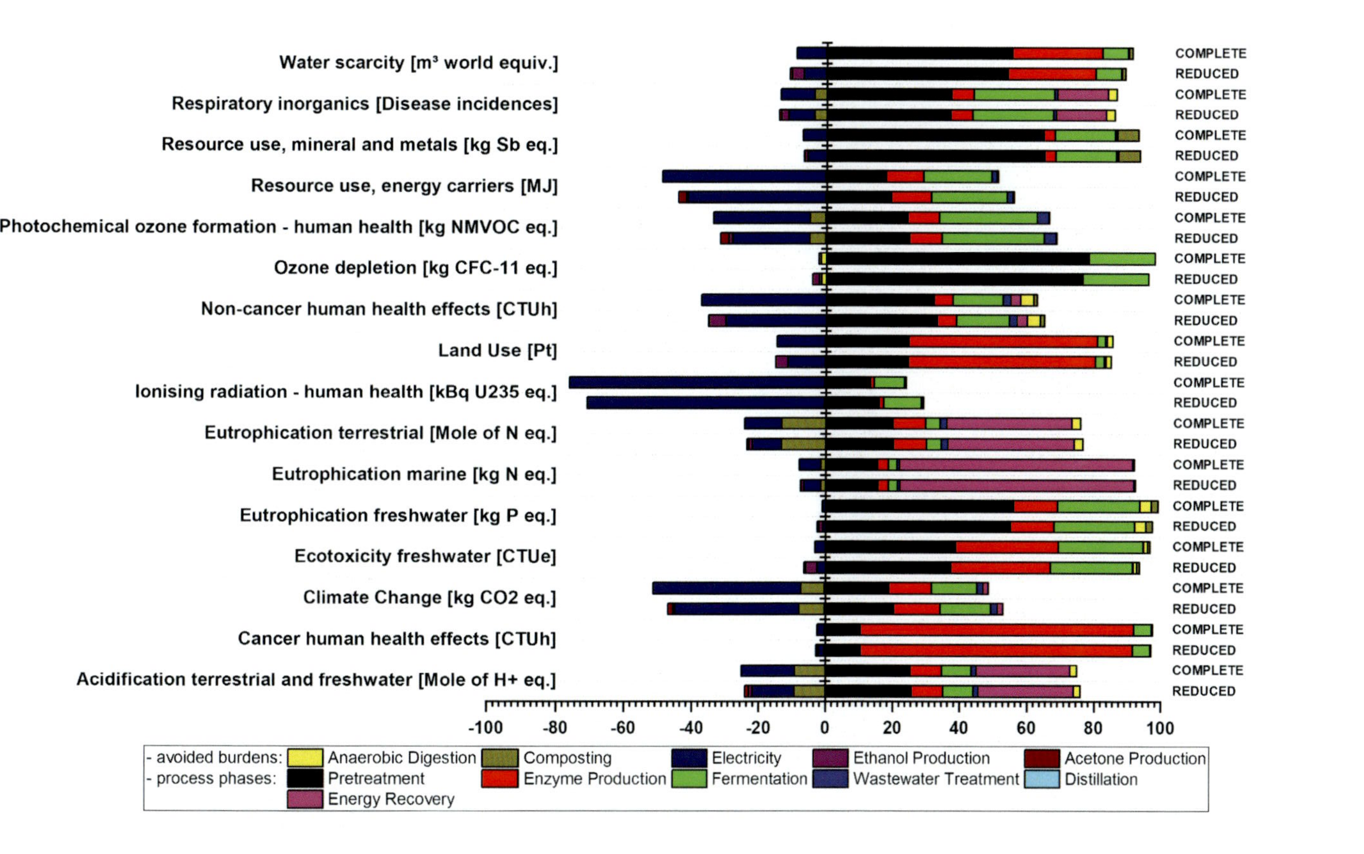

FIG. 4

Hotspot analysis of operational phase of biobutanol production.

The hotspot analysis highlights the importance of the credits in determining the environmental performance of biobutanol. The results are therefore dependent on the approach to allocate environmental impacts between different functions as well as on the technologies (or mix thereof) that we assume are being displaced; an example of this is provided by Tereshchenko and Nord (2015) for the case of heat and power cogeneration. Furthermore, if potato peel wastes achieve a positive market value, for example due to an expansion of waste utilisation technologies, the process of biobutanol production would not be credited for managing potato peel; rather the environmental impacts of producing the agri-food waste would need to be accounted for, like those of other resources and intermediate products.

Life cycle costing

Table 2 presents the total production cost (TPC), which ranges from 0.08 €/MJ in the case of a complete distillation sequence to 0.13 €/MJ for the reduced sequence. Table 2 also presents that nearly 90% of the TPC derives from the variable operating costs (VOC), with the total capital investment (TCI) and fixed operating costs (FOC) amounting to about 5%.

Going a bit further on the main cost, Fig. 5 reports a contribution analysis for the VOC for both distillation sequences. On the one hand, nearly half of the variable

Table 2 Total production cost and contributions for complete and reduced distillation sequences.

Distillation scenario	TCI	VOC	FOC	TCP, €/MJ
Reduced	5.6%	88.7%	5.7%	1.33E−01
Complete	5.4%	89.3%	5.3%	8.48E−02

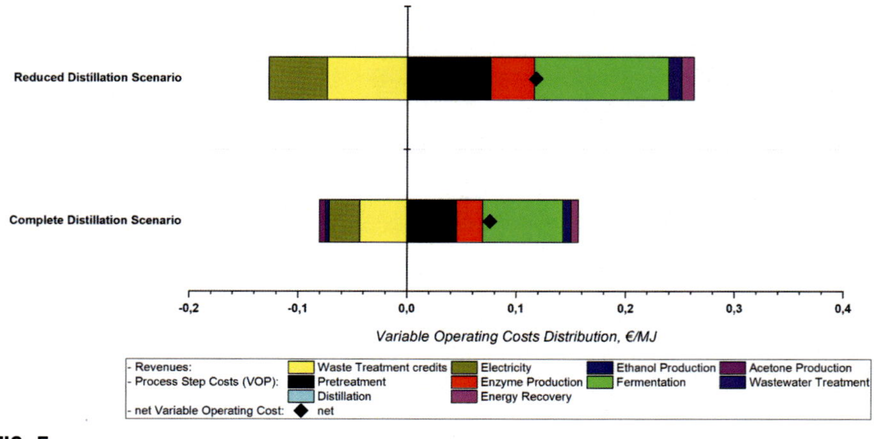

FIG. 5

Hotspot analysis for variable operating cost (VOC) for biobutanol production.

costs for producing biobutanol are associated with the fermentation step, whilst pre-treatment and enzyme production are responsible for around 29% and 15%, respectively. Interestingly, the analysis indicates that the fermentation step is far more relevant from an economic rather than an environmental standpoint. For example, pre-treatment, enzyme production, and agri-food waste treatment credits feature substantially larger contributions to climate change impacts. Two issues limit the economic performance of biobutanol production from potato peel: (i) the adsorbent and the citric acid cost used for butanol in situ recovery and the pre-treatment step, respectively, and (ii) very low butanol productivity. On the other hand, revenues account for nearly half of the costs in both process layouts. They are dominated by the sale of surplus electricity and by the avoided costs for disposal of agri-food waste with similar contributions. As for environmental credits, the revenues associated with ethanol and acetone for the complete distillation sequence option are negligible. Similar to the LCA analysis, the LCC demonstrates the importance of the revenues from selling surplus electricity and credits from potato peel waste treatment in determining the economic sustainability of biobutanol, and therefore that the total production cost is also very dependent on the future market value of agro-food wastes.

Comparative analysis

Fig. 6 shows results of the comparative analysis between biobutanol and the conventional fossil-based counterparts (i.e. gasoline and E85). The comparison is based on the total production cost (TPC), the environmental score obtained via normalisation

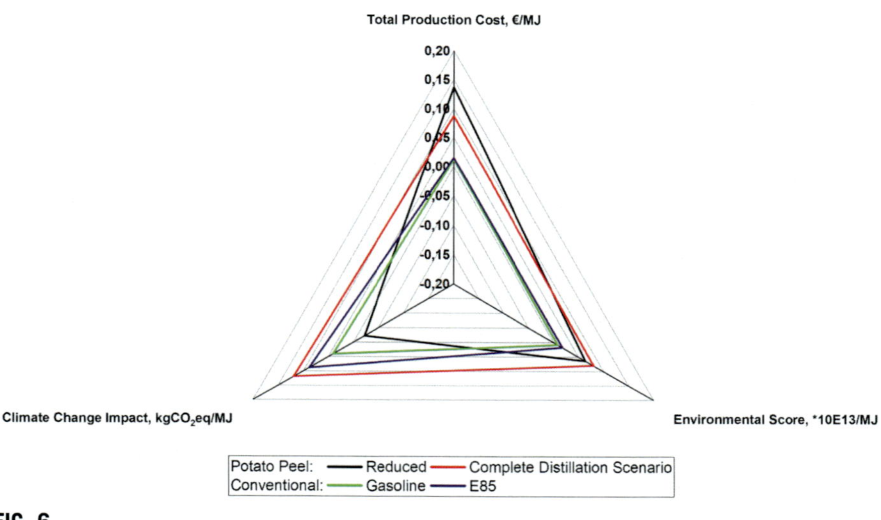

FIG. 6

Total cost of production, environmental score, and climate change impact of biobutanol, gasoline, and E85.

and weighting of environmental impacts, as well as climate change impacts; these, arguably, represent the main motivation behind the development of biofuels. The results show interesting trade-offs. Biobutanol has substantially higher total production costs (6–11 times higher when compared to gasoline and E85) and environmental score (from 3 to 9 times), but it yields significantly lower carbon emissions (for the reduced distillation scenario), with net savings compared to both gasoline and E85.

Our analysis indicates that biobutanol derived from potato peel waste represents a valid strategy to reducing carbon emissions from transportation fuels production. However, the environmental advantages are less straightforward when more environmental categories are considered, whilst it remains significantly disadvantageous for an economic standpoint, at least in the short term. It is expected that in the long term the economy of scale will contribute to lowering the production costs of biobutanol as well as its environmental impacts.

It must be underlined that the present analysis does not include the use phase of the fuels. This is particularly important for the comparison because greenhouse gas emissions from the use phase of biobutanol would be considered biogenic, and thus have significantly lower climate change impacts than those originating from gasoline and E85. Therefore it is expected that the environmental advantages—in particular those related to climate change—would be higher when the analysis is extended to the use phase. The inclusion of the use phase may also affect the economic comparison as carbon and other taxes may in future be levied on the use of fossil fuels.

5 Conclusions

The analysis reported in this chapter concerned the production of a biofuel via biotechnological route. The exploitation of renewable resources used as case study—potato peel waste—positions the production process in the circular economy scenario with the unquestionable advantages related to this industrial approach.

The results of the present analysis highlight advantages and drawbacks of the biotechnological production route and shed the light on the key critical stages of the production chain that mostly deserve optimisation from the economic and environmental standpoints. The total production cost of biobutanol is almost six times higher than the conventional gasoline under the best scenario. The environmental advantage in terms of climate change locates the process amongst the highly ranked ones with respect to the production processes typically proposed for the second-generation biofuels. Indeed the reduced scenario shows a greenhouse gas reduction of 158% compared to conventional gasoline. However, if all the impact categories are considered, the results show that biobutanol production from potato peel is still less competitive compared to conventional fuel production nearly seven times larger than conventional gasoline. As regards the economic balance, results highlight that the effort for improvement should be mostly directed to the sugar conversion and pre-concentration steps.

Acknowledgements

The case study presented in this chapter was developed as part of the project H2020-LCE-2015 Waste2Fuels 'Sustainable production of next-generation biofuels from waste streams' N. 654623.

References

Amiri, H., & Karimi, K. (2018). Pretreatment and hydrolysis of lignocellulosic wastes for butanol production: Challenges and perspectives. *Bioresource Technology*, *270*, 702–721.

Berndes, G., Youngs, H., Ballester, M. V. R., Cantarella, H., Cowie, A. L., Jewitt, G., et al. (2015). Soils and water (Chapter 18). In G. M. Souza, R. L. Victoria, C. A. Joly, & L. M. Verdade (Eds.), *Bioenergy and sustainability: Bridging the gaps. SCOPE 72* (pp. 618–659). Scientific Committee on Problems of the Environment.

Briens, C., Piskorz, J., & Berruti, F. (2008). Biomass valorization for fuel and chemicals production—A review. *International Journal of Chemical Reactor Engineering*, *6*(1). https://doi.org/10.2202/1542-6580.1674.

Brito, M., & Martins, F. (2017). Life cycle assessment of butanol production. *Fuel*, *208*, 476–482. https://doi.org/10.1016/J.FUEL.2017.07.050.

Clift, R., Doig, A., & Finnveden, G. (2000). The application of life cycle assessment to integrated solid waste management. Part 1 – methodology. *Process Safety and Environmental Protection*, *78*(4), 279–287. https://doi.org/10.1205/095758200530790.

Colling Klein, B., Bonomi, A., & Maciel Filho, R. (2018). Integration of microalgae production with industrial biofuel facilities: A critical review. *Renewable and Sustainable Energy Reviews*, *82*, 1376–1392. https://doi.org/10.1016/j.rser.2017.04.063.

Desta, M., Lee, T., & Wu, H. (2021). Well-to-wheel analysis of energy use and greenhouse gas emissions of acetone-butanol-ethanol from corn and corn Stover. *Renewable Energy*, *170*, 72–80. https://doi.org/10.1016/j.renene.2021.01.079.

European Environment Agency. (2020). *Bio-waste in Europe—Turwning challenges into opportunities (issue 04).*

European Union, E. (2008). *Directive 2008/98/EC of the European Parliament and of the Council of 19 November 2008 on waste and repealing certain directives (text with EEA relevance)* (pp. 3–30).

European Union, E. (2018). Directive (EU) 2018/2001 of the European Parliament and of the Council of 11 December 2018 on the promotion of the use of energy from renewable sources. *Official Journal of the European Union*, *5*, 82–209.

Ezeji, T. C., & Blaschek, H. P. (2010). Butanol production from lignocellulosic biomass. In *3. Biofuels from agricultural wastes and byproducts* (pp. 19–37). Wiley. https://doi.org/10.1002/9780813822716.ch3.

Fazio, S., Castellani, V., Salasa, S., Schau, E., Secchi, M., Zamporti, L., et al. (2018). *Supporting information to the characterisation factors of recommended EF life cycle impact assessment method.*

García, V., Päkkilä, J., Ojamo, H., Muurinen, E., & Keiski, R. L. (2011). Challenges in biobutanol production: How to improve the efficiency? *Renewable and Sustainable Energy Reviews*, *15*(2), 964–980. https://doi.org/10.1016/j.rser.2010.11.008.

Giacobbe, S., Pezzella, C., Lettera, V., Sannia, G., & Piscitelli, A. (2018). Laccase pretreatment for agrofood wastes valorization. *Bioresource Technology*, *265*(June), 59–65. https://doi.org/10.1016/j.biortech.2018.05.108.

Hanaki, K., & Portugal-Pereira, J. (2018). The effect of biofuel production on greenhouse gas emission reductions. In *Biofuels and sustainability* (pp. 53–71). Tokyo: Springer.

Hauschild, M. Z., Rosenbaum, R. K., & Olsen, S. I. (2017). *Life cycle assessment: Theory and practice*. Springer International Publishing. https://doi.org/10.1007/978-3-319-56475-3.

Hijosa-Valsero, M., Garita-Cambronero, J., Paniagua-García, A. I., & Díez-Antolínez, R. (2018). Biobutanol production from coffee silverskin. *Microbial Cell Factories*, *17*(1), 154.

Hijosa-Valsero, M., Paniagua-García, A. I., & Díez-Antolínez, R. (2017). Biobutanol production from apple pomace: The importance of pretreatment methods on the fermentability of lignocellulosic agro-food wastes. *Applied Microbiology and Biotechnology*, *101*(21), 8041–8052.

Hijosa-Valsero, M., Paniagua-García, A. I., & Díez-Antolínez, R. (2018). Industrial potato peel as a feedstock for biobutanol production. *New Biotechnology*, *46*, 54–60.

Humbird, D., Davis, R., Tao, L., Kinchin, C., Hsu, D., Aden, A., et al. (2011a). *Process design and economics for biochemical conversion of lignocellulosic biomass to ethanol: Dilute-acid pretreatment and enzymatic hydrolysis of corn Stover*. Golden, CO, United States: National Renewable Energy Lab. (NREL).

Humbird, D., Davis, R., Tao, L., Kinchin, C., Hsu, D., Aden, A., et al. (2011b). Process design and economics for biochemical conversion of lignocellulosic biomass to ethanol. *Renewable Energy*, *303*(May), 147. https://doi.org/10.2172/1013269.

ISO. (2006a). *Environmental management – life cycle assessment – principles and framework. EN ISO 14040:2006*.

ISO. (2006b). *Environmental management – life cycle assessment – requirements and guidelines. EN ISO 14044:2006*.

Jones, D. T., & Woods, D. R. (1986). Acetone-butanol fermentation revisited. *Microbiological Reviews*, *50*(4), 484.

JRC. (2018). *Product environmental footprint category rules guidance. Version 6.3*. JRC.

Kaza, S., Yao, L. C., Bhada-Tata, P., & Van Woerden, F. (2018). In World Bank (Ed.), *A global snapshot of solid waste management to 2050*. https://openknowledge.worldbank.org/handle/10986/30317.

Lettieri, P., & Chirone, R. (2019). *Technical report: Sustainable life cycle assessment final report. Project name: Sustainable production of next generation biofuels from waste streams (Waste2Fuels) – H2020, LCE-11-2015, developing next generation technologies for biofuels and sustainable*.

Levasseur, A., Bahn, O., Beloin-saint-pierre, D., Marinova, M., & Vaillancourt, K. (2017). Assessing butanol from integrated forest biorefinery: A combined techno-economic and life cycle approach. *Applied Energy*, *198*, 440–452. https://doi.org/10.1016/j.apenergy.2017.04.040.

Littlejohns, J., Rehmann, L., Murdy, R., Oo, A., & Neill, S. (2018). Current state and future prospects for liquid biofuels in Canada. *Biofuel Research Journal*, *5*(1), 759–779.

Macedo, I. C., Seabra, J. E. A., & Silva, J. E. A. R. (2008). Green house gases emissions in the production and use of ethanol from sugarcane in Brazil: The 2005/2006 averages and a prediction for 2020. *Biomass and Bioenergy*, *32*(7), 582–595. https://doi.org/10.1016/j.biombioe.2007.12.006.

Meng, F., Ibbett, R., De Vrije, T., Metcalf, P., Tucker, G., & Mckechnie, J. (2021). Process simulation and life cycle assessment of converting autoclaved municipal solid waste into butanol and ethanol as transport fuels. *Waste Management*, *89*(2019), 177–189. https://doi.org/10.1016/j.wasman.2019.04.003.

Muench, S., & Guenther, E. (2013). A systematic review of bioenergy life cycle assessments. *Applied Energy*, *112*, 257–273. https://doi.org/10.1016/j.apenergy.2013.06.001.

Nanda, S., Golemi-Kotra, D., McDermott, J. C., Dalai, A. K., Gökalp, I., & Kozinski, J. A. (2017). Fermentative production of butanol: Perspectives on synthetic biology. *New Biotechnology*, *37*, 210–221. https://doi.org/10.1016/j.nbt.2017.02.006.

Niemistö, J., Saavalainen, P., Isomäki, R., Kolli, T., Huuhtanen, M., & Keiski, R. L. (2013). Biobutanol production from biomass. In *Biofuel technologies* (pp. 443–470). Springer.

Niglio, S., Procentese, A., Russo, M. E., Sannia, G., & Marzocchella, A. (2019). Investigation of enzymatic hydrolysis of coffee silverskin aimed at the production of butanol and succinic acid by fermentative processes. *Bioenergy Research*, *12*(2), 312–324.

Papoutsakis, E. T. (2008). Engineering solventogenic clostridia. *Current Opinion in Biotechnology*, *19*(5), 420–429.

Pereira, L. G., Chagas, M. F., Dias, M. O. S., Cavalett, O., & Bonomi, A. (2014). Life cycle assessment of butanol production in sugarcane biore fi neries in Brazil. *Journal of Cleaner Production*, *96*, 557–568. https://doi.org/10.1016/j.jclepro.2014.01.059.

Pereira, L. G., Dias, M. O. S., Mariano, A. P., Filho, R. M., & Bonomi, A. (2015). Economic and environmental assessment of n-butanol production in an integrated first and second generation sugarcane biorefinery: Fermentative versus catalytic routes. *Applied Energy*, *160*(2015), 120–131. https://doi.org/10.1016/j.apenergy.2015.09.063.

Procentese, A., Raganati, F., Olivieri, G., Russo, M. E., de la Feld, M., & Marzocchella, A. (2017). Renewable feedstocks for biobutanol production by fermentation. *New Biotechnology*, *39*, 135–140.

Raganati, F., Procentese, A., Olivieri, G., Russo, M. E., Salatino, P., & Marzocchella, A. (2018). Bio-butanol separation by adsorption on various materials: Assessment of isotherms and effects of other ABE-fermentation compounds. *Separation and Purification Technology*, *191*, 328–339.

Renewable, I., & Agency, E. (2019). *Renewable power generation costs in 2019*.

Sala, S., Cerutti, A. K., & Pant, R. (2018). *Development of a weighting approach for the environmental footprint*. Luxembourg: Publications Office of the European Union.

Sala, S., Crenna, E., Secchi, M., & Pant, R. (2017). *Global normalisation factors for the environmental footprint and life cycle assessment*. https://doi.org/10.2760/88930.

Setlhaku, M., Heitmann, S., Górak, A., & Wichmann, R. (2013). Investigation of gas stripping and pervaporation for improved feasibility of two-stage butanol production process. *Bioresource Technology*, *136*, 102–108.

Tao, L., Tan, E. C. D., McCormick, R., Zhang, M., Aden, A., He, X., et al. (2014). Techno-economic analysis and life-cycle assessment of cellulosic isobutanol and comparison with cellulosic ethanol and n-butanol. *Biofuels, Bioproducts and Biorefining*, *6*(3), 246–256. https://doi.org/10.1002/bbb.

Tereshchenko, T., & Nord, N. (2015). Uncertainty of the allocation factors of heat and electricity production of combined cycle power plant. *Applied Thermal Engineering*, *76*, 410–422. https://doi.org/10.1016/j.applthermaleng.2014.11.019.

Turton, R., Bailie, R. C., Whiting, W. B., & Shaeiwitz, J. A. (2008). *Analysis, synthesis and design of chemical processes*. Pearson Education.

Ubando, A. T., Rivera, D. R. T., Chen, W. H., & Culaba, A. B. (2019). A comprehensive review of life cycle assessment (LCA) of microalgal and lignocellulosic bioenergy products from thermochemical processes. *Bioresource Technology*, *291*(May), 121837. https://doi.org/10.1016/j.biortech.2019.121837.

van Dyk, S., Su, J., McMillan, J. D., & Saddler, J.(. J.). N. (2019). *'DROP-IN' biofuels: The key role that co-processing will play in its production (issue January)*. IEA Bioenergy.

Walter, A., Seabra, J. E. A., Machado, P. G., de Barros Correia, B., & de Oliveira, C. O. F. (2018). Sustainability of biomass. In *Biomass and green chemistry* (pp. 191–219). Springer.

Wukovits, W., Kirchbacher, F., Miltner, M., & Friedl, A. (2018). Assessment of hybrid processes for bio-butanol purification applying process simulation. *Chemical Engineering Transactions*, *70*, 319–324.

Xue, C., Liu, F., Xu, M., Tang, I.-C., Zhao, J., Bai, F., et al. (2016). Butanol production in acetone-butanol-ethanol fermentation with in situ product recovery by adsorption. *Bioresource Technology*, *219*, 158–168. https://doi.org/10.1016/j.biortech.2016.07.111.

CHAPTER

Footprint assessment of solid waste management systems

19

Daniela Gavrilescu

Department of Environmental Engineering and Management, "Gheorghe Asachi" Technical University of Iasi, Iasi, Romania

1 Aims

The 4 case studies presented have as main objective the evaluation of solid waste management systems (SWMS), municipal solid waste (MSW), waste of electric and electronic equipment (WEEE), packaging waste (PW), and biowaste performance in terms of GHG emissions, by using aggregated models. The MSW case study has as specific objective to obtain the carbon footprint of a region in which landfilling is the predominant waste treatment option. The objective of the WEEE case study was to compare the carbon footprint of two EU member states (Romania and Italy) in different position regarding WEEE management performance over a longer period of time. The objective of the PW case study was to compare the PW management performance in terms of GHG emissions in four EU member states, over a longer period of time. The biowaste case study has as main objective the comparison of carbon footprint of 3 scenarios, in the current status versus implementation of waste treatment technologies.

2 State of the art

The topic of solid waste management system itself is of extreme significance because of the *Sustainable Development Goals* which state that all humans have the right to a proper sanitation service and to a clean environment in which waste is collected and treated in a sustainable manner. Furthermore, European Union underlines the concept of *circular economy* which means the change in the way wastes are perceived as materials or products which nobody wants to a valuable resource either for material recycling or energy valorisation.[a] Simultaneously, the *Zero Waste* concept

[a](https:/ec.europa.eu/environment/strategy/circular-economy-action-plan_enhttps:/ec.europa.eu/environment/strategy/circular-economy-action-plan_en).

Assessing Progress towards Sustainability. https://doi.org/10.1016/B978-0-323-85851-9.00016-X

emphasises the need for transition to closing material and energy loops for materials and energy before diverting waste streams to landfilling.[b]

Even though profound changes in the way solid waste management systems are organised can be foreseen in the future, these changes should not come without the analysis of their consequences in terms of environmental impact. In general, for the solid waste management systems, the main focus is on waste generation, waste composition and material flow analysis, different processes' configuration for a more efficient treatment, and more recently on the life cycle analysis of waste management system elements.

It is a well-known fact that improper waste management lead to environmental impacts such as increases in GHG emissions, soil contamination, groundwater and surface water pollution, as well as loss in biodiversity, just to name a few (Obersteiner, Gollnow, & Eriksson, 2021). While each one of them is important, most of the existing waste management policies and strategies are connected to strategies on global warming/climate change direction talking about reduction in GHG emissions. According to EU (2020), greenhouse gas emissions come from 4 sources: fuel combustion, agricultural activities, industrial processes, and solid waste. In line with the circular economy concept, selecting a proper treatment for the waste streams helps in reducing emissions and mitigate climate change phenomena.

Desired for the multitude of information offered by the resulting environmental profiles, the full life cycle assessment of solid waste management systems is a difficult approach because of many quality data input requirements, the lack of information on the local/regional conditions, depending on the scale of the study, the complex result interpretation, and the challenges in understanding the results in the case of non-practitioners.

A literature scan, performed on *Science Direct* engine using keywords such as 'solid waste environmental impact' gave 159,803 responses, 'solid waste and climate change' gave 63,345 responses, while 'solid waste carbon footprint' gave 16,202. This means that out of the environmental impacts, the hot topic is climate change caused by solid waste for almost a half of the studies, while out these, 25% are focused in determining actually the performance based on carbon footprint.

Furthermore, authors such as Bernstad Saraiva, Souza, and Valle (2017), Lee, Han, and Wang (2017), and Pérez, Manuelde Andrés, Lumbreras, and Rodríguez (2018) use the LCA methodology to calculate only the emissions of greenhouse gases (GHG) associated with the municipal solid waste treatments, thus evaluating the environmental impact in terms of climate change. Sankar Cheela, John, Biswas, and Dubey (2021), even though employing life cycle assessment, have found out that the global warming potential indicator has the highest potential to be reduced, by sound solid waste practices. Vázquez-Rowe, Ziegler-Rodriguez, Margallo, Kahhat, and Aldaco (2021) have attributed the predominant quantity of GHG emissions at country level (Peru) to the generation and disposal of wastes, such as food waste. Fan et al. (2021) have used only carbon footprint to compare the solid waste management systems performance in 5 neighbouring European member states. Malakahmad, Abualqumboz, Kutty, and Abunama (2017) have compared different solid waste treatment technologies mainly in terms of carbon footprint. Căilean and Teodosiu (2016) have performed an assessment of the Romanian solid waste

[b](https:/zerowasteeurope.eu/).

management system based on sustainable development indicators and within the environmental performance indicators, carbon footprint was analyzed. Nazmul Islam (2017) have used carbon footprint associated with solid waste management strategy for improvements in the urban metabolism in Bangladesh.

Among the existing instruments or methodologies to calculate the carbon footprint, aggregated models are used in the case studies presented in this chapter (Cifrian, Galan, Andres, & Viguri, 2012). An aggregated model is a model based on a single aggregated emission factor used to characterise the generation of GHG for a certain material type and a certain waste treatment option. If for the treatment options, the aggregated models give emission factors for material recycling, composting, anaerobic digestion, incineration, and landfilling, in the case of waste material composition, some aggregated models consider a more detailed material classification, while others do not consider any classification.

The primary data on waste generation require further processing in establishing material balances per each waste treatment option used in that waste management system.

The case studies presented in the next subsections display results for carbon footprint calculated based on the following aggregated models: Smith, Brown, Ogilvie, Rushton, and Bates (2001), (US EPA WARM, 2006, US EPA WARM, 2016), Chen and Lin (2008), Christensen et al. (2009) with minimum and maximum values. Out of these aggregated models, USEPA WARM methodology has been constantly updated, with adjusted aggregated emissions factors varying from version to version.

In the calculation step, the indicators that may be obtained are the carbon footprint for a certain material and waste treatment, the total carbon footprint estimated by the sum of each carbon footprint, and finally the GHG efficiency indicator (tonnes of CO_2e/tonnes of treated waste) which gives the performance of the solid waste treatment system.

The case studies and subsequent results are independent and must be treated as such, therefore no comparisons between the key waste streams under analysis are intended.

In terms of novelty, the MSW case study combines for the first time material flow analysis with carbon footprint to monitor the changes over time at the regional level of a Romanian SWMS, strongly dependent on landfilling. Related to the WEEE and PW case studies, there are few studies that deal with the entire management system at country level and enables inter-country comparisons in terms of carbon footprint. The biowaste case study models for the first time the waste management alternatives to be implemented in an EU member state (Romania), by comparing the present and mid-term changes in time.

3 Novelty

This chapter presents a new methodology which combines material flow analysis with emission factors per waste type and treatment stage to estimate greenhouse gas emissions related to the management stages of waste streams. Aggregated models are incorporated within the methodology, making it suitable for rapid evaluations of waste management systems environmental performance characterized by carbon footprint. The methodology has been applied for the first time at regional and national levels (Bulgaria, Hungary, Italy, Poland and Romania) and for waste streams such as: MSW, WEEE, biowaste, PW. The obtained results allow comparisons over past times series or within future scenarios and/or different geographical levels.

4 Case study description

4.1 Carbon footprint of regional/national municipal solid waste management systems

This case study discusses the environmental impact of the municipal solid waste system (MWSM) at regional level, expressed as GHG emissions calculated by the carbon footprint indicator (Ropcean, Gavrilescu, Teodosiu, & Fiore, 2019). MSW refers strictly to solid waste generated by households and similar waste coming from institutions, commercial units, and economic operators. This excludes sludges and construction and demolition waste. As mentioned before, this is a key waste stream, according to the European legislation, with mandatory monitoring and reporting obligations and one of the sources of GHG generations.

Primary data have been selected from *Romanian Environmental Status Reports* (2017) and refer to domestic and similar waste generated in North-East region of Romania, in both urban and rural areas in households, institutions, commercial units and from economic operators, street waste collected from public spaces, streets, parks, green areas. The investigated period of time is 2013–2015, because it reflects a transition time between two Romanian National Strategies for Waste Management implementation. The region includes 6 counties: Bacau, Botosani, Iasi, Neamt, Suceava, and Vaslui), having a surface area of 36,850 km^2, approximately 3.7 million inhabitants, the lowest GDP in Romania and an approximately waste collection rate of 0.6 million tonnes/year.

Investments on sustainable solid waste management practices are among the priorities of the Romanian NE Regional Development Strategy (2014–2020); however, there is a lack of scientific studies regarding the current environmental impact of solid waste management and how the various treatment options could diminish this environmental impact in terms of climate change (Ropcean et al., 2019).

Over the 2013–2015 period, at regional level, the treatment and disposal methods used for the MSW management were predominantly landfilling (87%), recycling (material recycling and composting) with 11%, and lastly incineration (2%).

In Fig. 1 the MSW management system flowsheet is depicted. The input in the system is the MSW collected, while the outputs are divided between refuse to enter the landfill, refuse-derived fuel sent to incineration, and recycled materials and biodegradable waste. Because the data on MSW mixed waste sorting and process efficiencies and selective collection are scarce, the calculations use the MSW collected quantities, MSW composition, and the information on the final quantities directed towards a certain treatment to estimate the carbon footprint.

By using the aggregated models and corresponding emission factors suggested by Smith et al. (2001), Chen and Lin (2008), Christensen et al. (2009), and US EPA WARM (2006), the carbon footprint values have been obtained. The sum of recycling, composting, incineration, and landfilling carbon footprints gives the Total carbon footprint (TCF) or the total GHG emissions derived from waste management practices and intermediary operations (waste collection, sorting, transportation, and storage) (Ropcean et al., 2019).

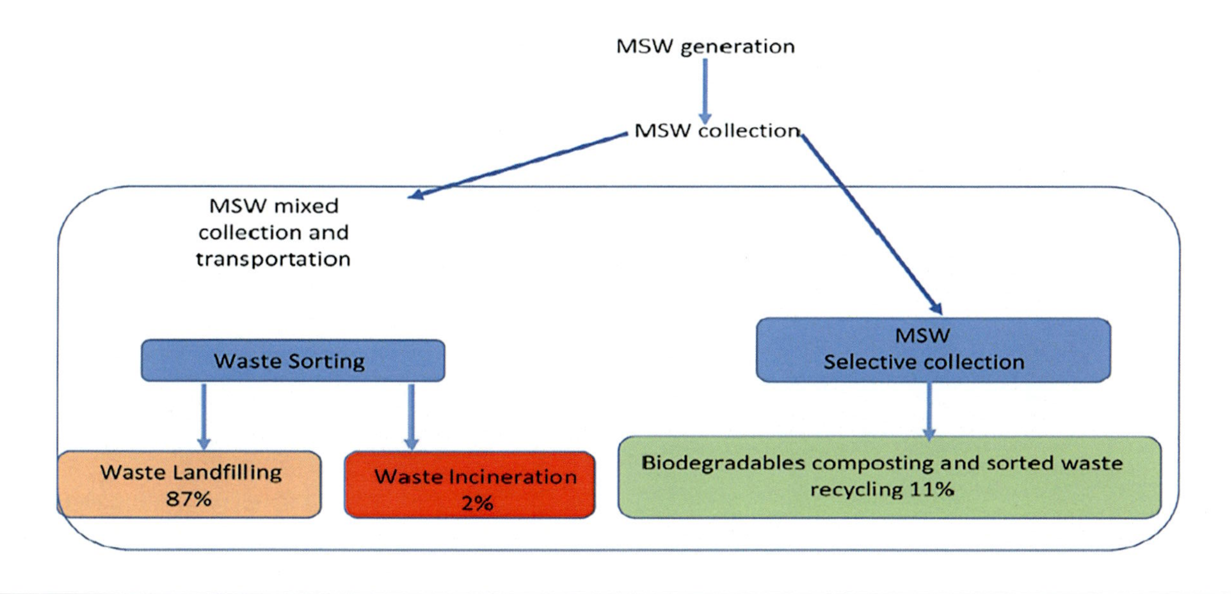

FIG. 1

MSW management system in Romania (2013–2015).

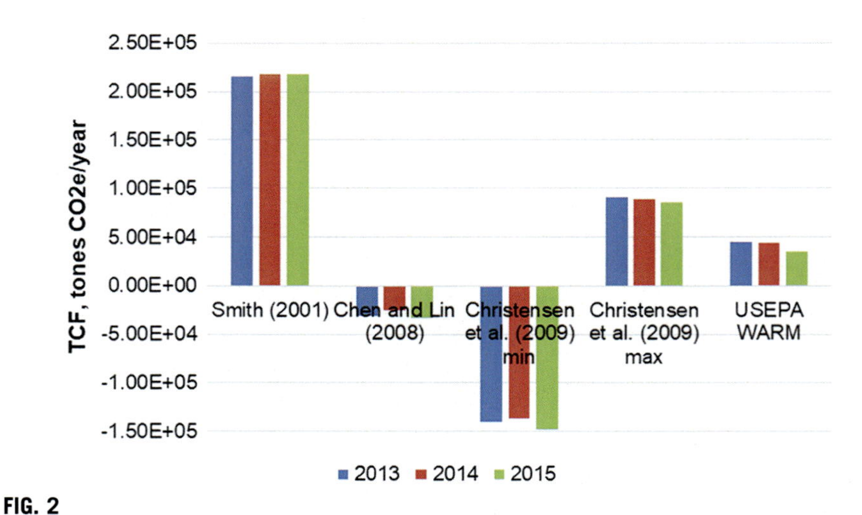

FIG. 2

Total carbon footprint at regional level from the MSWM system, NE region, Romania.

The results represented in Fig. 2 give the following information: (i) for 3 out of 5 models, namely Smith et al. (2001), Christensen et al. (2009)—maximum values, and US EPA WARM (2016), the TCF values are positive, meaning that the MSW system is generating GHG emissions; (ii) 2 models indicate negative values of the TCF, Christensen et al. (2009)—minimum values due to the negative values of the emission factors irrespective of the MSW composition and Chen and Lin (2008) which consider negative emission factors for paper and wood as subcategories of MSW; (iii) in the investigated period, the results from the Smith et al. model do not show a significant variation in time when compared to the other models; (iv) in this case waste composition influences mostly the TCF results obtained with Chen and Lin (2008) model.

The regional TCF results are the contribution of each TCF at local (in this case, county) level. The highest contributor to the regional TCF is Iasi, followed by Bacau, Suceava, Neamt, Vaslui, Botosani.

To summarise the main findings, it can be underlined that the strong dependency on landfilling generates the most significant amounts of GHG emissions; however, the situation has improved when 2015 is compared to the 2014 and 2013 values of the TCF at the regional level (Ropcean et al., 2019).

4.2 Carbon footprint of national waste electrical and electronic management systems

Worldwide, there is an increase in generation rates of waste electrical and electronic equipment (WEEE), a fact that is caused by several factors: enhanced technological developments, the smart functions incorporated in electrical devices and tools, the larger accessibility, pleasant design, and higher overall living standards (Gavrilescu, Enache, Ibănescu, Teodosiu, & Fiore, 2021). The increasing WEEE generation rates have placed the collection and treatment stages under pressure to develop sustainable ways to perform the corresponding actions (Ibanescu, Cailean, Teodosiu, & Fiore, 2018). It has to be underlined that WEEE is considered a key waste steam in European Union, requiring special monitoring and reporting activities and having as specific legislation Directive 2012/19/EU (2012).

The WEEE management system environmental profiles can be completely characterised by LCA studies. Depending on the defined scope, authors may obtain the environmental profile for the whole WEEE management and recycling system, partial WEEE management and recycling system, or WEEE treatment system. Efforts to obtain such environmental profiles have been intensified starting with 2013 (Ismail & Hanafiah, 2019). Out of environmental impact indicators, the contribution to climate change has been evaluated by using the carbon footprint of WEEE management systems in countries such as Romania (Ibanescu et al., 2018) and United Kingdom (Clarke, Williams, & Turner, 2019).

To perform the carbon footprint analysis of national WEEE management systems, in this case study, 2 countries were considered, namely Romania and Italy, similar in the efficiency of the solid waste management in general—meaning that both counties are considered to perform under average (Ibanescu et al., 2018), but with different economic situations. While according to the 2017 classification of the International Monetary Fund, Romania is considered a developing country and Italy a major developed economy (https:/www.imf.org). Romania has implemented the legislative framework on waste starting with 2007, while Italy implemented it since 2002 (Ibanescu et al., 2018).

For the carbon footprint calculation, an average composition of a general WEEE is considered and associated waste treatment options that are usually performed (Table 1) (Ibanescu et al., 2018). The WEEE generation rates are taken from the Eurostat database (https:/ec.europa.eu/eurostat). The case study covers the reference period 2007–2013.

The Total Carbon Footprint main results show that, over 2007–2013, the TCF values (tonnes CO_2e/year) of the Italian WEEE management system are one or

Table 1 WEEE composition and treatment options.

WEEE material type		Percentage (weight, %)	Treatment option
Metals	Ferrous	43.63	Recycling
	Non-ferrous	4.6	Recycling
Glass		10.01	Recycling
Plastics		14.61	Recycling
Wood		2	Recycling
Capacitors		0.1	Thermal treatment
Printed circuit boards		2.2	Thermal treatment
Non-recyclable		4	Landfilling
Refrigerants		0.02	Thermal treatment
Oil		0.1	Thermal treatment
Other components		18.71	Landfilling

two orders of magnitude higher than the corresponding values of the Romanian WEEE management system, due mainly to the higher Italian WEEE generation rates (Ibanescu et al., 2018).

The main findings in terms of national Total carbon footprints per capita calculated for Romania and Italy by using the aggregated models are depicted in Fig. 3A and B.

In the case of both countries, all results are negative, because of the contribution of the recycling carbon footprints, throughout the entire investigated period which is a clear indication that the WEEE management system organisation facilitates the reduction of GHG emission generation and from this indicator point of view the WEEE key stream is treated in a sustainable manner. According to the results, each Italian inhabitant has a higher contribution in terms of GHG emissions compared to a Romanian inhabitant (the reduction of GHG is higher for Romania than for Italy).

The GHG efficiency indicator was also calculated and the results show a change in the performance of the WEEE management system. If in 2007, the GHG efficiency indicator calculated for Italy was lower than the Romanian one, the 2013 shows a reversed situation. The 2013 situation reflects the improvements in the Romanian WEEE management system; however, in Italy the WEEE generation rates are 20–30 times higher than in Romania (Ibanescu et al., 2018) (Table 2).

4.3 Carbon footprint of national packaging waste management system

Packaging waste (PW) is another key waste stream in the European Union legislation, with generation rates up to approximately 80 million tonnes in 2018, in EU member states. The first PW specific legislation has been developed in 1985, continuing in 1994 with Directive 94/62/EC (n.d.) on packaging and packaging waste which has been modified and updated constantly to answer the changes and challenges. The legislation monitors the following packaging waste streams: paper and cardboard; plastics; wood; glass; metals, out of which steel and aluminium are the most important fractions (Gavrilescu, Bârjoveanu, & Teodosiu, 2017).

This case study proposes the evaluation of the Packaging Waste (PW) management system carbon footprint in 4 member states of the European union, namely Bulgaria (BG), Hungary (HU), Poland (PL), and Romania (RO), in the period 2007–2014. The selected period represents a full cycle of strategic planning in waste management options in European Union. All these countries have faced similar economic and societal challenges and most importantly, all have been classified as underperforming in terms of solid waste management. Hungary and Poland are 3-year period ahead of Romania and Bulgaria in gathering experience with the European legislation implementation in the solid waste field, a fact that should be considered in the results obtained in this assessment (Gavrilescu et al., 2017).

The organisation of the PW management system follows the life cycle stage, such as generation, collection, recycling for material recovery and thermal treatment for energy recovery (R1 type), and final disposal. This is the first study that evaluates the

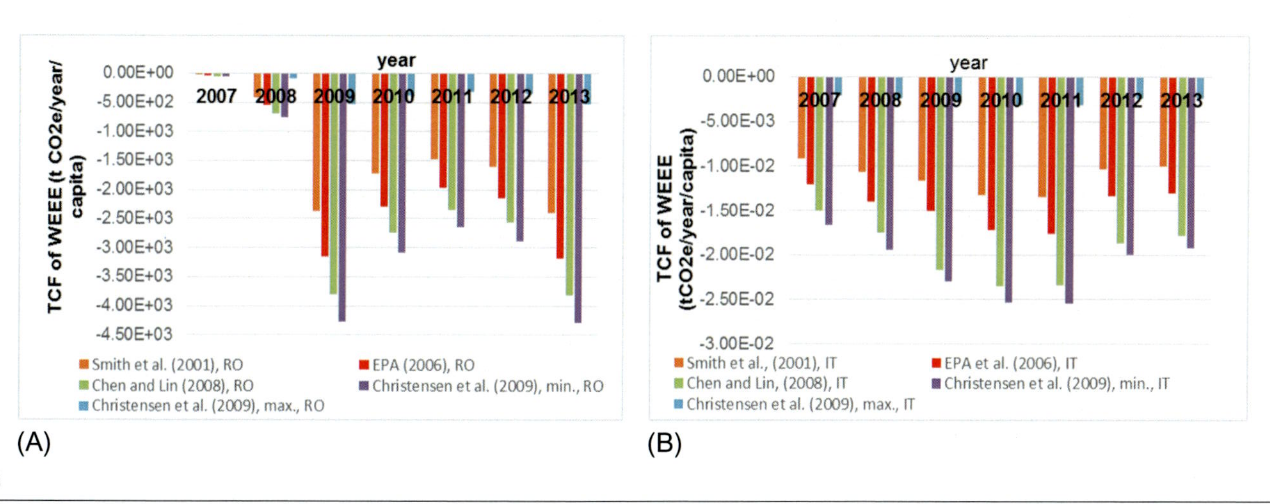

FIG. 3

(A) TCF per capita of WEEE in Romania. (B) TCF per capita of WEEE in Italy.

Table 2 GHG efficiency indicator for the WEEE stream in Romania and Italy.

Model	2007	2008	2009	2010	2011	2012	2013
Romania							
Smith et al. (2001)	−0.86	−1.12	−1.39	−1.41	−1.42	−1.40	−1.45
US EPA WARM (2006)	−1.11	−1.47	−1.85	−1.87	−1.88	−1.86	−1.92
Chen and Lin (2008)	−1.56	−1.89	−2.23	−2.25	−2.26	−2.24	−2.29
Christensen et al. (2009)—minimum value	−1.67	−2.08	−2.50	−2.52	−2.54	−2.51	−2.58
Christensen et al. (2009)—maximum value	−0.21	−0.26	−0.31	−0.31	−0.31	−0.31	−0.32
Italy							
Smith et al. (2001)	−1.24	−1.24	−0.81	−0.93	−1.00	−0.88	−0.94
US EPA WARM (2006)	−1.64	−1.63	−1.04	−1.21	−1.31	−1.14	−1.22
Chen and Lin (2008)	−2.04	−2.03	−1.50	−1.66	−1.74	−1.59	−1.67
Christensen et al. (2009)—minimum value	−2.27	−2.26	−1.59	−1.79	−1.89	−1.71	−1.80
Christensen et al. (2009)—maximum value	−0.28	−0.28	−0.20	−0.22	−0.23	−0.21	−0.22

environmental impact of the PW management system, based on the carbon footprint, by using aggregated models and comparing the greenhouse gas (GHG) emission variation during a seven-year period (2007–2014) (Gavrilescu et al., 2017).

The European PW composition (% in weight), for 2013, is paper and cardboard (41%), glass (20%), plastic (19%), wood (15%), and metal (6%) and displays very few variations in time. For example, the last available data from 2018 show that paper and cardboard percentage is 40.9%, glass 18.7%, plastic 19.1%, wood 16.1%, metal 5%, and 0.3% other type of packages (https:/ec.europa.eu/eurostat).

The PW generation rates and the recovery shares are presented in Table 3. Both indicators are below the EU average. The maximum increase in terms of recovery percentage (approximately 20%) is registered in Romania in 2014. Out of all recovery options, recycling is predominant and recycling rates follow the same national trends as the recovery ones, thus it is expected that the recycling carbon footprint will significantly impact the total carbon footprint results (Gavrilescu et al., 2017).

Fig. 4A–D presents the results obtained in the case of the total carbon footprint of the packaging waste in the 4 countries. Apart from Christensen et al. (2009) with maximum values, the aggregated models employed in the carbon footprint calculation indicate negative values, meaning that by using the current treatment options, reductions in the GHG emissions were registered. Over 2007–2014, the PW generation rates have fluctuated, while the recycling rates have been increasing, which lead to reductions of the national GHG coming from the PW management system (Gavrilescu et al., 2017).

The GHG efficiency indicator for the PW key stream is presented in Table 4. It is difficult to say which country is performing the best, because 2 models (Smith and US EPA WARM) give Bulgaria as the best performing country, while Chen and Lin (2008) as well as Christensen et al. (2009)—minimum indicate Romania. However, it can be said that the values of the GHG efficiency indicator for one model are within the same range for all the countries for both 2007 and 2014.

A comparison in the achievements of waste management policies in the beginning and the end of a full strategic planning period in EU can be compared by observing variations in terms of GHG efficiency indicator (Table 4).

4.4 Carbon footprint of biowaste management system

Biowaste or the municipal biodegradable stream consists of biodegradable fraction of household and similar waste, waste from trimmings and green areas (parks, gardens), as well as biodegradable street and market waste. According to the Romanian National Waste Management Plan for 2014–2020, biowaste generation predictions for 2020 account for 2477 million tonnes of biowaste and will reach 2161 t by 2025.

In terms of municipal solid waste infrastructure improvements at national level, in Romania, 3 alternatives have been proposed in order to improve the municipal solid waste system at national level and biowaste being the predominant fraction is, of course, directly influenced. Alternative 0 or the reference scenario considers

Table 3 Packaging waste generation rates (kg/inhabitant)-first line and recycling percentage of PW (%)-second line, 2007–2014 (https:/ec.europa.eu/eurostat).

Country	2007	2008	2009	2010	2011	2012	2013	2014
Bulgaria								
Waste generation	42.19	40.33	40.82	43.43	42.82	45	48.18	52.42
Recycling %	54.82	50.31	45.86	61.64	65.13	66.53	65.73	62.04
Hungary								
Waste generation	96.27	100.08	97.56	88.08	84.08	102.1	103.34	102.58
Recycling %	46.38	50.80	51.12	58.70	59.30	48.54	49.18	52.29
Poland								
Waste generation	82.21	109.69	99.08	112.85	121.14	122.69	126.88	127.49
Recycling %	48.16	42.91	36.85	38.87	41.24	41.37	36.06	55.38
Romania								
Waste generation	61.63	57	49.03	48.15	49.26	52.82	52.81	62.36
Recycling %	30.55	33.51	40.49	43.36	49.98	56.80	52.80	54.76
EU27 average								
Waste generation	164.27	164.05	153.68	157.23	160.05	157.29	158.21	163.55
Recycling %	59.16	60.53	62.45	63.53	63.79	64.72	65.36	65.53

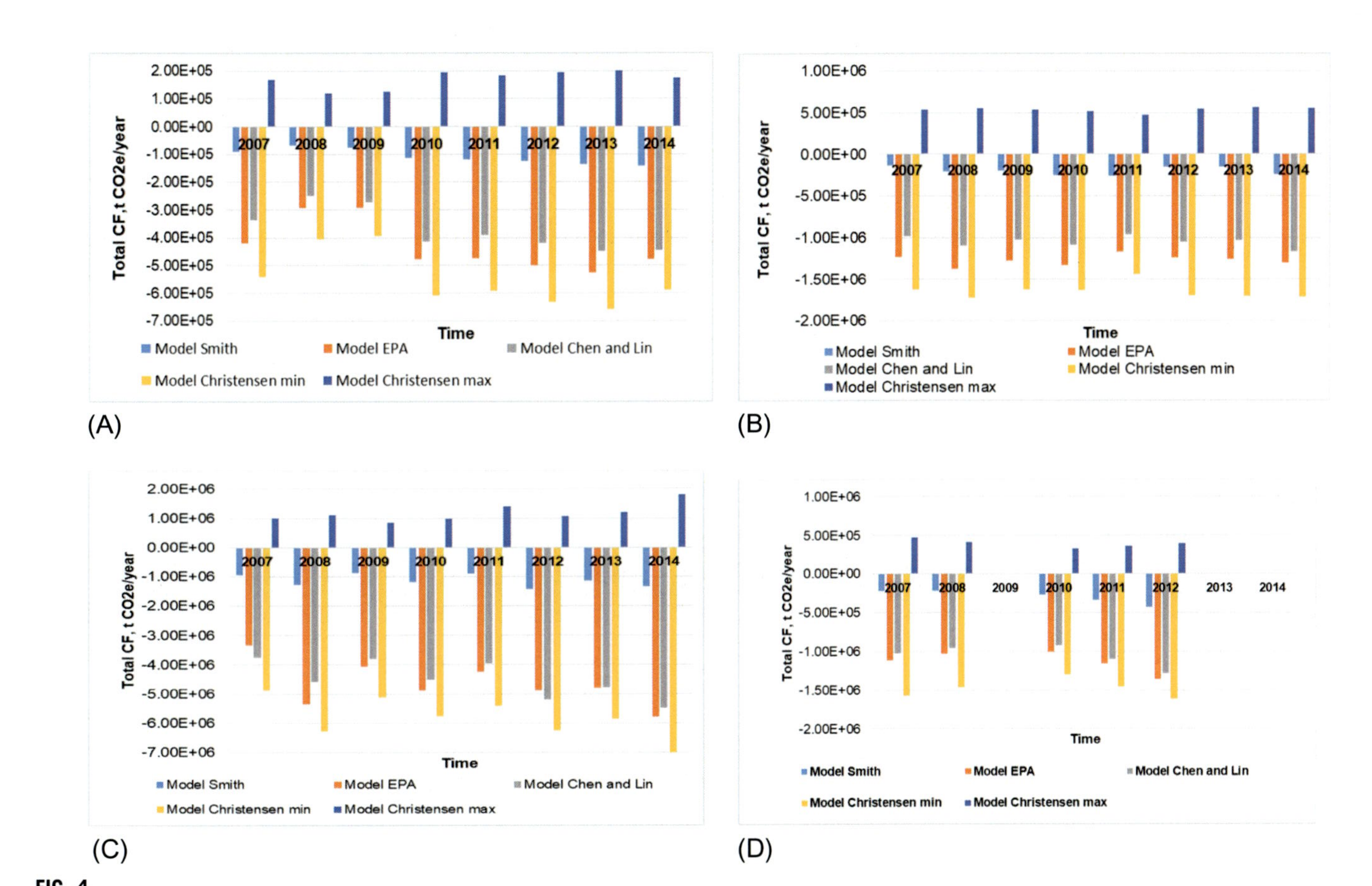

FIG. 4

(A) TCF of PW system in Bulgaria. (B) TCF of PW system in Hungary. (C) TCF of PW system in Poland. (D) TCF of PW system in Romania.

Table 4 GHG efficiency indicator for the PW system.

	Smith et al. (2001)		US EPA WARM (2006)		Chen and Lin (2008)		Christensen et al. (2009)—min		Christensen et al. (2009)—max	
Model applied	**2007**	**2014**	**2007**	**2014**	**2007**	**2014**	**2007**	**2014**	**2007**	**2014**
Bulgaria	−0.52	−0.56	−2.41	−2.25	−1.92	−1.88	−3.10	−2.85	0.97	0.87
Hungary	−0.26	−0.25	−2.33	−2.04	−1.86	−1.72	−3.09	−2.79	1.01	0.89
Poland	−0.50	−0.54	−1.77	−1.83	−1.99	−1.95	−2.60	−2.35	0.53	0.40
Romania	−0.47	−0.70	−2.36	−2.23	−2.16	−2.10	−3.34	−2.66	1.01	0.66

that all new waste sorting and treatment facilities are fully operational in 2020 and the waste collection service will cover 100% of population (Logigan, Gavrilescu, & Teodosiu, 2019).

Alternative 1 involves besides all the hypothesis formulated in Alternative 0, the extension of the separate collection system for recyclable waste, sorting facilities, and a separate biowaste collection and treatment system. All facilities for waste treatment will include green waste treatment by composting stations, anaerobic digestion facilities, and mechanical-biological treatment facilities (with biodrying) all fully operational by 2025, the opening of new cells in the case of the compliant landfills, and the closure of any non-compliant landfills (Logigan et al., 2019).

Alternative 2 considers, besides all the measures taken in Alternative 1, the extensions of the energy valorisation infrastructure with a massive municipal solid waste incineration facility with energy recovery, operating in Bucharest (capital city of Romania) able to treat 173.000 t/year, by 2025 (Logigan et al., 2019). The facility types and operating capacities are presented in Table 5.

Based on the material balances and the carbon footprint calculation methodology, in 2020, the environmental impact in terms of greenhouse gas emission derived from the biowaste management system can be observed in Table 6. In 2020 the treatment facilities existing in Alternatives 1 and 2 are not in operation, thus making the results identical in all alternatives (Logigan et al., 2019). However, by the end of 2025, a clear differentiation between the environmental impacts of the alternatives, in terms of GHG emissions, is observed. Based on the results indicated in Tables 6–9, the main conclusion is that by 2025, Alternative 1 ensures the lowest GHG emissions from the biowaste management system, because 4 out of 5 methodologies for the aggregated carbon footprint calculations have the lowest values.

5 Conclusions

A special interest regarding the link between SWMS and climate change in terms of GHG emissions generation and ways of reductions has been observed in conducting the literature review. The GHG emissions can be estimated by using the LCA methodology if sufficient quality data are available and the results are more relevant if specific factors are considered. However, the GHG emissions can be estimated by using aggregated methodologies as demonstrated in this chapter, thus overcoming to a certain point the data availability and level of detail issue. Aggregated models have proven to be a rapid assessment tool to obtain the desired information in terms of total or treatment specific emissions (CF of a treatment method or the TCF) or solid waste performance in treating the waste (GHG efficiency indicator) and/or to compare situations in different territorial levels (regional, country, or inter-country), or in different periods of time (past/current/future).

SWMS such as MSW, WEEE, PW, and biowaste have been investigated. For the MSW case study, the regional total carbon footprint is mainly influenced by the carbon footprint associated with landfilling, while the main contributor to the carbon footprint is Iasi (NE Region, Romania), a county with 25% of the population.

Table 5 Overall capacity of sorting and treatment facilities (tonnes), operational in 2020.

Alternative	Composting station	Sorting station[a]	TMB with biostabilisation	TMB with biodrying	Anaerobic digestion	Incineration
0	377,000	1,072,700	1,529,000	–	–	–
1	403,800	1,106,700	–	1,146,000	812,000	–
2	403,800	1,106,700	–	973,000	812,000	173,000

[a]*For recyclables (from separate collection).*

Table 6 Carbon footprint of various biowaste treatment options and total carbon footprint, in Romania, 2020.

Alternative	Biowaste treatment option	Smith et al. (2001)	US EPA WARM (2016)	Chen and Lin (2008)	Christensen et al. (2009)—min	Christensen et al. (2009)—max
A0=	Material recycling	−7229.02	−31,260.64	−39,075.80	−171,308.31	59,785.97
A1=	Incineration	−3420.43	−8291.95	−12,956.17	−52,809.34	2539.41
A2	Landfilling	279,496.18	76,574.30	45,944.58	−80,020.14	61,642.31
	Total CF	**268,846.73**	**37,021.71**	**−6087.39**	**−304,137.79**	**123,967.69**

Table 7 Carbon footprint of various biowaste treatment options and total carbon footprint, in Romania, predicted for 2025, Alternative 0.

Alternative	Biowaste treatment option	Smith et al. (2001)	US EPA WARM (2016)	Chen and Lin (2008)	Christensen et al. (2009)—min	Christensen et al. (2009)—max
A0	Material recycling	−6796.78	−29,391.47	−36,739.33	−161,065.24	56,211.18
	Incineration	−3215.91	−7796.15	−12,181.48	−49,651.70	2387.57
	Landfilling	1,026,548.36	281,246.13	168,747.68	−293,902.20	226,403.13
	Total CF	**1,016,535.67**	**244,058.51**	**119,826.87**	**−504,619.14**	**285,001.88**

Table 8 Carbon footprint of various biowaste treatment options and total carbon footprint, in Romania, predicted for 2025, Alternative 1.

Alternative	Biowaste treatment option	Smith et al. (2001)	US EPA WARM (2016)	Chen and Lin (2008)	Christensen et al. (2009)—min	Christensen et al. (2009)—max
A1	Material recycling	−19,339.26	−83,629.25	−104,536.56	−458,288.28	159,940.94
	Incineration	−16,079.55	−38,980.73	−60,907.38	−248,258.49	11,937.85
	Anaerobic digestion	−32,376.02	−32,376.02	−32,376.02	−32,376.02	−32,376.02
	Landfilling	585,908.30	160,522.82	96,313.69	−167,746.35	129,220.87
	Total CF	**518,113.46**	**5536.83**	**−101,506.27**	**−906,669.14**	**268,723.63**

Table 9 Carbon footprint of various biowaste treatment options and total carbon footprint, in Romania, predicted for 2025, Alternative 2.

Alternative	Biowaste treatment option	Smith et al. (2001)	US EPA WARM (2016)	Chen and Lin (2008)	Christensen et al. (2009)—min	Christensen et al. (2009)—max
A2	Material recycling	−7279.94	−31,480.83	−39,351.04	−172,514.97	60,207.09
	Incineration	−21,643.07	−52,468.06	−81,981.34	−334,155.93	16,068.34
	Anaerobic digestion	−32,376.02	−32,376.02	−32,376.02	−32,376.02	−32,376.02
	Landfilling	524,372.35	143,663.66	86,198.19	−150,128.52	115,649.24
	Total CF	**463,073.31**	**27,338.74**	**−67,510.21**	**−689,175.44**	**159,548.66**

The WEEE case study results demonstrated that Italians have a higher contribution in terms of GHG emissions than Romanians and that when 2013 values are compared with 2007, the GHG efficiency indicator is higher for Romania (up to 1.6 times). The PW case study gives an image of the magnitude of 4 national PWMS carbon footprints (between -7×10^6 and -7×10^5) tonnes/year, corresponding to Bulgaria, Hungary, Poland, and Romania and their variation in time. In both WEEE and PW case studies, there is a significant contribution of the recycling carbon footprint to the total carbon footprint. In the case of Biowaste case study, the results in terms of total carbon footprint indicate that a shift in the waste treatment options towards mechanical biological treatment with biodrying and anaerobic digestion in 5 years' time would be the key to reduce the GHG emissions.

Acknowledgements

This study was supported by the Romanian Ministry of Research and Innovation, CCCDI-UEFISCDI, project number 26PCCDI/01.03.2018, 'Integrated and sustainable processes for environmental clean-up, wastewater reuse and waste valorization' (SUSTENV-PRO), within PNCDI III.

References

Bernstad Saraiva, A., Souza, R. G., & Valle, R. A. B. (2017). Comparative lifecycle assessment of alternatives for waste management in Rio de Janeiro – Investigating the influence of an attributional or consequential approach. *Waste Management*, *78*, 701–710.

Căilean, D., & Teodosiu, C. (2016). An assessment of the Romanian solid waste management system based on sustainable development indicators. *Sustainable Production and Consumption*, *8*, 45–56.

Chen, T. C., & Lin, C. F. (2008). Greenhouse gases emissions from waste management practices using life cycle inventory model. *Journal of Hazarduous Materials*, *155*(1–2), 23–31.

Christensen, T., Gentil, E., Boldrin, A., Larsen, A. W., Weidema, B. P., & Hauschild, M. (2009). C balance, carbon dioxide emissions and global warming potentials in LCA modelling of waste management systems. *Waste Management and Research*, *27*, 707–715.

Cifrian, E., Galan, B., Andres, A., & Viguri, J. R. (2012). Materials flow indicators and carbon footprint for MSW management systems: Analysis and application at regional level. *Cantabria, Spain, Resources, Conservation and Recycling*, *68*, 54–66.

Clarke, C., Williams, I. D., & Turner, D. A. (2019). Evaluating the carbon footprint of WEEE management in the UK. *Resources, Conservation and Recycling*, *141*, 465–473.

Directive 2012/19/EU. (2012). *Directive 2012/19/EU on waste of electric and electronic equipment*.

Directive 94/62/EC. n.d. Directive 94/62/EC on packaging and packaging waste.

Eurostat. (2020). https:/ec.europa.eu/eurostat/web/products-eurostat-news/-/DDN-20200123-1.

Fan, Y. V., Jiang, P., Klemeš, J. J., Yen, P., Chew, L., & Lee, T. (2021). Integrated regional waste management to minimise the environmental footprints in circular economy transition, resources. *Conservation and Recycling*, *168*, 105292.

Gavrilescu, D., Bârjoveanu, G., & Teodosiu, C. (2017). Sustainability of the packaging waste management systems in Eastern European Countries. In *Conference paper, 9th international conference on environmental engineering and management ICEEM09, Bologna, Italy.*

Gavrilescu, D., Enache, A., Ibănescu, D., Teodosiu, C., & Fiore, S. (2021). Sustainability assessment of waste electric and electronic equipment management systems: Development and validation of the SUSTWEEE methodology. *Journal of Cleaner Production*, *306*, 127214.

Ibanescu, D., Cailean, D., Teodosiu, C., & Fiore, S. (2018). *Assessment of the waste electrical and electronic equipment management systems profile and sustainability in developed and developing European Union countries. vol. 73* (pp. 39–53).

Ismail, H., & Hanafiah, M. M. (2019). An overview of LCA application in WEEE management: Current practices, progress and challenges. *Journal of Cleaner Production*, *232*, 79–93.

Lee, U., Han, J., & Wang, M. (2017). Evaluation of landfill gas emissions from municipal solid waste landfills for the life-cycle analysis of waste-to-energy pathways. *Journal of Cleaner Production*, *166*, 335–342. https:/doi.org/10.1016/j.jclepro.2017.08.016.

Logigan, F., Gavrilescu, D., & Teodosiu, C. (2019). *Modelling the greenhouse gas emissions coming from the national municipal solid waste system.* Dissertation Thesis (in Romanian), supervisors Carmen Teodosiu and Daniela Gavrilescu.

Malakahmad, A., Abualqumboz, M. S., Kutty, S. R. M., & Abunama, T. J. (2017). Assessment of carbon footprint emissions and environmental concerns of solid waste treatment and disposal techniques; case study of Malaysia. *Waste Management*, *70*, 282–292.

Nazmul Islam, K. M. (2017). Greenhouse gas footprint and the carbon flow associated with different solid waste management strategy for urban metabolism in Bangladesh. *Science of the Total Environment*, *580*, 755–769.

Obersteiner, G., Gollnow, S., & Eriksson, M. (2021). Carbon footprint reduction potential of waste management strategies in tourism. *Environmental Development*, *39*, 100617.

Pérez, J., Manuelde Andrés, J., Lumbreras, J., & Rodríguez, E. (2018). Evaluating carbon footprint of municipal solid waste treatment: Methodological proposal and application to a case study. *Journal of Cleaner Production*, *205*, 419–431.

Romanian Environmental Status Reports. (2017).

Ropcean, G., Gavrilescu, D., Teodosiu, C., & Fiore, S. (2019). Environmental indicators in the assessment of municipal solid waste management systems. In *Case study: NE Region, Romania, conference paper, 10th international conference on environmental engineering and management ICEEM10, Iasi, Romania.*

Sankar Cheela, V. R., John, M., Biswas, W. I., & Dubey, B. (2021). Environmental impact evaluation of the current municipal solid waste treatment in India using life cycle assessment. *Energies*, *14*, 3133.

Smith, A., Brown, K., Ogilvie, S., Rushton, K., & Bates, J. (2001). *Waste management options and climate change*. Final report to the European Commission Luxembourg: DG Environment, Office for Official Publications of the European Communities, ISBN:92-894-1733-1.

US EPA WARM. (2006). In *Version 8*. Available at: https:/www.epa.gov/warm/versions-waste-reduction-model-warm.

US EPA WARM. (2016). In *Version 14*. Available at: https:/www.epa.gov/warm/versions-waste-reduction-model-warm.

Vázquez-Rowe, I., Ziegler-Rodriguez, K., Margallo, M., Kahhat, R., & Aldaco, R. (2021). Climate action and food security: Strategies to reduce GHG emissions from food loss and waste in emerging economies. *Resources, Conservation and Recycling*, *170*, 105562.

CHAPTER

How can we validate the environmental profile of bioplastics? Towards the introduction of polyhydroxyalkanoates (PHA) in the value chains

20

Alba Roibás-Rozas[a], Mateo Saavedra del Oso[a], Giulia Zarroli[a,b], Miguel Mauricio-Iglesias[a], Anuska Mosquera-Corral[a], Silvia Fiore[b], and Almudena Hospido[a]

[a]*CRETUS, Department of Chemical Engineering, Universidade de Santiago de Compostela, Santiago de Compostela, Spain* [b]*Department of Environment, Land and Infrastructure Engineering (DIATI), Politecnico di Torino, Torino, Italy*

1 Plastics and bioplastics

Plastic goods have been in our lives for no longer than one century. However, in about 70 years, they have become one of the greatest concerns and environmental threats worldwide. The growth of plastics market during the twentieth century was remarkably fast, with a growth rate only surpassed by some construction goods, such as steel and cement. Plastics production increased from 2 Mt in 1950 to 380 Mt in 2015, yielding in about 83,000 Mt of total produced virgin plastics in that period (Geyer et al., 2017). Moreover, the plastic sector employs about 1.5 M people in the European Union (EU) and generates around 340 billion € yearly (European Commission, 2018a, 2018b). However, its production and use are linked to the generation of extremely high amounts of wastes. It is estimated that, of all the plastics ever produced, only about 30% of them are still in use, so the remaining 70% of them were wasted. Of this enormous amount of waste, only 9% is recycled, 12% is incinerated, and 79% is accumulated in landfills or directly disposed in the natural environment with no treatment at all (Geyer et al., 2017).

The related impacts of plastics on ecosystems, health, and economy cannot be quantified yet due to the lack of consistent frameworks and actual data. However, the number of studies on this topic is growing very fast, and their results show that

Assessing Progress towards Sustainability. http://doi.org/10.1016/B978-0-323-85851-9.00010-9

the effects are serious and global (Halsband & Herzke, 2019; Saling et al., 2020). In fact, plastics are slowly decomposing and leaking in the environment. First, they affect terrestrial ecosystems (de Souza Machado et al., 2018), from where they are transferred into freshwater (Wagner et al., 2014). Finally, plastics reach the oceans in the form of macro, micro, and nanoplastics (Galloway et al., 2017; Geyer et al., 2017; Halsband & Herzke, 2019; Saling et al., 2020). Modern plastics are complex mixtures of polymers, residual monomers, and chemical additives, some of them being endocrine disruptors that alter metabolic and reproductive patterns (Galloway et al., 2017). Plastic debris accumulates in the environment and is transferred to many trophic levels through food chains, thereby affecting animal and human health (Akhbarizadeh et al., 2020). Moreover, unmanaged plastic waste also affects the economy, seriously impacting fishing, aquaculture, and tourism sectors, and it implies direct costs for governments and administrations, as an example, the Netherlands spend millions of euros each year on removing litter from beaches and coastal areas (Conejo-Watt & Luisetti, 2019).

Furthermore, around 50% of the produced plastics are single-use plastics (SUP) (Galloway et al., 2017), which constitute 49% of the marine litter in the EU (European Commission, 2018a). Moreover, it is estimated that about 95% of the economic value of plastic packaging materials is lost due to the single-use life cycles (European Commission, 2018a). Therefore the European Commission recently addressed this issue by publishing a directive on the reduction of the impact of plastics on the environment (European Commission, 2019a, 2019b) and by generating a new legal framework in the European Green Deal (European Commission, 2019a) and the new *Circular Economy Action Plan for a cleaner and more competitive Europe* (European Commission, 2020). Amongst the actions of this plan are to develop a framework on sourcing, labelling, and use of bio-based, biodegradable, and/or compostable plastics.

Whilst biodegradable plastics can have petrochemical origin, bioplastics include a broad number of renewable, biodegradable, and bio-based polymers that are expected to substitute conventional plastics, as they have similar physicochemical, thermal, and mechanical properties (Mannina et al., 2020). Nowadays, the annual production of bio-based and/or biodegradable plastics is less than 1% of the global plastic production (Geyer et al., 2017), polyhydroxyalkanoates (PHA), polylactic acid (PLA), and polybutylene succinate (PBS) being the biomaterials with more potential in the market. The three of them are produced involving microbiological processes. For PLA and PBS, the building blocks of the polymer (the monomers, lactic, succinic acid, and 1,4-butanediol, respectively) are produced by the activity of microorganisms, but monomer polymerisation is not a natural process. However, for PHA, polymerisation occurs naturally by the microorganisms, so the production is totally biological. Recently, PHA is gaining more attention, but its sustainability needs to be ensured before entering the value chains. Thus this chapter reviews the current state of the art regarding PHA environmental assessment and discusses the hotspots and bottlenecks that its production and processing are facing.

2 Polyhydroxyalkanoates: Feasible production and challenges along the value chain

PHAs are produced by some microorganisms under stress conditions. If feedstock and nutrients are available transiently, some microbes respond by accumulating extra carbon that could be used in the case of feeding deprivation (Valentino et al., 2017). That extra carbon is stored inside the cells in the form of polyesters, which, after extraction, can be employed to replace petrochemical plastics. Therefore in PHA production, carbon-rich feedstocks are transformed into value-added products by microorganisms, so they are an opportunity to fit the circular economy action plans by EU.

First, studies were based on pure substrates, mostly coming from dedicated crops such as sugar cane or palm oil, to feed pure cultures of microorganisms consisting in one single strain of bacteria able to store PHA. However, the costs were very high, reaching 45% of the total expenses due to high purity feedstocks (Kourmentza et al., 2017). Besides, pure culture processes require sterile conditions, which imply high energy costs for reactor sterilisation (Bengtsson et al., 2017). Finally, the use of dedicated crops involves an ethic conflict, as a global production of bioplastic based on agricultural feedstocks could alter the world's food supply chain, reducing the global availability of food and increasing the price of some basic nourishments, such as corn, sugar cane, or soy (Sabapathy et al., 2020). Even if this problem is partially faced by replacing these feedstocks by non-food crops, the land use would subliminally distress food production. Moreover, the production of PHA from dedicated substrates could provoke high impacts in categories such as eutrophication or acidification (Yadav et al., 2020).

This section presents the current challenges faced by the research community and industry on their way to achieve full-scale PHA production, as well as the main issues related to the following stages that form the biomaterial value chain.

2.1 Ongoing efforts on feasible PHA production

Research efforts are currently oriented to replace pure feedstocks by inexpensive substrates, such as waste streams. They are carbon rich, and sometimes nitrogen and nutrient rich, so they can also be used to feed the microorganisms in charge of PHA production. Moreover, waste streams are not competing at all with the food chain supplies or land use, and they are environmentally burden free (Yadav et al., 2020). When wastes are employed for PHA production, pure cultures are often replaced by mixed microbial cultures (MMCs), where a high diversity of microorganisms is present. Therefore a selection strategy is necessary to favour the growth of the desired strains and disfavour the thriving of the non-storing populations of microorganisms (Kourmentza et al., 2017).

Although full-scale PHA production is currently dominated by pure culture systems (Sabapathy et al., 2020), which are generally crop based (Yates & Barlow, 2013),

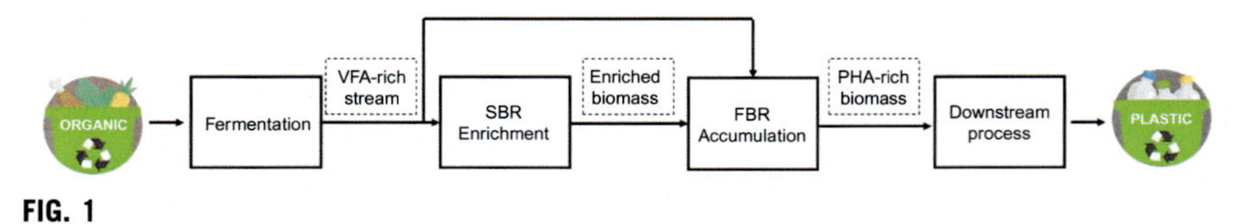

FIG. 1

Flowchart of a typical MMC-based PHA production process.

the pilot-scale projects reported in literature and operated for PHA production are nowadays working with MMCs (Rodriguez-Perez et al., 2018), showing a switch in trend in the culture type used. As seen in Fig. 1, this production pathway is generally performed in three stages (Valentino et al., 2017): (i) a pre-treatment where the organic sources present in the feedstock (like carbohydrates or proteins) are transformed into suitable forms of carbon for PHA production through anaerobic fermentation (i.e. volatile fatty acids (VFA), which will be naturally esterified and accumulated by bacteria (Kosseva & Rusbandi, 2018)); (ii) once the feedstock is fermented, the VFA-rich stream will be fed transiently to an open MMC that will be enriched in microorganisms with the capacity of accumulating PHA; and (iii) the outgoing stream, rich in storing biomass, will be fed with the same VFA-rich stream where the MMC accumulating capacity will be maximised (Sabapathy et al., 2020). The second stage is generally performed in a Sequencing Batch Reactor (SBR) and the third one in a Fed-Batch Reactor (FBR) (Fig. 1).

Nowadays, 1 kg of most petrochemical conventional plastics cost less than 1 €, whilst the commercial prices of pure culture PHA still range between 1.5 and 5.0 €/kg (Kumar et al., 2020), so the MMC-based route is a promising alternative to increase competitiveness of bioplastics production. However, there is still a great economical drawback due to lower overall yields of the typical three-stage process for MMC-based PHA (Valentino et al., 2017), as pure culture PHA processes reach 90% accumulation with 1.0–1.3 g PHA/(L·h) productivity, whilst MMC-based PHA reaches up to 75% and 1.2 g PHA/(L·h) (Mannina et al., 2020). Besides, there is an extra cost linked to the additional required fermentation unit (Kumar et al., 2020), and to the downstream operations needed to extract the PHA from the bacterial cells, which can be even more expensive for MMCs due to different cell wall fragility with respect to pure strains (Mannina et al., 2020).

2.2 Beyond PHA production: Recovery and purification

Downstream process (DSP) involves the procedures necessary to extract the PHA from the biomass and transform it into useful materials. It still accounts for 30%–50% of the total production costs (Colombo et al., 2020), being, economically and environmentally, the most expensive part of the whole MMC-based production process (Fernández-Dacosta et al., 2015). Nowadays, the existing DSPs are divided into two categories: the ones that separate biomass from PHA by dissolving the polymers, and the ones that dissolve or mechanically disrupt the cellular material to release the PHA (Jiang et al., 2018; Kosseva & Rusbandi, 2018; Yadav et al., 2020). Generally, the process starts with the concentration of the PHA-rich biomass (Pérez-Rivero et al., 2019). Then, pre-treatments are used to weaken the cell membrane before it is disrupted. Although the most applied pre-treatment involves freeze-drying the biomass (Rodriguez-Perez et al., 2018), it presents technical and economic difficulties for its full-scale application, so other techniques like thermal dehydration (Kourmentza et al., 2017) or osmotic shocks (Koller et al., 2013a, 2013b; Kourmentza et al., 2017; Rathi et al., 2013) are being researched.

Once the cell is weakened or broken up, the methods directly dissolving the biopolymeric material usually employ solvents that change the permeability of the cell membrane, so they contact with the PHA and dissolve it (Yadav et al., 2020). Here, the most used method is solvent extraction (generally using chlorinated compounds) (Li & Wilkins, 2020; Thakur et al., 2018). It provides a high-quality final product, but the use of chlorinated solvents is economically and environmentally detrimental (Fernández-Dacosta et al., 2015; Saavedra del Oso et al., 2020). Although novel approaches replace chlorinated compounds by non-halogenated solvents (Jiang et al., 2018) or by bio-based solvents (de Souza Reis et al., 2020), they are still energy intensive, highly expensive, and environmentally unfavourable for the process (Pérez-Rivero et al., 2019; Saavedra del Oso et al., 2020), compromising PHA's environmental benefits (Li & Wilkins, 2020).

The alternative to PHA direct solving is cell digestion, which is normally chemical (by acids or alkalis). However, in the last years, a new pathway was introduced, where the cell material is dissolved employing surfactants (Fernández-Dacosta et al., 2015; Mannina et al., 2019; Rathi et al., 2013), the sodium dodecyl sulphate (SDS) being the most employed one (Pérez-Rivero et al., 2019). This detergent enters the lipid membrane and increases the volume of the cell envelope until it breaks down, so the PHA is released. This method obtains good results when it is combined with other treatments, like alkali addition (Fernández-Dacosta et al., 2015; Rathi et al., 2013), so surfactant treatment is, economically and environmentally, the most promising alternative for low-grade PHA extraction (Fernández-Dacosta et al., 2015; Saavedra del Oso et al., 2020). After extraction, a separation process is performed (Li & Wilkins, 2020) and, finally, a purification step can be needed, mostly if the PHA is used for pharmaceutical or food grade applications. In this sense, depending on the final degree of purification, it has been recently proposed to classify the final product on low- and high-grade PHA (Saavedra del Oso et al., 2020).

2.3 Closing the loop: PHA compounding and shaping, use and end of life

Once the polymer is obtained, its final uses are multiple and depend on its properties (Roohi et al., 2018), which vary with the feedstock composition and the extraction method (Bengtsson et al., 2017; Melendez-Rodriguez et al., 2018), as monomer distribution and chain length are two of the main issues defining the final properties

(Raza et al., 2018). There is still a great way until the tailoring of the PHA materials is perfectly understood, although it seems likely that polyhydroxybutyrate (PHB) can substitute some polypropylene (PP) and polyethylene terephthalate (PET) applications (Roohi et al., 2018). Common PHA manufacturing techniques are moulding and injection (Bengtsson et al., 2017; Changwichan et al., 2018) and electrospinning (Melendez-Rodriguez et al., 2018). Besides, it has been stated that the combination of PHA and PLA fibres can produce materials with improved final properties (Bengtsson et al., 2017; Lamberti et al., 2020), and fully green composites formed by PHA, PLA, and PBS reinforced with non-woven flax fibres achieve the necessary requirements to be employed in the car industry (Pantaloni et al., 2020). Finally, poor-grade or wasted (after several use cycles) PHA films can be further valorised in feedstock recycling to produce crotonic acid and its oligomers, which are used as building blocks in the production of some chemicals, like paints or adhesives (Fernández-Dacosta et al., 2016; Lamberti et al., 2020).

Finally, these bio-based materials need to be properly disposed and managed at their end of life (EoL) (Fig. 2). Here, the waste management options are landfilling, incineration, anaerobic digestion, composting, and recycling (Yates & Barlow, 2013). Although PHAs are biodegradable, their degradation is complex and hard to model, as it is dependent on many factors (Laycock et al., 2017). In this point, there is some sort of paradox regarding bioplastics, as they need to be completely biodegradable, but their properties need to endure for some time, responding to products' shelf-life and life cycle uses. PHA readily biodegrades in compost, soil, and marine environments, although, in absence of microorganisms and ambient conditions (mainly UV radiation), it remains stable (Lamberti et al., 2020). Moreover, PHA shows a good anaerobic biodegradability (Battista et al., 2021) and it is completely composted after 10–22 months (Arcos-Hernandez et al., 2012). However, the EN 13432 standard (2005), which regulates the requirements for packaging recoverable through composting and biodegradation, states a time from three to 6 months of biodegradation time for a material to be defined as compostable, so the management of PHA as organic/compostable waste can be complex.

Despite the lack of experimental data, suitable EoL routes have been proposed. They start with several use cycles in which mechanical recycling is involved, as PHA is biodegradable but endurable, and its properties are maintained after several extrusion cycles (PHBV remains almost unaltered after five cycles), where the number of re-uses might be improved by combining PHA with PLA (Lamberti et al., 2020). Environmentally, the EoL has low contributions to the total impact in the life cycle (Yates & Barlow, 2013), and recent studies confirmed that mechanical recycling is the best alternative for EoL single-use bioplastics (Changwichan et al., 2018). Therefore after several use cycles, the circular economy loop can be closed by employing the deteriorated PHA to produce biogas through anaerobic digestion, compost through aerobic mineralisation, new PHA (as its carbon can be used as biomass feedstock), or to generate crotonic acid as a building block in the green chemistry industry by mild temperature pyrolysis (Lamberti et al., 2020).

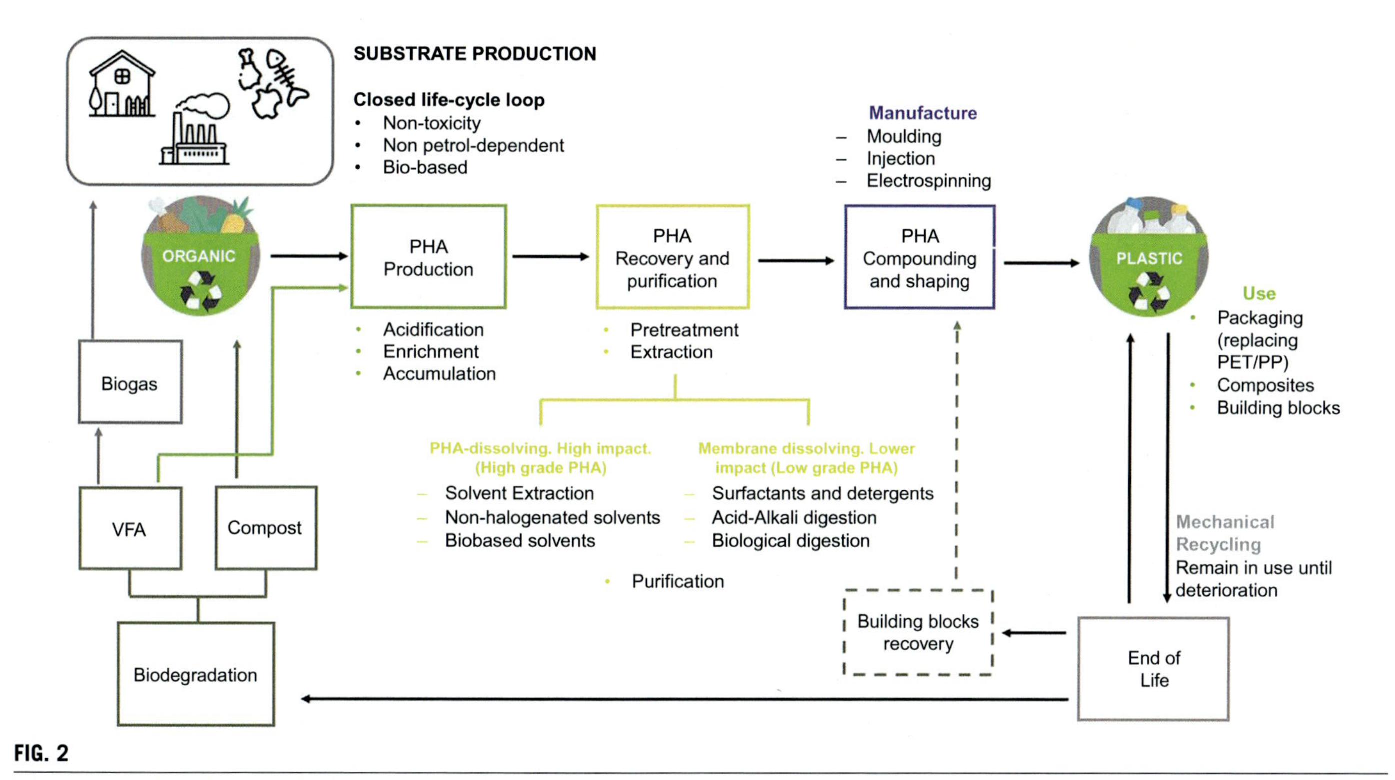

FIG. 2

Life cycle of a biopolymer in the frame of circular economy.

There is still a regulatory gap between the production and the application of bioplastics market, and specific regulations have not yet been implemented due to several factors, like the lack of full-scale demonstrations and the few data generated regarding the final properties and applications of the product. Nonetheless, policies are being proposed and the EU established a legal framework to begin shaping the general regulations which will rule bioplastic production, use, and management for the next years (European Commission, 2018a). Here, the European Commission engages to develop specific policies to manage the use of bio-based and biodegradable plastics and the gradual change to a circular industrial and consumption mode. Therefore express regulations will surely be claimed in the next years. The proposed policies must provide supportive environments which will let bioeconomy growth by employing different tools, like funding research and innovation, or supporting the early market (for example, by tax incentives) (OECD, 2013).

3 Review strategy: Current literature regarding LCA of PHA

This review was performed using the Scopus search engine, where the following key words and acronyms were selected: life cycle assessment (LCA), polyhydroxyalkanoates (PHA), bioplastics, sustainable process index (SPI), waste reduction algorithm (WRA), techno-economic analysis (TEA), and sustainability. The results of the review will be discussed according to (i) LCA methodological issues, (ii) process issues, and (iii) the studies' outputs. Regarding LCA methodological issues, a standardised protocol to perform LCA studies of PHA and bioplastics has been recently proposed by the European Commission and the Joint Research Centre of the European Commission (JCR-EC) to conduct comparative LCA studies for bioplastics and fossil-based polymers (Nessi et al., 2020), which is based on several widely recognised guidance documents, such as ISO or EN standards. The main principles of this guidelines framed the revision and discussion presented later. Finally, besides the discussion of the main outputs, the referred process issues classification is linked to relevant PHA production features, explicitly:

- **Substrate employed: ad hoc produced material vs waste stream.** Substrate selection has economic, social, ethical, and environmental implications, but it also partially defines the PHA production process (wastes are generally preferred for MMC-based systems). Moreover, the relevant environmental categories for each system type are linked to the substrate employed (in crop-based systems, land use or eutrophication is of special importance). Therefore reviewed articles will be classified as waste streams or dedicated crops.
- **Culture type: mixed culture vs pure culture.** Although there is not explicit correlation between culture type and feedstock selection, these two issues are often linked. Besides, the culture type defines the process configuration and type, so the presented studies will be also categorised regarding the microbiological features of the process, as mixed and pure cultures.

4 Critical review of the current LCA studies for PHA production

A selection of 60 papers where the environmental profile of PHA production was assessed resulted from the literature search. These papers will be discussed in the following sections, whilst the ones not explicitly addressed but included amongst the 60 documents are included in the Further Reading section. The results for LCA methodological issues and process issues are quantitatively summarised in Table 1, and further discussed in Sections 4.1 and 4.2, whilst the studies' outputs will be qualitatively discussed in Section 4.3.

The first document was published back in 1998 and, since then, two studies were published per year on average, with an important growth in the last 2 years (i.e. 15 documents in 2019/2020). About 70% of the documents were research papers, and around 20% were reviews where LCA of biopolymers was the main issue, or where the PHA state of the art was assessed and LCA was included. The rest of the documents were four books or book chapters (Gupta et al., 2019; Martinez-Hernandez et al., 2018; Medina-Martos et al., 2019; Plackett, 2011), one report (Bengtsson et al., 2017), and two methodological documents: one assessing carbon displacement of bioproducts in comparison to conventional materials (Lynd & Wang, 2003), and the JCR-EC report (Nessi et al., 2020). Nevertheless, one of the reviews also addressed LCA methodological issues and a case study (Heimersson et al., 2014).

4.1 Review outputs regarding methodological issues

Motivation of the studies, functional unit, and system boundaries

In the first years when LCA-PHA was researched, studies aimed to compare the environmental profile of PHA with the profiles of polystyrene (PS), polyethylene (PE), or PP. Nonetheless, in the last years, studies aim to investigate the optimal routes to treat industrial streams, so the obtained results provide information about the routes with lower energy requirements or Greenhouse Gas (GHG) emissions.

On the other hand, the question of the system boundaries (SB) is linked to the functional unit (FU) definition. The most common SB were the cradle-to-gate type (about 65% of the studies), which means that the EoL of the product is not considered. Thus only the production process is included in the study, as very few data is yet available about bioplastics EoL. However, as seen previously, JCR-EC method recommends cradle-to-grave SB for LCA of PHA, as, when EoL of the product is not included, biodegradability is disregarded, and the obtained results are just partial. It is also truth that fair comparison cannot be established between PHA and petrochemical plastics with the current LCA methodology, as the actual effects of plastic pollution are not yet modelled. Thus biodegradability and energy recovery from bioplastics are disregarded in the study if the LCA has cradle-to-gate SB, and the devastating effects of marine and terrestrial plastic pollution are dismissed if

Table 1 Main quantitative results of the critical review.

Classification of the studies attending to LCA methodological issues									
Document type	**NS**	**Assessment type**	**NS**	**Functional unit**	**NS**	**System boundaries**	**NS**	**Data source**[a]	**NS**
Paper	40	LCA	30	Polymer mass	25	Cradle-to-gate	27	Secondary	16
Review	13	LCA+Economics	9	Feedstock	7	Cradle-to-grave	7	Primary*	26
Book chapter	4	SPI	4	Consumer product	6	Only extraction	3	***Primary source type**	
Methodology	2	Social +Environmental +Economic	3	Other	3	Gate-to-gate	3	Personal communication	6
Report	1	TM-LCA	2	Not clear	2	Only EoL	1	Simulation or mass balance	9
TOTAL	**60**	Other	2			Cradle-to-cradle	1	Lab or pilot reactors	14

Classification of the studies attending to process issues			
Substrate type	**NS**	**Culture type**	**NS**
Pure (dedicated)	18	Pure	25
Waste	28	MMC	11
Synthetic/not clear	3	Not specified	4

Results are expressed as the total number of studies (NS) and, therefore, some of the values include more than one of the considered features (for example, some studies compared MMC and pure culture performance, thus, they are accounted for within both culture types).

[a]*The data is expressed as studies only based on previous literature works (secondary sources) and studies based on data generated explicitly for the assessment (primary sources).*

cradle-to-grave approach is applied. This means that, possibly, most of the studies performed to date suffer from bias in favour of petrochemical plastics, so results might not show the actual net benefits of PHA. Moreover, the scientific community must orient their efforts to gather data about the EoL of PHA and the effects of plastic leaks in the environment, as addressed in recent works and projects (Changwichan et al., 2018; MarILCA, 2021).

About 75% of the studies considered a FU linked to the final product (normally, mass of polymer or mass of final product, as plastic bags or boxes). Cradle-to-gate SB are enough if the studied systems have the same gate (a certain mass of PHA), and the aim of the study is to assess different bioplastic production routes (Heimersson et al., 2014). However, for assessments with a FU considering the mass of polymer, and where the aim of the evaluation is to compare the environmental performance of bio-based and petrochemical materials, the issue of the SB becomes relevant and hard to solve. Moreover, PHA properties are different than the ones of petrochemical plastics, and they depend on the extraction protocol, the monomer proportion distribution, and so on. Therefore the assumption generally made that 1 kg of PHA will directly replace 1 kg of PE, PP, or PET is not very accurate and can lead to wrong results. Unfortunately, most of the current studies reviewed follow this approach, and this might be one of the reasons why the obtained results are highly variable and sometimes conflicting. This question needs to be considered and properly addressed, at least until more information about PHA use and EoL and plastic pollution is provided. Meanwhile, the lack of data can be managed by applying conservative replacement ratios of PHA to other polymers (Vega et al., 2019) and/or by performing sensitivity analysis (Roibás-Rozas et al., 2020).

On the other hand, a FU referred to the feedstock appears to be the most suitable option to compare different treatment/valorisation systems to produce the same polymer. This means assessments where the aim is not to compare the environmental profile of biopolymers to petrochemical polymers, but to study the performance of different process schemes. This influent-FU approach was considered for the first time more than a decade ago (Gurieff & Lant, 2007). However, the standard approach was the opposite (except for assessments only considering DSP) until it was used again a few years ago (Morgan-Sagastume et al., 2016), and then more research works used it in their assessments (Fernández-Braña et al., 2019; Pérez et al., 2020; Roibás-Rozas et al., 2020; Vega et al., 2019, 2020). This FU switch is linked to the process approach in the studied systems, which were initially pure cultures fed by pure substrates, so the aim was to compare the performance of the produced PHA to a conventional material. Consequently, when more process approaches appeared (not only MMCs, but also new production or extraction methods) and wastes started to be the main considered option for PHA production, the trend switched, and the chosen FU was referred to the feedstock, as the aim was not only to produce PHA but also to treat a waste. In fact, 5 out of 10 studies reported in the last 2 years considered the influent as a FU, in opposition to the approach followed in the past decades, also indicating that the goal of these novel process approaches is not only to produce a polymer, but also to establish processes that fit circularity by valorising waste streams.

Impact categories

The most assessed impact category is Global Warming Potential (GWP), as it is expected that bioproducts can mitigate the effect of climate change. Non-renewable Energy Use (NREU) or Fossil Resource Depletion is also commonly evaluated as bioplastics would replace petrochemical materials. Acidification and Eutrophication Potential were also commonly studied, and just a few papers reviewed an important number of the categories included in the ReCiPe impact methodology (Fernández-Braña et al., 2019; Harding et al., 2007; Roibás-Rozas et al., 2020; Vega et al., 2020; Vogli et al., 2020).

Again, the lack of data regarding EoL affects the selection of impact categories, as it is not completely clear how the effects linked to polymer production, use, and disposal are transferred on the environmental compartments. On the other hand, the JRC-EC method states that the whole 16 categories addressed in the Product Environmental Footprint Guidelines (Manfredi et al., 2012) need to be considered, so each of them must be discussed in LCA studies from now on. In any case, it has been stated that, when the LCA is performed for polymers processed from crops, categories such as Land Use, Acidification, and Eutrophication must be mandatorily considered (Heimersson et al., 2014).

Assessment type and methodology

More than 60% of the assessments used pure LCA (meaning according to ISO 14040 and 14044), where the most employed software was SimaPro (Gupta et al., 2019). Around 20% combined environmental (i.e. LCA) and cost assessment by adding any economic evaluation (Changwichan et al., 2018; Gurieff & Lant, 2007; Leong et al., 2016, 2017; Martinez-Hernandez et al., 2018) or implementing TEA (Fernández-Dacosta et al., 2015; Medina-Martos et al., 2019; Saavedra del Oso et al., 2020; Vega et al., 2020). Almost 5% of the reviewed papers used Territorial Metabolism-LCA (TM-LCA) (Vega et al., 2019, 2020), which considers regional aspects of the environmental assessments, and almost 10% used SPI (Koller et al., 2013a; Kourmentza et al., 2017; Rathi et al., 2013), which is an index that normalises impacts according to the availability of planet surface (it calculates the necessary area for industrial processes).

Recently, studies started including other sustainability aspects by linking environmental impacts with social and economic effects (Chen et al., 2020; Pérez et al., 2020; Talan et al., 2020). In fact, the most recent findings have pointed out that the assessment of bioproducts cannot be linked only to environmental aspects, as suggested by the triple-bottom line of sustainability, which includes environment, economy, and society. It makes sense that, if one wants to evaluate the feasibility of bioeconomy implementation, the concept of sustainability needs to be linked not only to the environmental issues but also to economic aspects, which are necessarily related to society. Finally, the remaining studies combined LCA with green design principles by scoring products according to proposed Ecodesign metrics (Tabone et al., 2010) or with WAR algorithm, which links waste minimisation with a decrease in ecotoxic impacts (Leong et al., 2016).

Source of data. Is there enough information to perform accurate assessments?

The first literature review of PHA sustainability assessments was published about a decade ago (although it also included PLA and starch-based polymers) (Yates & Barlow, 2013). They reported that the studies published until that date showed unclear data sources, as some LCAs were performed using confidential information or personal communications (Hermann et al., 2007; Kim & Dale, 2005; Pietrini et al., 2007) and the reader cannot know the features of the studied system. Moreover, assessments performed in the last decade are relying on data provided by studies carried out 20 years ago. To see some examples, see Fig. 3, where the studies reported on Table 1 based on secondary sources (dashed lines) were tracked to find the origin of the information (continuous lines).

LCA practitioners need to stop relying on data generated 20 years ago to provide results referring today's processes. New inventories need to be generated and adapted to the novel strategies for PHA production (MMCs, waste streams, or strategies to improve PHA yield). Otherwise, LCA of PHA is halted and results are going to repeat the same outputs generated in the past, reproducing the same biases. Fortunately, some of the most recent studies are based on the inventories generated explicitly for the assessments, and based on up-scaled laboratory experiences (Nitkiewicz et al., 2020; Pérez et al., 2020; Righi et al., 2016; Roibás-Rozas et al., 2020; Vogli et al., 2020) or pilot-scale experiences (Bengtsson et al., 2017; Morgan-Sagastume et al., 2016; Vega et al., 2019, 2020).

4.2 Classification of the studies regarding process type

Amongst the research works published until 2010, only few considered the use of any waste feedstock. However, in the last decade the trend switched, and most of the studies focused on the valorisation of waste streams, where only two studies addressed PHA production from dedicated crops (Changwichan et al., 2018; Kookos et al., 2019) using secondary data from literature (Akiyama et al., 2003), and (Khoo & Tan, 2010), respectively, see Fig. 3) for the production of the substrate.

Regarding culture type, only two studies considered MMC-based processes until 2015 (Fernández-Dacosta et al., 2015; Gurieff & Lant, 2007). Nevertheless, and considering the recent trends in PHA production pathways, most of the papers published in the last years evaluated and processed based on MMC and/or waste streams. Moreover, it has been pointed out that data sources need to be clear and transparent (Yates & Barlow, 2013), as sometimes data sources are unclear and it is hard to understand features as important as the culture type or the feedstock employed. This becomes an important issue especially for old references and works relying on them (see Fig. 3), as some of these studies were based on personal communications or confidential information provided by companies.

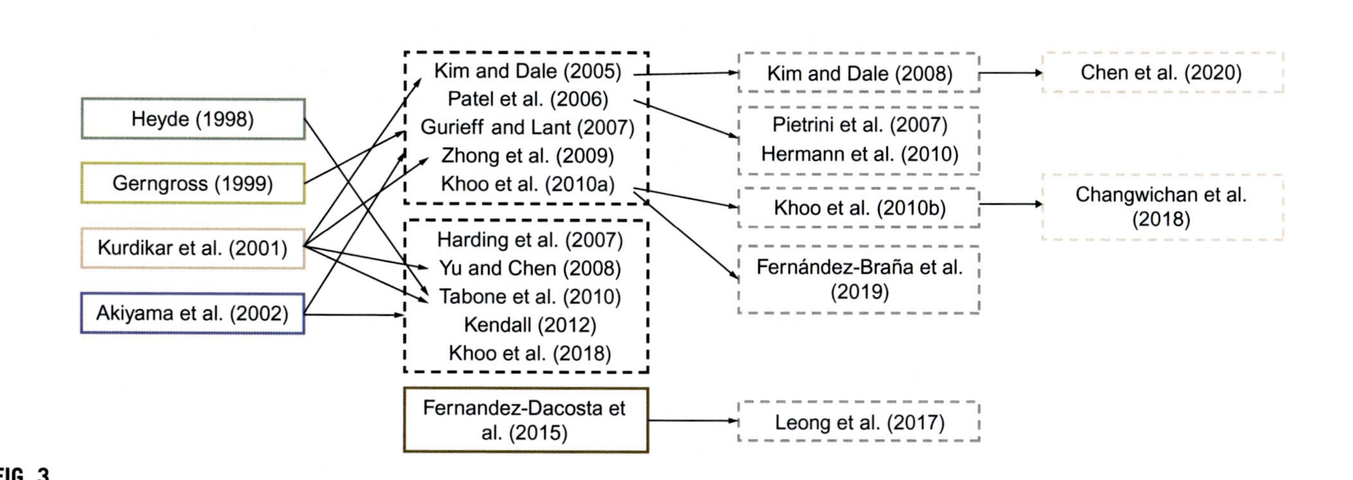

FIG. 3

Relationship amongst some of the published LCA-PHA studies and their sources. References framed in *continuous lines* represent studies presenting original inventories (using primary and/or secondary data), and *dashed lines* represent the studies derived from these inventories.

4.3 Main outputs

As mentioned, the results obtained from PHA-LCA studies were sometimes contradictory or conflicting amongst them, and the reasons might be linked to biases related to the lack of information for the last stages of polymers' life cycle (bio-based and petrochemical) and allocation choices (Heimersson et al., 2014; Kookos et al., 2019). However, studies agreed at indicating that the environmental performance of the process can be improved by introducing waste streams in the PHA production process (Chanprateep, 2010; Kim & Dale, 2005, 2008; Koller et al., 2013b; Shahzad et al., 2013), and, about a decade ago, it was explicitly proved for the first time that using them is more favourable than employing dedicated crops (Kendall, 2012).

Moreover, the energy source was also important, as green energy usage can switch the environmental profile of the bioproduct (Khoo et al., 2010; Kurdikar et al., 2000; Zhong et al., 2009), suggesting that impacts might be regionally dependent on the energy mix employed (Shahzad et al., 2013). Finally, the most recent outputs highlight especially that the success of the process can only be complete if several aspects of sustainability (economic, environmental, and social) are considered (Chen et al., 2020; Koller et al., 2017; Kookos et al., 2019; Pérez et al., 2020; Talan et al., 2020; Yadav et al., 2020). Results point out that this success will be reached if an integral biorefinery approach is applied (Medina-Martos et al., 2019; Nitkiewicz et al., 2020; Yadav et al., 2020), where all streams are valorised (for example, the generated sludge is converted into biogas by anaerobic digestion). In fact, the latest outputs proved that economic and environmental viability of PHA production facilities can be reached if part of the waste carbon used for biopolymer generation is derived for on-site for energy generation, like the case of anaerobic digestion (Dietrich et al., 2017; Kookos et al., 2019; Martinez-Hernandez et al., 2018; Pérez et al., 2020; Vega et al., 2019, 2020; Vogli et al., 2020).

Full value chain vision: What do we know about the use and EoL stages?

As mentioned, little is known about the last stages of PHA life cycle. On the one hand, it has been considered that impacts derived from the use stage are the same as for petrochemical plastics, so they can be disregarded for comparative LCA with conventional materials (Hermann et al., 2007). On the other hand, it is especially urgent to generate information about the EoL as just 9 out of the 40 papers revised considered it and only 4 papers compared different management routes for the (bio) plastic materials once used.

Regarding EoL, some authors concluded that waste bioplastic management might determine the global sustainability of the process (Heyde, 1998), whilst others highlighted that this stage usually has a low impact in relation to the full PHA life cycle (Khoo & Tan, 2010). The options considered in these studies were landfilling, composting, and incineration, where landfilling showed the worst results and the success of the remaining options depended on factors like energy origin. Other research works compared AD, incineration, and home/industrial composting to find that AD had generally the best performances and industrial composting the worst (Hermann et al., 2011). Finally, a more recent study considered landfilling, composting, incineration, and mechanical recycling, where recycling clearly showed the best results (Changwichan et al., 2018). These outputs indicate that new research works including modern approaches and recent findings are needed, as composting (the most typically considered EoL route) might not be necessarily the optimal alternative, and the introduction of several cycles of mechanical recycling could improve the bio-based material life cycle performance.

4.4 Lessons learned

PHAs are a promising alternative to solve the serious challenges we are currently facing regarding plastic pollution and climate change. However, PHA full-scale production is hindered by the economic feasibility and the lack of consistency concerning its environmental validation, which, as most studies indicated, are linked to (i) the use of waste streams as feedstock, (ii) employing clean or renewable energy, and (iii) applying an integral biorefinery vision, where each stream is valorised (energy production on-site is encouraged). In this sense, the main aspects that need to be urgently solved for this validation are as follows:

- Lack of data. The uncertainties regarding plastic leaks in the environment for petrochemical plastics and EoL pathways for biopolymers are generating a bias in favour of conventional materials, as cradle-to-gate SB exclude the benefits generated from valorising wasted bioplastics and cradle-to-grave SB disregard the serious effects of uncontrolled plastic leaks in the environment. These gaps generate conflicting or contradictory results depending, amongst other factors, on the allocation choices and the hypothesis formulated when trying to fill those data requirements up.
- Sources of information. Although the number of studies performed based on new inventories is increasing, there is still a big number of assessments carried out relying on old information, as seen in "Source of data. Is there enough information to perform accurate assessments?" section (about 60% of the research works used only secondary data sources from previous studies). This provokes results that are stuck in old technologies and process approaches, as novel strategies for PHA production are not considered, where the reproduction of biases is aggravated.
- Process scale. There is still no full-scale market for PHA, so studies are based on up-scaled laboratory data or pilot-scale systems, where only 19 MMC-based pilot experiences were reported till date (Estévez-Alonso et al., 2021). This is another issue that can increase the uncertainties of the assessments, as, when the environmental performance of a conventional petrochemical plastic is compared to the one of PHA, the comparison is performed between a process widely implemented at full scale and a process with just a few pilot-scale working facilities. Although several assessment approaches have been proposed in order

to fill this gap from pilot to full scale in emerging technologies (Bergerson et al., 2019; Hung et al., 2020; Piccinno et al., 2016), to the best of our knowledge, none of them has been applied yet to PHA.

In any case, the success of the process is not only linked to the environmental validation, and future studies must mandatorily include economic and/or social aspects in the assessment.

5 Future outlook, challenges, and key players

The problem of plastic pollution is serious and real, and the necessity of looking for a suitable solution as soon as possible is a fact. However, MMC-based full-scale production of PHA from waste streams is not a reality yet due to some economic and environmental bottlenecks. Basically, the research community needs to provide information about the effects of plastic leaks in the environment and about the last stages of PHA life cycle (manufacture, use, and EoL). Then, a robust framework can be established to determine the main features of the PHA value chain and, finally, that information can be employed to develop policies and regulations that fit the principles of circular economy.

Although this information is not available yet, the scientific community and agents such as the Life Cycle Initiative are gathering efforts to generate this framework, and several groups and projects deserve special attention in this scenario (MarILCA, 2021; Plastic Leak Project, 2021). The objective of the Plastic Leak Project is to conceive a methodology to identify an inventory for plastic leakage in the environment, focusing on important sectors such as transport and textiles, and on goods like plastic products, packaging materials, or plastic pellets. The mentioned project MarILCA aims to determine the effects of marine litter, so they are gathering information not only regarding inventories, but also impacts, as they are trying to develop a model to describe how the release of plastic affects the different environmental compartments, and how these plastics are transferred amongst them.

The introduction of biodegradable plastics in the value chains is a necessity, but it is still not a reality due to several drawbacks that need to be urgently addressed. As economic issues are the main barrier in this pathway, new process configurations and strategies need to be explored, so less energy-intense routes can be proposed. However, in a bioeconomy framework, financial viability and environmental validation are two sides of the same coin. In this sense, aspects such as energy efficiency optimisation will not only yield in monetary savings but also in environmental credits.

6 Conclusions

The environmental validation of PHA was studied for the first time in 1998, and, since then, approximately two research works were published yearly. Although the first studies performed assessed systems based on pure cultures, the current trend is to evaluate MMC-based processes, as they are supposed to be less expensive. Moreover, current evaluations generally study systems fed with waste feedstocks, as the use of dedicated crops might not fit the circular economy principles.

The main issue found is linked to the FU and the allocation employed, as excluding EoL of the system boundaries disregards the benefits derived from PHA biodegradability and including it dismisses the effects of plastic leaks in the environment and plastic pollution. Therefore new information regarding these topics needs to be urgently generated. In this sense, it is also necessary to provide data regarding pilot- and full-scale processes, as current information is based on secondary sources of data, simulations, or up-scaled lab results. Moreover, LCA practitioners need to be especially careful when using these secondary sources of information, as we found that many current references still use inventories generated 20 years ago. This is necessarily going to generate conflicting results and is not going to provide information in concordance to today's processes, so the creation of inventories for current process approaches is encouraged.

Finally, LCA of PHA need to consider each stage of the life cycle to complete the value chain, and cradle-to-grave approach is encouraged. Furthermore, it is recommended that assessments do not include only environmental aspects, but also other issues linked to bioeconomy and sustainability, as social or economic aspects, to get a full overview of the evaluation that addresses the integral approach linked to circularity.

References

Akhbarizadeh, R., Dobaradaran, S., Nabipour, I., Tajbakhsh, S., Darabi, A. H., & Spitz, J. (2020). Abundance, composition, and potential intake of microplastics in canned fish. *Marine Pollution Bulletin*, *160*. https://doi.org/10.1016/j.marpolbul.2020.111633.

Akiyama, M., Tsuge, T., & Doi, Y. (2003). Environmental life cycle comparison of polyhydroxyalkanoates produced from renewable carbon resources by bacterial fermentation. *Polymer Degradation and Stability*, *80*(1), 183–194. https://doi.org/10.1016/S0141-3910(02)00400-7.

Arcos-Hernandez, M. V., Laycock, B., Pratt, S., Donose, B. C., Nikolič, M. A. L., Luckman, P., et al. (2012). Biodegradation in a soil environment of activated sludge derived polyhydroxyalkanoate (PHBV). *Polymer Degradation and Stability*, *97*(11), 2301–2312. https://doi.org/10.1016/j.polymdegradstab.2012.07.035.

Battista, F., Frison, N., & Bolzonella, D. (2021). Can bioplastics be treated in conventional anaerobic digesters for food waste treatment? *Environmental Technology and Innovation*, *22*. https://doi.org/10.1016/j.eti.2021.101393.

Bengtsson, S., Werker, A., Visser, C., & Korving, L. (2017). *PHARIO – stepping stone to a sustainable value chain for PHA bioplastic using municipal activated sludge*.

Bergerson, J. A., Brandt, A., Cresko, J., Carbajales-Dale, M., MacLean, H. L., Matthews, S. H., et al. (2019). Life cycle assessment of emerging technologies: Evaluation techniques at different stages of market and technical maturity. *Journal of Industrial Ecology*, 11–25. https://doi.org/10.1111/jiec.12954.

Changwichan, K., Silalertruksa, T., & Gheewala, S. H. (2018). Eco-efficiency assessment of bioplastics production systems and end-of-life options. *Sustainability*, *10*(4). https://doi.org/10.3390/su10040952.

Chanprateep, S. (2010). Current trends in biodegradable polyhydroxyalkanoates. *Journal of Bioscience and Bioengineering*, *110*(6), 621–632. https://doi.org/10.1016/j.jbiosc.2010.07.014.

Chen, W., Oldfield, T. L., Cinelli, P., Righetti, M. C., & Holden, N. M. (2020). Hybrid life cycle assessment of potato pulp valorisation in biocomposite production. *Journal of Cleaner Production*, 122366. https://doi.org/10.1016/j.jclepro.2020.122366.

Colombo, B., Pereira, J., Martins, M., Torres-Acosta, M. A., Dias, A. C. R. V., Lemos, P. C., et al. (2020). Recovering PHA from mixed microbial biomass: Using non-ionic surfactants as a pretreatment step. *Separation and Purification Technology*, *253*. https://doi.org/10.1016/j.seppur.2020.117521.

Conejo-Watt, H., & Luisetti, T. (2019). *Tackling marine litter in the Atlantic area.*

de Souza Machado, A. A., Kloas, W., Zarfl, C., Hempel, S., & Rillig, M. C. (2018). Microplastics as an emerging threat to terrestrial ecosystems. *Global Change Biology*, *24*(4), 1405–1416. https://doi.org/10.1111/gcb.14020.

de Souza Reis, G. A., Michels, M. H. A., Fajardo, G. L., Lamot, I., & de Best, J. H. (2020). Optimization of green extraction and purification of PHA produced by mixed microbial cultures from sludge. *Water*, *12*(4). https://doi.org/10.3390/W12041185.

Dietrich, K., Dumont, M. J., Del Rio, L. F., & Orsat, V. (2017). Producing PHAs in the bioeconomy—Towards a sustainable bioplastic. *Sustainable Production and Consumption*, *9*, 58–70. https://doi.org/10.1016/j.spc.2016.09.001.

Estévez-Alonso, Á., Pei, R., van Loosdrecht, M. C. M., Kleerebezem, R., & Werker, A. (2021). Scaling-up microbial community-based polyhydroxyalkanoate production: Status and challenges. *Bioresource Technology*, *327*. https://doi.org/10.1016/j.biortech.2021.124790.

European Commission. (2018a). *A European strategy for plastics in a circular economy.*

European Commission. (2018b). *Reducing marine litter: Action on single use plastics and fishing gear.*

European Commission. (2019a). *Directive (EU) 2019/904 of the European Parliament and of the council of 5 June 2019 on the reduction of the impact of certain plastic products on the environment.*

European Commission. (2019b). *The European green deal.*

European Commission. (2020). *A new circular economy action plan for a cleaner and more competitive Europe.*

Fernández-Braña, Á., Feijoo-Costa, G., & Dias-Ferreira, C. (2019). Looking beyond the banning of lightweight bags: Analysing the role of plastic (and fuel) impacts in waste collection at a Portuguese city. *Environmental Science and Pollution Research*, *26*(35), 35629–35647. https://doi.org/10.1007/s11356-019-05938-w.

Fernández-Dacosta, C., Posada, J. A., Kleerebezem, R., Cuellar, M. C., & Ramirez, A. (2015). Microbial community-based polyhydroxyalkanoates (PHAs) production from wastewater: Techno-economic analysis and ex-ante environmental assessment. *Bioresource Technology*, *185*, 368–377. https://doi.org/10.1016/j.biortech.2015.03.025.

Fernández-Dacosta, C., Posada, J. A., & Ramirez, A. (2016). Techno-economic and carbon footprint assessment of methyl crotonate and methyl acrylate production from wastewater-based polyhydroxybutyrate (PHB). *Journal of Cleaner Production*, *137*, 942–952. https://doi.org/10.1016/j.jclepro.2016.07.152.

Galloway, T. S., Matthew, C., & Ceri, L. (2017). Interactions of microplastic debris throughout the marine ecosystem. *Nature Ecology & Evolution*. https://doi.org/10.1038/s41559-017-0116.

Geyer, R., Jambeck, J. R., & Lavender Law, K. (2017). Production, use, and fate of all plastics ever made. *Science Advances*, e1700782. https://doi.org/10.1126/sciadv.1700782.

Gupta, J., Rathour, R., Medhi, K., Tyagi, B., & Thakur, I. S. (2019). Microbial-derived natural bioproducts for a sustainable environment: A bioprospective for waste to wealth. In *Refining biomass residues for sustainable energy and bioproducts: Technology, advances, life cycle assessment, and economics* (pp. 51–85). Elsevier. https://doi.org/10.1016/B978-0-12-818996-2.00003-X.

Gurieff, N., & Lant, P. (2007). Comparative life cycle assessment and financial analysis of mixed culture polyhydroxyalkanoate production. *Bioresource Technology*, *98*(17), 3393–3403. https://doi.org/10.1016/j.biortech.2006.10.046.

Halsband, C., & Herzke, D. (2019). Plastic litter in the European arctic: What do we know? *Emerging Contaminants*, *5*, 308–318. https://doi.org/10.1016/j.emcon.2019.11.001.

Harding, K. G., Dennis, J. S., von Blottnitz, H., & Harrison, S. T. L. (2007). Environmental analysis of plastic production processes: Comparing petroleum-based polypropylene and polyethylene with biologically-based poly-β-hydroxybutyric acid using life cycle analysis. *Journal of Biotechnology*, *130*(1), 57–66. https://doi.org/10.1016/j.jbiotec.2007.02.012.

Heimersson, S., Morgan-Sagastume, F., Peters, G. M., Werker, A., & Svanström, M. (2014). Methodological issues in life cycle assessment of mixed-culture polyhydroxyalkanoate production utilising waste as feedstock. *New Biotechnology*, *31*(4), 383–393. https://doi.org/10.1016/j.nbt.2013.09.003.

Hermann, B. G., Blok, K., & Patel, M. K. (2007). Producing bio-based bulk chemicals using industrial biotechnology saves energy and combats climate change. *Environmental Science and Technology*, *41*(22), 7915–7921. https://doi.org/10.1021/es062559q.

Hermann, B. G., Debeer, L., De Wilde, B., Blok, K., & Patel, M. K. (2011). To compost or not to compost: Carbon and energy footprints of biodegradable materials' waste treatment. *Polymer Degradation and Stability*, *96*(6), 1159–1171. https://doi.org/10.1016/j.polymdegradstab.2010.12.026.

Heyde, M. (1998). Ecological considerations on the use and production of biosynthetic and synthetic biodegradable polymers. *Polymer Degradation and Stability*, *59*(1–3), 3–6. https://doi.org/10.1016/s0141-3910(97)00017-7.

Hung, C. R., Ellingsen, L. A. W., & Majeau-Bettez, G. (2020). LiSET: A framework for early-stage life cycle screening of emerging technologies. *Journal of Industrial Ecology*, *24*(1), 26–37. https://doi.org/10.1111/jiec.12807.

Jiang, G., Johnston, B., Townrow, D. E., Radecka, I., Koller, M., Chaber, P., et al. (2018). Biomass extraction using non-chlorinated solvents for biocompatibility improvement of polyhydroxyalkanoates. *Polymers*, *10*(7). https://doi.org/10.3390/polym10070731.

Kendall, A. (2012). A life cycle assessment of biopolymer production from material recovery facility residuals. *Resources, Conservation and Recycling*, *61*, 69–74. https://doi.org/10.1016/j.resconrec.2012.01.008.

Khoo, H. H., & Tan, R. B. H. (2010). Environmental impacts of conventional plastic and bio-based carrier bags: Part 2: End-of-life options. *International Journal of Life Cycle Assessment*, *15*(4), 338–345. https://doi.org/10.1007/s11367-010-0163-8.

Khoo, H. H., Tan, R. B. H., & Chng, K. W. L. (2010). Environmental impacts of conventional plastic and bio-based carrier bags. *International Journal of Life Cycle Assessment*, *15*(3), 284–293. https://doi.org/10.1007/s11367-010-0162-9.

Kim, S., & Dale, B. E. (2005). Life cycle assessment study of biopolymers (polyhydroxyalkanoates) derived from no-tilled corn. *International Journal of Life Cycle Assessment*, *10*(3), 200–210. https://doi.org/10.1065/lca2004.08.171.

Kim, S., & Dale, B. E. (2008). Energy and greenhouse gas profiles of polyhydroxybutyrates derived from corn grain: A life cycle perspective. *Environmental Science and Technology*, *42*(20), 7690–7695. https://doi.org/10.1021/es8004199.

Koller, M., Niebelschütz, H., & Braunegg, G. (2013a). Strategies for recovery and purification of poly[(R)-3-hydroxyalkanoates] (PHA) biopolyesters from surrounding biomass. *Engineering in Life Sciences*, *13*(6), 549–562. https://doi.org/10.1002/elsc.201300021.

Koller, M., Sandholzer, D., Salerno, A., Braunegg, G., & Narodoslawsky, M. (2013b). Biopolymer from industrial residues: Life cycle assessment of poly(hydroxyalkanoates) from whey. *Resources, Conservation and Recycling*, *73*, 64–71. https://doi.org/10.1016/j.resconrec.2013.01.017.

Koller, M., Maršálek, L., de Sousa Dias, M. M., & Braunegg, G. (2017). Producing microbial polyhydroxyalkanoate (PHA) biopolyesters in a sustainable manner. *New Biotechnology*, *37*, 24–38. https://doi.org/10.1016/j.nbt.2016.05.001.

Kookos, I. K., Koutinas, A., & Vlysidis, A. (2019). Life cycle assessment of bioprocessing schemes for poly(3-hydroxybutyrate) production using soybean oil and sucrose as carbon sources. *Resources, Conservation and Recycling*, *141*, 317–328. https://doi.org/10.1016/j.resconrec.2018.10.025.

Kosseva, M. R., & Rusbandi, E. (2018). Trends in the biomanufacture of polyhydroxyalkanoates with focus on downstream processing. *International Journal of Biological Macromolecules*, *107*, 762–778. https://doi.org/10.1016/j.ijbiomac.2017.09.054.

Kourmentza, C., Plácido, J., Venetsaneas, N., Burniol-Figols, A., Varrone, C., Gavala, H. N., et al. (2017). Recent advances and challenges towards sustainable polyhydroxyalkanoate (PHA) production. *Bioengineering*, *4*(2). https://doi.org/10.3390/bioengineering4020055.

Kumar, M., Rathour, R., Singh, R., Sun, Y., Pandey, A., Gnansounou, E., et al. (2020). Bacterial polyhydroxyalkanoates: Opportunities, challenges, and prospects. *Journal of Cleaner Production*, *263*. https://doi.org/10.1016/j.jclepro.2020.121500.

Kurdikar, D., Fournet, L., Slater, S. C., Paster, M., Gruys, K. J., Gerngross, T. U., et al. (2000). Greenhouse gas profile of a plastic material derived from a genetically modified plant. *Journal of Industrial Ecology*, *4*(3), 107–122. https://doi.org/10.1162/108819800300106410.

Lamberti, F. M., Román-Ramírez, L. A., & Wood, J. (2020). Recycling of bioplastics: Routes and benefits. *Journal of Polymers and the Environment*, *28*(10), 2551–2571. https://doi.org/10.1007/s10924-020-01795-8.

Laycock, B., Nikolić, M., Colwell, J. M., Gauthier, E., Halley, P., Bottle, S., et al. (2017). Lifetime prediction of biodegradable polymers. *Progress in Polymer Science*, *71*, 144–189. https://doi.org/10.1016/j.progpolymsci.2017.02.004.

Leong, Y. K., Show, P. L., Lin, H. C., Chang, C. K., Loh, H. S., Lan, J. C. W., et al. (2016). Preliminary integrated economic and environmental analysis of polyhydroxyalkanoates (PHAs) biosynthesis. *Bioresources and Bioprocessing*, *3*(1). https://doi.org/10.1186/s40643-016-0120-x.

Leong, Y. K., Show, P. L., Lan, J. C. W., Loh, H. S., Lam, H. L., & Ling, T. C. (2017). Economic and environmental analysis of PHAs production process. *Clean Technologies and Environmental Policy*, *19*(7), 1941–1953. https://doi.org/10.1007/s10098-017-1377-2.

Li, M., & Wilkins, M. R. (2020). Recent advances in polyhydroxyalkanoate production: Feedstocks, strains and process developments. *International Journal of Biological Macromolecules*, *156*, 691–703. https://doi.org/10.1016/j.ijbiomac.2020.04.082.

Lynd, L. R., & Wang, M. Q. (2003). A product-nonspecific framework for evaluating the potential of biomass-based products to displace fossil fuels. *Journal of Industrial Ecology*, *7*(3–4), 17–32. https://doi.org/10.1162/108819803323059370.

Manfredi, S., Allacker, K., Chomkhamsri, K., Pelletier, N., & Souza. (2012). *Product environmental footprint (PEF) guide*. https://eplca.jrc.ec.europa.eu//EnvironmentalFootprint.html.

Mannina, G., Presti, D., Montiel-Jarillo, G., & Suárez-Ojeda, M. E. (2019). Bioplastic recovery from wastewater: A new protocol for polyhydroxyalkanoates (PHA) extraction from mixed microbial cultures. *Bioresource Technology*, *282*, 361–369. https://doi.org/10.1016/j.biortech.2019.03.037.

Mannina, G., Presti, D., Montiel-Jarillo, G., Carrera, J., & Suárez-Ojeda, M. E. (2020). Recovery of polyhydroxyalkanoates (PHAs) from wastewater: A review. *Bioresource Technology*, *297*. https://doi.org/10.1016/j.biortech.2019.122478.

MarILCA. (2021). *Marine Impacts in Life Cycle Assessment*. https://marilca.org/.

Martinez-Hernandez, E., Ng, K. S., Amezcua Allieri, M. A., Aburto Anell, J. A., & Sadhukhan, J. (2018). Value-added products from wastes using extremophiles in biorefineries: Process modeling, simulation, and optimization tools. In *Extremophilic microbial processing of lignocellulosic feedstocks to biofuels, value-added products, and usable power* (pp. 275–300). Springer International Publishing. https://doi.org/10.1007/978-3-319-74459-9_14.

Medina-Martos, E., Istrate, I. R., & Dufour, J. (2019). Technoeconomic and environmental review of value-added products from wastewater: Bioplastic production and algal cultivation for biofuels. In J. A. Olivares, J. A. Melero, D. Pujol, & J. Dufour (Eds.), *Wastewater treatment residues as resources for biorefinery products and biofuels* (pp. 435–454). Elsevier. https://doi.org/10.1016/B978-0-12-816204-0.00019-9.

Melendez-Rodriguez, B., Castro-Mayorga, J. L., Reis, M. A. M., Sammon, C., Cabedo, L., Torres-Giner, S., et al. (2018). Preparation and characterization of electrospun food biopackaging films of poly(3-hydroxybutyrate-co-3-hydroxyvalerate) derived from fruit pulp biowaste. *Frontiers in Sustainable Food Systems*. https://doi.org/10.3389/fsufs.2018.00038.

Morgan-Sagastume, F., Heimersson, S., Laera, G., Werker, A., & Svanström, M. (2016). Techno-environmental assessment of integrating polyhydroxyalkanoate (PHA) production with services of municipal wastewater treatment. *Journal of Cleaner Production*, *137*, 1368–1381. https://doi.org/10.1016/j.jclepro.2016.08.008.

Nessi, S., Sinkko, T., Bulgheroni, C., Garcia-Gutierrez, P., Giuntoli, J., Konti, A., et al. (2020). *Comparative life cycle assessment (LCA) of alternative feedstock for plastics production*.

Nitkiewicz, T., Wojnarowska, M., Sołtysik, M., Kaczmarski, A., Witko, T., Ingrao, C., et al. (2020). How sustainable are biopolymers? Findings from a life cycle assessment of polyhydroxyalkanoate production from rapeseed-oil derivatives. *Science of the Total Environment*, *749*. https://doi.org/10.1016/j.scitotenv.2020.141279.

OECD. (2013). Policies for bioplastics in the context of a bioeconomy. In *OECD Science, Technology and industry policy papers*. https://doi.org/10.1787/5k3xpf9rrw6d-en.

Pantaloni, D., Shah, D., Baley, C., & Bourmaud, A. (2020). Monitoring of mechanical performances of flax non-woven biocomposites during a home compost degradation. *Polymer Degradation and Stability*, *177*. https://doi.org/10.1016/j.polymdegradstab.2020.109166.

Pérez, V., Mota, C. R., Muñoz, R., & Lebrero, R. (2020). Polyhydroxyalkanoates (PHA) production from biogas in waste treatment facilities: Assessing the potential impacts on economy, environment and society. *Chemosphere*, *255*. https://doi.org/10.1016/j.chemosphere.2020.126929.

Pérez-Rivero, C., López-Gómez, J. P., & Roy, I. (2019). A sustainable approach for the downstream processing of bacterial polyhydroxyalkanoates: State-of-the-art and latest developments. *Biochemical Engineering Journal*, *150*. https://doi.org/10.1016/j.bej.2019.107283.

Piccinno, F., Hischier, R., Seeger, S., & Som, C. (2016). From laboratory to industrial scale: A scale-up framework for chemical processes in life cycle assessment studies. *Journal of Cleaner Production*, *135*, 1085–1097. https://doi.org/10.1016/j.jclepro.2016.06.164.

Pietrini, M., Roes, L., Patel, M. K., & Chiellini, E. (2007). Comparative life cycle studies on poly(3-hydroxybutyrate)-based composites as potential replacement for conventional petrochemical plastics. *Biomacromolecules*, *8*(7), 2210–2218. https://doi.org/10.1021/bm0700892.

Plackett, D. (2011). Preface. In *Biopolymers - new materials for sustainable films and coatings* (pp. xiii–xiv). https://doi.org/10.1002/9781119994312.

Plastic Leak Project. (2021). *Plastic Leak Project, 2020. Tackling plastic pollution: A prioneering methodology to measure plastic leakage and identify its pathways into the environment*. https://quantis-intl.com/strategy/collaborative-initiatives/plastic-leak-project/.

Rathi, D. N., Amir, H. G., Abed, R. M. M., Kosugi, A., Arai, T., Sulaiman, O., et al. (2013). Polyhydroxyalkanoate biosynthesis and simplified polymer recovery by a novel moderately halophilic bacterium isolated from hypersaline microbial mats. *Journal of Applied Microbiology*, *114*(2), 384–395. https://doi.org/10.1111/jam.12083.

Raza, Z. A., Abid, S., & Banat, I. M. (2018). Polyhydroxyalkanoates: Characteristics, production, recent developments and applications. *International Biodeterioration and Biodegradation*, *126*, 45–56. https://doi.org/10.1016/j.ibiod.2017.10.001.

Righi, S., Baioli, F., Samorì, C., Galletti, P., Stramigioli, C., Tugnoli, et al. (2016). A life-cycle assessment of poly-hydroxybutyrate extraction from microbial biomass using dimethylcarbonate. *10th Italian LCA Conference: 10th Convegno dell'Associazione Rete Italiana LCA 2016*.

Rodriguez-Perez, S., Serrano, A., Pantión, A. A., & Alonso-Fariñas, B. (2018). Challenges of scaling-up PHA production from waste streams. A review. *Journal of Environmental Management*, *205*, 215–230. https://doi.org/10.1016/j.jenvman.2017.09.083.

Roibás-Rozas, A., Mosquera-Corral, A., & Hospido, A. (2020). Environmental assessment of complex wastewater valorisation by polyhydroxyalkanoates production. *Science of the Total Environment*, *744*. https://doi.org/10.1016/j.scitotenv.2020.140893.

Roohi, Zaheer, M. R., & Kuddus, M. (2018). PHB (poly-β-hydroxybutyrate) and its enzymatic degradation. *Polymers for Advanced Technologies*, *29*(1), 30–40. https://doi.org/10.1002/pat.4126.

Saavedra del Oso, M., Mauricio-Iglesias, M., & Hospido, A. (2020). Evaluation and optimization of the environmental performance of PHA downstream processing. *Chemical Engineering Journal*. https://doi.org/10.1016/j.cej.2020.127687.

Sabapathy, P. C., Devaraj, S., Meixner, K., Anburajan, P., Kathirvel, P., Ravikumar, Y., et al. (2020). Recent developments in Polyhydroxyalkanoates (PHAs) production—A review. *Bioresource Technology*, *306*. https://doi.org/10.1016/j.biortech.2020.123132.

Saling, P., Gyuzeleva, L., Wittstock, K., Wessolowski, V., & Griesshammer, R. (2020). Life cycle impact assessment of microplastics as one component of marine plastic debris. *International Journal of Life Cycle Assessment*, *25*(10), 2008–2026. https://doi.org/10.1007/s11367-020-01802-z.

Shahzad, K., Kettl, K. H., Titz, M., Koller, M., Schnitzer, H., & Narodoslawsky, M. (2013). Comparison of ecological footprint for biobased PHA production from animal residues utilizing different energy resources. *Clean Technologies and Environmental Policy*, *15*(3), 525–536. https://doi.org/10.1007/s10098-013-0608-4.

Tabone, M. D., Cregg, J. J., Beckman, E. J., & Landis, A. E. (2010). Sustainability metrics: Life cycle assessment and green design in polymers. *Environmental Science and Technology*, *44*(21), 8264–8269. https://doi.org/10.1021/es101640n.

Talan, A., Kaur, R., Tyagi, R. D., & Drogui, P. (2020). Bioconversion of oily waste to polyhydroxyalkanoates: Sustainable technology with circular bioeconomy approach and multidimensional impacts. *Bioresource Technology Reports*, *11*. https://doi.org/10.1016/j.biteb.2020.100496.

Thakur, S., Chaudhary, J., Sharma, B., Verma, A., Tamulevicius, S., & Thakur, V. K. (2018). Sustainability of bioplastics: Opportunities and challenges. *Current Opinion in Green and Sustainable Chemistry*, *13*, 68–75. https://doi.org/10.1016/j.cogsc.2018.04.013.

Valentino, F., Morgan-Sagastume, F., Campanari, S., Villano, M., Werker, A., & Majone, M. (2017). Carbon recovery from wastewater through bioconversion into biodegradable polymers. *New Biotechnology*, *37*, 9–23. https://doi.org/10.1016/j.nbt.2016.05.007.

Vega, G. C., Sohn, J., Bruun, S., Olsen, S. I., & Birkved, M. (2019). Maximizing environmental impact savings potential through innovative biorefinery alternatives: An application of the TM-LCA framework for regional scale impact assessment. *Sustainability*, *11*(14). https://doi.org/10.3390/su11143836.

Vega, G. C., Voogt, J., Sohn, J., Birkved, M., & Olsen, S. I. (2020). Assessing new biotechnologies by combining TEA and TM-LCA for an efficient use of biomass resources. *Sustainability*, *12*(9). https://doi.org/10.3390/su12093676.

Vogli, L., Macrelli, S., Marazza, D., Galletti, P., Torri, C., Samorì, C., et al. (2020). Life cycle assessment and energy balance of a novel polyhydroxyalkanoates production process with mixed microbial cultures fed on pyrolytic products of wastewater treatment sludge. *Energies*, *13*(11). https://doi.org/10.3390/en13112706.

Wagner, M., Scherer, C., Alvarez-Muñoz, D., Brennholt, N., Bourrain, X., Buchinger, S., et al. (2014). Microplastics in freshwater ecosystems: What we know and what we need to know. *Environmental Sciences Europe*, *26*(1), 1–9. https://doi.org/10.1186/s12302-014-0012-7.

Yadav, B., Pandey, A., Kumar, L. R., & Tyagi, R. D. (2020). Bioconversion of waste (water)/residues to bioplastics—A circular bioeconomy approach. *Bioresource Technology*, *298*. https://doi.org/10.1016/j.biortech.2019.122584.

Yates, M. R., & Barlow, C. Y. (2013). Life cycle assessments of biodegradable, commercial biopolymers—A critical review. *Resources, Conservation and Recycling*, *78*, 54–66. https://doi.org/10.1016/j.resconrec.2013.06.010.

Zhong, Z. W., Song, B., & Huang, C. X. (2009). Environmental impacts of three polyhydroxyalkanoate (pha) manufacturing processes. *Materials and Manufacturing Processes*, *24*(5), 519–523. https://doi.org/10.1080/10426910902740120.

References included in the critical review but not explicitly discussed in the text

Cristóbal, J., Matos, C. T., Aurambout, J. P., Manfredi, S., & Kavalov, B. (2016). Environmental sustainability assessment of bioeconomy value chains. *Biomass and Bioenergy*, *89*, 159–171. https://doi.org/10.1016/j.biombioe.2016.02.002.

Gerngross, T. U. (1999). Can biotechnology move us toward a sustainable society? *Nature Biotechnology*, *17*(6), 541–544. https://doi.org/10.1038/9843.

Halley, P. J., & Dorgan, J. R. (2011). Next-generation biopolymers: Advanced functionality and improved sustainability. *MRS Bulletin*, *36*(9), 687–691. https://doi.org/10.1557/mrs.2011.180.

Hottle, T. A., Bilec, M. M., & Landis, A. E. (2013). Sustainability assessments of bio-based polymers. *Polymer Degradation and Stability*, *98*(9), 1898–1907. https://doi.org/10.1016/j.polymdegradstab.2013.06.016.

Kettl, K. H., Shahzad, K., Eder, M., & Narodoslawsky, M. (2012). Ecological footprint comparison of biobased PHA production from animal residues. *Chemical Engineering Transactions*, *29*, 439–444. https://doi.org/10.3303/CET1229074.

Koller, M., Gasser, I., Schmid, F., & Berg, G. (2011). Linking ecology with economy: Insights into polyhydroxyalkanoate-producing microorganisms. *Engineering in Life Sciences*, *11*(3), 222–237. https://doi.org/10.1002/elsc.201000190.

Patel, M., Crank, M., Dornburg, V., Hermann, B., Roes, Husing, B., et al. (2006). *Medium and long-term opportunities and risks of the biotechnological production of bulk chemicals from renewable resources the BREW project*.

Ramesh, P., & Vinodh, S. (2020). State of art review on life cycle assessment of polymers. *International Journal of Sustainable Engineering*, *13*(6), 411–422. https://doi.org/10.1080/19397038.2020.1802623.

Rostkowski, K. H., Criddle, C. S., & Lepech, M. D. (2012). Cradle-to-gate life cycle assessment for a cradle-to-cradle cycle: Biogas-to-bioplastic (and back). *Environmental Science and Technology*, *46*(18), 9822–9829. https://doi.org/10.1021/es204541w.

Titz, M., Kettl, K. H., Shahzad, K., Koller, M., Schnitzer, H., & Narodoslawsky, M. (2012). Process optimization for efficient biomediated PHA production from animal-based waste streams. *Clean Technologies and Environmental Policy*, *14*(3), 495–503. https://doi.org/10.1007/s10098-012-0464-7.

Urtuvia, V., Villegas, P., González, M., & Seeger, M. (2014). Bacterial production of the biodegradable plastics polyhydroxyalkanoates. *International Journal of Biological Macromolecules*, *70*, 208–213. https://doi.org/10.1016/j.ijbiomac.2014.06.001.

Wampfler, B., Ramsauer, T., Rezzonico, S., Hischier, R., Köhling, R., Thöny-Meyer, L., et al. (2010). Isolation and purification of medium chain length poly(3-hydroxyalkanoates) (mcl-PHA) for medical applications using nonchlorinated solvents. *Biomacromolecules*, *11*(10), 2716–2723. https://doi.org/10.1021/bm1007663.

Yu, J., & Chen, L. X. L. (2008). The greenhouse gas emissions and fossil energy requirement of bioplastics from cradle to gate of a biomass refinery. *Environmental Science and Technology*, *42*(18), 6961–6966. https://doi.org/10.1021/es7032235.

CHAPTER

Conclusions

21

Silvia Fiore[a], Almudena Hospido[b], and Carmen Teodosiu[c]

[a]*Department of Environment, Land and Infrastructure Engineering (DIATI), Politecnico di Torino, Torino, Italy* [b]*CRETUS, Department of Chemical Engineering, University of Santiago de Compostela, Santiago de Compostela, Spain* [c]*Department of Environmental Engineering and Management, "Gheorghe Asachi" Technical University of Iasi, Iasi, Romania*

Back in 1966, the economist, educator, peace activist, and interdisciplinary philosopher Kenneth Ewart Boulding deplored in front of the international community the 'cowboy economy' (i.e. unlimited ethics and resources, with success measured by production and consumption) and proclaimed the need to a paradigm shift to a 'spaceship economy', where resources are limited and pollution must be contained, with success measured by the total capital stock, including human health. Twenty years later, after tragic disasters as Bhopal and Chernobyl, and many oil spills, the Brundtland Report (1987) defined the concept of Sustainable Development, stressing the vital importance of improving quality of life rather than consumption for current and future generations. Since then, Sustainable Development timeline is full of milestones, which take us in 2015 to the Agenda 2030 adopted by all United Nations Member States and its related 17 Sustainable Development Goals (SDGs) together with a 15-year plan to achieve them (Chapter 2). The SDG actions are currently progressing at a slow pace, affected from 2020 by the coronavirus pandemic with its devastating consequences. Now, even more than ever, there is a clear consensus on the importance of 'following the science' when defining and implementing actions, strategies, and policies, so there seems to be no doubt that there is an urgent need to strengthen the contributions of science, which will be decisive to renew and support the development process towards the SDG implementation.

The concept of Circular Economy (2016), i.e. "*an alternative to a traditional linear make-use- dispose economy in which we keep resources in use for as long as possible*" (Chapter 3), potentially supports the SDGs, particularly SDG12—Sustainable production and consumption, but needs to be more present in others, such as SDG8—Decent work and economic growth, as the SGDs ignore the distinction between economic growth and sustainable development, and the performance economy approach of generating more jobs without increasing resources consumption.

Assessing Progress towards Sustainability. http://doi.org/10.1016/B978-0-323-85851-9.00021-3

The Agenda 2030 was defined within and for a planet in which world population is approaching 8 billion people, and the related needs for food, water, and energy are exacerbating the pressure on the Earth Natural Capital, i.e. "any resource (including plants, animals, minerals, end ecosystems) that provides functions that produce ecosystem goods and services." And therefore, it is essential to identify how to reduce the multiple impacts affecting the food, energy, and water sectors and their multiple and strong interconnections, as well as to define and apply a consistent approach in their assessment. Such role is played by the food–energy–water nexus (Chapter 4), which provides an overarching framework that helps investigating the interrelations between different natural resources, resource sectors, and policy fields.

Moving down to the policy makers level, the European Green Deal (2019) (Chapter 5) is an example of how science-based policies can promote sustainability at multi-country level, providing an ambitious action plan for transforming the European Union economy into a sustainable one. So, it took half a century after Kenneth Boulding's statement for the European Commission to announce an integrated action plan for decoupling economic growth from resource consumption and environmental impacts, as well as acting on production and consumption at system level to stay within the planetary boundaries. In any case, it is never too late and finally the European Union is aware of its good position to lead the global transition towards sustainable development.

Not only in Europe, but all over the world, policy makers create plans and strategies to steer the future direction of business, government, and society. In recent years, policy makers are increasingly understanding that policies in a particular sector may have unintended knock on environmental, social, and economic impacts in other sectors or areas. In this sense Life Cycle Thinking appears to be the right "framework, based on a system's approach, to measure the environmental, social and economic impacts of a product, process or service over its entire life cycle," as "it provides a long-term perspective of the multiple impacts, including burden-shifting, affecting a system, showing the global picture to ensure a global solution" (Chapter 6). The life cycle based toolbox, including Life Cycle Assessment (LCA, focused on the environmental sustainability), Life Cycle Costing (looking at the economic domain of sustainability), Social LCA (focused on the social dimension), and more recently the Life Cycle Sustainability Assessment Framework (aiming at integrating the triple bottom line model of sustainability) can be used to tackle different questions and get the needed answers. In parallel, and sharing in most cases the system approach that characterised the life cycle based tools, a variety of footprints (Chapter 7) have been developed to better understand the impacts of human activities on the environment. Without questioning their usefulness and attractiveness, some researchers support the development of integrated and harmonised tools to overcome the challenges associated with single issue footprints, whereas others lean towards classification to retain the value of simpler easy-to-understand footprints as part of a footprint family. The latter approach is particularly important when new urgent issues that impact on sustainability arise.

Entailing a set of indicators (i.e. impact categories) but focusing on the environmental element of the triple bottom line model of sustainability vector, LCA has been able to show a significant level of versatility by being combined to other tools or by adapting itself to be applied to territories. Firstly, the combined use of LCA and Data Envelopment Analysis (DEA) (Chapter 8) has been employed over the past years to assess eco-efficiency thanks to the more holistic environmental evaluation of production systems achieved by its joint implementation, which enables the quantification of environmental benchmarks and cleaner production thresholds. Significant progress has been made in the combined application of both methodologies, but there are still open challenges to be faced, such as uncertainty and robustness, and undoubtfully standardisation is highly necessary. Secondly, Territorial LCA (Chapter 9), e.g., the adaptation of the conventional LCA approach for the assessment of territories (being them a wide range of subnational systems such as cities, metropolitan areas, agricultural areas, or regions), allows to identify pollution transfers and to quantify the environmental impacts of trajectories regarding a bouquet of services rendered. Being territories are a key element in the necessary ecological transition of our societies, the review presented demonstrated the relevance and need of this approach as well as pointed out some of the open questions to be addressed soon, such as the need to consider absolute eco-efficiency with global and regional planetary boundaries and to include a safe operating space to broaden LCA utility for decision support.

Going back again to the past century, Environmental Impact Assessment (EIA) was introduced in 1969, and still now is one of the important tools of environmental management used in environmental policy and planning worldwide. At present integrative approaches consider EIA together with risk assessment and other tools (Chapter 10), aiming at developing evaluation methods that are more objective and offer precise information about the likely impacts and risks associated to certain activities or projects, to generate robust results and reduce uncertainties. Closing the section on tools and methodologies, Multi-Criteria Decision Making (Chapter 11) provides a scientifically sound decision methodology to combine different types of inputs with stakeholder views and cost/benefit information; this approach is required in complex decision-making problems, where in general, there is no obvious and/or unique solution, to match the stakeholders' preferences and be plain about the critical aspects amongst different elements (economy, environment, society). Its application has grown and developed significantly over the past decades, but there is still room for further evolution including integrated sustainability assessments of the social, environmental, and economic areas in the final vision of the SDGs.

The selection of case studies aims at providing examples of real applications of the above-mentioned frameworks and tools in different fields (food and personal care products, bioplastics, bioenergy and biorefineries, buildings, water/wastewater systems, solid waste management).

Eco-design, i.e. the integration of environmental aspects into the product development process, and LCA combined application is especially relevant at the very early stages of product development (lab-scale synthesis and testing) as they can bring important information related to the environmental profiles of the new products. This was the case of the technical and environmental evaluation of several organic/inorganic composites and biosorbents for the removal of heavy metals in water (Chapter 12).

The LCA for the design of a pilot recovery plant of brewery by-products for aquaculture feed ingredients (Chapter 13) was used for the decision making from the early stage design of the building, focusing mainly on two strategies: on the one hand, the reduction of heating and cooling energy consumption during use phase, and on the other hand, the selection of construction materials with lower environmental impacts. The optimisation of active strategies as well as renewable generation onsite should be included in further research to assess the energy saving potential and the repercussion of these systems and materials on the life cycle impacts.

In a different field, the environmental performance of various kinds of riceberry rice products (food and personal care products) was assessed coupling LCA and Eco-Efficiency (Chapter 14), aiming at identifying the most appropriate alternatives for developing innovative value-added products, which turned to be the recommendation towards non-food products because of their higher value added and eco-efficiency.

The environmental and economic sustainability assessment of cocoa production in West sub-Saharan Africa (Chapter 15) revealed that Cocoa represents a *luxury* food of consumption in developed countries, but it is also a fundamental agricultural product in the poorer tropical countries where it is cultivated. Although West sub-Saharan Africa represents the core region (nearly 70%) of global production, the lack of industries and the low local demand make this area the main exporter worldwide. The lack of adequate infrastructures for crop processing significantly questions the economic sustainability of cocoa production. The economic interest behind cocoa imports in the European market and to the US unavoidably challenges the environmental sustainability of cocoa production in the study areas. Most of the production increase over the past decades was driven by land extensification at the expense of forests. Cocoa plantations in these areas are entirely fed by rainfall. Hence, dry periods can significantly threaten cocoa's productivity and, thus, the economic revenues coming from the sale of cocoa beans.

Wastewater treatment plants, specifically conventional ones, are well documented through LCA studies, and mechanistic models are available for the estimation of missing inventory data when direct measurements are not available. However, limited inventories are available for non-conventional wastewater treatment plants and nature-based solutions, which require to manage multi-functionality (i.e. wastewater treatment, energy recovery from sludge, reuse of wastewater and recovered nutrients in agriculture, etc.). Besides, when moving up to the whole urban water systems (Chapter 16) LCA studies are by far scarcer, both at the complete cycle level, but also at the level of individual elements (drinking water production, wastewater reuse, etc.). It is necessary to harmonise the application of LCA to urban water systems, providing recommendations in the different steps: goal and scope and functional unit, inventory, impact assessment and interpretation.

When moving to energy production systems (Chapter 17), the economic aspect has been widely studied and, for the renewable electricity production, is usually supported by the presence of subsidy framework, whilst the social aspect remains less investigated, as in many other areas, due to the lack of a widely accepted assessment methodology. The LCA and LCC of second-generation biobutanol production (Chapter 18) highlighted the advantages and drawbacks of the biotechnological production route and shed the light on the key critical stages of the production chain that mostly deserve optimisation from the economic and environmental standpoints.

Carbon footprint was applied to solid waste management systems (Chapter 19), covering several types of waste (i.e. municipal solid waste (MSW), waste of electric and electronic equipment (WEEE), packaging waste (PW), and biowaste), as well as different methodologies for the estimation of GHG emissions, concluding that aggregated models were proven to be a rapid alternative to obtain the desired information in terms of total or treatment specific emissions as well as to compare situations in different territorial levels (regional, country, or intercountry) or in different periods of time (past/current/future).

Finally, the analysis of the LCA of bioplastics (specifically Polyhydroxyalkanoates, PHA) (Chapter 20) demonstrated that literature generally focus on waste feedstocks, as the use of dedicated crops might not fit the circular economy principles. Still, the exclusion of the end-of-life phase within the scope of the studies disregards the expected benefits derived from PHA biodegradability whilst its inclusion dismisses the effects of plastic leaks in the environment and plastic pollution. Further modelling on this is required, as it is also needed to provide data at pilot and full-scale, as current information is based on secondary sources, simulations, or up-scaled lab results. Besides, there is also a need of expanding the scope of the assessments by covering not only environmental aspects, but also other issues linked to bioeconomy and sustainability, as social or economic aspects, to get a full overview of the evaluation that addresses the integral approach linked to circularity.

The idea of this book derived from the desire to provide a comprehensive approach to the evaluation of sustainability progress, considering the main frameworks and tools, and their application in case studies describing different sectors (food and personal care products, bioplastics, bioenergy and biorefineries, buildings, water/wastewater systems, solid waste management). We consider the case studies a fundamental element of the book, essential to appreciate and discuss the application of the different frameworks and tools in real contexts in an international perspective. Our readers find in one book valuable contributions from well-known professionals and academics in the fields of environmental sciences and engineering, with the common goal of analysing the opportunities to reduce environmental impacts and to increase resources and energy efficiency. Our specific focus was to highlight the connections amongst the challenges posed by production–consumption–end-of-life phases in different geographical boundaries, the international policies on Sustainable Development and Circular Economy, the food–energy–water nexus, and the key assessment tools. The case studies describing the application of the assessment tools were carefully selected based on the above-mentioned specific criterion and a quantitative engineering approach.

A bend in the road is not the end of the road…unless you fail to make the turn

Helen Keller

Index

Note: Page numbers followed by *f* indicate figures, *t* indicate tables, and *b* indicate boxes.

F

N

O

P

R

S

Printed in the United States
by Baker & Taylor Publisher Services